Biomedical Measurement Systems and Data Science

Discover the fundamental principles of biomedical measurement design and performance evaluation with this hands-on guide. Whether you develop measurement instruments or use them in novel ways, this practical text will prepare you to be an effective generator and consumer of biomedical data. Designed for both classroom instruction and self-study, it explains how information is encoded into recorded data and can be extracted and displayed in an accessible manner. The text describes and integrates experimental design, performance assessment, classification, and system modeling. It combines mathematical concepts with computational models, providing the tools needed to answer advanced biomedical questions. In addition, MATLAB® scripts throughout the text help readers model all types of biomedical systems. The text also contains numerous homework problems, with a solutions manual available online.

This is an essential text for advanced undergraduate and graduate students in bioengineering, electrical and computer engineering, computer science, and medical physics, and for anyone preparing for a career in biomedical sciences and engineering.

Michael Insana is a Donald Biggar Willett Professor in Engineering at the University of Illinois at Urbana-Champaign. He is a Fellow of the AIMBE, the IEEE, and the IoP.

Biomedical Measurement Systems and Data Science

MICHAEL INSANA

University of Illinois at Urbana-Champaign

CAMBRIDGE
UNIVERSITY PRESS

University Printing House, Cambridge CB2 8BS, United Kingdom

One Liberty Plaza, 20th Floor, New York, NY 10006, USA

477 Williamstown Road, Port Melbourne, VIC 3207, Australia

314–321, 3rd Floor, Plot 3, Splendor Forum, Jasola District Centre, New Delhi – 110025, India

79 Anson Road, #06–04/06, Singapore 079906

Cambridge University Press is part of the University of Cambridge.

It furthers the University's mission by disseminating knowledge in the pursuit of education, learning, and research at the highest international levels of excellence.

www.cambridge.org
Information on this title: www.cambridge.org/9781107179066
DOI: 10.1017/9781316831823

First published 2021

A catalogue record for this publication is available from the British Library.

ISBN 978-1-107-17906-6 Hardback

To my family, my inspiration – Debbie, Rebecca, Mark, Annie, and Greg

Contents

Preface

When I told colleagues I was writing this book, the most thoughtful responses were some version of "Why?" There are already many wonderful texts describing measurements and data, some are accessible electronically from libraries, and some publishers will synthesize elements from several texts into a single volume for classroom instruction. A little web surfing reveals a wealth of knowledge from educators and scientific organizations offering many course materials and homework problems for free. So why add another text to the mix?

The short answer is to connect fundamental analytical methods that every engineer should know to computational models enabling data visualization. This book was developed as a first course in a bioengineering graduate program. It reviews core skills in mathematics and statistics in the context of linear systems analysis. It describes straightforward models illustrating the principles of measurement design and evaluation. The hope is that the MATLAB scripts embedded throughout will seed the reader's imagination for exploring solutions to more complex problems. Part of the transition from successful undergraduate to practicing professional is the scale and depth of problems encountered. In consequence, many of the chapter problems found in this book are mini-projects. Specialty functions in MATLAB are avoided to encourage readers to code equations themselves so that they will become familiar with the limiting assumptions so critical to successful analysis.

This book aims to connect topics that often appear disjoint to undergraduate engineering students. The reflexive approaches to formulating solutions that come naturally to practicing professionals have yet to be activated in most newly minted graduates. The biomedical applications discussed in these pages often begin with data being synthesized. Armed with data in which every detail is known, readers can learn to classify and estimate biomedical properties as well as to evaluate task performance from results. Depth of knowledge comes from hands-on interactions.

I also see this book as enabling self-learning among practicing engineers who wish to expand their toolbox. I encourage readers to read this material with a pencil and a laptop handy to keep the engagement active. Senior undergraduates and beginning graduate students will be able to quickly take charge of the material under the guidance of an experienced instructor, who can help them navigate more challenging topics. Experts and novices alike tend to see measurement system design and evaluation through sets of equations. However, those equations and the data they generate speak to experts as stories and mental pictures not yet obvious to a novice. Visualizing results while playing with model parameters is essential when learning to extract knowledge from data. These activities build confidence in the findings that inform medical decisions.

The basic skills for analyzing measurements and mining data for information are linear algebra, statistics, ODEs, and computational methods. Assuming readers have been introduced to these topics, the following materials build upon that base knowledge. I purposely did not cover machine learning techniques in this text, because there are already many sources on these rapidly evolving methods. It is also important to understand the strengths and weaknesses of standard methods to know when machine learning techniques can and should be applied for greater performance.

A solutions manual for the chapter problems is offered to instructors who request it at www.cambridge.org/9781316831823 or who write to me directly. If you find errors or have suggestions for improvements, please let me know. I will appreciate hearing about your experiences with this book.

Finally, I wish to thank my many colleagues and students for their comments and suggestions during the years that helped my class notes evolve into a textbook. I am especially grateful to Fan Lam, Craig Abbey, Cameron Hoerig, Yang Zhu, and Frank Brooks for their helpful comments and suggestions. Most of all, I am grateful to the Grainger College of Engineering at UIUC that immerses its faculty and students in a nurturing environment of innovation and excellence. It is a truly remarkable place for scholarship.

1 Introduction to Measurement Systems

1.1 Information Flow

We scientists and engineers develop approaches to problem solving early in our training that stick with us a lifetime. The most influential training occurs through one-on-one interactions with mentors during graduate and postdoctoral years. That was my situation. My first job after graduate school was to work with Robert F. Wagner and David G. Brown (Bob and David) at the Center for Devices and Radiology Health (CDRH), Food and Drug Administration (FDA), in Rockville, Maryland. After arriving in July 1984, I soon learned that one inspiration behind Bob and David's seminal 1985 paper on the analysis of medical imaging[1] systems [114] was the famous papers of Claude Shannon that introduced information theory to the world [95]. A famous diagram from Shannon's papers is reproduced in Figure 1.1a. It shows a source transmitting information through a "channel" from which information is received and interpreted by an observer at the destination. This channel model could represent the process of making a telephone call without considering details of the associated instrumentation. It is also a basis for defining fundamental concepts such as *information* and *channel capacity*, which are central to any communication process. This model suggests that any well-designed measurement maximizes the flow of expected information through the noisy channel despite the distortions that inevitably occur to the transmitted signals during the process.

Bob and David viewed the process of medical-image formation similarly; they saw imaging devices as noisy information channels through which stochastic[2] task information flows from patient to physician, generally as electronic signals. They posited that the design of any device must begin by identifying a medical *task* – the reason for acquiring patient information. For example, if the task is to image a nondisplaced bone fracture using projection radiography, it is important to evaluate the performance of the system as part of that specific undertaking. Moreover, since realistic signal and noise processes are stochastic, we should employ statistical

---

[1] Bob told me he wanted to title the paper "Grand Unified ANalysis Of medical imaging systems (GUANO)," but the editor of *Physics in Medicine and Biology* objected. Probably a wise decision – but it taught me that one can and should try to have some fun while working!

[2] For the most part, *random* and *stochastic* are applied interchangeably to describe limits on our ability to predict the response of systems and data associated with the system. Some reserve the term *random* for data variables and the term *stochastic* for the systems that act on those variables.

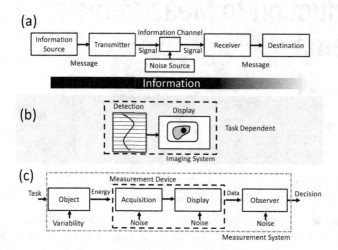

Figure 1.1 (a) Schematic diagram of a general communication system from Shannon's 1948 paper [95]. (b) A schematic diagram of a general medical imaging system consisting of detection and display stages, from Wagner and Brown's 1982 paper [113]. Depicted is a sinogram from CT acquisition followed by a reconstructed image. (c) A general representation of the biomedical measurement systems discussed in this book as built upon the foundations of communications and imaging theory. The object/patient, the device, and the observer are components of the total measurement system.

decision-making methods [10, 70, 110] to evaluate performance. So the goal of medical imaging is to maximize observer performance [47] even if that involves discarding some task information during the process of forming images. Of course, efforts to improve diagnostic performance must also consider overall cost and patient risk, so optimization is multifactorial and often complicated.

Wagner and Brown identified two stages of an imaging device (Figure 1.1b) that I believe generalize to any biomedical measurement. First is an *acquisition stage*, which for x-ray imaging includes the photon energy spectrum that irradiates patients as well as properties of the photon-collection geometry and radiation detectors. Detection is followed by a *display stage* that includes all processes that prepare the recorded data for presentation to an observer, including reconstruction, image processing, and display-monitor feature settings. The principal job of the acquisition stage is to gather and store as much information about the task from the patient as possible while minimizing overall risk. The principal job of the display stage is to format the stored data into a form such that task information is readily accessible by observers. Around these ideas, the field of medical image science was developed and adopted [54].

Although the earliest concepts of image science grew from the photographic and television industries [93], many approaches remained fragmented until they were unified in a comprehensive framework by Harrison H. Barrett and Kyle J. Myers [10]. Harry and Kyle provided a comprehensive mathematical theory from which the field of imaging has coalesced into a rigorous science. The foundational principles of image science generalize to all types of biomedical measurements.

The development of information-theoretic approaches to medical devices is encouraged by the CDRH, as their mission statement includes assuring "patients and providers have timely and continued access to safe, effective, and high-quality medical devices." Through the work of the four imaging scientists mentioned earlier and a great many colleagues from academia and industry, diagnostic performance standards were formulated from the principles of statistical decision theory [54]. The interdisciplinary foundation for that construction, which is labeled by some as *computational data science*, involves methods for extracting from recorded data and effectively presenting specific medical information. This was Bob and David's vision for the unification of medical imaging as described in [114], and it is a goal shared broadly by anyone making biomedical measurements.

1.2 Design of the Book

As a reference book, the primary objective of this text is to describe the engineering principles underlying biomedical measurements and the extraction of knowledge from the data these measurement systems produce. The approach is to leverage the numerous advances made over the last seventy-plus years by pioneers of information theory and photographic, television, and medical-image sciences for this purpose. Since all engineers working in biomedical sciences generate or apply measurement data regularly, there is a core need to understand how measurements are designed for transferring task information gathered from objects for presentation to decision makers (Figure 1.1c).

As a textbook, this book's goal is to remind advanced engineering students of their training in mathematics and engineering methods by applying both to the exploration of measurement systems and data science. Just as Claude Shannon abstracted the concept of a communication channel in his diagram (Figure 1.1a), we will describe principles and techniques for measurement analysis with occasional reference to specific instrumentation, biology, and the underlying physics of radiation interactions with matter. Consequently, this book complements courses in instrumentation and medical imaging, as the ideas described herein build on that basic knowledge. Electrical engineering students concentrating in signal processing or communications and bioengineering students concentrating in imaging and sensing will have been exposed to many of these ideas. However, any senior undergraduate or graduate student with training in engineering or physical science disciplines will be able to follow the discussion by applying the background materials and references provided. Although many of the worked-out examples are biomedical in nature, these concepts apply broadly to any linear-system measurement process. The problem sets found at the end of each chapter either provide problem-solving practice or introduce additional methods and applications not discussed in the text.

Computational modeling in MATLAB is a large part of this approach. Discussions begin by modeling measurement processes in order to simulate data. We employ *forward-problem approaches* that begin with physical descriptions of objects/patients and instruments to generate measurement data. Then we employ *inverse-problem approaches*, whereby measurement data are reduced to describe object properties. Mathematical and computational models are essential for understanding measurements systems and the biological processes to which measurements are applied. Most examples are simplified to allow for concise analytical solutions and short modeling scripts. More complex problems are provided in the end-of-chapter problems.

One impediment to learning technical material is inconsistent notation and lack of conceptual visualization. To the extent possible, I followed the notation provided by standard texts [5, 10, 13, 42] (summarized in Section 1.3). I hope the notational similarities motivate some readers to refer to those references, and to dig deeper and reach for other perspectives. Analysis (mathematics) and computation are different sides of the same modeling coin. Sometimes ideas that appear murky on one side become clearer when viewed through the other. Computation enables visualization, which is absolutely essential for understanding the complex processes that dominate modern approaches to biomedical measurements and data analysis.

Measurement scientists must be knowledgeable about probability and statistics (Chapters 2, 3, and 8) and linear algebra (Appendix A). Elements essential to measurement science are reviewed in these chapters. When I teach graduate courses from this material, I provide a handful of MATLAB coding exercises to knock the rust off of my students' coding skills. The discussions of linear systems (Chapter 4) and basis decompositions (Chapters 5 and 6) describe core engineering principles of measurement systems and data science. These methods are applied to medical imaging in Chapter 7 and to flow cytometry in Chapter 9. Both chapters will greatly benefit those readers with some knowledge of the physics of photon–tissue interactions. Chapter 9 focuses on the utility of standard statistical pattern recognition methods. Biological systems and sensors are modeled with differential equations in Chapters 10 and 11, respectively. There, important differences between simple and complex systems are emphasized so that readers will appreciate the value of modeling at different levels.

1.3   Notation

Modeling the measurement process requires mathematical representations of objects, signals, and devices using functions, matrices, and operators. I have diligently sought to keep the notation consistent throughout the chapters. It is, of course, impossible to use symbols uniquely; e.g., the symbol x is used many times for many different quantities. However, the use of a symbol in a specific context combined with its units can help readers determine the symbol's meaning. It is essential to visualize representations at each step in the measurement process to truly understand the concepts involved.

A continuous function of one continuous variable will be written as $f(t)$ when the independent variable is time and as $f(x)$ for a spatial variable. Scalar functions, f, of 3-D vector position, boldface $\mathbf{x}$, will be written $f(\mathbf{x}) = f(x_1, x_2, x_3)$. The quantity $h(\mathbf{x}, t; \boldsymbol{\theta})$ represents a 4-D continuous function of space-time that is parameterized by vector $\boldsymbol{\theta}$; e.g., $h(\mathbf{x}, t; a) = A(x_1, x_2, x_3) \exp(-at)$ has four independent variables and one constant parameter a.

Lowercase boldface quantities are column vectors, e.g., $\mathbf{y} = \begin{pmatrix} y_1 \\ y_2 \\ y_3 \end{pmatrix}$, and $\mathbf{y}^\top = (y_1, y_2, y_3)$ is its transpose given by a row vector. Matrices are indicated by uppercase boldface quantities. For example, an $M \times N$ matrix of real values, $\mathbf{H}$, is a 2-D array of elements with M rows and N columns, where the mth row and nth column are indicated by $H[m, n]$ or, equivalently, H_{mn}. Applying shorthand notation, a matrix described as $\mathbf{H} \in \mathbb{R}^{M \times N}$ and its transpose as $\mathbf{H}^\top \in \mathbb{R}^{N \times M}$ tell us the elements are a 2-D array of real numbers. The complex $M \times N$ matrix $\mathbf{A}$ is indicated using $\mathbf{A} \in \mathbb{C}^{M \times N}$. We indicate its complex-conjugate transpose as $\mathbf{A}^\dagger \triangleq (\mathbf{A}^*)^\top \in \mathbb{C}^{N \times M}$, where " $x \triangleq y$ " means "x is defined as y." The inverse of matrix $\mathbf{B}$ is $\mathbf{B}^{-1}$ and its pseudoinverse is $\mathbf{B}^+$. Lowercase letters i, j, k, ℓ, m, and n usually represent integers.

Continuous functions can be expressed as vectors and matrices. For example, the real function $y(t)$ may be represented as a $\infty \times 1$ column vector, $\mathbf{y} \in \mathbb{R}^{\infty \times 1}$. If y is sampled on the interval T that remains constant over the recording time of the measurement T_0, then all values of $y(t)$ are set to zero except those found at $y(mT)$ for integers $m = 0, 1, 2, \ldots, M - 1$, where $T_0 = (M - 1)T$ (Figure 1.2). In that case,

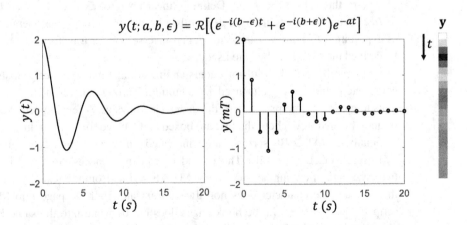

$$y(t; a, b, \epsilon) = \mathcal{R}\left[\left(e^{-i(b-\epsilon)t} + e^{-i(b+\epsilon)t}\right)e^{-at}\right]$$

Figure 1.2 Three representations of one function. $y(t)$ is a continuous function of continuous time with parameters $a = 2, b = 1, \epsilon = 0.05$. $y(mT)$ is the same continuous function of continuous time but now only samples at $t = mT$ are preserved. Twenty-one-element column vector $\mathbf{y}$ describes the continuously varying nonzero values of $y(t)$ on a grayscale display.

$y(mT)$ is still a continuous function of time, as indicated by the curved parentheses around the independent variable $t = mT$. However, we can discard all values of $y(t)$ except those at mT and sequence them as $M \times 1$ vector $\mathbf{y}$ to enable use of matrix algebra or arrays on which MATLAB and similar coding languages are based.

Elements of data column vector $\mathbf{g}$ are represented as $g[m]$, where the square brackets indicate a discrete-time variable at integer m. We may equivalently use $[\mathbf{g}]_m$ or g_m, depending on how specific we need to be or how cumbersome a form may be in each circumstance. If the amplitude of the function is continuous and time is quantized, we refer to $\mathbf{g}$ as a *discrete-time function*. If both amplitude and time are quantized, we use the same notation but refer to $\mathbf{g}$ as a *discrete function*. Note that discrete-time functions are useful analytically, but measurement data suitable for computation are discrete, which means that *quantization noise* is an element of acquisition noise. The continuous-time axis of a discrete-time function may be recovered using one of many types of interpolation [77].

Applying analogous notation, 2-D functions of space $g(x_1, x_2)$ that are sampled on the rectangular interval (X_1, X_2) appear as $g(mX_1, nX_2)$ or as matrix $\mathbf{G}$ with elements G_{mn}, where $0 \le m \le M - 1$ and $0 \le n \le N - 1$.

We use $G(u)$ to describe the 1-D Fourier transform of $g(t)$, where t and u are *conjugate variables* associated with the transformation. The symbol u indicates frequency, either temporal or spatial, while $\Omega = 2\pi u$ is radial frequency. Similarly, $G(s)$ is the Laplace transform of $g(t)$, with s and t as conjugate variables. Thus, uppercase and lowercase letters can have multiple meanings that we help readers negotiate at the appropriate time.

We may *apply* operator $\mathcal{A}$ to function $f(t)$ by writing $\mathcal{A}\{f(t)\}$ or simply $\mathcal{A}f(t)$. Operators are represented by bold, uppercase, calligraphy letters. Application of an operator is not necessarily multiplication. For example, if we define quadratic operator $\mathcal{A} \triangleq (\cdot)^2$, then $\mathcal{A}f(t) = f^2(t)$. Defining linear operator $\mathcal{B} \triangleq d/dx$, then $\mathcal{B}y(x) = dy/dx$. Often operators are used as shorthand notation for common operations, such as expectation $\mathcal{E}$, Fourier transform $\mathcal{F}$, and inverse Fourier transform $\mathcal{F}^{-1}$, but must be defined the first time they are used.

Units can be quite helpful for understanding strange mathematical quantities like comb, step, and Dirac delta used in sampling. Throughout this text, we describe units to guide our interpretation of the underlying physics. A few important equations frequently cited throughout the text are boxed and labeled to aid in finding them.

Finally, MATLAB scripts are indicated in the `courier` font, e.g., `fftshift(fft(g))*dt`. Those readers who have access to the e-book have the advantage of cutting and pasting MATLAB scripts from the text. If you do, note that at least one character does not always translate from the page into MATLAB scripts. Apostrophe (') in the book pages does not translate into the same keyboard character in MATLAB scripts. Characters that cannot be interpreted show up in red in MATLAB, making them easy to find. I developed all of the code using MATLAB R2016a, but I've tried parts of the code in later versions up to R2020a. Intrinsic functions in MATLAB are avoided when there is value in writing our own.

2   Probability

Probability is a branch of mathematics widely used in science and engineering to understand and manage uncertainty. For example, if you apply a known input stimulus to a measurement system and it responds exactly the same way every time the stimulus is applied, you can say with confidence that the object and the measurement device are both deterministic – there will be no uncertainty in measuring that specific object with that specific device. Of course, such systems don't exist in practice. All objects have some degree of randomness in the property that couples to measurement energy, the energy fields themselves are stochastic, and every measurement device introduces one or more sources of *noise*. In addition, interactions among acquisition-stage elements of a measurement device can compound the uncertainty. In later chapters, we describe how display-stage processing modifies the acquisition-stage probability models, and how displayed-data uncertainty can influence observer decisions.

2.1   Introduction

You might be surprised to hear that some experts remain sharply divided on the interpretation of "probability." Despite the randomness it describes and the differing definitions, probability theory is an exact science built on fundamental principles, some of which we describe in this chapter. The challenge is always to interpret the rules and then correctly apply them to each problem. This description might sound like a job for a lawyer, but effective engineering very often results from having a strong intuitive sense of how to apply basic probability theory.

Frequentists define probability as the frequency of event occurrences during an experiment relative to the opportunity for those events to occur. For them, probability is determined empirically: Data tell the whole story. In contrast, *Bayesians* view probability as the degree of belief in a state. They define a prior probability before taking data and then update that belief with a posterior probability after viewing the experimental data. As you might imagine, each view has its strengths and weaknesses. The biggest distinction is that frequentists consider probabilities as objective states of an experiment, whereas Bayesians consider probabilities as subjective states of the experimenter. We shall remain agnostic here, using each definition as it suits us.

Let's introduce the frequentist position with an example. Assume water is exposed to sunlight that absorbs photons depending on their wavelength and the molecular

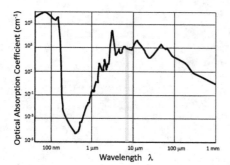

Wavelength λ

Figure 2.1 Example of the optical absorption coefficient in water across the solar spectrum (after [33]).

properties of water. Photon absorption is a *stochastic process* that can be *known statistically* but is unknown exactly. That is, the exact number of events occurring at a specific time and location is unknown, but the likelihood of absorption can be estimated if we can specify a representative probability model based on the physics of optical absorption. To illustrate, assume the spectrum in Figure 2.1 represents all possible wavelengths that can be incident on the water volume. We set up the experiment so that if a photon is absorbed, it can only have a wavelength between 50 nm and 1 mm. The probability that every absorbed photon has a wavelength in this range is 1 or 100%. We can estimate the probability that the absorbed photon has a wavelength in the narrow band near 8 μm as the ratio of shaded area to the total area under the curve of Figure 2.1 between 50 nm and 1 mm.[1] Let's look at this more closely.

Define probability $\Pr(E)$ as a measure of the chance that event E occurs during an experiment, and S as the *event space* in which all events occur. In Figure 2.1, E indicates a photon within the 8 μm-wavelength bin is absorbed by water, while S specifies the number N and wavelength range of all photons illuminating that volume of water during the experiment time. We measure N' photons emerging from the water volume, although to keep things simple we ignore absorption by air, the container, and all scattered photons. Hence, $n(E) = N - N'$ is the number of absorption events occurring during the experiment time.

DEFINITION 2.1.1   *The probability of E within event space S is defined as*

$$\Pr(E) \triangleq \lim_{N \to \infty} \frac{n(E)}{N} \quad \text{for } E \in \mathsf{S}. \tag{2.1}$$

It is important to specify S because it defines the N opportunities for a photon to encounter an absorbing molecule. This simple equation is packed with implied information about the physics of the experiment, which is why many find it intuitive but not always practical. Notice that the frequentist probability of (2.1) is defined in the

---

[1] The area must be computed from a linear plot. Figure 2.1 is a log–log plot of the absorption coefficients.

limit of infinite experimental evidence, so any practical measurement of probability must be an approximation. Questions about whether the ratio converges in the limit using experimental data are a topic we leave to others to discuss.

Let's examine a simpler experiment, one that is a clear favorite in probability discussions: The coin toss. We all know a "fair coin" offers a 50–50 chance of measuring a heads or a tails with each toss. Let E be the event of obtaining heads, and let its *complement* E^c (not E) indicate that a tails was obtained. Event space S includes influences from the coin, the person doing the tossing, and the surfaces a coin may encounter once tossed. These details determine whether tosses are "fair," in which case (2.1) gives $\Pr(E) = 0.5 = \Pr(E^c)$ and $\Pr(S) = \Pr(E) + \Pr(E^c) = 1$.

Suppose we want to determine if a coin is fair. We toss it once and find heads. All we know from this one event is that we can eliminate the possibility that $\Pr(E) = 0$. So we toss it $N = 10$ times and estimate from this sample of 10 trials that $\Pr(E) \simeq 0.3$ and $\Pr(E^c) \simeq 0.7$. Can we predict the next toss any better? Well, yes, but not especially well. The event space may still be unfair, but we don't have enough experience to say for sure. Tossing the coin a thousand times and finding $\Pr(E) \simeq 0.503$ gives us more confidence that the experiment involves a fair coin. Just exactly how fair the coin can be determined to be depends on your patience for experimentation, but you will never know exactly because "exact" is found as $N \to \infty$.

2.2 Probability Axioms

Probability theory is built upon the following three axiomatic statements:

DEFINITION 2.2.1

Axiom 1 : $\Pr(E) \in \mathbb{R}$, $\Pr(E) \geq 0$ *(probabilities are nonnegative real numbers)*

Axiom 2 : $\Pr(S) = 1$ *(one of the events in S is certain to occur)*   (2.2)

Axiom 3 : $\Pr\left(\bigcup_{i=1}^{M} E_i\right) = \sum_{i=1}^{M} \Pr(E_i)$ *(for mutually exclusive events)*

The first two axioms are fairly obvious, but the third axiom needs some explanation. Assume $E_1, E_2, \ldots, E_M$ are mutually exclusive events – a photon with wavelength λ_1 cannot also have wavelength λ_2. Then the third axiom states that the probability of at least one of these events occurring is the sum of their respective probabilities.

$E \cup F$ denotes the union of events (E or F where $\cup \triangleq$ "or"). Generalizing, $\bigcup_{i=1}^{M} E_i$ denotes the union of M events, E_1 or E_2 or $\ldots$ or E_M. Also, the product EF and $E \cap F$ both denote the intersection of events (E and F where $\cap \triangleq$ "and").

Example 2.2.1. *Playing with dice: A fair die with six sides generates events $E_1 = \{1\}$, $E_2 = \{2\}\ldots$ with probability $\Pr(E_i) = 1/6$. Note $E_1 = \{1\}$ indicates the set of all ones rolled in S. From (2.1) and Axiom 3, since the events are mutually exclusive (each* roll can generate only one value), a fair die yields the following probability of rolling an odd number: $\Pr(\{odd\}) = \Pr(\{1,3,5\}) = 1/6+1/6+1/6 = 1/2$. This result should match your intuition.

2.3   Consequences of the Axioms

- If E^c is the *complement* of event E, then the axioms yield

$$\Pr(S) \triangleq 1 = \Pr(E \cup E^c) = \Pr(E) + \Pr(E^c) \text{ so that } \Pr(E^c) = 1 - \Pr(E).$$

- If the space of event E is *contained* within the space of event F, then $\Pr(E)$ is less than or equal to $\Pr(F)$, or

$$\text{if } E \subset F, \text{ then } \Pr(E) \le \Pr(F).$$

- The probability of the union of two events $E \cup F$ that are not mutually exclusive is

$$\Pr(E \cup F) = \Pr(E \cup E^c F) = \Pr(E) + \Pr(E^c F).$$

From Axiom 3 and Figure 2.2, we see that

$$\Pr(F) = \Pr(EF \cup E^c F) = \Pr(EF) + \Pr(E^c F).$$

Combining the two equations,

$$\Pr(E \cup F) = \Pr(E) + \Pr(F) - \Pr(EF). \tag{2.3}$$

Equation (2.3) is always true but necessary only for dependent events because $\Pr(EF) = 0$ when E and F are mutually exclusive.

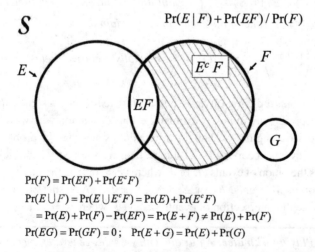

$$\Pr(F) = \Pr(EF) + \Pr(E^c F)$$
$$\Pr(E \cup F) = \Pr(E \cup E^c F) = \Pr(E) + \Pr(E^c F)$$
$$= \Pr(E) + \Pr(F) - \Pr(EF) = \Pr(E+F) \ne \Pr(E) + \Pr(F)$$
$$\Pr(EG) = \Pr(GF) = 0; \quad \Pr(E+G) = \Pr(E) + \Pr(G)$$

Figure 2.2 Venn diagram illustrating various probabilities associated with non–mutually exclusive events E and F and mutually-exclusive event G, all within event space S.

Example 2.3.1. *Probability of being sick: Let E be the event that a patient is sick with heart disease; F be the event that a patient is sick with cancer; and EF be the event that a patient is sick with both diseases, i.e., $E \cap F$ (poor guy!). Further, $n(E) = 5$ patients have heart disease out of $N = 9$ total patients, $n(F) = 5$ patients have cancer, and $n(EF) = 1$ patient has both diseases. Clearly, $\Pr(EF) = 1/9$ for the S defined. Equation (2.3) gives the intuitive probability that patients in this event space who are sick with either disease is equal to 1: $\Pr(S) = \Pr(E \cup F) = 5/9 + 5/9 - 1/9 = 1$. The last term subtracts the probability of patients with both diseases to eliminate double counting.*

2.4 Conditional Probability

Dice seem to offer the simplest examples. Let E be the event that the sum of two throws of a die is 7 and F be the event that the first throw is 3. The probability that E occurs given that (or conditional upon) event F has occurred is indicated by $\Pr(E|F)$.

We find that $\Pr(F) = 6/36 = 1/6$ because $\{F\} = \{(3,1)(3,2)(3,3)(3,4)(3,5)(3,6)\}$ are the six possibilities within a sample space S with $6^2 = 36$ possibilities. Also, $\{E\} = \{(1,6)(6,1)(2,5)(5,2)(3,4)(4,3)\}$, so $\Pr(E) = 6/36 = 1/6$. Looking at the set $\{E\}$, the intersection EF is just one event, $(3,4)$. Hence $\Pr(EF) = 1/36$ even though $\Pr(E|F) = 1/6$; that latter is found by limiting the choices from the whole event space to include only those in set $\{F\}$. Conditioning statements modify the event space, the denominator of (2.1).[2]

2.4.1 Pr(EF) versus Pr(E|F)

To emphasize the difference between these probabilities, consider the example of Figure 2.3. Follow along by counting squares. This event space has $N = 56$ elements on a 7×8 grid. E and F are events in two shaded regions that overlap. Counting, we see that $\Pr(E) = \Pr(F) = 16/56$. Also, $\Pr(E^c) = \Pr(F^c) = (56 - 16)/56$ and $\Pr(EF) = 4/56$. Finally,

$$\Pr(E|F) = \frac{\Pr(EF)}{\Pr(F)} = (4/56)/(16/56) = 1/4, \qquad (2.4)$$

which is defined only when $\Pr(F) \neq 0$. Equation (2.4) shows that the probability of events E and F occurring simultaneously equals the probability of E given F times $\Pr(F)$. Note that $\Pr(E|E) = 1$. Conditioning limits the event space.

---

[2] The most amusing statements of conditions placed on S are from those seeking to hold a "world record." For example, Jane Doe holds the world record for most somersaults in 20 minutes while wearing boots on a Tuesday afternoon during a thunderstorm. ... If you restrict the event space enough, everyone can be a star!

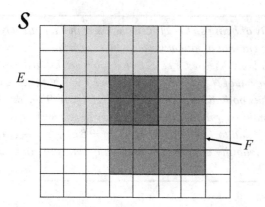

Figure 2.3   Graphical representation of an event space.

2.5   Bayes' Formula

From (2.4), $\Pr(EF) = \Pr(E|F)\Pr(F)$ and $\Pr(FE) = \Pr(F|E)\Pr(E)$. Because $\Pr(EF) = \Pr(FE)$,

$$\Pr(E|F)\Pr(F) = \Pr(F|E)\Pr(E)$$

$$\Pr(E|F) = \frac{\Pr(F|E)\Pr(E)}{\Pr(F)}, \qquad (2.5)$$

which is *Bayes' formula*. Decomposing $\Pr(F)$,

$$\Pr(F) = \Pr(FE) + \Pr(FE^c)$$

$$= \Pr(F|E)\Pr(E) + \Pr(F|E^c)\Pr(E^c) \qquad (2.6)$$

and generalizing to include N different events in S, viz., $E_1, E_2, \ldots, E_N$, we have

$$\Pr(F) = \sum_{i=1}^{N} \Pr(F|E_i)\Pr(E_i).$$

Combining this result with (2.5) yields a more general form,

$$\Pr(E_j|F) = \frac{\Pr(F|E_j)\Pr(E_j)}{\sum_{i=1}^{N}\Pr(F|E_i)\Pr(E_i)}. \qquad \text{(Bayes' formula)} \qquad (2.7)$$

Example 2.5.1. *Assessing probabilities: A patient enters a clinic with flu symptoms. All patients coming to the clinic live in three cities A, B, and C, such that $\Pr(S) = \Pr(A) + \Pr(B) + \Pr(C)$. Based on census data, $\Pr(A) = 0.2$, $\Pr(B) = 0.7$, and $\Pr(C) = 0.1$. The conditional probabilities that patients from each city have the flu are $\Pr(F|A) = 0.01$, $\Pr(F|B) = 0.001$, and $\Pr(F|C) = 0.05$. What is the probability this new patient is from each city?*

A. *The probability this flu patient is from City A is*

$$\Pr(A|F) = \frac{\Pr(A \cap F)}{\Pr((A \cap F) \cup (B \cap F) \cup (C \cap F))} = \frac{\Pr(AF)}{\Pr(AF) + \Pr(BF) + \Pr(CF)}$$

$$= \frac{\Pr(F|A)\Pr(A)}{\Pr(F|A)\Pr(A) + \Pr(F|B)\Pr(B) + \Pr(F|C)\Pr(C)}$$

$$= \frac{(0.01)(0.2)}{(0.01)(0.2) + (0.001)(0.7) + (0.05)(0.1)} = 0.26.$$

Similarly, we find that $\Pr(B|F) = 0.09$ *and* $\Pr(C|F) = 0.65$. *Note that* $\Pr(A|F) + \Pr(B|F) + \Pr(C|F) = \Pr(S)$. *This patient is most likely from City C.*

2.6   Statistically Independent Events

Random events E and F are *statistically independent* if

$$\Pr(EF) = \Pr(E)\Pr(F). \tag{2.8}$$

In the example based on Figure 2.2, if it was found that $\Pr(E) = \Pr(F) = 1/3$ and $\Pr(EF) = 1/6$, we would immediately know that E and E were dependent events. N different events are independent if

$$\Pr\left(\prod_i^N E_i\right) = \prod_i^N \Pr(E_i).$$

EXAMPLE 2.6.1   *Explaining terms: What are some of the differences between disjoint sets, mutually exclusive events, and statistically independent events?*

- *If the two sets containing events E and events F are disjoint, their intersection is the empty set; i.e., $E \cap F = \emptyset$. Empty sets have no elements.*
- *If sets E and F are mutually exclusive, then $\Pr(EF) \triangleq \Pr(E \cap F) = 0$. Consequently, (2.3) results in Axiom 3, $\Pr(E + F) = \Pr(E) + \Pr(F)$.*
- *The difference between disjoint sets and mutually exclusive events is that the former is a statement about set membership and the latter one of probability. Often the statements are true simultaneously, but not always. Consider $E = \{0\}$ and $F = \{0\}$, where each set consists of just one element, zero. These sets are mutually exclusive, $\Pr(EF) = 0$, but they are not disjoint, $E \cap F \neq \emptyset$, because zero is a member of the intersection of sets.*
- *Disjoint sets contain mutually exclusive events. However, sets of mutually exclusive events are not necessarily disjoint.*
- *If E and F are not mutually exclusive but they are statistically independent, then, from (2.4) and (2.8), $\Pr(E|F) = \Pr(EF)/\Pr(F) = \Pr(E)\Pr(F)/\Pr(F) = \Pr(E)$, telling us that conditioning has no effect in this situation.*
- *If E and F are mutually exclusive but dependent, $\Pr(E|F) = \Pr(EF)/\Pr(F) = 0$.*

2.7   Problems

2.1 As a new graduate student, you work with your advisor to shop and find just the right research instrument. It is a little fragile, and you're concerned it might not stay fully functional for the four or five years you'll be a student. So you ask the manufacturer, What is the probability that the device will remain functional for five years? The manufacturer responds that the probability that the device will function properly for t years before needing service is $\Pr(T \geq t) = \exp(-t/10)$ for $t > 0$.

(a) On the day of your purchase, what are the chances the device will be fully functional for five years?

(b) Three years later, you've used the device without failure. Yeah! At this point in time, what is the chance the device will break down sometime in the next two years?

2.2 Find $\Pr(E)$, $\Pr(F)$, $\Pr(G)$, and $\Pr(H)$ from the following facts (if it is possible). Then draw a Venn diagram illustrating the relationships among these event sets.

- Events E and F are independent and events F and G are independent
- $\Pr(E \cup F) = 11/20$
- $\Pr(F \cup G) = 19/40$
- $\Pr(EF) = 1/10$
- $\Pr(EG) = \Pr(HG) = \Pr(GH) = 0$
- $\Pr(EH) = \Pr(H)$

2.3 A population of N patients has Medicare and/or private insurance. Let $\Pr(P)$ be the probability that a patient has private insurance, $\Pr(M) = 0.625$ be the probability that a patient has Medicare, and $\Pr(MP) = 0.025$ be the probability that a patient has both types of insurance.

All members of this group of patients have been diagnosed with diabetes and/or atherosclerosis, with the probabilities being $\Pr(D) = 0.6875$ and $\Pr(A) = 0.875$.

(a) Find the number of patients who have both diabetes and atherosclerosis.

(b) If 50 patients have Medicare insurance, how many patients have atherosclerosis?

(c) Given that a patient has diabetes, what is the chance the patient also has atherosclerosis?

(d) What is the probability that a patient in this group has diabetes given that the patient has atherosclerosis?

2.4 Let $S = \{1, 2, 3\}$. The probabilities for event E are $\Pr(E = 1) = 0.1$, $\Pr(E = 2) = 0.5$, and $\Pr(E = 3) = 0.4$, while the probabilities for event F are $\Pr(F = 1) = 0.8$ and $\Pr(F = 2) = 0.2$.

The conditional probabilities are $\Pr(E = 1|F = 1) = 0.1$, $\Pr(E = 2|F = 1) = 0.5$, $\Pr(E = 3|F = 1) = 0.4$, $\Pr(E = 1|F = 2) = 0.1$, $\Pr(E = 2|F = 2) = 0.5$, and $\Pr(E = 3|F = 2) = 0.4$.

(a) Are E and F independent random events? Why?

(b) Find $\Pr(E \neq 3 \cap F)$.

3 Statistics of Random Processes

This chapter introduces variables associated with random processes in the context of biomedical measurements. Consider Figure 3.1, which describes $X(t)$ as a measurement drawn from a set of vectors, (b)→(a), and as a simulation drawn from a distribution function, (c)→(a).

Experiments. Let event space S include all possible measurements of serum cholesterol from patient blood samples. Let E be a set of measurements in a subspace of S originating from a device that your company manufactures. Let F be a set of measurements from a competitor's device. Because both devices can be applied to the same patient, these subspaces overlap. Within E, let G define a smaller subspace for measurements made using your device under specific experimental testing conditions. Measurements drawn from G during testing define *random variable X*, which represents all voltage sequences that can be recorded by this device under these experimental conditions.

If X depends on time, $X(t)$, we refer to it as a *random process*. Among the N time series recorded, the nth realization of process $X(t)$ is indicated using $x_n(t)$ for any $n \in N$. $X(t) = \{x_n(t)\}$ is the set of measurements acquired. If testing is expanded to include blood samples from patients located in different cities over time, then subspace G must expand to become a function of both space and time, e.g., $X(z,t)$. Alternatively, the data set may include $J = 2$ different measurements within the same event space. Now $\mathbf{X} \in \mathbb{R}^{J \times 1}$ is a random vector, and its nth realization is $\mathbf{x}_n = (x_{1n}, x_{2n})^\top$. To discuss the probability of acquiring a time series, we must take into consideration the event space for that experiment.

Simulations. From the physical properties of the experiment, we can sometimes identify the underlying distribution that characterizes random process $X(t)$. For example, if the signal in Figure 3.1a is a recording dominated by acquisition noise in an electronic measurement device, the physics of that process suggests $X(t)$ is a continuous random variable that is well represented by a normal distribution. Once we know the distributions of measurement variables defining a process and can identify the associated parameters, we are able simulate experimentally measured vectors like those drawn from set G for testing and development. We now examine continuous and discrete distributions commonly used to represent random measurement variables.

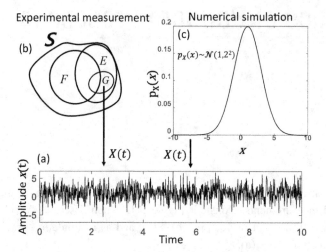

Figure 3.1 Random process $X(t)$ plotted in (a) is represented experimentally as vectors X drawn from set $G \in E \in S$ in (b). These data can be modeled numerically from the physics that determines the probability distribution, $p_X(x(t))$ – in this case, a normal distribution with parameters $\mu = 1, \sigma = 2$ in (c). (b)→(a) describes the acquisition of experimental data as drawing vectors from a set without direct reference to the underlying physics. (c)→(a) describes simulation of data based on known physical properties that are represented as parameters of a probability distribution.

3.1   Univariate Continuous Random Variables

3.1.1   Probability Density Function, pdf

DEFINITION 3.1.1   *X is a continuous random variable if there exists a nonnegative function p defined for all $x \in \mathbb{R}$ with the property that for any set of real numbers B,*

$$\Pr(X \in B) = \int_{B \in S} dx \; p_X(x). \tag{3.1}$$

Equation (3.1) states that the probability of X being in event subspace B is found by integrating $p_X(x)$ over B. If B does not extend over the range of x, then $\Pr(X \in B) < 1$. We will write $p(x)$ to indicate $p_X(x)$ except in situations where it is unclear that x is a realization of variable X.

Because measurement variable X is continuous, $p(x)$ has units of $[x]^{-1}$, which explains why we call it a *probability density function (pdf)* for continuous random variable X. For example, if $p(v)$ is the pdf for signal voltage variable V, then $p(v)$ has units [volts]$^{-1}$ and thus probability $\Pr(a < V < b) = \int_a^b dv\, p(v)$ is unitless, as it must be.

It is natural to ask, What is the probability that variable X equals a specific realization x? Mathematically, the answer is $\Pr(X = x) = 0$ because X is continuous and there are an infinite number of potential realizations between any two points.

Examining $\Pr(X)$ over a small continuous interval dx centered at $X = x$ over which $p(x)$ is constant gives

$$\Pr(x - dx/2 \leq X \leq x + dx/2) = \int_{x-dx/2}^{x+dx/2} dx' \, p(x')$$

$$= p(x) \int_{x-dx/2}^{x+dx/2} dx' \simeq dx \, p(x). \qquad (3.2)$$

In words, $dx \, p(x)$ is a *measure* on a set in the event space that tells us how likely we are to find X *near* x. While $p(x)$ is not a probability, $dP(x) = dx \, p(x)$ is a probability and what we mean descriptively by $\Pr(X = x)$ for a continuous random variable.

3.1.2 Cumulative Distribution Function, cdf

Any integral of a pdf yields a probability provided the integral exists. The *cumulative distribution function (cdf)* of X is defined as

$$P(x) \triangleq \Pr(X \leq x) = \int_{-\infty}^{x} dx' \, p(x'). \qquad (3.3)$$

The pdf is returned from the cdf by differentiating:

$$\left. \frac{dP(x')}{dx'} \right|_{x'=x} = p(x).$$

3.1.3 Normal Random Variable, $\mathcal{N}(\mu, \sigma^2)$

The most commonly used continuous probability distribution is based on a *normal or Gaussian probability density function (pdf)* [90]. If X is one normal random variable, its univariate pdf is expressed as

$$X \sim \mathcal{N}(\mu, \sigma^2) = p(x; \mu, \sigma^2) = \frac{1}{\sigma\sqrt{2\pi}} e^{-(x-\mu)^2/2\sigma^2} \quad \text{for } X \in \mathbb{R}^\infty. \quad \text{(UVN pdf)}$$

$$(3.4)$$

One form uses operator notation, $X \sim \mathcal{N}(\mu, \sigma^2)$, indicating parameters μ, σ that entirely define this pdf. Another form, $p(x; \mu, \sigma^2)$, explicitly indicates random variable X and the pdf parameters separated by a semicolon.

The second axiom of probability states that the probability associated with the entire event space must be 1. Consequently, we must check whether the pdf integrates to 1. Changing variables in (3.4), $y = (x - \mu)/\sigma$, we obtain $dy = dx/\sigma$ and the integration limits remain unchanged:

$$dI = dy \, p(y) = \frac{1}{\sqrt{2\pi}} dy \, e^{-y^2/2}.$$

The square of the integral is

$$I^2 = \int_{-\infty}^{\infty} dI \int_{-\infty}^{\infty} dI' = \frac{1}{2\pi} \int_{-\infty}^{\infty} dy \int_{-\infty}^{\infty} dy'\, e^{-(y^2+y'^2)/2}.$$

Transforming to polar coordinates, $y' = r\cos\theta$, $y = r\sin\theta$, $dy\,dy' = r\,dr\,d\theta$, and changing the integration limits as required gives

$$I^2 = \frac{1}{2\pi} \int_0^{2\pi} d\theta \int_0^{\infty} dr\, r e^{-r^2/2} = -\frac{2\pi}{2\pi} e^{-r^2/2}\Big|_0^{\infty} = 1.$$

Since $I = 1$, we have shown that $\int dx\, p(x) = 1$. The scale factor $1/\sigma\sqrt{2\pi}$ in (3.4) is essential to satisfying the axioms of probability. In general, it is helpful to know that

$$\int_{-\infty}^{\infty} dy\, e^{-ay^2} = \sqrt{\pi/a}. \tag{3.5}$$

Then you only need to perform a change of variable.

3.1.4   Ensemble Operator and Moments

Ensemble operator $\mathcal{E}$ is defined as $\mathcal{E}\{\cdot\} \triangleq \int_{-\infty}^{\infty} dx\{\cdot\}\, p(x)$. Applying $\mathcal{E}$ to an integer power of X^m, we find the mth *moment* of X. This form applies to any continuous random variable. Let's examine two cases.

DEFINITION 3.1.2   *The mth* noncentral moment *of univariate X is*

$$\mathcal{E}X^m = \int_{-\infty}^{\infty} dx\, x^m\, p(x). \tag{3.6}$$

The mth central moment *of X is*

$$\mathcal{E}\{(X - \mathcal{E}X)^m\} = \int_{-\infty}^{\infty} dx\, (x - \mathcal{E}X)^m\, p(x). \tag{3.7}$$

For a normal distribution, the first noncentral moment is the mean given by parameter μ, and the second central moment is the variance given by σ^2.

Let's show that μ is the *mean* of X.

$$\text{mean}(X) \triangleq \mathcal{E}X = \int_{-\infty}^{\infty} dx\, x\, p(x) \tag{3.8}$$

$$= \frac{1}{\sigma\sqrt{2\pi}} \int_{-\infty}^{\infty} dx\, x\, e^{-(x-\mu)^2/2\sigma^2}$$

$$= \frac{1}{\sigma\sqrt{2\pi}} \left[\int_{-\infty}^{\infty} dx\, (x - \mu) e^{-(x-\mu)^2/2\sigma^2} + \mu \int_{-\infty}^{\infty} dx\, e^{-(x-\mu)^2/2\sigma^2} \right]$$

$$= \frac{1}{\sigma\sqrt{2\pi}} \int_{-\infty}^{\infty} dy\, y\, e^{-y^2/2\sigma^2} + \mu \int_{-\infty}^{\infty} dx\, p(x) = \mu.$$

The first integral in the last line is zero because $y\,e^{-y^2/2\sigma^2}$ is antisymmetric about zero, where $y = (x - \mu)$. The second integral is from the second axiom of probability.

3.1.5 Expectation of Functions of Random Variables

Let $q(x)$ be a function of random variable X. The mth noncentral moment is

$$\mathcal{E}q^m = \int_{-\infty}^{\infty} dx\, q^m(x)\, p(x). \tag{3.9}$$

3.1.6 Finite Ensembles and Sample Means

We employ different types of means depending on the questions we have about the data. The ensemble mean for X defined in (3.8) sums all possible realizations $x(t)$ after they are weighted by $p(x(t))$. The result is a function of time, $\mathcal{E}X(t)$, which we now indicate explicitly. In practice, only N measured time series are acquired, as illustrated in Figure 3.2. The sample ensemble mean $\bar{X}(t)$ for a finite data set is

$$\bar{X}(t) = \mathcal{E}X(t; N) = \sum_{n=1}^{N} x_n(t)\, p(x_n(t)) \underset{N \to \infty}{=} \mu(t). \tag{3.10}$$

This estimate of the ensemble mean for the population, $\mu(t)$, is found using N realizations at each time t. We indicate the limited sample number with a parameterized ensemble operator, $\mathcal{E}X(t; N)$. It is possible that signals like those in Figure 3.1a might have a pdf that varies with each instant of time. Notice that the sum in (3.10) is over waveform realizations via n, and not over time. If our assumptions about $p(x)$

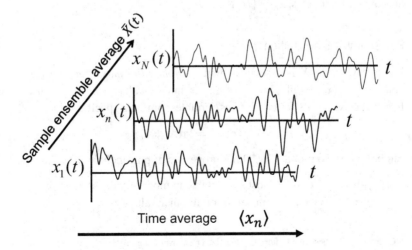

Figure 3.2 Illustration of three of N realizations in the set $\{x_n(t)\}$. Also shown are the axes over which the sample ensemble average $\bar{X}(t)$ and time average $\langle x_n \rangle$ are obtained.

are correct and N becomes large, the sample mean will asymptotically approach the population mean $\mu(t)$. (See Problem 3.4.)

We may also be interested in a time average for the nth realization of X,

$$\langle x_n \rangle = \frac{1}{T_0} \int_0^{T_0} dt\, x_n(t). \tag{3.11}$$

Assuming values in the univariate pdf are equally likely, we apply the factor $1/T_0$ to equally weight each time value in $x_n(t)$. By this definition, the time-averaged value is not a function of time,[1] but rather a function of the waveform, as indicated by subscript n. Sometimes the ensemble and temporal means are equal, but to describe those conditions we need to first describe stationary and ergodic processes.

3.1.7   Standard Normal Random Variable

In MATLAB, z=randn(100,1); gives a 100×1 column array of realizations for *standard normal* $Z \sim \mathcal{N}(0,1)$. The conversion between standard normal Z and general normal random variable X is $x = \sigma z + \mu$. In MATLAB, x=s*randn(M,N)+m generates a 2-D array of normal random values having mean m and variance s^2.

3.2   Discrete Random Variables

If X can take on at most a countable number of values in event space S, then X is a *discrete random variable*. As discussed in Section 2.1, for example, X may be the number of photons absorbed by a solution exposed to N photons or the number of heads following N coin tosses. The range of possible discrete values X can take on is given by the set $\{x_i\}$. In the case of photon absorption, $\{0 \le x_i \le N\}$ for integer N. In the case of a coin toss, $x_i = \{H, T\}$.

3.2.1   Probability Mass Function, pmf

Discrete variable X is associated with a *probability mass function (pmf)* using the notation[2] $p_X[x_i] = p[x_i]$ for $x_i \in$ S. For integer-valued x_i, we may use i, i.e., $p[i]$. In contrast with pdfs for continuous variables, pmf at a specific value,

$$p[x_i] \triangleq \Pr(X = x_i), \tag{3.12}$$

can be a nonzero probability. Other properties of pmf include the following:

- $p[x_i]$ is positive for at most a countable number of values of x_i; i.e., $p[x_i] > 0$ $i = 1, 2, \ldots$ and zero for all other values of x_i.

---

[1] Chapter 5 describes signal filtering, a localized time-averaging process that remains a function of time.

[2] Although I use p for pdf and pmf, the exact meaning can be adjusted without ambiguity in context.

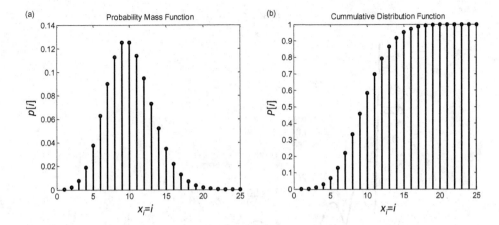

Figure 3.3 (a) Probability mass function $p[i]$ and (b) cumulative distribution function $P[i]$ for Poisson random variable X, for input range $1 \leq i \leq 25$ and $\lambda = 10$. Notice that $p[i]$ is slightly asymmetric.

- $\sum_{\forall i \in S} p(x_i) = 1$.
- The cumulative distribution function (cdf) is $P[x_i] = \sum_{i'=-\infty}^{i} p[x_{i'}]$.

Examples of pmf $p[i]$ and cdf $P[i]$ for the Poisson process defined in Section 3.2.2 are found in Figure 3.3. These curves were generated using MATLAB functions (Figure 3.3a) p=poisspdf(x,10); and (Figure 3.3b) P=poisscdf(x,10); for parameter $\lambda = 10$. Another useful function is x=poissrnd(lambda,M,N);, which generates an $M \times N$ matrix of "uncorrelated" Poisson pseudo-random numbers for parameter λ. Note that the symbol λ has many uses in this book.

3.2.2   Poisson Random Variable

The event space of a Poisson random variable lies within the nonnegative integers, $S \ni i \geq 0$, and is completely defined by one unitless parameter, $\lambda > 0$. Equivalent representations of a Poisson process include

$$X \sim \mathcal{P}(\lambda) = p[i] = \Pr(X = i) = e^{-\lambda} \frac{\lambda^i}{i!} \quad \text{for } i = 0, 1, 2, \ldots \in S. \quad \text{(Poisson pmf)}$$

$$(3.13)$$

The Poisson operator $\mathcal{P}(\lambda)$ applied to λ completely specifies the pmf. The dependence of $p[i]$ on λ is shown in Figure 3.4.

Let's check that $p[i]$ sums to 1, as it must to represent $\Pr(X = i)$.

$$P(\infty) = \sum_{i=0}^{\infty} p[i] = e^{-\lambda} \sum_{i=0}^{\infty} \frac{\lambda^i}{i!} = e^{-\lambda} e^{\lambda} = 1.$$

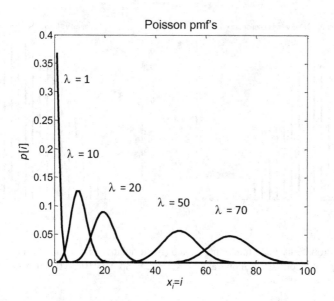

Figure 3.4 Poisson probability mass functions for a range of λ values.

The result requires that we know the series expansion $e^x = \sum_{i=0}^{\infty} x^i / i!$. Examining Figure 3.4, we see that as λ increases, the pmf broadens and the peak decreases so the area under the pmf maintains a value of 1.

Poisson variables are appropriate for describing many common discrete stochastic processes. For example, they are widely used to model photon noise (microscopy, x-rays, nuclear imaging) since photons are positive, countable, discrete-number events. Other common phenomena modeled as Poisson processes include the following:

- The number of cells in a unit volume of tissue
- The number of radioactive decays per second for an isotope sample
- The number of photons absorbed in a detector area per second

Evans ([34], chapter 26) illustrates the principles underlying a *Poisson process* by deriving its frequency distribution through a consideration of radioactive nuclear decay. Let's modify his example to consider biological cell proliferation. In the following example, the goal is to enumerate assumptions required to derive the pmf and to show how they enter into the derivation.

Example 3.2.1. *Cell numbers:* *Let event X be the number of cells that undergo division in a culture sample during measurement time interval T, where* S ∋ *positive integers. We assume that*

1. *The rate of cellular proliferation is the same for all cells in an experiment.*
2. *Each cell proliferates independently of the others.*

3. *Cell proliferation rate is constant over the measurement time. That generally means the reproductive lifetime of cells should be much longer than the observation time of the experiment.*
4. *The number of cells and the total observational time intervals are large, so that we can obtain statistical averages that approach the ensemble statistics.*

If positive constant λ' is the average rate of cell division for a given culture, then $\lambda'T$ is the probability that a cell will divide during time interval T, viz., $\Pr(X(T) = 1) = \lambda'T$. Reducing T to an infinitesimal dt, $\Pr(X(dt) = 1) = \lambda' dt$, we can say that $\lambda' dt \ll 1$. Also the probability of observing two or more cells dividing during dt becomes much less than that of observing one division; i.e., $\Pr(X(dt) = 1) \gg \Pr(X(dt) = 2)\ldots$ Therefore, a good approximation to the probability of observing no cells dividing during interval dt is

$$\Pr(X(dt) = 0) = 1 - \Pr(X(dt) = 1) = 1 - \lambda' dt.$$

The chance of finding n cells dividing during $t + dt$, i.e., $\Pr(X(t + dt) = i)$, may be expressed as a combination of the probabilities of finding $i - 1$ divisions in time t and one division during dt, i.e., $\Pr(X(t) = i - 1) \times \Pr(X(dt) = 1)$, or i divisions during t and no divisions during dt, i.e., $\Pr(X(t) = i) \times \Pr(X(dt) = 0)$. Mathematically, this statement reads

$$\Pr(X(t + dt) = i) = \Pr(X(t) = i) \Pr(X(dt) = 0)$$
$$+ \Pr(X(t) = i - 1) \Pr(X(dt) = 1)$$
$$= \Pr(X(t) = i)(1 - \lambda' dt) + \Pr(X(t) = i-1) \lambda' dt$$
$$\frac{\Pr(X(t + dt) = i) - \Pr(X(t) = i)}{dt} = \lambda'\big(\Pr(X(t) = i - 1) - \Pr(X(t) = i)\big)$$
$$\frac{d \Pr(X(t) = i)}{dt} = \lambda' \left(\Pr(X(t) = i - 1) - \Pr(X(t) = i)\right). \quad (3.14)$$

The solution to this first-order differential equation is

$$\Pr(X(t) = i) = p[i] = \frac{(\lambda' t)^i}{i!} e^{-\lambda' t}, \quad (3.15)$$

which may be verified by substituting (3.15) into (3.14).

The expressions for the Poisson pmf of (3.13) and (3.15) differ in several important ways. First, λ' is not the unitless parameter λ, but rather a rate constant that has the units of time^{-1}. In fact, $\lambda \leftrightarrow \lambda' t$. Second, the X in (3.15) is a time-dependent random process. The associated distributions of X in (3.13) and (3.15) are essentially the same except that the time dependence in (3.15) can be important for interpretation. Note that integer i is a function of time for this random process $X(t) = \{i(t)\}$.

One reason to go through the Poisson pmf derivation is to reinforce the point that the analysis tools used in research are strictly valid only under stated assumptions. We must be careful to ensure the assumptions hold for the problem at hand before we put any faith in what the tool tells us. In the cell-growth problem, if the observation time

of the experiment is on the order of or greater than the lifetime of the cells, we violate
the third condition necessary for our cell proliferation experiment to be modeled as a
strict Poisson process. It turns out that we can remedy this situation by adding a cell-
death term in the differential equation (see Section 10.4). Ultimately, it is up to you
to decide how egregious any violations of the assumptions are when modeling your
experiment as a Poisson variable.

3.2.3 Mean of a Poisson Random Variable

The mean of Poisson variable X is found from the first moment as follows:

$$\mathcal{E}X(t) \triangleq \sum_{i:p[i]>0} i\, p[i] = \sum_{i=0}^{\infty} i e^{-\lambda} \frac{\lambda^i}{i!}$$

$$= \lambda e^{-\lambda} \sum_{i=1}^{\infty} \frac{\lambda^{(i-1)}}{(i-1)!} = \lambda e^{-\lambda} \sum_{j=0}^{\infty} \frac{\lambda^j}{j!} = \lambda e^{-\lambda} e^{\lambda} = \lambda(t), \tag{3.16}$$

where we made a change of variable, $j = i - 1$. The population ensemble mean of a
Poisson variable is generally a function of time.

The population ensemble variance of a Poisson variable is

$$\text{var}\{X(t)\} = \mathcal{E}\left\{(X - \mathcal{E}X)^2\right\} = \lambda(t), \tag{3.17}$$

where "var," denoting variance, is applied as an operator. (See Problem 3.3.)

Notational Comparison of Discrete and Continuous Variables
Pr is probability; P is cdf; p denotes pdf or pmf.

Discrete Random Variable X	Continuous Random Variable X
$\Pr(X)$	$\Pr(X)$
$P_X[x_i] = P[i] \triangleq \Pr(X \le x_i)$	$P_X(x) = P(x) \triangleq \Pr(X < x)$
$p_X[x_i] = p[i] \triangleq \Pr(X = x_i)$	$p_X(x) = p(x) = dP(x')/dx'\|_{x'=x}$
$p[i] = P[i] - P[i-1]$	$dx\, p(x) = \Pr(x - dx/2 \le X \le x + dx/2) = dP(x)$
$\Pr(X \in (-\infty,\infty)) = P(\infty) = \sum_{i \in S} p[i] = 1$	$\Pr(X \in (-\infty,\infty)) = P(\infty) = \int_S dx\, p(x) = 1$
$\Pr(X = a) = \sum_{i=a}^{a} p[i] = p[a]$	$\Pr(X = a) = \int_a^a dx\, p(x) = 0$

3.3 Jointly Distributed Random Variables

Probabilities associated with two continuous random variables X and Y are described
using the joint cumulative distribution function,

$$\Pr(X \le x, Y \le y) = P_{XY}(x,y) = \int_{-\infty}^{y} dy' \int_{-\infty}^{x} dx'\, p_{XY}(x',y') \qquad \text{for } x, y \in \mathbb{R}.$$

Analogous to the univariate case,

$$\Pr(x < X < x + dx, y < Y < y + dy) = \int_y^{y+dy} dy' \int_x^{x+dx} dx' \, p(x', y')$$

$$\simeq dx \, dy \, p(x, y) = \partial^2 P(x, y),$$

which states the relationships among probability, pdf, and cdf for *bivariate* distributions.

The *marginal cdf*, $P_X(x)$, is found from the joint bivariate distribution using

$$P_{XY}(x, y = \infty) = \Pr(-\infty < X \le x, -\infty < Y \le \infty)$$

$$= \int_{-\infty}^x dx' \int_{-\infty}^\infty dy' \, p_{XY}(x', y') = \int_{-\infty}^x dx' \, p_X(x') = P_X(x).$$

Notice the integration limits in this equation. We can apply a similar procedure to find $P_Y(y)$. Dropping the subscripts, the *marginal probability density functions* are

$$p(x) = \frac{dP(x)}{dx} = \int_{-\infty}^\infty dy' \, p(x, y') \quad \text{and} \quad p(y) = \frac{dP(y)}{dy} = \int_{-\infty}^\infty dx' \, p(x', y).$$

J-variate continuous random variables are compactly expressed as column vectors $\mathbf{X}^\top = (X_1 \ldots X_j \ldots X_J) \in \mathbb{R}^{J \times 1}$, where $X_j = \{x_j\}$ for $1 \le j \le J$. The associated cdf is $P(\mathbf{x}) = \Pr(X_1 \le x_1, \ldots, X_J \le x_J)$ and the pdf is $p(\mathbf{x}) = \partial^N P(\mathbf{x})/\partial x_1 \ldots \partial x_N$. Also, the marginal density of the jth component is

$$p(x_j) = \int_{-\infty}^\infty dx_J \cdots \int_{-\infty}^\infty dx_{j+1} \int_{-\infty}^\infty dx_{j-1} \cdots \int_{-\infty}^\infty dx_1 \, p(\mathbf{x}). \quad (3.18)$$

From (3.18), the mean vector, its jth component, and a mean vector conditioned on state θ_i are, respectively,

$$\mu = \mathcal{E}\{\mathbf{X}\} = \int_{\mathbf{X} \in \mathbb{S}} d\mathbf{x} \, \mathbf{x} \, p(\mathbf{x})$$

$$\mu_j = \mathcal{E}\{X_j\} = \int_{\mathbf{X} \in \mathbb{S}} d\mathbf{x} \, x_j \, p(\mathbf{x}) = \int_{-\infty}^\infty dx_j \, x_j \, p(x_j)$$

$$\mu_i = \mathcal{E}\{\mathbf{X}|\theta_i\} = \int_{\mathbf{X} \in \mathbb{S}} d\mathbf{x} \, \mathbf{x} \, p(\mathbf{x}|\theta_i),$$

where $\int_{\mathbf{X} \in \mathbb{S}} d\mathbf{x}$ indicates a J-dimensional integral over the entire J-D event space. Elements of random vector $\mathbf{X}$ are generally dependent, but when they are independent, $p(\mathbf{x}) = p(x_1)p(x_2) \ldots p(x_J)$.

3.4   Data Statistics

Descriptive statistics summarize a sample of data drawn from a population. Statistical moments of data recorded from device output can describe object and/or measurement system properties in equilibrium. For example, the sample mean and variance of core temperature, systemic blood pressure, and heart rate measurements acquired from a specific group of adults inform physicians about that group's state of health through the lens of the measurement devices when compared with population statistics.

   Inferential statistics analyze sample data to infer properties of the greater population from which the sample was drawn. When these statistics are used, the goal is to reach conclusions about the whole from a random sampling of the population. Hypothesis testing, confidence intervals, and regression analysis help us infer population properties from statistical estimates, often called *test statistics*.

   The following discussion of descriptive statistics helps to set up an exploration of inferential statistics at the end of this chapter and in Chapter 8.

3.4.1   Population and Sample Statistics

We now apply the moment equations of (3.6) and (3.7) to various data sets.

Ensemble mean.
Equation (3.10) states that the N-sample ensemble mean for continuous process $X(t)$ is

$$\bar{X}(t) = \mathcal{E}X(t; N) = \sum_{n=1}^{N} x_n(t)\, p(x_n(t)) = \frac{1}{N} \sum_{n=1}^{N} x_n(t). \text{ Sample ensemble mean}$$

(3.19)

The last form is equivalent to assuming a uniformly distributed pdf, viz., $p(x_n(t)) = 1/N$. Without knowledge of the pdf, it is appropriate to assume the *maximum entropy* state (see Section 3.14).

   As $N \to \infty$ while increment $\Delta x = x_n - x_{n-1} \to 0$, the discrete measure $\Delta x\, p(x_n)$ becomes continuous $dx\, p(x)$, and we find the sample ensemble mean approaches the population ensemble mean from (3.8):

$$\mathcal{E}X(t) = \int_{-\infty}^{\infty} dx\, x(t)\, p(x(t)). \qquad \text{Population ensemble mean} \qquad (3.20)$$

Problems 3.4 and 3.5 describe *maximum likelihood estimates* of population mean and variance from sample data.

Ensemble variance.
Beginning with the second central moment from (3.7), the population ensemble variance is

$$\text{var}(X(t)) = \mathcal{E}\{(X(t) - \mathcal{E}X(t))^2\} = \int_{-\infty}^{\infty} dx \, (x(t) - \mathcal{E}X(t))^2 \, p(x(t)). \qquad (3.21)$$

Population ensemble variance

For finite sample size N and uniform pdf over S, (3.21) and (3.19) combine to give the sample ensemble variance:

$$\widehat{\text{var}}(X(t)) = \frac{1}{N-1} \sum_{n=1}^{N} (x_n(t) - \bar{X}(t))^2. \qquad \text{Sample ensemble variance} \qquad (3.22)$$

Since the sample mean is computed from the same data as the sample variance is, the number of degrees of freedom is reduced by one. Consequently, the unbiased estimate is scaled by $1/(N-1)$ instead of $1/N$.

Temporal moments.
The definition of a temporal mean from (3.11) is intuitive but not very practical when describing sampled data from a continuous random process. Consequently, we now discuss temporal moments of temporally sampled waveforms.

We introduce the basic sampling operator in Section 5.7, but apply it now to estimate $dx \, p_s(x(t))$, the probability of a sampled waveform. The ensemble operator shown here samples $x(t)$ in time before computing the moment. For notational simplicity, define $dPR(X(t)) \triangleq \Pr(x(t) - dx/2 \leq X \leq x(t) + dx/2)$ from (3.2). Applying (5.23), the probability of the sampled waveform becomes

$$dx \, p_s(x(t)) = \mathbf{S}^\dagger \mathbf{S}\{dx \, p(x(t))\} = \sum_{j=-\infty}^{\infty} dPR(X(t)) \, \delta(x(t) - x(jT)),$$

where index j indicates the time sample along the waveform, $x(jT)$. Sampling operator $\mathbf{S}^\dagger \mathbf{S}$ converts acquired data $x(t)$ from the continuous-time domain t to the discrete-sample domain j and back to the continuous domain via jT. Although the sum of sampling deltas is not bounded, the acquired data to which the sampling operator is applied are bounded in time.

From (3.6), the mth time-average moment for the sampled waveform is

$$\langle x^m \rangle = \int_{-\infty}^{\infty} x^m(t) \, dx \, p_s(x(t)) = \int_{-\infty}^{\infty} x^m(t) \sum_{j=-\infty}^{\infty} dPR(X(t)) \, \delta(x(t) - x(jT))$$

$$= \sum_{j=-\infty}^{\infty} \int_{-\infty}^{\infty} x^m(t) \, dPR(X(t)) \, \delta(x(t) - x(jT))$$

$$= \sum_{j=1}^{J} x^m(jT) \, \Pr(X = x(jT)) = \sum_{j=1}^{J} x^m(jT) \, p(x(jT)). \qquad (3.23)$$

That's lots of work but doesn't lead to any surprises.

The mechanics of applying a sampling operator and integrating over Dirac deltas are handled in Section 5.7. The importance of (3.23) for this chapter is the expression for computing temporal moments from sampled data acquired from a continuous random process.

Following the ideas leading to (3.19) and applying (3.23), the sample temporal mean for the nth waveform is

$$\langle x_n \rangle = \frac{1}{J} \sum_{j=1}^{J} x_n(jT), \qquad \text{Sample temporal mean} \qquad (3.24)$$

which is constant over time index j. Comparing (3.24) and (3.19), you will notice that one is a sum over n (ensemble samples) and the other is a sum over j (time samples).

Following the ideas leading to (3.22) and carefully applying (3.23) for the second central moment, we find the sample temporal variance:

$$s_n^2 = \frac{1}{J-1} \sum_{j=1}^{J} (x_n(jT) - \langle x_n \rangle)^2, \qquad \text{Sample temporal variance} \qquad (3.25)$$

which strictly applies only to the nth waveform. In Section 3.8, we discuss situations in which the temporal and ensemble moments are approximately equal.

3.4.2   Mean-Squared Error

Random errors are well represented by central moments such as *variance*, which describe measurement *precision*. Systematic errors add *bias* to measurement uncertainty, which is a statement of measurement *accuracy*. Bias is defined as the tendency of a measurement to generate a sample mean that does not converge to the population mean as the sample size increases. Suppose there are legitimate reasons to expect $X(t) \sim \mathcal{N}(\mu, \sigma^2)$, where the sample ensemble mean is time invariant, $\bar{X}(t) = \bar{X}$. Bias in the sample temporal mean $\langle x_n \rangle$ is $b(x_n(jT)) \triangleq \langle x_n \rangle - \mu$, which can be positive or negative.

Total measurement uncertainty is the deviation of measurements about a constant population mean. It is quantified by *mean squared error* for the nth time series, MSE_n,

$$\text{MSE}_n \triangleq \frac{1}{J} \sum_{j=1}^{J} (x_n(jT) - \mu)^2. \qquad (3.26)$$

MSE_n resembles sample variance, except that μ is used instead of $\langle x_n \rangle$ so there are now J degrees of freedom. These differences are important.

Beginning with (3.26), we complete the square to find

$$
\begin{aligned}
\mathrm{MSE}_n &= \left[\frac{1}{J}\sum_{j=1}^{J} x_n^2(jT)\right] - \left[\frac{2\mu}{J}\sum_{j=1}^{J} x_n(jT)\right] + \mu^2 \\
&= \left[\frac{1}{J}\sum_{j=1}^{J} x_n^2(jT)\right] - 2\langle x_n\rangle\mu + \mu^2 + \langle x_n\rangle^2 - \langle x_n\rangle^2 \\
&= \frac{1}{J}\left[\left(\sum_{j=1}^{J} x_n^2(jT)\right) - J\langle x_n\rangle^2\right] + \langle x_n\rangle^2 - 2\langle x_n\rangle\mu + \mu^2 \\
&= \left[\frac{1}{J}\sum_{j=1}^{J}\left(x_n(jT) - \langle x_n\rangle\right)^2\right] + \left(\langle x_n\rangle - \mu\right)^2 \simeq s_n^2 + b_n^2. \quad (3.27)
\end{aligned}
$$

MSE_n is the sum of sample variance and squared bias for the nth waveform. It displays random and systematic errors, as illustrated in Figure 3.5. The *root-mean-square error* (RMSE) is $\sqrt{\mathrm{MSE}_n}$. Although the sample temporal variance, s_n^2, is scaled by $1/J$, that estimation bias is negligible if $J \gg 1$.

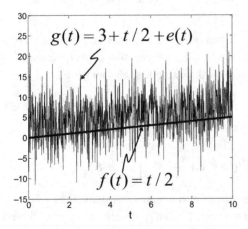

Figure 3.5 Object function $f(t) = t/2$ represents the population mean for a data set. Data sampled from the same object are described by $g_n(jT) = 3 + (jT)/2 + e(tT)$, which includes significant additive zero-mean white-Gaussian noise $e(jT) \sim \mathcal{N}(0, 5^2)$ and a temporally constant bias term $b(g_n) = 3$. From (3.27), $\mathrm{MSE}_n = s_n^2 + b^2(g_n) = 25 + 9 = 34$. While the standard deviation describing random error is 5, the RMSE is 17% larger, 5.8, because of the addition of systematic error (bias).

3.4.3   Moment-Generating Functions

In the preceding discussion, we stated that the physics of a measurement situation should dictate the probability distribution selected to represent the random process under investigation. For example, the positive values, event independence, and discrete nature of photon or cell counting suggest that these quantities may be represented by a Poisson variable. In contrast, the thermal agitation of charge carriers in electronic components generates normally distributed additive acquisition noise. Earlier, we saw that distribution parameters could be estimated from the moments of measurement data, and for Poisson/normal processes all we need are one or two moments to completely represent the data.

In practice, a variety of distributions are needed to fully represent data capturing biological processes [35]. The moments of these distributions are functions of the distribution parameters (see Problem 3.16). *Moment-generating functions* are a simple method for finding the mth moment for any population distribution to which you can compute its mth-order derivative.

Consider dependent random variable X and independent nonrandom variable s. The moment-generating function of X, $M_X(s)$, is defined from the expectation of an exponential function:

$$M_X(s) \triangleq \mathcal{E}e^{sX} \tag{3.28}$$

$$= \int_{-\infty}^{\infty} dx \; e^{sx} \; p_X(x) \qquad \text{For continuous random variables}$$

$$= \sum_{x_n : p(x_n) > 0} e^{sx_n} \; p_X(x_n). \qquad \text{For discrete random variables}$$

Variable $s = \sigma + i\Omega$ is complex and interpretable as a frequency variable[3] with units of $[x]^{-1}$. Statistical moments are found by successive differentiation of $M(s)$ with respect to s, followed by setting $s = 0$. For the first moment, we have

$$M_X^{(1)}(0) \triangleq \left[\frac{d}{ds} \mathcal{E}e^{sX} \right]_{s=0} = \mathcal{E}\left[\frac{d}{ds} e^{sX} \right]_{s=0} = \mathcal{E}X.$$

More generally, $\quad M_X^{(m)}(0) \triangleq \mathcal{E}\left[\frac{d^m}{ds^m} e^{sX} \right]_{s=0} = \mathcal{E}X^m.$

Superscript (m) in parentheses identifies the number of derivatives taken with respect to s, while $\mathcal{E}X^m$ identifies the moment. We can reverse the order of differentiation and expectation (most of the time) because both are linear operators. Another approach is to consider the power series expansion of the exponent, $e^x = \sum_{n=0}^{\infty} x^n / n!$:

---

[3] Actually s is the Laplace frequency variable whose imaginary part is the Fourier frequency variable $\Omega = 2\pi u$. The moment-generating function has the form of a two-sided, conjugate-Laplace transform of the probability density or mass function. That is, $M_X(s) = \mathcal{E}e^{sX} = \int_{-\infty}^{\infty} dx \; p_X(x) \, e^{sx} = \mathcal{L}\{p_X(x)\}_{s \to -s} = \mathcal{L}^* p_X(x)$. We return to this topic in Chapter 11.

$$M_X(s) = \mathcal{E}e^{sX} = \mathcal{E}\left\{ \sum_{k=0}^{\infty} \frac{s^k}{k!} X^k \right\} = \sum_{k=0}^{\infty} \frac{s^k}{k!} \mathcal{E}X^k$$

$$M_X^{(m)}(0) = \left. \frac{d^m M_X}{ds^m} \right|_{s=0} = \mathcal{E}X^m \left. \sum_{k=m}^{\infty} \frac{s^{k-m}}{(k-m)!} \right|_{s=0} = \mathcal{E}X^m, \quad m > 0. \quad (3.29)$$

Let's apply this method to Poisson and standard normal distributions.

Example 3.4.1. *Poisson moment: Find the first moment of Poisson distribution*

$$\mathcal{P}(\lambda) = \Pr(X = i) = p_X(i;\lambda) = \lambda^i e^{-\lambda}/i!$$

using moment-generating functions, where i is the Poisson variable.

$$M_X(s) = \mathcal{E}e^{sX} = \sum_{i:p_X(i)>0} e^{si} p_X(i) = \sum_{i=0}^{\infty} e^{si} \frac{\lambda^i}{i!} e^{-\lambda} = e^{-\lambda} \sum_{i=0}^{\infty} \frac{(\lambda e^s)^i}{i!}$$

$$= e^{-\lambda} \exp(\lambda e^s) = \exp(\lambda(e^s - 1))$$

$$M_X^{(1)}(0) = \lambda e^s \exp(\lambda(e^s - 1))\big|_{s=0} = \lambda \quad \left[\text{chain rule } \frac{d}{dx} e^{f(x)} = e^{f(x)} \frac{df}{dx} \right]$$

Example 3.4.2. *Normal moments: Find the first three moments of a standard normal distribution*

$$p_Z(z;0,1) = \mathcal{N}(0,1) = \frac{1}{\sqrt{2\pi}} e^{-z^2/2}.$$

From (3.28),

$$M_Z(s) = \mathcal{E}e^{sZ} = \frac{1}{\sqrt{2\pi}} \int_{-\infty}^{\infty} dz \, e^{sz} e^{-z^2/2} = \frac{1}{\sqrt{2\pi}} \int_{-\infty}^{\infty} dz \, e^{-(z^2-2sz)/2}$$

$$= \frac{e^{s^2/2}}{\sqrt{2\pi}} \int_{-\infty}^{\infty} d(z-s) \, e^{-(z-s)^2/2} = e^{s^2/2} \quad (3.30)$$

$$M_Z^{(1)}(0) = s \, e^{s^2/2}\Big|_{s=0} = 0$$

$$M_Z^{(2)}(0) = \left[(s^2 + 1) \, e^{s^2/2} \right]_{s=0} = 1$$

$$M_Z^{(3)}(0) = \left[(s^3 + 3s) \, e^{s^2/2} \right]_{s=0} = 0.$$

We found that the mean $= M_Z^{(1)}(0) = 0$ and the variance $= M_Z^{(2)}(0) - \left(M_Z^{(1)}(0) \right)^2 = 1$ as expected. The scaled, central, third moment is called *skewness* in statistics. It is equal to $\mathcal{E}(X - \mu)^3/\sigma^3$ generally, or to $\mathcal{E}Z^3$ for the standard normal process, where $Z = (X - \mu)/\sigma$. The third moment indicates the symmetry of the pdf about the mean. Since normal distributions are symmetric, it makes sense that the skewness measure is zero.

If $M(s)$ can be found and if it is continuously differentiable to order m, the mth population moment for $p_X(x)$ or $p_X[i]$ can be found. This is an important tool for predicting statistical moments from measured data when verifying the model distribution selected and when estimating model parameters from moments.

3.4.4   Characteristic Functions

Although we have not yet described Fourier analysis (this topic is covered in Chapter 5), here we will discuss *characteristic functions* – a type of Fourier transform applied to probability density and mass functions. Specifically, C_X is the conjugate Fourier transform of $p_X(x)$,

$$C_X(\Omega) \triangleq \int_{-\infty}^{\infty} dx\, p_X(x)\, e^{i\Omega x} = \left[\int_{-\infty}^{\infty} dx\, p_X(x)\, e^{-i\Omega x} \right]^* = \mathcal{F}^*\{p_X(x)\} = \mathcal{E} e^{i\Omega X},$$

(3.31)

where Y^* is the complex conjugate of Y. The same result can be obtained from moment-generating functions using the relation $C_X(\Omega) = M_X(s)_{s=i\Omega}$, where the imaginary part of complex $s = \sigma + i\Omega$ is radial frequency $\Omega = 2\pi u$. The *conjugate transform* described in (3.31) and indicated by $\mathcal{F}^*\{p(x)\}$ is also given by the expectation of the Fourier basis, $\mathcal{E} e^{i2\pi u x}$.

Example 3.4.3. *Normal distribution: (a) From the moment-generating function computed in Example 3.4.2, compute the characteristic function for standard normal* $Z \sim \mathcal{N}(0, 1)$.

$$C_Z(\Omega) = M_Z(s)_{s=i\Omega} = e^{-\Omega^2/2}$$

(b) Find $C_X(\Omega)$ for the more general form $X \sim \mathcal{N}(\mu, \sigma^2)$.

$$C_X(\Omega) = \mathcal{E} e^{i\Omega X} = \mathcal{F}^*\{p_X(x)\} = \frac{1}{\sigma\sqrt{2\pi}} \int_{-\infty}^{\infty} dx\, e^{-(x-\mu)^2/2\sigma^2}\, e^{i\Omega x}$$

$$= e^{i2\pi u \mu}\, e^{-2\pi^2 u^2 \sigma^2} = e^{i\Omega \mu - \Omega^2 \sigma^2/2}$$

Can you see how the result of (b) becomes the result for (a) using the standard normal distribution? In (b), we considered the problem as the conjugate Fourier transform of the normal pdf, in which case the theorems and functions of Appendix D apply.

The characteristic function for discrete, integer-valued random variable $X[n]$ that is periodic over time T_0 is given by a variation of the Fourier series expression:

$$C_X(\Omega) = \mathcal{E} e^{i\Omega X} = \sum_n p_X(n)\, e^{i2\pi u n}.$$

In this light, $C_X(\Omega)$ may be interpreted as coefficients that, when appropriately weighted and summed, reconstruct the original pdf:

$$p(n) = \Pr(X = n) = \frac{2}{T_0} \int_{-T_0/2}^{T_0/2} du \, C_X(\Omega) \, e^{-i2\pi un}.$$

What is different is that complex conjugates of the Fourier operators are involved. The sum is over all $n \in S$ for which $p(n) > 0$.

An important application of characteristic functions is in describing the pdf of the sum of probability density (or mass) functions. Note that convolution is discussed in Chapter 4.

Example 3.4.4. *Sum of random variables: Let X and Y be independent, identically distributed (iid) variables with densities $p_X(x)$ and $p_Y(y)$. If $Z = X + Y$, find $p_Z(z)$ in terms of the other two pdfs.*

To find a general expression, begin with the cdf for Z and note that $Y = Z - X$:

$$P_Z(z) = \Pr(X + Y \le z) = \int_{-\infty}^{\infty} dx \, p_X(x) \int_{-\infty}^{z-x} dy \, p_y(y)$$

$$= \int_{-\infty}^{\infty} dx \, p_X(x) P_Y(z - x).$$

Differentiating with respect to Z, the cdfs are converted into pdfs and we find

$$p_Z(z) = \int_{-\infty}^{\infty} dx \, p_X(x) \frac{d P_Y(z - x)}{dz} = \int_{-\infty}^{\infty} dx \, p_X(x) \, p_Y(z - x). \tag{3.32}$$

Hence the pdf of the sum of two independent variables is given by the convolution of the component pdfs, $p_Z(z) = [p_X * p_Y](z)$.

Now we see an application for characteristic functions. To find $p_Z(z)$, first find $C_X(\Omega)$ and $C_Y(\Omega)$, take their product $C_Z(\Omega) = C_X(\Omega) C_Y(\Omega)$, and convert $C_Z(\Omega) \rightarrow p_Z(z)$ using Fourier transforms. See Section 5.6 for details. For example,

$$p_X(x) = \frac{1}{\sqrt{2\pi}} \exp(-x^2/2) \qquad \text{and} \qquad p_Y(y) = \frac{1}{\sqrt{2\pi}} \exp(-y^2/2)$$

$$C_X(u) = \exp(-2\pi^2 u^2) \qquad \text{and} \qquad C_Y(u) = \exp(-2\pi^2 u^2)$$

$$C_Z(u) = C_X(u) C_Y(u) = \exp(-4\pi^2 u^2)$$

$$p_Z(z) = \frac{1}{\sqrt{2\pi}\sqrt{2}} \exp(-z^2/2(2)) = \frac{1}{2\sqrt{\pi}} \exp(-z^2/4).$$

The last line requires that you know the inverse conjugate Fourier transform of a Gaussian function,

$$(\mathcal{F}^{-1})^* \{\exp(a^2 u^2)\} = \frac{\sqrt{\pi}}{a} \exp(-\pi^2 x^2/a^2),$$

which is also the inverse transform. It is very helpful to remember (or keep in your back pocket)

$$\mathcal{F}\left\{\frac{1}{\sqrt{2\pi}}\exp\left(-(t-t_0)^2/2\sigma^2\right)\right\} = \exp(-i2\pi t_0 u)\exp(-2\pi^2\sigma^2 u^2).$$

You can see that for N iid normal variables, X_j, the pdf of their sum is

$$p_Z(z) = \frac{1}{\sqrt{2\pi N}}\exp(-z^2/2N), \quad \text{for } Z = \sum_{j=1}^{N} X_j \text{ and}$$

$$p_{X_j}(x) = \frac{1}{\sqrt{2\pi}}\exp(-x^2/2).$$

The sum of independent normal random variables is another normal random variable. The key word here is *independent*.

3.5   Second-Order Statistics: Matrix Forms

To this point, we have been examining *first-order point statistics*, where random variable $X(t)$ is analyzed one point at a time. Means and variances are examples of first-order statistics. In contrast, covariance and correlation are examples of *second-order statistics* that apply to univariate or multivariate processes, e.g., $\mathbf{X}(t) \sim p(\mathbf{x}(t)) = p(x_1, x_2, \ldots)$, where $\mathbf{X}$ is a vector of random variables that may each vary in time. Second-order temporal statistics describe how variations in the jth variable at time t, $x_j(t)$, couple to those at other times, $x_j(t')$, e.g., $\mathcal{E}\{x_j(t)\,x_j(t')\}$. Other second-order measures describe how two variables co-vary at one point in time, e.g., $\mathcal{E}\{x_j(t)\,x_{j'}(t)\}$. Both cases are second order in variable x.

Consider the situation where we are testing a new blood pressure-regulating medicine in a large group of paid volunteers. Each patient's blood pressure (BP) is measured at the time the medicine is administered and then every hour for 24 hours. Consequently, $t = nT, 0 \le n \le N-1, T = 1$ hr, $N = 25$. To study coupled variability among volunteers at various times, we examine the *covariance* of the BP variable.

3.5.1   Population Covariance

Population covariance is a quantity assumed a priori or estimated from a joint pdf. Let's consider the covariance between the first and the last BP measurements. The two relevant variables are ensembles of patient measurements made at the initial and the final times: $y_1 = \{x(t = 0\,\text{hr})\}$ and $y_2 = \{x(t = 24$ hr$)\}$. These form data vector $\mathbf{Y} = \begin{pmatrix} y_1 \\ y_2 \end{pmatrix}$, which is characterized by joint pdf $\mathbf{Y} \sim p(y_1, y_2)$, which is assumed to be multivariate normal (see Section 3.9). The full pdf, $p(y_1, y_2)$, may be approximated by normalizing a 25×25 histogram of BP measurements from a large patient ensemble.

Applying the population ensemble mean vector, $\mu = \begin{pmatrix} \mu_1 \\ \mu_2 \end{pmatrix} = \begin{pmatrix} \mathcal{E}y_1 \\ \mathcal{E}y_2 \end{pmatrix}$, we define the *population covariance matrix* as the second central moment:

$$
\begin{aligned}
\mathbf{K} &= \mathcal{E}\{(\mathbf{Y} - \mu)(\mathbf{Y} - \mu)^\top\} \qquad \text{(Population Covariance Matrix)} \\
&= \begin{pmatrix} \mathcal{E}(y_1 - \mu_1)^2 & \mathcal{E}\{(y_1 - \mu_1)(y_2 - \mu_2)\} \\ \mathcal{E}\{(y_2 - \mu_2)(y_1 - \mu_1)\} & \mathcal{E}(y_2 - \mu_2)^2 \end{pmatrix} \\
&= \begin{pmatrix} \text{var}(Y_1) & \text{cov}(Y_1 Y_2) \\ \text{cov}(Y_2 Y_1) & \text{var}(Y_2) \end{pmatrix}.
\end{aligned}
\tag{3.33}
$$

Variances $\text{var}(Y_1)$ and $\text{var}(Y_2)$ are given by (3.21). The population ensemble covariances are equal, $\text{cov}(Y_1 Y_2) = \text{cov}(Y_2 Y_1)$, and are given by

$$
\text{cov}(Y_1 Y_2) = \int_{-\infty}^{\infty} dy_2 \int_{-\infty}^{\infty} dy_1 (y_1 - \mu_1)(y_2 - \mu_2) p(y_1, y_2).
\tag{3.34}
$$

The matrix described by (3.33) is actually quite simple. The work lies in calculating the matrix elements, and is not described here. By indicating operators as we have done here, we hide the computational details of (3.33), making the result more conceptual than practical.

In regard to correlation, covariance, and independence as determined from population statistics:

- Y_1 and Y_2 are statistically independent random variables when $p(y_1, y_2) = p(y_1) p(y_2)$. This expression represents an extension of the discussion in Section 2.6 to density functions.
- Y_1 and Y_2 are uncorrelated random variables when $\text{cov}(Y_1, Y_2) = 0$. Since $\text{cov}(Y_1, Y_2) = \mathcal{E}\{Y_1 Y_2\} - \mathcal{E}Y_1 \, \mathcal{E}Y_2$, then $\mathcal{E}\{Y_1 Y_2\} = \mathcal{E}Y_1 \, \mathcal{E}Y_2$ for uncorrelated variables.
- Statistically independent variables are uncorrelated, but not all uncorrelated variables are independent. All uncorrelated normal variables are statistically independent.
- The degree to which two variables are correlated is quantified by

$$
\rho = \frac{\text{cov}(Y_1, Y_2)}{\sqrt{var(Y_1)var(Y_2)}}. \qquad \text{Correlation coefficient} \tag{3.35}
$$

For perfectly correlated variables, $\rho = 1$ and $\text{cov}(Y_1, Y_2) = \text{cov}(Y_1, Y_1) = \text{var}(Y_1) = \text{var}(Y_2)$. If $\rho = -1$, the variables are perfectly anti-correlated: $\text{cov}(Y_1, Y_2) = -\text{var}(Y_1)$. We now illustrate practical solutions through a numerical example.

Example 3.5.1. *Statistics estimated from a joint pdf: Let* $y_1 = (3, 2, 1)$ *and* $y_2 = (-1, 1, 0)$ *indicate the first and last blood pressure measurements over time for just* $M = 3$ *subjects. The numbers are not recognizable as pressures, but calculational*

Table 3.1 The joint pdfs for Case 1 (left), Case 2 (center), and Case 3 (right). The y_1-axis values are along the vertical dimension of the table and the y_2-axis values are along the horizontal dimension. Note that the joint and marginal pdfs, $p(y_1, y_2)$, $p(y_1)$, $p(y_2)$, must each sum to 1.

Case 1:

$p(y_1,y_2)$	1	2	3	$p(y_1)$
1	1/9	2/9	0	1/3
2	0	2/9	1/9	1/3
3	0	1/9	2/9	1/3
$p(y_2)$	1/9	5/9	1/3	1

Case 2:

$p(y_1,y_2)$	1	2	3	$p(y_1)$
1	1/9	1/9	1/9	1/3
2	1/9	1/9	1/9	1/3
3	1/9	1/9	1/9	1/3
$p(y_2)$	1/3	1/3	1/3	1

Case 3:

$p(y_1,y_2)$	1	2	3	$p(y_1)$
1	1/3	0	0	1/3
2	0	1/3	0	1/3
3	0	0	1/3	1/3
$p(y_2)$	1/3	1/3	1/3	1

simplicity is important for clarity. Initially, let y_1 and y_2 be unpaired (different volunteers) so the sample space consists of all possible y_1, y_2 pairs:

$$S = \{(3, -1), (3, 1), (3, 0), (2, -1), (2, 1), (2, 0), (1, -1), (1, 1), (1, 0)\}.$$

We do not know the population means, so we settle for the sample means $\bar{Y}_1 = 2$ and $\bar{Y}_2 = 0$. Assume the joint pdf is given by the Case 1 values in Table 3.1. The last row and the last column in the table are marginal probabilities, e.g., $p(y_2) = \int_{y_1 \in S} dy_1\, p(y_1, y_2)$.

(a) Estimate the population statistics $\mathrm{cov}(Y_1, Y_2)$, $\mathrm{var}(Y_1)$, and $\mathrm{var}(Y_2)$ for the nonuniform, unpaired, joint pdf of Case 1. State whether Y_1 and Y_2 are correlated and/or statistically independent. (b) Repeat for the uniform, unpaired, joint pdf of Case 2 in Table 3.1. (c) Repeat for the uniform, paired, joint pdf of Case 3 in Table 3.1.

Solution

(a) Interpreting (3.34) for discrete data and given the preceding data, we find

$$\mathrm{cov}(Y_1, Y_2) = \sum_{m=1}^{3} \sum_{m'=1}^{3} (y_{1m} - \bar{Y}_1)(y_{2m'} - \bar{Y}_2)\, p(y_{1m}, y_{2m'}) \tag{3.36}$$

$$= \frac{1}{9}(3-2)(-1-0) + \frac{2}{9}(3-2)(1-0) + 0(3-2)(0-0)$$

$$+ 0(2-2)(-1-0) + \frac{2}{9}(2-2)(1-0) + \frac{1}{9}(2-2)(0-0)$$

$$+ 0(1-2)(-1-0) + \frac{1}{9}(1-2)(1-0) + \frac{2}{9}(1-2)(0-0) = 0.$$

The covariance between Y_1 and Y_2 is zero. Consequently, $\rho = 0$, and we find

$$\mathrm{var}(Y_1) = \sum_{m=1}^{3} (y_{1m} - \bar{Y}_1)^2 p(y_{1m}) = (3-2)^2\frac{1}{3} + (2-2)^2\frac{1}{3} + (1-2)^2\frac{1}{3} = 2/3$$

$$\mathrm{var}(Y_2) = (-1-0)^2\frac{1}{9} + (1-0)^2\frac{5}{9} + (0-0)^2\frac{1}{3} = 2/3.$$

These variances are not the same as unbiased sample variance estimates: $s_1^2 = ((3-2)^2 + (2-2)^2 + (1-2)^2)/2 = 1$ and $s_2^2 = ((-1-0)^2 + (1-0)^2 + (0-0)^2)/2 = 1$, as M is small. Also, although $p_{Y_1 Y_2}(2, 3) = p_{Y_1}(2) p_{Y_2}(3)$, we find $p_{Y_1 Y_2}(1,1) = 1/9 \neq p_{Y_1}(1) p_{Y_2}(1) = 1/27$. *Consequently, Y_1, Y_2 are dependent for Case 1, where* $\mathbf{K}_I = \begin{pmatrix} 2/3 & 0 \\ 0 & 2/3 \end{pmatrix}$.

(b) Repeating the calculation for the uniformly distributed $p(y_1, y_2)$ for Case 2, we find again that $\mathrm{var}(Y_1) = 2/3 = \mathrm{var}(Y_2)$, $\mathrm{cov}(X, Y) = 0$ and $\mathbf{K}_{II} = \mathbf{K}_I$, but now we find these variables are statistically independent. Since Y_1, Y_2 are unpaired (not necessarily from the same volunteers), it makes sense that the data are uncorrelated.

(c) For Case 3, $\mathrm{var}(Y_1) = 2/3 = \mathrm{var}(Y_2)$, $\mathrm{cov}(Y_1, Y_2) = -1/3$, and $\mathbf{K}_{III} = \begin{pmatrix} 2/3 & -1/3 \\ -1/3 & 2/3 \end{pmatrix}$. Volunteer measurements Y_1, Y_2 are now paired, and only diagonal elements of $p(y_1, y_2)$ in Table 3.1 are nonzero. We see that as y_2 increases, y_1 tends to decrease, although the correlation is inexact: $\rho = -0.5$. This covariance tells us that the effect of the medicine is to reduce blood pressure more in patients with higher blood pressures. We also see that $p(y_1, y_2) \neq p(y_1) p(y_2)$, so the variables are statistically dependent, which follows from the nonzero covariance. This bite-sized example is fine for a hands-on illustration, but it is unrealistic since we rarely have the pdf and a small sample size. If we have only a data set available, we can estimate the sample covariance matrix.

3.5.2 Sample Covariance

To estimate the *sample covariance* from available data, we again consider the covariance between time samples acquired during the first (0 hr) and last (24 hr) blood pressure measurement times. $\mathbf{X} \in \mathbb{R}^{M \times N}$ contains paired data ($N = 2$) from $M = 3$ volunteers. The sample means are $\bar{X}_1 = \sum_{m=1}^{M} X[m,1]/M$ and $\bar{X}_2 = \sum_{m=1}^{M} X[m,2]/M$. Define $\mathbf{Z} \in \mathbb{R}^{M \times N}$ with column vectors $z_1 = X[m,1] - \bar{X}_1$ and $z_2 = X[m,2] - \bar{X}_2$. The unbiased sample covariance matrix is

$$
\hat{\mathbf{K}} = \frac{1}{M-1} \mathbf{Z}^T \mathbf{Z} \qquad \text{(Sample covariance matrix)}
$$

$$
= \frac{1}{M-1} \begin{pmatrix} \sum_{m=1}^{M}(X[m,1] - \bar{X}_1)^2 & \sum_{m=1}^{M}(X[m,1] - \bar{X}_1)(X[m,N] - \bar{X}_N) \\ \sum_{m=1}^{N}(X[m,N] - \bar{X}_N)(X[m,1] - \bar{X}_1) & \sum_{m=1}^{M}(X[m,N] - \bar{X}_N)^2 \end{pmatrix}
$$

$$
= \begin{pmatrix} \widehat{\mathrm{var}}(X_1) & \widehat{\mathrm{cov}}(X_1 X_N) \\ \widehat{\mathrm{cov}}(X_N X_1) \widehat{\mathrm{var}}(X_N) & \widehat{\mathrm{var}}(X_N) \end{pmatrix} = \begin{pmatrix} \hat{\sigma}_1^2 & \rho \hat{\sigma}_1 \hat{\sigma}_N \\ \rho \hat{\sigma}_1 \hat{\sigma}_N & \hat{\sigma}_N^2 \end{pmatrix}. \tag{3.37}
$$

The last form relabels $\widehat{\mathrm{var}}(X_i) = \hat{\sigma}_i^2$ and applies (3.35). In (3.37), we have a sample covariance matrix composed of sample variances and covariances instead of the population statistics from the last section. Do you see how this specific matrix product averages over the M cases to give the temporal covariance estimate? Even though $\mathbf{Z}$ is $M \times N$, the sample covariance is $N \times N$ because there are N "variables."

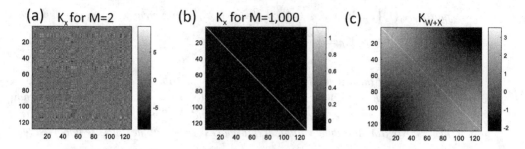

Figure 3.6 (a) Sample covariance $\hat{\mathbf{K}}$ for an average of two 128-point standard-normal time series, $X(2, 128)$. (b) $\hat{\mathbf{K}}$ for $X(1, 000, 128)$. (c) $\hat{\mathbf{K}}$ for sum $W + X$, where rows of W are the same Gaussian function and $X(1, 000, 128)$. Details are provided in the code shown later in this section.

If $\mathbf{Z}$ is $(M = 1, 000) \times (N = 2)$, $\hat{\mathbf{K}}$ would remain 2×2 and asymptotically approach the population covariance, $\mathbf{K}$ from (3.33), as $M \to \infty$. The use of $\mathbf{Z}^\top \mathbf{Z}$ or $\mathbf{Z}\mathbf{Z}^\top$ in (3.37) depends on whether we orient the M cases along the columns or the rows of $\mathbf{Z}$.

We can illustrate the convergence of $\hat{\mathbf{K}} \to \mathbf{K}$ for uncorrelated normal noise data because we know $\mathbf{K}$. The sample covariance approaches the population covariance by averaging $\mathbf{X}$ for large M. In the following code, I first average two realizations of a standard normal random variable $X(2, 128)$, and find the very messy result in Figure 3.6a. We know $\hat{\mathbf{K}} \xrightarrow[M \to \infty]{} \mathbf{I}_{128}$, the identity matrix. The result in Figure 3.6b, found using $X(1, 000, 128)$, is much closer to the expected result. Adding a deterministic function to the same random process, we obtain the sum of the two covariances, shown in Figure 3.6c.

```
N=128;M=2;x=randn(M,N);K=x'*x/(M-1);      % cov of std norm sequence; ave 2
subplot(1,3,1);imagesc(K);colormap(gray);axis square;colorbar
M=1000;X=randn(M,N);Ka=X'*X/(M-1);        % cov of std norm sequence; ave 1000
subplot(1,3,2);imagesc(Ka);colormap(gray);axis square;colorbar
wp=3*exp(-(0:127).^2/(2*50^2));w=wp-mean(wp); % deterministic function
Y=zeros(M,N);                              % initialize
for n=1:M
    Y(n,:)=w+randn(size(w));               % sum det function and X(1000,128)
end
Kw=Y'*Y/(M-1);subplot(1,3,3);imagesc(Kw);colormap(gray);axis square;colorbar
```

If in the third line of the code we substituted `M=128;X=randn(M,N);Ka=X*X'/(M-1);`, you would find that transposing the second copy of the data matrix instead of the first forms a $1, 000 \times 1, 000$ covariance matrix. Can you explain this new result? (Hint: It's the covariance of standard normal realizations averaged over time.)

Computing $\hat{\mathbf{K}}$ from Example 3.5.1, `y1=[3;2;1];y2=[-1;1;0];cov([y1-mean(y1) y2-mean(y2)])` gives $\hat{\mathbf{K}} = \begin{pmatrix} 1 & -0.5 \\ -0.5 & 1 \end{pmatrix}$. The equivalent population covariance is $\mathbf{K}_3$ above, which should match. We find that $\hat{\mathbf{K}} = \mathbf{K}_3$ once bias is eliminated by adjusting for the use of sample means instead of population means. Specifically, $\mathbf{K}_3$ should be multiplied by $M/(M-1) = 3/2$ to remove the bias. Bias is significant with small sample sizes.

3.6 Stationary Random Processes

Ensemble averages of random time series generally remain functions of time, whereas time averages do not. However, if we find that ensemble averages of a stochastic process are constant over time, then we have a temporally *stationary process*. If we then find a case where the ensemble and temporal averages are equal, that stationary process is also an *ergodic process*.

This special situation provides experimentalists with many options. For example, overall cost and risk can be reduced by measuring a drug response on just a few in vivo subjects at hundreds of time points instead of measuring hundreds of subjects a few times each to generate sufficient statistical power to make a strong statement about efficacy (see Section 8.2.2). Ergodic processes are rare, unfortunately. Let's first look closely at the more general stationary processes.

DEFINITION 3.6.1 *For random process* $X(t)$, *let* $p_X(x[1], \ldots, x[j], \ldots, x[J])$ *be the joint pdf at time samples* $t = jT$, $1 \le j \le J$. $X(t)$ *is a stationary random process if, for any value of integer* ℓ *such that* $x[j + \ell] \in X(t)$, *we find*

$$p(x[1], \ldots, x[j], \ldots, x[J]) = p(x[1 + \ell], \ldots, x[j + \ell], \ldots, x[J + \ell]).$$

That is, if we shift the time for each sample in the pdf by constant ℓ *and find that the distribution does not change, the process is* strictly stationary.

The covariance matrix of $X(t)$ in Figure 3.6b is strictly stationary (to the extent the pseudo-random numbers are truly random) because each value in the time series is drawn from the same random number generator without reference to time. There are only two nonzero moments, $\mathcal{E}X(t) = \mu$ and $\text{var}(X) = \sigma^2$, and both are *time invariant*. Any shift in the time axis produces the same joint pdf.

It is difficult to establish stationarity for the same reason it is difficult to estimate a pdf: You need lots of annotated data and must show the pdf converges as additional data are added. A more practical method examines moments to see if they change as the time axis is shifted. If the *first two moments* of a process are time invariant, then for the population mean we find $\mathcal{E}X(jT) = \mathcal{E}X((j + \ell')T)$ and for the population covariance matrix we find for diagonal elements $\mathbf{K}_{\ell\ell} = \mathbf{K}_{\ell+\ell',\ell+\ell'}$ for any $\ell' \in \mathsf{S}$. When this is true, the process is *wide-sense stationary (WSS)*. Wide-sense stationary random processes have covariance matrices with a Toeplitz structure (see Appendix A).

Example 3.6.1. *Nonstationary covariance: Let's examine the covariance matrix for an obviously nonstationary random process. Using the MATLAB script shown here, a standard normal process has a variance that increases with time because of* $\sigma(jT)$, *as illustrated in Figure 3.7. The effects on the covariance matrix show this is still an uncorrelated process (diagonal covariance matrix) but the diagonal variance values clearly increase with time.* $\mathbf{K}$ *is not Toeplitz, although it is still square, Hermitian, and positive semi-definite.*

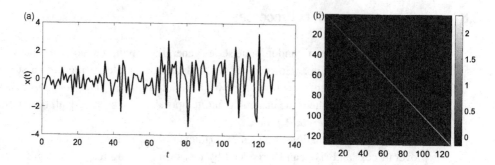

Figure 3.7 (a) One realization of a nonstationary, uncorrelated normal random time series. (b) The corresponding covariance matrix for an ensemble $N = 1,000$ realizations shows that the variance (diagonal values) increases with time. From Example 3.6.1.

```
%%%%%  Code for Example 3.6.1 yielding the results of Figure 3.7  %%%%%
r=64/128:1/128:191/128;xp=randn(128,1000);x=zeros(size(xp));  % standard normal x'
for j=1:1000;
  x(:,j)=r'.*xp(:,j);  % scale the variance of x' as a function of time
end;
plot(x(:,1));figure;K=x*x'/999;imagesc(K);colormap gray;axis square;colorbar
```

3.7   Continuous-Time Covariance and Correlation

We computed moments of sampled, continuous-time data via $p_s(t)$ in (3.23), enabling the use of matrix calculations. These methods are practical for estimating statistical moments from data. We may also consider moments of continuous-time functions without sampling to approximate underlying population properties – the topic of this section.

The *autocovariance function* for stationary $X(t)$ defined over $0 \leq t, \tau \leq T_t$ is

$$K_X(t, t - \tau) = \mathcal{E}\left\{(x(t) - \mu_x)(x(t - \tau) - \mu_x)\right\} \tag{3.38}$$

$$= \int_{x \in S} dx(t) \int_{x \in S} dx(t - \tau)(x(t) - \mu_x)$$

$$\times (x(t - \tau) - \mu_x)\, p_X(x(t), x(t - \tau)).$$

For strictly stationary X, the pdf is time invariant, i.e., $p_X(x(t), x(t - \tau)) = p_X(x(t - t_0), x(t - \tau - t_0))\ \forall\ t_0 \in S$. Applying the linear property of the ensemble operator to (3.38), we find

$$K_X(\tau) = \mathcal{E}\{x(t)\,x(t - \tau)\} - \mathcal{E}\{x(t)\}\mu_x - \mu_x\mathcal{E}\{x(t - \tau)\} + \mu_x^2$$

$$= \mathcal{E}\{x(t)x(t - \tau)\} - \mu_x^2$$

$$= \left[\int_{x \in X} dx(t) \int_{x \in X} dx(t - \tau)\left(x(t)\,x(t - \tau)\right) p_X(x(t), x(t - \tau))\right] - \mu_x^2$$

$$= \phi_X(\tau) - \mu_x^2, \tag{3.39}$$

where $\phi_X(\tau)$ is the ensemble *autocorrelation function* for stationary X. Also, for $X \in S$ and $Y \in S$, the *cross-covariance function* is

$$K_{XY}(\tau) = \phi_{XY}(\tau) - \mu_x \mu_y$$
$$\text{where} \quad K_X(\tau) = \phi_X(\tau) - \mu_x^2$$
$$K_Y(\tau) = \phi_Y(\tau) - \mu_y^2 .$$

The *cross-correlation function* is

$$\phi_{XY}(\tau) = \int_{x \in S} dx(t) \int_{y \in S} dy(t-\tau) \left(x(t) y(t-\tau)\right) p_{XY}(x(t), y(t-\tau)). \quad (3.40)$$

It makes sense to discuss continuous-time functions when talking conceptually. However, we must use discrete forms when processing data.

3.8 Ergodic Processes

Consider WSS random process $X(t)$.

DEFINITION 3.8.1 *An ergodic process is a stationary process in which ensemble averages equal time averages. For example, $X(t)$ is first-order ergodic if sample ensemble mean $\bar{X}(t)$ and sample temporal mean $\langle x_n \rangle$ are equal. If $\lim_{N \to \infty} \bar{X}(t) = \lim_{T_t \to \infty} \langle x_n \rangle = \mu$, the process is ergodic and $\bar{X}$ is a consistent estimate of μ.*

The process is second-order ergodic if the ensemble correlation function $\phi_X(\tau)$ in (3.39) can be represented by a time-averaged form, $\varphi_X(\tau)$, given by

$$\varphi_X(\tau) \triangleq \lim_{T_t \to \infty} \frac{1}{T_t} \int_0^{T_t} dt\, x(t) x(t-\tau) = \phi_X(\tau), \quad (3.41)$$

where T_t is the total duration of X.

When a process is ergodic, you may substitute averages over the independent variables, usually time and/or space, in place of ensemble averages when ensembles are more difficult to obtain experimentally.

Specifically, consider $x_n(jT)$ as a sample waveform acquired from the ensemble $X(t) = \{x_n(t)\}$, where n denotes the waveform in the ensemble and j indexes time. $X(t)$ is *first-order ergodic* if

$$\bar{x} = \langle x_n \rangle$$

$$\frac{1}{N} \sum_{n=1}^{N} x_n(jT) \simeq \frac{1}{J} \sum_{j=1}^{J} x_n(jT),$$

where equality is achieved asymptotically. Clearly, $X(t)$ must be at least WSS for this statement to make any sense, because $\bar{x}(t)$ is generally a function of time while $\langle x_n \rangle$ is not.

All ergodic processes are stationary, but not all stationary processes are ergodic. Also, it is possible for a process to be first-order ergodic and *not* second-order ergodic, as shown in the following example from [13].

Example 3.8.1. *Ergodicity: Consider random process $G(t)$, where the nth sinusoidal waveform realization at frequency u_0 is*

$$g_n(t) = A_n \sin(2\pi u_0 t + \theta_n). \tag{3.42}$$

$G(t)$ is a random process because its amplitude A_n and phase θ_n for the nth waveform are both drawn from uniform *distributions; e.g.,*

$$p(A) = \begin{cases} \frac{1}{A_2 - A_1} & \text{for amplitude range } A_1 \le A \le A_2 \\ 0 & \text{otherwise.} \end{cases}$$

Examples of four such waveforms are shown in Figure 3.8a.

Ideally, $G(t)$ is a first-order ergodic process because the time-averaged (summed over integer number of periods) and ensemble-averaged (summed over $N \rightarrow \infty$ random waveforms) means are both asymptotically zero. However, there are practical considerations when considering this statement using experimental data. While the time-averaged mean is zero for every waveform, the ensemble-averaged mean converges to zero slowly as $1/\sqrt{N}$. This is shown in Figure 3.8b. It does not mean $G(t)$ is first-order nonergodic if only 100 waveforms are available, but it does mean that sample moments computed using 100 or fewer waveforms converge more quickly to population means using time averaging for this first-order ergodic process. It is difficult to make strong statements about ergodicity from measurements where few data samples are available. As with pdf estimation, evidence for ergodicity is obtained analytically from knowledge of the physics and not numerically from measurements.

We can analytically show that $G(t)$ is not second-order ergodic. The time-averaged correlation is found by substituting (3.42) into (3.41),

$$\varphi_G(\tau, n) = \frac{A_n^2}{2} \cos(2\pi u_0 \tau),$$

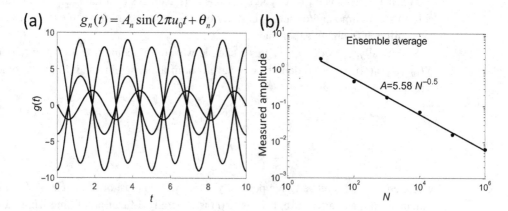

(a) $g_n(t) = A_n \sin(2\pi u_0 t + \theta_n)$ **(b)**

Ensemble average

$A = 5.58\ N^{-0.5}$

Figure 3.8 (a) $N = 4$ sine waves $g_n(t)$ with random amplitude and phase. (b) Plot of the ensemble mean maximum amplitude as a function of ensemble size N. $\bar{g}(t) \rightarrow 0$ as $N \rightarrow \infty$ in a manner $\propto 1/\sqrt{N}$.

which is a function of the exact waveform used via the index n. By comparison, the ensemble-averaged correlation $\phi_G(\tau)$ is not a function of a specific waveform, so $G(t)$ is not second-order ergodic. Hence, if you need to compute the covariance matrix, for example, you need to obtain many realizations of waveforms and cannot use time-averaged estimates.

If the waveform amplitude is constant, even if the phase remains random, then $\phi_G(\tau,n) = \frac{A^2}{2}\cos(2\pi u_0\tau) = \phi_G(\tau)$ and so $G(t)$ is both first- and second-order ergodic. This example serves to alert readers to some of the issues related to the determination of ergodicity.

3.9 Multivariate Normal Density

In Example 3.6.1, each sample in the time series had a different normal parameter σ, which meant the process was non-WSS. Generally, there is no reason to expect an ensemble for an experimentally recorded time series to emerge from a pdf with fixed parameters. Thinking back to the drug study described in Example 3.5.1, we might find that the effects of the drug administered vary systematically between men and women over time. The pdf might be the same, but the parameters may vary or the data may be correlated. We now introduce equations for modeling realistic processes using a *multivariate normal (MVN)* pdf.

Real time series $\mathbf{X} = \begin{pmatrix} x_1 \\ \vdots \\ x_J \end{pmatrix}$, expressed as a column vector, has normally distributed elements. The multivariate normal pdf is

$$p(\mathbf{X}) = \mathcal{N}(\boldsymbol{\mu},\mathbf{K}) = [(2\pi)^N \det \mathbf{K}]^{-1/2} \exp\left[-\frac{1}{2}(\mathbf{X}-\boldsymbol{\mu})^\top \mathbf{K}^{-1}(\mathbf{X}-\boldsymbol{\mu})\right]. \quad \text{(MVN pdf)}$$

(3.43)

The population mean is $\boldsymbol{\mu}_X = \mathcal{E}\mathbf{X} = \begin{pmatrix} \mathcal{E}x_1 \\ \vdots \\ \mathcal{E}x_J \end{pmatrix} = \begin{pmatrix} \mu_1 \\ \vdots \\ \mu_J \end{pmatrix}$ and the covariance is

$$\mathbf{K} = \mathcal{E}\{(\mathbf{X}-\boldsymbol{\mu})(\mathbf{X}-\boldsymbol{\mu})^\top\} \quad (3.44)$$

$$= \mathcal{E}\left\{\begin{pmatrix} x_1-\mu_1 \\ \vdots \\ x_J-\mu_J \end{pmatrix}((x_1-\mu_1) \cdots (x_J-\mu_J))\right\} = \begin{pmatrix} \sigma_{11}^2 & \cdots & \sigma_{1j'}^2 & \cdots & \sigma_{1J}^2 \\ & \ddots & & & \\ \vdots & & \sigma_{jj'}^2 & & \vdots \\ & & & \ddots & \\ \sigma_{J,1}^2 & & \cdots & & \sigma_{JJ}^2 \end{pmatrix}.$$

We can legitimately use $\sigma_{jj'}^2 = \mathcal{E}\{(x_j - \mu_j)(x_{j'} - \mu_{j'})\}$ as shorthand notation for variances and covariances because we are modeling a population for which parameters are assumed or determined by fits with data.

For stationary process $X(t)$ with diagonal $\mathbf{K}$ (uncorrelated variables), the exponent in (3.43) reduces to

$$\frac{1}{2}(\mathbf{X} - \boldsymbol{\mu})^{\top} \mathbf{K}^{-1}(\mathbf{X} - \boldsymbol{\mu})$$

$$= \frac{1}{2}\left((x_1 - \mu_1) \cdots (x_J - \mu_J)\right) \begin{pmatrix} 1/\sigma_{11}^2 & & & 0 \\ & 1/\sigma_{22}^2 & & \\ & & \ddots & \\ 0 & & & 1/\sigma_{JJ}^2 \end{pmatrix} \begin{pmatrix} x_1 - \mu_1 \\ \vdots \\ x_J - \mu_J \end{pmatrix},$$

so that (3.43) becomes

$$p(\mathbf{x}) = \prod_{j=1}^{J} \frac{1}{\sigma_j \sqrt{2\pi}} e^{-(x_j - \mu_j)^2/2\sigma_j^2}$$

$$= (2\pi)^{-J/2} \sigma^{-J} \exp\left[-\frac{1}{2}\sum_{j=1}^{J}(x_j - \mu_j)^2/\sigma^2\right]. \tag{3.45}$$

Being able to diagonalize the covariance matrix greatly simplifies the MVN expressions. Appendix A shows that square, symmetric, positive-semi-definite matrices, such as $\mathbf{K}$, are diagonalizable. Approximate forms may be applied to manage very large matrices.

Covariance matrix $\mathbf{K}$ can be written in terms of the ensemble correlation matrix ϕ for real data as follows:

$$\mathbf{K} = \mathcal{E}\{(\mathbf{X} - \boldsymbol{\mu})(\mathbf{X} - \boldsymbol{\mu})^{\top}\} = \mathcal{E}\{\mathbf{XX}^{\top}\} - \mathcal{E}\{\mathbf{X}\}\boldsymbol{\mu}^{\top} - \boldsymbol{\mu}\mathcal{E}\{\mathbf{X}^{\top}\} + \boldsymbol{\mu}\boldsymbol{\mu}^{\top}$$

$$= \mathcal{E}\{\mathbf{XX}^{\top}\} - \boldsymbol{\mu}\boldsymbol{\mu}^{\top} = \phi - \boldsymbol{\mu}\boldsymbol{\mu}^{\top}. \tag{3.46}$$

Finally, we can partition the covariance matrix into diagonal *standard deviation matrix* $\mathbf{S}$ and *correlation coefficient matrix* $\mathbf{R}$,

$$\mathbf{K} = \mathbf{SRS}, \tag{3.47}$$

where

$$\mathbf{S} = \begin{pmatrix} \sigma_1 & 0 & \cdots & 0 \\ 0 & \sigma_2 & \cdots & 0 \\ \vdots & & \ddots & \vdots \\ 0 & \cdots & & \sigma_J \end{pmatrix} \quad \text{and} \quad \mathbf{R} = \begin{pmatrix} 1 & \rho_{12} & \cdots & \rho_{1J} \\ \rho_{21} & 1 & \cdots & \rho_{2J} \\ \vdots & & \ddots & \vdots \\ \rho_{J1} & \cdots & & 1 \end{pmatrix}.$$

3.9.1   Glossary

Terms used is Chapters 2 and 3 are summarized here:

- $\Pr, P, p$, WSS: probability, cdf, pmf/pdf, and wide-sense stationary
- $\mathcal{E}X^m(t)$: mth moment; $\mathcal{E}X^m$: the mth moment for a stationary process
- $\mathcal{E}X(t)$: population ensemble mean; $\mathcal{E}X$: population ensemble mean for WSS process
- $\mathbf{K}$: population covariance matrix and $\boldsymbol{\phi}$: correlation matrix
- $\mathcal{N}(\mu, \sigma^2)$: UVN pdf; $\mathcal{N}(\boldsymbol{\mu}, \mathbf{K})$: MVN pdf
- $\mathcal{P}(\lambda)$: univariate Poisson pmf
- $\phi_X(\tau)$: ensemble correlation function
- $\varphi_X(\tau)$: temporal correlation function
- s_n^2: sample temporal variance for nth waveform
- $\mathbf{S}, \mathbf{R}$: standard deviation and correlation coefficient matrices
- $\text{var}(X(t))$: population ensemble variance; $\text{var}(X)$: population ensemble variance for WSS process
- $\widehat{\text{var}}(X(t))$: sample ensemble variance; $\widehat{\text{var}}(X)$: sample ensemble variance for WSS process
- $\bar{X}(t)$: sample ensemble mean; $\bar{X}$: sample ensemble mean for WSS process
- $\langle x_n \rangle$: sample temporal mean for nth waveform
- If $\bar{X} = \langle X_n \rangle$ and $\widehat{\text{var}}(X) = s_n^2$, then X is a second-order ergodic process

The equations used to process data that result in property estimates or model parameters are called *estimators*.

- **Estimators** are mathematical expressions or algorithms that input data and output statistics associated with a parametric model of the process generating the data.
- The **efficient estimator** is the one that yields the smallest MSE for a specific parameter.
- **Consistent estimators** yield sample estimates that approach population values asymptotically.

The *maximum-likelihood estimator* (see Problems 3.4 and 3.5) of sample variance is biased. Therefore it is not an efficient estimator, but it is consistent since $\lim_{N \to \infty} \mathcal{E}\{s_n^2\} = \sigma^2$. Unbiased estimates of sample variance are efficient estimators.

3.10   Mixture Densities

If an N-dimensional event space consists of L classes of data, θ_i, $i = 1, ..., L$, the density for the ith class is given by the multivariate conditional pdf $p_i(\mathbf{X}) \triangleq p(\mathbf{X}|\theta_i)$. For example, classes might include truly healthy and sick patients as verified clinically. Conditional densities are used to keep track of classes within event space S.

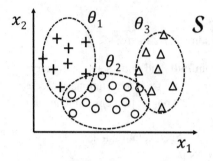

Figure 3.9 N = two-dimensional measurements, (x_1, x_2), of $L = 3$ classes of data θ_i.

Let $\Pr(\theta_i)$ be the *prior probability* that a data point in S belongs to class θ_i, where $\sum_{i=1}^{L} \Pr(\theta_i) = 1$. Then the *mixture density*, i.e., the unconditional pdf $p(\mathbf{X})$, is

$$p(\mathbf{X}) = \sum_{i=1}^{L} p_i(\mathbf{X}) \Pr(\theta_i). \tag{3.48}$$

From Section 2.5, the posterior probability $\Pr(\theta_i|\mathbf{X})$ combines the likelihood $p_i(\mathbf{X})$ and prior probability $\Pr(\theta_i)$ as follows:

$$\Pr(\theta_i|\mathbf{X}) = \frac{p_i(\mathbf{X}) \Pr(\theta_i)}{p(\mathbf{X})}. \tag{3.49}$$

Marginal densities $p(x_n)$ may be found by integrating over all dimensions except n using

$$p(x_n) = \int_{-\infty}^{\infty} dx_1 \ldots \int_{-\infty}^{\infty} dx_{n-1} \int_{-\infty}^{\infty} dx_{n+1} \ldots \int_{-\infty}^{\infty} dx_N \, p(\mathbf{x}). \tag{3.50}$$

$p(x_n)$ is the pdf of any one measurement dimension.

3.11   Functions of Random Variables

Consider an object with stochastic properties that vary in space and time. For example, the optical absorption coefficient α [cm^{-1}] for tissue describes the probability of absorbing an incident photon of a particular wavelength per unit absorber thickness Δx. The incremental change in the number of incident photons N_0 from absorption by incremental tissue thickness dx is $dN = -\alpha \, dx \, N$.[4] From α, the distribution of N, and the function describing the measurement process (described in Chapter 4), we can compute the measurement distribution.

---

[4] The probability of N_a photons being absorbed by tissue thickness Δx is $\Pr(N_a) = N_a/N_0 = 1 - \exp(-\alpha\Delta x)$. The probability of absorption over incremental dx is $1 - 1 + \alpha \, dx - (\alpha \, dx)^2/2 + \ldots \simeq \alpha \, dx$.

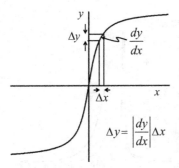

Figure 3.10 Example of an invertible probability transformation between variables X and Y from (3.52).

3.11.1 Univariate Probability Transformations

Let $X(t)$ be a time-varying continuous process where $X(t) = \{x(t)\}$, and let $Y(t) = f(X(t))$ be a one-to-one monotonic transformation of $X(t)$ by function f (see Figure 3.10). Since one value of $x(t)$ is associated with one value of $y(t)$, the inverse $x = f^{-1}(y)$ is well defined. Suppose the probability in region Δx, $\Pr(x - dx/2 \le X \le x + dx/2)$, is approximated by the probability for the region Δy of Y,

$$p_Y(y)\Delta y = p_X(x)\Delta x. \tag{3.51}$$

Figure 3.10 illustrates the linear relationship between the two variables, $\Delta y \simeq |dy/dx|\Delta x$, where $|dy/dx|$ is the *Jacobian* of the transformation. Combining this result with (3.51) yields

$$p_Y(y) = p_X(x)\frac{\Delta x}{\Delta y} = p_X(x)\left|\frac{dx}{dy}\right| = p_X(f^{-1}(y))\left|\frac{df^{-1}(y)}{dy}\right|. \tag{3.52}$$

The last form of (3.52) is a bit strange. It instructs us to express the *original* pdf and the Jacobian in terms of the new variable $y(t)$. The following examples illustrate use of (3.52).

Example 3.11.1. *Linear transformation:* *Let $X \sim p_X(x) = \exp(-x)$ for $x > 0$. The challenge is to find $p_Y(y)$ where $y = f(x) = ax + b$. That is, what is the pdf after X undergoes this simple linear transformation? Computing terms in the last form of (3.52), $x = f^{-1}(y) = (y - b)/a$ and $|df^{-1}(y)/dy| = |a|^{-1}$, we find*

$$p_Y(y) = p_X\left(\frac{y - b}{a}\right)|a|^{-1}$$

$$= \frac{1}{|a|}e^{-(y-b)/a} \quad \text{for} \quad y > b \text{ and } a > 0.$$

Examples for two different values of (a,b) are shown in Figure 3.11a. Notice that when x is scaled, the Jacobian is not equal to 1, which ensures that the resulting pdf integrates to 1.

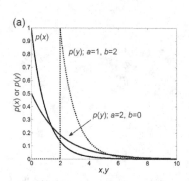
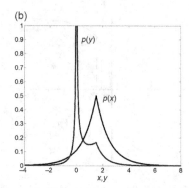

Figure 3.11 (a) Illustration of Example 3.11.1 where X follows a unit exponential pdf and the output undergoes linear amplification and a base offset via $y = ax + b$ for $a = 2, 1$ and $b = 0, 2$. (b) Results of Example 3.11.2 where $A = 1$ and $a = 1.5$. Carefully note the location of the origin.

Transformations that are not functions (one-to-one) do not have well-defined inverses $x = f^{-1}(y)$, although they may be well defined when considered piecewise. In that case, by partitioning $f(x)$ into M monotonic segments, we find that (3.52) extends to

$$p_Y(y) = \sum_{m=1}^{M} p_X(f_m^{-1}(y)) \left| \frac{d f_m^{-1}(y)}{dy} \right|. \tag{3.53}$$

Example 3.11.2. *Piecewise transformation:* *Let the transformation be* $y = Ax^2$, *where* $x = f^{-1}(y) = -\sqrt{y/A}$ *for* $x \le 0$ *and* $+\sqrt{y/A}$ *for* $x > 0$. *This nonlinear transformation is not one-to-one, so we need to partition X into halves at the origin. We find that* $|df^{-1}(y)/dy| = (2\sqrt{Ay})^{-1}$ *for all y except at the origin. Therefore*

$$p_Y(y) = \frac{p_X(-\sqrt{y/A}) + p_X(\sqrt{y/A})}{2\sqrt{Ay}}. \tag{3.54}$$

Let's try the following problem. Select a two-sided exponential $p_X(x) = 0.5 \exp(-|x - a|)$ *as the original pdf and apply transformation* $y = Ax^2$ *via (3.54). Both $p_X(x)$ and $p_Y(y)$ are shown in Figure 3.11b. You see hints of $p_X(x)$ in $p_Y(y)$, but the $(Ay)^{-1/2}$ scaling term tends to dominate near the origin.*

I found it rather tricky to interpret everything correctly, so I constructed the following MATLAB code for the transformation plotted, assuming $A = 1$ and $a = 1.5$. The code is purposely inefficient; that is, it is broken up into parts for illustration. Note that x1 *equals* $-\sqrt{y}$ *for $x < 0$ and* x2 *equals* $\sqrt{y}$ *for $x > 0$. I avoided $x = 0$.*

```
x=-4:0.01:8;px=0.5*exp(-abs(x-1.5));plot(x,px);hold on %choose -4 < x < 8
x1=-4:0.01:-0.01;py1=0.5*exp(-abs(x1-1.5))./(2*abs(x1));
x2=0.01:0.01:8;py2=0.5*exp(-abs(x2-1.5))./(2*x2);
plot(x1,py1,x2,py2);hold off
```

3.11.2   Functions of Multivariate Random Variables

Experimental data are often represented as functions of multivariate processes. Extending the univariate discussion from the preceding subsection, we have $\mathbf{y} = f(\mathbf{x})$ where $\mathbf{x}$ and $\mathbf{y}$ are, respectively, $N \times 1$ and $M \times 1$ column vectors of random variables. For the current discussion, assume $M = N$ and $\mathbf{x} = f^{-1}(\mathbf{y})$ exists. Then, expanding (3.52) and using $\mathbf{x}(\mathbf{y}) \triangleq f^{-1}(\mathbf{y})$,

$$p_Y(\mathbf{y}) = p_X(\mathbf{x}(\mathbf{y}))/|\det(\partial\mathbf{y}/\partial\mathbf{x})|, \tag{3.55}$$

where

$$\det\left(\frac{\partial\mathbf{y}}{\partial\mathbf{x}}\right) = \begin{vmatrix} \frac{\partial y_0}{\partial x_0} & \frac{\partial y_0}{\partial x_1} & \cdots & \frac{\partial y_0}{\partial x_{N-1}} \\ \frac{\partial y_1}{\partial x_0} & \frac{\partial y_1}{\partial x_1} & \cdots & \frac{\partial y_1}{\partial x_{N-1}} \\ \vdots & \vdots & \ddots & \vdots \\ \frac{\partial y_{M-1}}{\partial x_0} & \frac{\partial y_{M-1}}{\partial x_1} & \cdots & \frac{\partial y_{M-1}}{\partial x_{N-1}} \end{vmatrix}.$$

Notice that the scalar nature of determinants gives $\det(\partial\mathbf{x}/\partial\mathbf{y}) = 1/\det(\partial\mathbf{y}/\partial\mathbf{x})$, which was used in (3.55). The absolute value in (3.55) ensures $\det(\partial\mathbf{y}/\partial\mathbf{x})$ (and therefore p_Y) is positive. Jacobians describe scaling and rotations of data caused by a linear transformation.

Example 3.11.3. *Bivariate normal:* x_0 and x_1 *are iid standard normal random variables. Their joint density is*

$$p_X(\mathbf{x}) = p_X(x_0, x_1) = p(x_0)p(x_1) = \frac{1}{2\pi}e^{-(x_0^2+x_1^2)/2}. \tag{3.56}$$

We are interested in two new variables, the sum and different $y_0 = x_0 + x_1$ *and* $y_1 = x_0 - x_1$. *Find* $p_Y(\mathbf{y})$.

Solution
The Jacobian from (3.55) is

$$\det\left(\frac{\partial\mathbf{y}}{\partial\mathbf{x}}\right) = \det\begin{pmatrix} 1 & 1 \\ 1 & -1 \end{pmatrix} = -2$$

and $\mathbf{x}(\mathbf{y}) = \begin{pmatrix} (y_0 + y_1)/2 \\ (y_0 - y_1)/2 \end{pmatrix}$. *Combining (3.55) and (3.56),*

$$p_Y(\mathbf{y}) = \frac{1}{2\pi \times 2}\exp\left(-\frac{1}{2}\left(\frac{(y_0 + y_1)^2}{4} + \frac{(y_0 - y_1)^2}{4}\right)\right) = \frac{1}{4\pi}e^{-(y_0^2+y_1^2)/4}.$$

Y_0 *and* Y_1 *remain separable and thus independent variables. Because the width of the 2-D normal pdf is broader, the amplitude must be reduced so that the integral remains 1. This Jacobian called for a twofold reduction in transformed signal amplitude.*

3.11.3 Linear Transformation of MVN Vectors

Assume $\mathbf{x} \in \mathbb{R}^{N \times 1}$ are drawn from a MVN process. That is, the elements of $\mathbf{x}$ are each UVN; $\{x_1\}, \{x_2\}, \ldots$ Let $\mathbf{x}$ be linearly transformed into $\mathbf{y} \in \mathbb{R}^{N \times 1}$ via $\mathbf{y} = \mathbf{B}\mathbf{x}$, where matrix $\mathbf{B} \in \mathbb{R}^{N \times N}$ is nonsingular. From (3.43),

$$p(\mathbf{x}) = [(2\pi)^N \det \mathbf{K}_x]^{-1/2} \exp\left[-\frac{1}{2}(\mathbf{x} - \boldsymbol{\mu}_x)^\top \mathbf{K}_x^{-1}(\mathbf{x} - \boldsymbol{\mu}_x)\right].$$

To write $p(\mathbf{y})$ in terms of $p(\mathbf{x})$, we first relate the means and covariance matrices using $\mathcal{E}\mathbf{y} = \mathcal{E}\{\mathbf{B}\mathbf{x}\} = \mathbf{B}\boldsymbol{\mu}_x = \boldsymbol{\mu}_y$ and

$$\begin{aligned}
\mathbf{K}_y &= \mathcal{E}\{(\mathbf{y} - \boldsymbol{\mu}_y)(\mathbf{y} - \boldsymbol{\mu}_y)^\top\} \\
&= \mathbf{B}\mathcal{E}\{(\mathbf{x} - \boldsymbol{\mu}_x)(\mathbf{x} - \boldsymbol{\mu}_x)^\top\}\mathbf{B}^\top \\
&= \mathbf{B}\mathbf{K}_x\mathbf{B}^\top .
\end{aligned}$$

$$\text{Also,} \quad \mathbf{K}_y^{-1} = (\mathbf{B}\mathbf{K}_x\mathbf{B}^\top)^{-1} = (\mathbf{B}^\top)^{-1}\mathbf{K}_x^{-1}\mathbf{B}^{-1} \triangleq \mathbf{B}^{-\top}\mathbf{K}_x^{-1}\mathbf{B}^{-1}$$

$$\det \mathbf{K}_y = \det(\mathbf{B}\mathbf{K}_x\mathbf{B}^\top) = (\det\mathbf{B})^2 \det\mathbf{K}_x.$$

Then

$$\begin{aligned}
\Delta_y^2 &= (\mathbf{y} - \boldsymbol{\mu}_y)^\top \mathbf{K}_y^{-1}(\mathbf{y} - \boldsymbol{\mu}_y) = (\mathbf{x} - \boldsymbol{\mu}_x)^\top \mathbf{B}^\top \mathbf{B}^{-\top} \mathbf{K}_x^{-1} \mathbf{B}\mathbf{B}^{-1}(\mathbf{x} - \boldsymbol{\mu}_x) \\
&= (\mathbf{x} - \boldsymbol{\mu}_x)^\top \mathbf{K}_x^{-1}(\mathbf{x} - \boldsymbol{\mu}_x) = \Delta_x^2, \quad (3.57)
\end{aligned}$$

where scalar Δ^2 – that is, the (generalized squared) *Mahalanobis distance* [17] between points in data space and the mean vector per variance – is invariant under any nonsingular linear transformation. Consistent with (3.55), we find

$$\begin{aligned}
p(\mathbf{y}) &= [(2\pi)^N \det \mathbf{K}_y]^{-1/2} \exp\left[-\frac{1}{2}(\mathbf{y} - \boldsymbol{\mu}_y)^\top \mathbf{K}_y^{-1}(\mathbf{y} - \boldsymbol{\mu}_y)\right] \\
&= [(2\pi)^N \det \mathbf{K}_y]^{-1/2} e^{-\Delta_y^2/2} \\
&= \left|(\det \mathbf{B})^{-1}\right| [(2\pi)^N \det \mathbf{K}_x]^{-1/2} e^{-\Delta_x^2/2} = \frac{p(\mathbf{x})}{|\det \mathbf{B}|}. \quad (3.58)
\end{aligned}$$

$\det\mathbf{B}$ is the Jacobian of the transformation; $|\det\mathbf{B}|$ is the absolute values of the scalar quantity.

3.12 Regression Analysis

We recently acquired a linear optical sensor that the manufacturer assures us responds with output signal $\hat{y}$ proportional to input light levels x. The input–output relationship is $\hat{y} = x + e$, such that[5] $y(x) \triangleq \mathcal{E}\hat{y}(x) = x$ and acquisition noise $e \sim \mathcal{N}(0, \sigma^2)$. Before using the sensor, we run a test in which we input precisely determined light levels to the device and measure the output values. The data are

---

[5] Chapter 4 tells us that $\hat{y} = \mathcal{H}x = cx + e$. Here, c has units of [volts]/[flux] and magnitude 1.

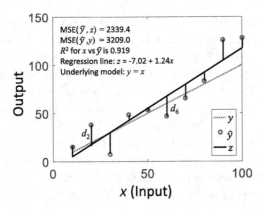

Figure 3.12 Optical sensor calibration experiment.

$$x = (10.000 \ 20.000 \ 30.000 \ 40.000 \ 50.000 \ 60.000 \ 70.000 \ 80.000 \ 90.000 \ 100.000)$$
$$\hat{y} = (15.377 \ 38.339 \ 7.412 \ 48.622 \ 53.188 \ 46.923 \ 65.661 \ 83.426 \ 125.784 \ 127.694)$$

These values are plotted as points in Figure 3.12. We have reason to believe the system responds linearly, so it makes sense to test that contention.

The process of fitting measurement data $\hat{y}_j$ to a linear function is referred to as *linear regression*. The process is implemented by the *method of least squares*, where we adopt the linear equation $z_j = P(1)x_j + P(2)$ and search for parameter vector $\mathbf{P} = (P(1), P(2))^{\top}$. The vector elements selected minimize the *objective function* Θ, which is the sum of squared deviations d_j^2 between the data $\hat{y}_j$ and the corresponding linear model values z_j for all N sample values:

$$\Theta = \sum_{j=1}^{N} d_j^2 = \sum_{j=1}^{N} \left(\hat{y}_j - z_j\right)^2 = \sum_{j=1}^{N} \left(\hat{y}_j - P_1 x_j - P_2\right)^2. \tag{3.59}$$

We minimize Θ by differentiating (3.59) with respect to each parameter and setting the results to zero:

$$\frac{\partial \Theta}{\partial P_1} = -2 \sum_{j=1}^{N} (\hat{y}_j - P_1 x_j - P_2) x_j = 0 \quad \text{and} \quad \frac{\partial \Theta}{\partial P_2} = -2 \sum_{j=1}^{N} \hat{y}_j - P_1 - P_2 = 0.$$

Rearranging and solving for P_1 and P_2, we find model parameters that satisfy the least-squares condition:

$$P_1 = \frac{N \sum_{j=1}^{N} \hat{y}_j x_j - \sum_{j=1}^{N} x_j \sum_{j=1}^{N} \hat{y}_j}{N \sum_{j=1}^{N} x_j^2 - \left(\sum_{j=1}^{N} x_j\right)^2}$$

$$P_2 = \frac{\sum_{j=1}^{N} x_j^2 \sum_{j=1}^{N} \hat{y}_j - \sum_{j=1}^{N} x_j \sum_{j=1}^{N} \hat{y}_j x_j}{N \sum_{j=1}^{N} x_j^2 - \left(\sum_{j=1}^{N} x_j\right)^2}.$$

Entering the data for x and $\hat{y}$, we find $P_1 = 1.24$ and $P_2 = -7.02$. Thus, the regression line is $z_j = 1.24x_j - 7.02$, which is plotted in Figure 3.12. In MATLAB we find the same result using two lines of code: `P=polyfit(x,yh,1);z=polyval(P,x);`. If you type `help polyfit`, you will see that we are fitting to a first-order (linear) model and that `P=(P1, P2)` is a parameter vector containing model slope `P1` and intercept `P2`. Function `polyval` combines `P` and x to generate the regression line. Defining the mean-square error for this purpose as `MSE=mean((yh-b).^2)`, where `yh=`$\hat{y}$ and `b` is either y or z, we find the values listed in Figure 3.12. Pearson's correlation coefficient R^2 between x and $\hat{y}$ is found by using `corrcoef(x,yh)` and examining the off-diagonal term in the resulting 2×2 matrix.

What can we conclude from these results? Figure 3.12 shows that the least-squares fit z to data $\hat{y}$ is not the underlying model y, but it is close. At first it may be puzzling that $MSE(\hat{y}, z) < MSE(\hat{y}, y)$, because how can anything fit the data better than the underlying process? Then we realize that just ten noisy measurements acquired over an input scale that spans a decade are not sufficient to uniquely determine the model. Given the relatively low ratio of signal energy to noise energy (Section 4.2), we observe that the model reflects the noise as well as the signal. All that least-squares methods ensure is that model parameters `P` minimize $MSE(\hat{y}, z)$ for the data provided. $R^2 = 0.919$ tells us that variations in $\hat{y}$ are explained almost entirely by changes in x, so fortunately the two variables are highly correlated. In practice, we won't know values for the noise-free function y, and we can be confident that z is a good representation of the process given the noisy data we have. From these results we may conclude the manufacturer's claim of a linear optical sensor is valid, at least over the range of input values tested.

3.12.1   Nonlinear Regression

We can adapt linear regression methods to analyze some nonlinear regression models. In this procedure, we first try to linearize the model so that linear regression methods apply. When *linearization* is not feasible, we can select a *basis* for expanding the model function into a sum of orthogonal polynomials. Other nonlinear methods are available as well. Here, we will examine the simplest and most common *basis*, a power series.

Example 3.12.1. *Linearized equations: We have a small sample of photon-emitting radioisotope for which we are told the photon energy is 140 keV and the activity is 100 Bq (1 Bq of activity for this isotope generates 1 photon/s). While the risk is low, we need to find a thickness of lead (Pb) to surround the isotope to protect personnel in the room from ionizing radiation exposure. We know that photon attenuation by Pb is a simple first-order differential equation $dN/N_0 = -\mu\,dx$, which tells us that the probability of photon absorption for an incremental layer is $\mu\,dx$, where μ is the linear attenuation coefficient $[cm^{-1}]$, whose value depends on photon energy and the electron density of the absorber. For Pb at 140 keV, $\mu_{Pb}(h\nu = 140\,keV) = 27.08\,cm^{-1}$.*

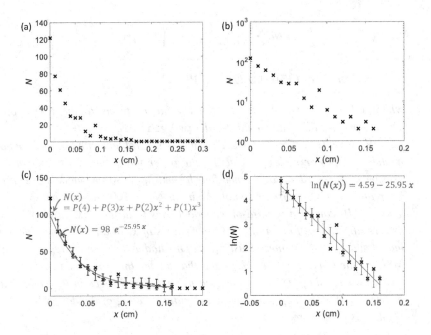

Figure 3.13 Results from Example 3.12.1.

The solution to the equation is $N(x) = N_0\, e^{-\mu x}$. The half-value layer (HVL) of Pb is the thickness that reduces the photon number by 50%. It equals $N(x = HVL)/N_0 = 0.5 = e^{-\mu \times HVL}$, or HVL= $\ln 2/\mu$. If we shield the sample with 5 HVL of Pb, we can assume negligible risk. What thickness of lead do we need?

Solution

If we are sure of the parameters described in the problem, the answer is straight-forward: 5 HVL = 0.128 cm of Pb. Unfortunately, all we can be sure of is the data acquired from the sample, where we count the number of photons detected per second while adding 10-μm-thick layers of lead between measurements. In lieu of experimental data, assume the following code simulates the data; plots of these data appear in Figures 3.13a and 3.13b on linear and semilog axes, respectively.

```
close all;x=0:0.01:0.3;rng(140);              %rng fixes random # gen seed.  x = lead thickness
N0=10^2;Np=N0*exp(-27.08*x);N=poissrnd(Np);   %Beer's law -> mean N, poissrnd -> measurement
plot(x,N,'x');figure;semilogy(x,N,'x')        %plot results
```

First, we linearize the exponential model, $\ln N(x) = \ln N_0 - \mu x$, so it can be fit to the log of the measurements, $\ln \hat{N}$. The resulting regression line with error bars is shown in Figure 3.13d. Converting back to linear coordinates, the results are plotted in Figure 3.13c.

Second, we expand the exponential model into a power series in x that we approximate at the third order using

$$N(x) = N_0 e^{-\mu x} = N_0 \sum_{k=0}^{\infty} \frac{(-\mu x)^k}{k!} = N_0 \left(1 - \mu x + \frac{\mu^2}{2} x^2 - \frac{\mu^3}{6} x^3 + O(4) \right)$$

$$\simeq P(4) + P(3)x + P(2)x^2 + P(1)x^3,$$

where $O(4)$ indicates the accumulated value of terms at the fourth order and above. Applying these data to third-order polyfit, *we find* P = (-7.39   2.48   -0.28   0.011) x 1.0e4. *The immediate problem with treating the nonlinear model as a power series is selecting the order that truncates the summation. We selected a third-order model since fourth- and higher-order models overfit the data by treating Poisson noise as signal. We input $N_0 = 100$ photons, but quantum noise gave us 110 counts. The third-order model tries to fit that first value at $x = 0$ and, as a result, it tends to oscillate at larger values of x. In this example, linearizing the model and applying a linear least-squares method gives superior results, even though MSE for the linearized fit is 47.30 and that for the third-order linear fit is 31.98. Be careful when using MSE as a decision metric! You will see in the code that values of $N(x) = 0$ were suppressed when computing the log of the data. My code for this solution is given here:*

```
close all;x=0:0.01:0.3;rng(140);              %rng fixes random # gen seed.  x = lead thickness
N0=10^2;Np=N0*exp(-27.08*x);N=poissrnd(Np); %Beer's law -> mean N, poissrnd -> measurement
plot(x,N,'x');figure;semilogy(x,N,'x')        %plot results
for j=1:length(x)              %Only include photon counts > 0 to avoid ln(x<=0)
    if N(j) > 0;
        M(j)=N(j);xp(j)=x(j);Np=Np+1;
    end
end                            %Try the linearized model first
[P,S]=polyfit(xp,log(M),1);  %S is a structure function used in polyval to give D
[Y,D]=polyval(P,xp,S);        %D are estimated uncertainties indicated in the plots as error bars
figure(2);plot(x,log(N));hold on;errorbar(xp,Y,D);hold off
figure(1);hold on;plot(xp,exp(Y))
[P,S]=polyfit(xp,M,3);        %Try the power law approximation to the exponential model
[Y,D]=polyval(P,xp,S);
errorbar(xp,Y,D);hold off
```

   The quick calculation of 5 HVL given earlier suggested that 0.128 cm of Pb is needed for adequate shielding. The linearized solution from the simulated data gives $5\,HVL = 5\ln 2/25.95 = 0.134$ cm. If we use 2 mm of Pb, we'll be fine. We now have more confidence in the suggestion that we have 100 Bq of a 140-keV photon emitter because the experimental data verify this finding within 5%.

   I urge you to try other orthogonal polynomial bases in place of the simple polynomial in x (e.g., Legendre). The basis choice should depend somewhat on details of the fitting problem you are addressing. Note that MATLAB provides lsqcurvefit, *which offers a least-squares solution to a general model function of your choice, F. If x are the input data and $\hat{y}$ are the corresponding output data to be fit, then* lsqcurvefit *finds K parameters in vector P that minimize objective function Θ,*

$$\Theta = \underset{P}{\arg\min} \sum_{k=0}^{K} (F(P, x_k) - \hat{y}_k)^2.$$

3.13   Central Limit Theorem

To this point, we have focused on Poisson and normal processes, even though there are a great number of other processes that represent the stochastic properties of measurements. It turns out that a powerful fundamental idea in probability theory, known as the central limit theorem, magnifies the importance of normal processes in many practical situations.

The *central limit theorem* states that the distribution of the sum of a large number of iid variables will be approximately normal regardless of the underlying distribution of each variable. Provided an experiment satisfies the assumptions associated with the theorem, one can model measurement processes as normal. The derivation of the central limit theorem can be found in many texts, so we will not repeat it here. Instead, we focus on assumptions required before invoking the theorem using measurement examples.

Consider a sensor that counts the number of photons absorbed in a specified energy band and over an adjustable interval of time T. The distribution of samples $g(mT)$ follows a Poisson random process. However, under the special condition that a "large number" of photons are captured by the sensor in each time interval T, we say that $g(mT)$ is asymptotically normal. What is important in this case is that we are summing counts and studying the distribution of the sum as the measured quantity.

If the photon fluence rate is reduced to a time average of one or two photons per measurement interval T, the photon-count distribution will more closely follow a Poisson distribution than a normal process. However, as the mean fluence rate increases, the Poisson distribution of net counts converges to a normal process, although the process remains a single-parameter distribution. That is, $\mathcal{P}(\lambda) = p_X(g[m]; \lambda) = \lim_{N \to \infty} p_X((g[m]; \lambda, \lambda) = \mathcal{N}(\lambda, \lambda)$. This is interesting because, under the central limit theorem, X may behave as a normal variable but there remains a fixed relationship between the mean and the variance as expected for a Poisson process.

Another important problem, which we will address in depth in Chapter 8, is the statistical behavior of observers viewing data to make decisions. In observer-performance experiments, human or computational observers view large numbers of data samples, such as image pixels or time series samples, as they generate a random variable we call their diagnostic decision. Decisions can be either binary (yes or no) or graded to indicate the degree of confidence in the decision (on a 1–10 scale). Provided the task required of the observer includes "many" iid random samples to be considered before making each decision, we can be reasonably sure that the decision variable will be normally distributed.

How many samples qualify as a "large number"? Well, that depends on the details of the physical processes and the measurement. We find numerical demonstrations of convergence to be much more instructive than rules of thumb. Problem 3.6 challenges readers to examine convergence rates for a uniformly distributed variable. As with many statistical rules of thumb, you must gain experience with data before you can understand how well the central limit theorem applies to your experiment. When the theorem applies with acceptable confidence, you have a distribution ready for data modeling.

3.14   Information

If we are to evaluate measurement performance based on the device's ability to capture and display task information, we must first ask, What is information? *Information* is a measurable statistical property of data that enables knowledge generation.

The *deficit of information* in a message is defined probabilistically by the scalar quantity *entropy*, H [28, 100]. Specifically, $H(X) \geq 0$ measures the uncertainty in random variable X, so it is the average amount of information required to specify that variable.

For a discrete-data sequence with N degrees of freedom drawn from random variable X,

$$H(X) = -\sum_{i=1}^{N} \Pr(X = i) \log \Pr(X = i) = -\sum_{i=1}^{N} p(x_i) \log p(x_i). \qquad (3.60)$$

For example, if X is a digital signal voltage that can assume 16 values (4 bits), there are $N = 16$ states or *degrees of freedom* in the event space for X. While the base of the logarithm in (3.60) is flexible, we will stick with convention and use log to mean $\log_2$ so that H is measured in bits. $p(x_i)$ is a measure of the certainty of occurrence of x_i in event space S. Its inverse is a measure of the uncertainty, so $-\log p(x_i)$ measures uncertainty in bits. $H(X)$ quantifies the mean uncertainty in X after considering the N degrees of freedom in the event space. $H(X)$ places a lower bound on the average number of bits required to represent X [28]. The more uncertainty regarding an event, the more information that is needed to identify X and the more capacity that X has for answering questions. Selecting one sample from random variable X, on average we need to ask a minimum of $H(X)$ binary questions to find, with certainty, the value of x_i.

If the task is to predict the outcome of flipping a *fair* coin, you need ask only one binary question after each flip, such as Is the outcome heads? After each flip, the state of the coin is completely determined. The event space includes $X \in \{h, t\}$ and the probabilities associated with each event are equal, $p(h) = p(t) = 0.5$. Consequently, $H = -0.5 \log 0.5 - 0.5 \log 0.5 = 1$ bit. Because $p(h) = p(t)$, this 1-bit system has *maximum entropy* (is the most uncertain it can be). Notice that we must know probabilities $p(h)$ and $p(t)$ if we are to compute the uncertainty of the coin-toss variable.

Let's modify the experiment by using a *biased* coin that always comes up heads. The probabilities associated with X are now $p(h) = 1$ and $p(t) = 0$, so $H = -1 \log 1 - 0 \log 0 = 0$ bits. The second term is zero because $\log x_i \rightarrow -\infty$ more slowly than $x_i \rightarrow 0$. Zero entropy tells us that no questions are needed to determine the state of the coin. You may not know what it is, but there is no uncertainty in this coin's state.

Finally, consider a *less-biased* coin, where now $p(h) = 0.3, p(t) = 0.7$, and $H = 0.88$ bits. We still need to ask one binary question to determine state, but there is less randomness in this coin than in a fair coin.

3.14.1   Definitions and Examples

The key terms introduced here are most clearly illuminated by examples. Let X be a discrete random variable with pmf $p(x) = \Pr(X = x)$, $x \in$ S. The entropy of X is the average uncertainty,

$$H(X) = \mathcal{E}_x \log(1/p(x)) = -\sum_{x \in S} p(x) \log p(x) \geq 0. \qquad (3.61)$$

where the shorthand notation for expectation is $\mathcal{E}_x\{\cdot\} \triangleq \sum_{x \in S}\{\cdot\}p(x)$.

Example 3.14.1. *Computing univariate entropies:* Let $X = \{x_i\}_{i=1}^5$ and $p(x_1) \triangleq \Pr(X = x_1) = 1/9$, $p(x_2) = 2/9$, $p(x_3) = 3/9$, $p(x_4) = 2/9$, and $p(x_5) = 1/9$. Find $H(X)$.

Solution

$$H(X) = -\frac{1}{9}(\log 1/9 + 2 \log 2/9 + 3 \log 3/9 + 2 \log 2/9 + \log 1/9) = 2.20 \text{ bits}$$

Example 3.14.2. *Computing jointly distributed entropies:*
(a) *Find the entropies for the joint and marginal distributions of Case 1 in Table 3.1.*

The joint distribution is reproduced here:

Y_1, Y_2	$y_2 = 1$	$y_2 = 2$	$y_2 = 3$
$y_1 = 1$	1/9	2/9	0
$y_1 = 2$	0	2/9	1/9
$y_1 = 3$	0	1/9	2/9

Note that $p(y_1, y_2) \triangleq \Pr(Y_1 = y_1, Y_2 = y_2)$ *and* $\mathcal{E}_{y_1 y_2}\{\cdot\} \triangleq \sum_{y_1, y_2 \in S}\{\cdot\}p(y_1, y_2)$.

Solution

$$H(Y_1, Y_2) = -\mathcal{E}_{y_1 y_2}\{\log p(y_1, y_2)\}$$
$$= -\sum_{y_1 \in S}\sum_{y_1 \in S} p(y_1, y_2) \log p(y_1, y_2) = -3/9(\log 1/9 + 2 \log 2/9) = 2.50$$

$$H(Y_1) = -\sum_{y_1 \in S} p(y_1) \log p(y_1) = -1/3(3 \log 1/3) = 1.59$$

$$H(Y_2) = -\sum_{y_2 \in S} p(y_2) \log p(y_2) = -1/9(\log 1/9 + 5 \log 5/9 + 3 \log 3/9) = 1.35$$

(b) *Find the conditional entropy* $H(Y_1|Y_2)$.

Solution

$$H(Y_1|Y_2) = -\mathcal{E}_{y_1, y_2} \log p(y_1|y_2) = -\sum_{y_1 \in S}\sum_{y_2 \in S} p(y_1, y_2) \log p(y_1|y_2)$$

$$= -\sum_{y_2 \in S} p(y_2) \sum_{y_1 \in S} p(y_1|y_2) \log p(y_1|y_2) \qquad (3.62)$$

$$= \sum_{y_2 \in S} p(y_2)\, H(Y_1|Y_2 = y_2).$$

Since conditioning sets the denominator of the probability,

$$p(y_1|y_2 = 1) = \left(\frac{1}{9},0,0\right)\Big/\frac{1}{9} = (1,0,0)$$

$$p(y_1|y_2 = 2) = \left(\frac{2}{9},\frac{2}{9},\frac{1}{9}\right)\Big/\frac{5}{9} = \left(\frac{2}{5},\frac{2}{5},\frac{1}{5}\right)$$

$$p(y_1|y_2 = 3) = \left(0,\frac{1}{9},\frac{2}{9}\right)\Big/\frac{1}{3} = \left(0,\frac{1}{3},\frac{2}{3}\right)$$

$$p(y_2) = \left(\frac{1}{9},\frac{5}{9},\frac{1}{3}\right)$$

$$H(Y_1|Y_2) = \frac{1}{9}H(1,0,0) + \frac{5}{9}H\left(\frac{2}{5},\frac{2}{5},\frac{1}{5}\right) + \frac{1}{3}H\left(0,\frac{1}{3},\frac{2}{3}\right)$$

$$= \frac{1}{9}(0) + \frac{5}{9}(1.522) + \frac{1}{3}(0.918)$$

$$= 0 + 0.846 + 0.306 = 1.15.$$

In this case, we find that conditioning reduces entropy since there is less uncertainty about Y_1 if we know Y_2; i.e., $H(Y_1|Y_2) < H(Y_1)$.

To arrive at (3.62), we applied the relation $p(x,y) = p(x|y)p(y)$. Taking the log of this expression, changing the sign, and computing means gives the *chain rule* for entropy:[6]

$$-\mathcal{E}_{x,y}\log p(x,y) = -\mathcal{E}_{x,y}\log p(x|y) - \mathcal{E}_y \log p(y)$$

$$H(Y_1,Y_2) = H(Y_1|Y_2) + H(Y_2) \tag{3.63}$$

$$2.50 = 1.15 + 1.35.$$

The last line numerically validates the results of Example 3.14.2. The chain rule for J variables is [28]

$$H(Y_1, Y_2, \ldots, Y_J) = H(Y_1) + \sum_{j=2}^{J} H(Y_j|Y_{j-1}\ldots Y_1) \tag{3.64}$$

e.g., $H(Y_1,Y_2,Y_3,Y_4) = H(Y_1) + H(Y_2|Y_1) + H(Y_3|Y_2,Y_1) + H(Y_4|Y_3,Y_2,Y_1).$

Similarly,

$$H(Y_1,Y_2|Y_3) = H(Y_1|Y_2,Y_3) + H(Y_2|Y_3). \tag{3.65}$$

---

[6] $\sum_{y_1,y_2 \in S} p(y_1) \triangleq \sum_{y_2 \in S}\sum_{y_1 \in S} p(y_1) = \sum_{y_1 \in S} p(y_1)$

Since $H(Y_1, Y_2) = H(Y_2, Y_1)$,

$$H(Y_1|Y_2) + H(Y_2) = H(Y_2|Y_1) + H(Y_1). \tag{3.66}$$

The following MATLAB function computes $H(Y_1, Y_2)$ from $p(y_1, y_2)$:

```
function e = entrop(p)              % p can be one or two dimensional
%
N=size(p);e=0;
for i=1:N(1)
    for j=1:N(2)
        if p(i,j)~= 0               % avoid p(i,j) = 0
            e = e - p(i,j)*log2(p(i,j));
        end
    end
end
```

The native MATLAB function `entropy` finds H from an array of data. It first creates a histogram of the data to estimate a joint distribution. The preceding script is useful if you already have the distribution. I used `histcounts2` in Problem 3.19 (with appropriate normalization) to estimate $p(y_1, y_2)$ from a list of samples.

Relative entropy, $D(p_1 \| p_2)$, is given by the *Kullback–Leibler divergence* [63],[7]

$$D(p_1 \| p_2) = \mathcal{E}_1 \log \frac{p_1(x)}{p_2(x)} = \sum_{x \in S} p_1(x) \log \frac{p_1(x)}{p_2(x)} \geq 0, \tag{3.67}$$

for pmfs $p_1(x)$ and $p_2(x)$. D measures the separation[8] between p_1 and p_2, which is at its minimum at $p_1 = p_2$.

Example 3.14.3. *Relative entropy calculations:*
(a) Compute D from (3.67) for $\mathcal{P}(\lambda)$ and $\mathcal{P}(\lambda + d)$ in the range $0 \leq d \leq 10$.
(b) Compute D for $\mathcal{N}(\mu, \sigma^2)$ and $\mathcal{N}(\mu + d, \sigma^2)$ for $0 \leq d \leq 15$ and $\sigma = 5$.
(c) Repeat (b) for $\sigma = 6$ and compare these results with detectability index $d' = d/\sigma$. The ranges for d are selected so that $d' \lesssim 3$.

Solution
The following script generated the Poisson results in (a) that are shown in Figure 3.14. It is easily adapted for normal variables by changing the pdf function.

```
%%%%%%%%%%%%%%%%%%%%%%%%% Script is for part (a) of Example 3.14.3 %%%%%%%%%%%%%%%%%%%%%%%%%%%%%%%%%%
x=0:100;N=length(x);N1=16;          % initialize random variable and parameters
D=zeros(1,N1);p1=zeros(N1,N);p2=p1; % initialize D and pmfs
for k=1:N1;                         % Compute D between delta mean: 0 <= d <= 15
    p1(k,:)=pdf('poiss',x,20);p2(k,:)=pdf('poiss',x,20+k-1);   % compute Poisson variables
    for j=1:N                       % for normal variables, change line above to 'norm' & add sigma
        if (p1(k,j)>1e-6) && (p2(k,j)>1e-6)          % avoid pmf=0 values
            D(k) = D(k) + p1(k,j)*log2(p1(k,j)/p2(k,j));   % find D
        end
    end
end
figure;subplot(1,3,1);plot(x(1:2:end),p1(1,(1:2:end)),'k');hold on
plot(x(1:2:end),p2(1,(1:2:end)),'k');hold off;title('\lambda_2-\lambda_1 = 0')
xlabel('x');ylabel('p(x)')
```

---

[7] The convention for $D = \sum p_1 \log(p_1/p_2)$ is $D = 0$ when $p_1 = 0, 0 \leq p_2 \leq 1$, and $D = \infty$ when $0 < p_1 \leq 1, p_2 = 0$. Carefully note the inequalities.
[8] Some references label (3.67) as a "distance." Although this metric does not have the symmetry of a true distance [10], it vanishes when $p_1(x) = p_2(x)$.

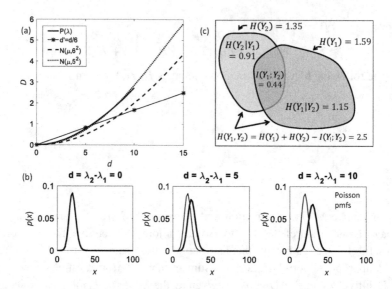

Figure 3.14 (a) Relative entropies $D(p_1 \| p_2)$ are plotted as a function of the difference in means d for one Poisson distribution and two normal distributions. $d' = d/\sigma$ for $\sigma = 6$. d' is a metric introduced in Chapter 7; it is plotted for the $\mathcal{N}(\mu, 6^2)$ results. (b) Three of the Poisson pmf pairs are shown: $d = 0, 5, 10$. (c) Venn diagram illustrating the relationships between mutual information I and entropy H. Numbers are from Examples 3.14.2 and 3.14.3.

```
subplot(1,3,2);plot(x(1:2:end),p1(6,(1:2:end)),'k');hold on
plot(x(1:2:end),p2(6,(1:2:end)),'k');hold off;title('\lambda_2-\lambda_1 = 10')
xlabel('x');ylabel('p(x)')
subplot(1,3,3);plot(x(1:2:end),p1(11,(1:2:end)),'k');hold on
plot(x(1:2:end),p2(11,(1:2:end)),'k');hold off;title('\lambda_2-\lambda_1 = 15')
xlabel('x');ylabel('p(x)')
figure;plot(0:10,D,'k','linewidth',2);xlabel('d');ylabel('D')    % plot D up to d'~=3, or d=sqrt(10)
```

The shape of the pmf influences the shape of D in Figure 3.14a, although the normal and Poisson results are similar for $d' \leq 3$ and $\sigma = 5$.

$D(p_1 \| p_2)$ is an *information-theoretic* measure of distribution separation. It differs from the parametric metric d' (detectability index) often applied in hypothesis testing in that D applies regardless of the distribution, whereas d' is a normal distribution metric. D is meaningful only when the distributions overlap, which is why the range of D is limited to $d' \lesssim 3$. This is not much of a limitation because we are most interested in measuring the separation between similar distributions. $D(p_1 \| p_2)$ is a measure of the information lost when we use $p_2(X)$ to describe X and the true distribution is $p_1(X)$.

Mutual information, $I(Y_1; Y_2)$, is the final quantity we introduce. It is closely related to relative entropy $D(p_1 \| p_2)$ [28]:

$$I(Y_1; Y_2) = \mathcal{E}_{y_1, y_2} \log \frac{p(y_1, y_2)}{p(y_1)p(y_2)} = D((p(y_1, y_2) \| p(y_1)p(y_2)) \geq 0$$

$$= \sum_{y_1, y_2 \in \mathsf{S}} p(y_1, y_2) \log \frac{p(y_1, y_2)}{p(y_1)p(y_2)}. \tag{3.68}$$

$I(Y_1; Y_2)$ quantifies the amount of information in bits that is shared between the two variables. If Y_1 and Y_2 are statistically independent, then $p(y_1, y_2) = p(y_1)p(y_2)$ and there is no mutual information, i.e., $I(Y_1; Y_2) = 0$.

From (3.68),

$$I(Y_1; Y_2) = \sum_{y_1, y_2 \in S} p(y_1, y_2) \log \frac{p(y_1 | y_2)}{p(y_2)}$$

$$= \sum_{y_2 \in S} p(y_2) \sum_{y_1 \in S} p(y_1 | y_2) \log p(y_1 | y_2) - \sum_{y_1 \in S} p(y_1) \log p(y_1)$$

$$= -H(Y_1 | Y_2) + H(Y_1). \tag{3.69}$$

Combining (3.63) and (3.69), we find $H(Y_1, Y_2) = H(Y_1) + H(Y_2) - I(Y_1; Y2)$. These relationships are illustrated in Figure 3.14c using the numerical values computed in Examples 3.14.2 and 3.14.3.

3.15 Problems

3.1 Let's expand on the simple Example 2.2.1. Assume two fair dice are thrown together many times, and let A and B be random variables describing the possible outcomes on each die. Consider A and B to be independent, where $\Pr(A = m) = \Pr(B = m) = 1/6$ for the event space $A, B \in S$, where $m = 1...6$. Now define random variable $C = |B - A|$; the possible outcomes for C are $\{0, 1, 2, 3, 4, 5\}$.

(a) Complete the following table of joint probabilities $\Pr(A = m, C = n)$ and marginal probabilities $\Pr(A = m)$ and $\Pr(C = n)$. (Hint: There is no formula, you just need to consider the possibilities.)

(b) Compute the statistics $\mathcal{E}\{A\}, \mathcal{E}\{C\}, \mathcal{E}\{AC\}$, var_A, var_C, and $\text{cov}_{AC} = \mathcal{E}\{(A - \mathcal{E}A)(C - \mathcal{E}C)\}$ from elements in the table.

$\Pr(A = m, C = n)$	$C: n = 0$	$n = 1$	$n = 2$	$n = 3$	$n = 4$	$n = 5$	$\Pr(A = m)$
$A:m = 1$	1/36	1/36	1/36	1/36	1/36	1/36	1/6
$m = 2$	1/36	2/36			1/36	0	
$m = 3$	1/36				0	0	
$m = 4$	1/36						
$m = 5$	1/36						
$m = 6$	1/36						
$\Pr(C = n)$	6/36				4/36	2/36	1

3.2 From the definition of moments, show how normal parameter σ is related to variance of the normal distribution.

3.3 Show that the variance of a Poisson variable is equal to λ
(a) in the manner of (3.16) and
(b) using moment-generating functions.

3.4 Derive maximum-likelihood estimates (MLEs) for the mean and the standard deviation assuming iid MVN data $\mathbf{g} \sim \mathcal{N}(\mu, \sigma^2)$.

Vector $\mathbf{g} \in \mathbb{R}^{N \times 1}$ contains N measurements in time, i.e, $g(t[n]) = g_n$. The likelihood function for these data and the parameter vector, $\boldsymbol{\theta} = (\mu, \sigma)^\top$, is the joint pdf $L = p(\mathbf{g}; \boldsymbol{\theta}) = p(g_1, g_2, ..., g_N; \mu, \sigma)$, which for iid samples becomes the product $\prod_{n \in N} p(g_n; \mu, \sigma^2)$. The MLE is the argument at the peak of the likelihood function. For example, the MLE of the mean can be found using $\hat{\mu} = \max_\mu \ln L(\mu) = \partial(\ln p(\mathbf{g}; \boldsymbol{\theta}))/\partial \mu \big|_{\hat{\mu} = \mu} = 0$. In this case, we take the derivative of the log-likelihood because the peak value is not affected by a monotonic transformation.

3.5 What is the maximum-likelihood estimate of the number of photons counted by a photodiode?

3.6 Conduct a numerical experiment that tests convergence of the central limit theorem.

(a) Start with a MATLAB uniform random variable routine and generate a matrix of $1,000 \times 10,000$ univariate samples. Create a histogram of the first row of 10,000 samples. Then sum the first two rows and create a histogram of the resulting 10,000 summed values. Repeat for the first 10 rows and for all 1,000 rows. Use a MATLAB routine to run a test for normality. In a 2×2 subplot, plot the histograms and best-fit normal pdf curves.

(b) Repeat part (a) using a log-normal rv. Also plot the 2×2 matrix of histograms and fits. What can you conclude about differences in convergence?

3.7 You are studying genes, and wonder what the chance of seeing a specific-size oligomer (a short sequence of base pairs) in a 10Kb (kilobase pair) sequence. There are four distinct base pairs A, C, T, G, so the chance of seeing a specific k-mer is 4^{-k}. For example, there are $4^8 = 65,536$ possible 8-mers and the chance of seeing one of them is $4^{-8} = 1.53 \times 10^{-5}$.

(a) Estimate the probability of finding at least one 8-mer in a 10Kb DNA segment.

(b) Compute the entropy for all k-mers between $1 \leq k \leq 25$ bps.

(c) Assuming this is the maximum entropy state, what does it tell you if you find H is less than the value found in part (b)?

3.8 (a) You decide that any new patient measurements must be part of the healthy population if those measurements fall within $\pm 3\sigma$ of the healthy-population mean. Those falling outside of the healthy range are considered positive for illness. What false-positive error α are you accepting by adopting this criterion?

(b) An instrument yields normally distributed data via $\mathcal{N}(\mu = 4V, \sigma^2 = 4V^2)$ when operated within design specifications. You hear of someone measuring a "high value," and you're concerned because it is outside the 5% error allowed by the manufacturer. In what range can you expect the measured value to be found?

3.9 Let (A, B) be a discrete random vector, where $S = \{(1,1),(1,2),(2,1),(2,2)\}$. The joint pmf for this event space is

$$p_{AB}(a,b) = \begin{cases} 1/4 & \forall \text{ combinations of } a,b \in S \\ 0 & \text{otherwise.} \end{cases}$$

Are random variables A and B independent?

3.10 Show that X and Y are both independent and uncorrelated given the joint pdf,

$$p_{XY}(x,y) = \frac{1}{2\pi\sigma^2} \exp\left[-\frac{(x-\mu_x)^2 + (y-\mu_y)^2}{2\sigma^2}\right].$$

3.11 This problem probes differences between the information capacity of 8-bit and 4-bit images for data from uniform and normal distributions.

(a) Generate a $1,000 \times 1,000$ image of uniformly distributed random values. The command in MATLAB is x=(2^8-1)*rand(1000), and the result is a 2-D array filled with double-precision floating point numbers between 0 and 255. Convert the image values to unsigned 8-bit integers using y=uint8(x) and generate a histogram from the results. Repeat the simulation using normally distributed data via the combination: uint8(39*randn(1000)+127). The shifting and scaling of the standard normal distribution place values in the range of an 8-bit unsigned integer. Now the two 8-bit data sets have 256 degrees of freedom. Plot histograms for both results, compute their entropies, and explain any differences.

(b) Compress the amplitudes of both data sets to have only 16 degrees of freedom. Even though both sets remain as unsigned 8-bit integers, the number of degrees of freedom (possible values that the amplitudes can assume) is reduced. Form and display their histograms, compute the associated entropies, and explain the differences.

3.12 The following code simulates data measured for a patient while tracking the flow of a blood tracer through a fixed point in an artery. The tracer is infused until it reaches an equilibrium body concentration of $C_1 + C_2$, where C_1 is the equilibrium concentration of tracer in the vasculature and C_2 is the corresponding concentration for the tissues. At $t = 0$, the infusion stops and the kidneys begin to clear the tracer, with the tracer concentration's exponential decay given by the simulated data. However, the signal is the sum of two exponential decay components. The majority of the tracer is found in the vasculature, which is cleared three times faster than the more slowly decaying tracer leaking from the tissues back into the blood. The model is $g(t) = C_1 \exp(-t/\tau_1) + C_2 \exp(-t/\tau_2)$, where $\tau_2 = 3\tau_1$. The problem asks you to run the code to find $g(t)$. Then apply regression analysis to estimate C_1, C_2, τ_1, and τ_2. $g(t)$ is acquired over $0 \leq t \leq 30$ min, where the sampling increment $T = 0.5$ min.

```
t=0:0.5:30;rng(140);   %30 min time axis.  Fix random sequence for additive noise
N=length(t);tau1=3;tau2=9; %Set time constants
e=20*randn(size(t));     %Generate the noise
g=1400-t/0.003-400*t/9; %Generate the noise-free data
for j=2:100              %Use power series for exp to generate data.
    g=g+1000*((-t/tau1).^j)/factorial(j) + 400*((-t/tau2).^j)/factorial(j);
end
g=g+e;                   %signal plus noise
plot(t,g,'or')           %Plot the results.
```

3.13 Find an expression for the characteristic function of a Poisson process. Plot your results for $\lambda = 1, 2, 5$ for frequencies $|\Omega| \leq 3$.

3.14 The N-variate normal pdf expressed in (3.43) shows that the scaling factor necessary to ensure the pdf integrates to 1 is $[(2\pi)^N \det \mathbf{K}]^{1/2}$. Show this is true.

3.15 Example 3.12.1 describes the equation for x-ray absorption (Beer's law) as $N(x) = N_0 e^{-\mu x}$. Give the cumulative distribution function (cdf) and the probability density function (pdf) describing the probability of photon attenuation as a function of absorber thickness x. Check that your solutions have the appropriate properties.

3.16 (a) Compute the moment-generating function for an exponential pdf, $X \sim p(x; \lambda) = \lambda e^{-\lambda x}$ with region of support $X = [0, \infty)$.
  (b) Find the mean and variance of X.
  (c) Show that $M^{(1)}(0) = \int_0^\infty dx \, x \, p(x)$.

3.17 Claude Shannon's original investigation into the nature of information sought to understand the influence of noise on a communication channel. He quickly realized these ideas could be applied to evaluate information in written language [96].
  Consider the frequencies of letters observed in 4.5×10^9 characters of English-language text.[9] From a histogram of letter occurrence, we estimate the following pmf:

```
a=0.0855, b=0.0160, c=0.0316, d=0.0387, e=0.1210, f=0.0218,
g=0.0209, h=0.0496, i=0.0733, j=0.0022, k=0.0081, l=0.0421,
m=0.0253, n=0.0717, o=0.0747, p=0.0207, q=0.0010, r=0.0633,
s=0.0673, t=0.0894, u=0.0268, v=0.0106, w=0.0183, x=0.0019,
y=0.0172, z=0.0011
```

As the text grows in size, the frequencies converge toward probabilities of finding a letter, e.g., $\Pr(X = a) = 0.0855$. Since $\sum_{i=1}^{26} p(x_i) = 1$, we are ignoring capital letters, punctuation, spaces, etc.
  (a) What is the minimum average number of binary questions we must ask to find an arbitrary letter? Demonstrate your solution for the letter 'b'.
  (b) If the 26 letters were equiprobable, $p(x_i) = 1/26$, does the strategy change?

3.18 Compute the mutual information for a bivariate normal distribution as a function of the correlation coefficient $0 \leq \rho \leq 1$. Explain the results at each limit.

---

[9] http://practicalcryptography.com/cryptanalysis/letter-frequencies-various-languages/english-letter-frequencies/

4 Spatiotemporal Models of the Measurement Process

4.1   Introduction

The first step in the measurement process is data *acquisition*, in which task-related properties of the object are reversibly encoded into data. The second step is data *display*, which prepares the acquired data for presentation to the observer/decision maker. This preparation considers the strengths and limitations of the observer when providing the information in accessible form.

Let $f(t')$ represent time-varying properties of the object that we wish to record. Properties $f(t')$ interact with the measurement device to acquire data $g(t)$ and are displayed as $\hat{f}(t')$. Representing the acquisition operator as $\mathcal{H}$, the acquisition process is expressed as $g(t) = \mathcal{H} f(t')$. Representing data display by operator $\mathcal{O}$, the preparation of acquired data for observation is written as $\hat{f}(t') = \mathcal{O}g(t)$. The overall measurement process is diagrammed in Figure 4.1.

For example, consider the pulse oximeter of Figure 4.2. This device measures the blood-oxygen concentration $f(t')$ in a finger, in a noninvasive manner. At the acquisition stage, light-emitting diodes embedded in a clip-on device transmit light through blood-perfused tissue at wavelengths of 660 nm and 940 nm. Any transmitted light sensed by a photodiode generates a voltage signal $g(t)$ proportional to the light absorbed. At the display stage, the differential voltage is interpreted via the oxyhemoglobin and deoxyhemoglobin absorption spectra to display estimates of the concentration of oxygen in the blood $\hat{f}(t')$. Section 4.2 analyzes measurement processes for devices classified as linear systems.

4.2   Measurement Equations

4.2.1   Acquisition

Let $f(t')$ be a time-varying *object function*, like blood-oxygen concentration. It is a function of patient time t'. Blood flow is also a function of location within the finger, $\mathbf{x} = (x, y, z)$, but we choose to ignore spatial features for the task at hand. Voltage waveform $g(t)$ is the time-varying *data function* at measurement time t, with a clock that resets at the beginning of each measurement.

Acquisition transforms the object functions from one vector space into data functions in another vector space. Vector spaces contain sets of vectors that are

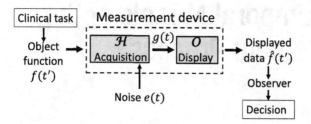

Figure 4.1 Illustration of the biomedical measurement process, $\hat{f}(t') = \mathcal{O}\{\mathcal{H} f(t')\}$.

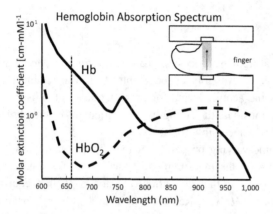

Figure 4.2 Plots of the molar extinction coefficients describing optical absorption for oxygenated (HbO$_2$) and deoxygenated (Hb) blood. Absorption measurements are made at $\lambda_1 = 660$ nm and $\lambda_2 = 940$ nm. The data are replotted from [46]. The inset shows a pulse oximeter on an index finger.

closed under vector addition and scalar multiplication conditions (see Appendix A) – but what does that mean in the pulse oximeter example? Object and data functions describe quantities with different physical units that run on different clocks, which makes them incommensurate. Thus, it does not make sense to add absorption spectra to voltage waveforms directly without first transforming one of them into the other's vector space. Simple conversions can be performed by calibration procedures. Here we focus on *linear operators* that approximate the exact acquisition process. Modeling the process mathematically and computationally is fundamental to understanding for purposes of engineering design and performance assessment.

The acquisition equation is[1]

$$g(t) = \mathcal{H} f(t') = \left(\int_{-\infty}^{\infty} dt'\, h(t,t')\, f(t') \right) + e(t), \quad \text{(Acquisition)}$$

$$\text{where } \mathcal{H} \triangleq e(t) + \int_{-\infty}^{\infty} dt'\, h(t,t'). \tag{4.1}$$

---

[1] Operator $\int dx \ldots$ integrates all terms to the right except where limited by parentheses.

Notice that $\mathcal{H}$ is "applied" to $f(t')$; it is not necessarily multiplied. The *integral operator* maps $f(t')$ into $g(t)$ while adding acquisition noise $e(t)$. The equation holds provided that each element can be modeled as a continuous function of time. The core of the operation is *impulse response function* $h(t,t')$, which translates object functions into data functions. Hence h must be a function of both object time t' and measurement time t. If the units of f are [obj] and the units of g and e are [data], then the units of h are [data/(time×obj)].

Note that g is a *function* of time because a one-to-one relationship exists between dependent variable g and independent variable t. However, g is a *functional* of f, i.e., $g(f(t))$, because the operator maps the entire function f onto each point of g. Equation (4.1) involves a *linear operation* because, for constants a and b, we find

$$\mathcal{H}\{af_1(t') + bf_2(t')\} = a\mathcal{H}f_1(t') + b\mathcal{H}f_2(t'). \tag{4.2}$$

It is important to clearly visualize (4.1) because of the central role this relation plays in modeling any linear measurement process. Figure 4.3 shows the consequences of measuring f using a device whose properties do not change with measurement time t (Figure 4.3a) and with a device whose properties change significantly with t (Figure 4.3b). The time-invariant measurement device has the same impulse response at each measurement time. Consequently, for a sinusoidal input, the measurement waveform is also a sinusoid with the same frequency but potentially a different overall amplitude and phase. In contrast, the time-varying measurement device has an impulse response that varies with t, producing a variable measurement amplitude at the same frequency.

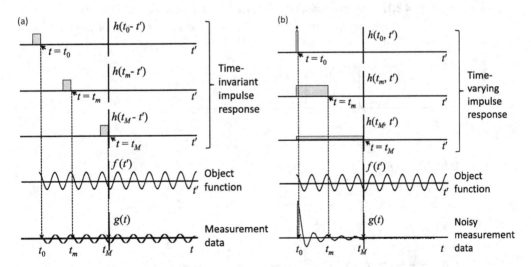

Figure 4.3 Illustration of linear time-invariant (LTI, a) and linear time-varying (LTV, b) measurements of the same object function $f(t') = \cos(2\pi u_0 t + \phi)$ producing data $g(t)$. (a) h is constant for all measurement times; (b) h varies significantly with t but has constant area. Plots in the top three rows show how the different impulse responses appear at measurement times $t = t_0, t_m, t_M$.

Because the time-varying impulse response changes with measurement time, $h(t', t)$ is a function of both object time t' and measurement time t. However, the linear time-invariant (LTI) device is only a function of the time difference, $h(t, t') = h(t - t')$. As we will see in the next subsection, acquisition using LTI devices may be described as a convolution.

4.2.2   Display

Can we fix acquisition distortions? Sometimes, if we have high signal-to-noise ratio. Distortion correction can be included within the *display-stage operator* applied to g (see Section 7.6). The goal is to display task-related features of the original object function – let's call it $\hat{f}(t')$. That process is generally represented as

$$\hat{f}(t') = \mathcal{O}\{g(t)\} = \mathcal{O}\{\mathcal{H}\{f(t')\}\}. \quad \text{(Display-stage equation)} \qquad (4.3)$$

Ideally, we would like to select $\mathcal{O} = \mathcal{H}^{-1}$, which tells us that the display operator completely "undoes" the effects of acquisition (e.g., blur, noise) to reveal task features of the object undistorted; i.e., $f(t')$ is perfectly rendered by the display. Unfortunately, the presence of acquisition noise pretty much guarantees that $\mathcal{H}^{-1}$ cannot be found (see Section 6.4). Example 4.2.1 offers a glimpse into the display-stage challenges that will be discussed in more depth in Section 7.6.

Example 4.2.1. *Deconvolution: Assume we have the oximeter with the blood-flow time series $x(t')$ shown at the top of Figure 4.4, which is nonzero only during three short*

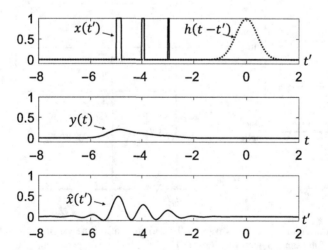

Figure 4.4   (top) Object function $x(t)$ and LTI impulse response $h(t - t')$ are illustrated. (middle) The measurement of x using system h yields acquired data $y(t)$, which poorly represents x. (bottom) Decorrelation methods were applied to partially restore the object, yielding $\hat{x}(t')$. The restoration is incomplete but acceptable if we can achieve the task goal.

time periods. Let's use x in place of f for this object and y in place of g for these data. The task is to measure the number and duration of the blood-flow periods. The Gaussian-shaped impulse response is LTI, $h(t - t')$, and responds very slowly compared with the duration of flow bursts in $x(t')$. This slow response means it could be difficult to achieve the task. Using this low-cost device also leads to a poor temporal resolution. We make the measurement and find the acquired data $y(t)$ are, in fact, blurred, making it nearly impossible to see the three flow bursts.

Using the following code, we create a time-domain deconvolution method to restore the data as best we can for display. This approach is feasible only because we have an LTI measurement device. Clearly, the object function is not fully restored in $\hat{x}(t')$, but the real question is whether we can achieve the task. If we have reason to assume blood flow is constant during the three bursts, the amplitude of peaks in $\hat{x}(t')$ may be interpreted in terms of the duration of the pulse. This information would become part of the display processing, allowing us to achieve the task even though restoration is incomplete. Readers may not follow the development of this code right now, but we will circle back to this solution. Try it while changing the pulse parameter (now set at $\sigma = 0.4$) to get a feel for the process.

```
%%%%%%%%%%%%%%%%%%%%  Code to generate results in Fig. 4.4 from Example 4.2.1  %%%%%%%%%%%%%%%%%%%%%%%
%
close all;T0=20;T=0.01;                                    % 20 s time series, 10 ms sampling interval
t=-T0/2:T:T0/2;N=length(t);                                % time axis & number of points
x=zeros(size(t));x(501:520)=1;x(601:610)=1;x(701:705)=1;  % LTI object function
subplot(3,1,1);plot(t,x);hold on                          % Display object function
h=exp(-(t.^2)/(2*0.4^2))/(0.4*sqrt(2*pi));plot(t,h);hold off % impulse response
y=conv(h,x,'same')*T;subplot(3,1,2);plot(t,y)             % Generate measurement data
hp=fftshift(h);                                           % Prep impulse response and...
H(1,:)=hp;                                                % ...load into first row of H
for j=2:N                                                 % Create circulant 2D H from 1D h
   H(j,1:j-1)=hp(N-j+2:N);
   H(j,j:end)=hp(1:N-j+1);
end
xp=H'/(H*H'+eye(N))*y'/T;                                 % Deconvolve the impulse response
figure(1);subplot(3,1,3);plot(t,xp)                       % Display results
```

Display-stage operator $\mathcal{O}$ is always needed to remap the acquired data back into the original vector space of objects. If the object is a function of space and time, $\mathcal{O}$ must include spatiotemporal remapping.[2] The temporal remapping in Example 4.2.1 was trivial because we synched the clocks during simulation. A display operator may also adjust any spatiotemporal distortions imposed during acquisition, as in the amplitude-decay problem for the time-varying impulse response depicted in Figure 4.3 and the blur problem illustrated in Figure 4.4. Such an operator might also minimize the effects of zero-mean noise, although no additive noise was introduced in the Example 4.2.1 simulations. However, restoration and noise reduction should be applied only after it is determined that these practices will truly improve task performance.

---

[2] This is the function of a scan converter in sonography and back-projection in computed tomography (CT).

4.2.3   Signal-to-Noise Ratio

Noise is often described in terms of noise energy because it can be compared to signal energy via a *signal-to-noise ratio*, SNR = (signal energy)/(noise energy), where variance is proportional to energy. We might measure noise variance in practice by acquiring data without an input signal – the equivalent of a control experiment in biology. Then, introducing a stochastic signal to the input, we can obtain the signal+noise energy. For signal-independent noise, the signal+noise variance is $\text{var}(g) = \text{var}(\mathcal{H}f(t))$ and

$$\frac{\text{var}\left(\int dt' h(t,t')\, f(t')\right) + \text{var}(e(t))}{\text{var}(e(t))} = \text{SNR} + 1 \tag{4.4}$$

$$\text{SNR} = \frac{\text{var}(g(t))}{\text{var}(e(t))} - 1. \qquad \text{Linear-scale SNR}$$

Remember, this formulation assumes that we can measure signal-in-noise g and noise-only e. Each term in (4.4) is in the data space and, as such, is a function of measurement time. Because we are interested in energy ratios, and because linear SNR can vary over many decades, results are often reported in decibels [dB]:

$$\text{SNR [dB]} = 10 \log_{10}\left(\frac{\text{var}(g)}{\text{var}(e)} - 1\right). \qquad \text{Log-scale SNR} \tag{4.5}$$

With sample variances and small sample sizes (which never produce a good estimate), it might be necessary to take the absolute values before computing the log. The energy in the measurement of a deterministic object function with additive noise, like that of $x(t')$ in Figure 4.4, is $\int_{-\infty}^{\infty} dt\, |y|^2$.

SNR = 0 dB indicates that the signal energy and noise energy are equal to each other. SNR = 20 dB indicates that the signal energy is 100 times the noise energy, and the average signal amplitude is 10 times that of the noise. SNR = −20 dB indicates that the noise energy is 100 times the signal energy. This last condition clearly describes a low-SNR measurement.

Example 4.2.2. *Signal-to-noise ratio: Adapt the initial parts of the* MATLAB *code from Example 4.2.1 to predict the SNR and then numerically verify the prediction using signal $x(t)$ from the example. Since that example was noise free, we must generate and add the zero-mean noise to the deterministic signal $x(t)$. Assuming $\sigma^2 = 0.01$, we find $e(t) \sim \mathcal{N}(0, 0.01)$ in* MATLAB *using* 0.1*randn(size(t))*. Further, assume an LTI device where $\mathcal{H}f(t')$ is approximated in* MATLAB *by* conv(h,f,'same'). *Further assume that noise is an ergodic process (Section 3.8) such that the ensemble and temporal moments are equal. We need to obtain $y(t)$ and $e(t)$ numerically. Finally, change the width of impulse response h from 0.4 to 0.1. These modifications change the task from being limited by temporal resolution to now being noise limited.*

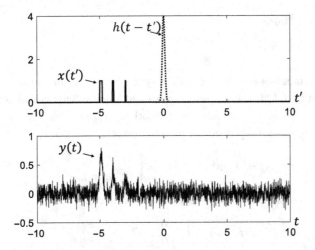

Figure 4.5 Results of the SNR calculation in Example 4.2.2. The impulse response was narrowed to make the task noise limited. $h(t - t')$ is centered at $t = 0$ with area 1.

Solution

Object $x(t')$ is composed of three pulses, each of amplitude 1, that are 0.2 s, 0.1 s, and 0.05 s in duration, respectively. The impulse is normalized to have area 1, so the energy of the object is the same as that of the noise-free measurement $y(t) - e(t)$. Squaring and summing $x(t')$ gives a noise-free signal energy of 0.35 V^2. The noise variance is 0.01 V^2, so SNR = 35 or 15.4 dB.

Numerically, we can approximate $\int_{-\infty}^{\infty} dt \, |y(t)|^2$ in MATLAB as `sum(y.^2)*T` *and the noise energy via* `var(e)`. *The result is averaged over 100 ensembles to provide more convergent sample moments. The measured result is SNR = 15.1 dB, which compares well with the prediction of 15.4 dB. The code that produces the results in Figure 4.5 is shown here.*

```
%%%%%%%%%%%%%%%%%%%%%%%%%%%%  Solution to Example 4.2.2  %%%%%%%%%%%%%%%%%%%%%%%%%%%%%%%%%%%%%%%%
close all;T0=20;T=0.01;                               % 20 s time series, sampling interval=10 ms
t=-T0/2:T:T0/2;N=length(t); snr=zeros(1,100);         % time axis, parameter initialization
x=zeros(size(t));x(501:520)=1;x(601:610)=1;x(701:705)=1; % object function
subplot(2,1,1);plot(t,x);hold on
h=exp(-(t.^2)/(2*0.1^2))/(0.1*sqrt(2*pi));plot(t,h);hold off % impulse response h
hp=fftshift(h); % position the impulse response appropriately
for j=1:100
    e=0.1*randn(size(t));                             % standard normal noise with std scaled by 0.1
    y=conv(h,x,'same')*T + e;                         % measure object w/noise to find y
    snr(j)=sum(y.^2)*T/var(e) - 1;                    % SNR from (4.4) for each noise realization
end
SNR=10*log10(mean(snr))                               % average 100, convert to dB via (4.5)
subplot(2,1,2);plot(t,y)                              % show one of the 100 measurements
```

Despite the assumptions, we measured the SNR values within 0.3 dB of the prediction. The bottom trace in Figure 4.5 is an example of noise at 15 dB SNR. You can build intuition by changing σ in $e(t)$ (set it to 0.1 in the code) and then observing the waveforms and SNR values. We now develop more analytical tools.

4.3   Signal Modeling Tools

4.3.1   Dirac Delta

The Dirac delta (impulse) is indicated by the symbol $\delta(t)$ for continuous variable t. It is a strange but useful "function," having zero width and infinite height at the point where its argument is zero. It can be tricky to work with $\delta(t)$ unless we find it within an integral. Among its useful properties, we find deltas integrate to 1,

$$\int_a^b dt \, \delta(t - t_0) = \int_a^b dt \, \delta(t_0 - t) = \begin{cases} 0 & t_0 < a \text{ or } t_0 > b \\ 1 & a < t_0 < b \end{cases}, \qquad (4.6)$$

and we can sift through continuous functions to select specific values,

$$\int_a^b dt \, f(t) \, \delta(t - t_0) = \begin{cases} 0 & t_0 < a \text{ or } t_0 > b \\ f(t_0) & a < t_0 < b \end{cases}. \qquad (4.7)$$

The sifting property means that

$$\int_{-\infty}^{\infty} dt \, f(t) \, \delta(t - t_0) = \int_{-\infty}^{\infty} dt \, f(t_0) \, \delta(t - t_0). \qquad (4.8)$$

Equation (4.7) is dimensionally correct when you consider that the units of a Dirac delta equal the inverse of the units of its argument. Like the probability density functions from Chapter 3, the product $dt \, \delta(t)$ is unitless; consequently, if dt has units of [s], then $\delta(t)$ has units $[\text{s}^{-1}] = [\text{Hz}]$.

The sifting property of Dirac deltas is helpful for representing a function of time sampled on the constant interval T, where $t = nT$ for integer n. Graphically, a sequence of Dirac deltas is indicated by upward-pointed arrows positioned along the t-axis at time intervals T. Writing a sequence of time-shifted deltas as $\delta(t - nT)$ and noticing that they do not overlap, we can sum the sequence to create a *comb* for sampling (see Section 5.7):

$$\text{comb}(t/T) \triangleq \sum_{n=-\infty}^{\infty} \delta\left(\frac{t}{T} - n\right) = T \sum_{n=-\infty}^{\infty} \delta(t - nT). \qquad (4.9)$$

Looking closely, you can see that combs have no units.

Derivatives for Dirac deltas are defined (Appendix B), but $\delta(t)$ is not a function, much less continuously differentiable. Limiting forms helps us see the origin of its units. To illustrate, let b denote a width parameter and a denote a shift parameter (Figure 4.6) for exponential and Gaussian functions. Limiting approximations include

$$\delta(t - a) = \begin{cases} \lim_{b \to 0} \frac{1}{2b} e^{-|t-a|/b} \\ \lim_{b \to 0} \frac{1}{b\sqrt{\pi}} e^{-(t-a)^2/b^2} \end{cases}. \qquad (4.10)$$

The scale factors, $1/2b$ for exponentials and $1/(b\sqrt{\pi})$ for Gaussians, are obtained from integrals of the respective functions (before taking the limit) over all time.[3]

---

[3] It is good to know that $\int_{-\infty}^{\infty} dt \, \exp(-a^2 t^2) = \sqrt{\pi/a^2}$ and $\int_{-\infty}^{\infty} dt \, \exp(-|t|/b) = 2b$.

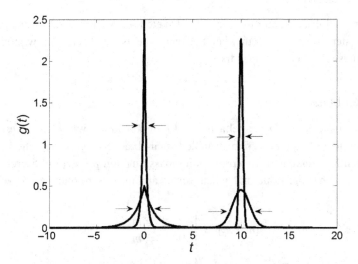

Figure 4.6 Plots of exponential (left) and Gaussian (right) functions for two different width parameters b, where one is five times the other. In the limit of $b \to 0$, both functions become Dirac deltas; see (4.10). We set the shift parameter $a = 0$ for the exponential functions, and $a = 10$ for the Gaussian functions just to separate them. Arrows indicate FWHM estimates.

These factors ensure the delta integrates to 1. You can always figure out the units of the constants by noting that units must all cancel in the arguments of transcendental functions. Consequently, if t in (4.10) has units of [ms], then the units of a and b must also be [ms].

The *full-width-at-half-maximum (FWHM) value* is $2b(\ln 2)$ for the exponential function and $2b(\sqrt{\ln 2})$ for the Gaussian function. These values are found by setting each function (before taking the limit) to half of its peak amplitude and computing its width. In both cases, as $b \to 0$, the width of the function goes to zero as the amplitude goes to infinity at $t = a$ so that the areas under the curves remain unity.

Multidimensional Dirac deltas are represented by N-dimensional vector arguments, $\delta(\mathbf{x}) = \delta(x_1)\delta(x_2)\ldots\delta(x_n)\ldots\delta(x_N)$. If x_n has units of length, e.g., [cm], and $N = 3$, then this delta has units of inverse volume, e.g., $[\text{cm}^{-3}] = [\text{mL}^{-1}]$.

Related quantities include *step*,

$$\text{step}(x) \triangleq \int_{-\infty}^{x} dy\, \delta(y) = \begin{cases} 0; & x < 0 \\ 1; & x > 0 \end{cases},$$

and rectangular "functions" with constants x_0 and X,

$$\text{rect}\left(\frac{x - x_0}{2X}\right) \triangleq \text{step}(x - x_0 + X) - \text{step}(x - x_0 - X)$$

$$= \begin{cases} 1; & x_0 - X < x < x_0 + X \\ 0; & \text{otherwise} \end{cases}.$$

To understand these equations, you should sketch them for yourself. Remember that delta and step transitions occur when those arguments equal zero. A few properties of Dirac deltas are listed in Appendix B.

4.3.2   Kronecker Delta

A close cousin of the Dirac delta is the *Kronecker delta*, which is represented by symbols $\delta[n]$ and δ_n for discrete-variable (square bracket) arguments. The Kronecker delta is unitless, analogous to a probability mass function $p_X[x_n]$ for discrete random variable X. It has the value of 1 whenever its argument is zero and zero everywhere else:

$$\delta_{mn} = \delta[n-m] = \begin{cases} 1 & n = m \\ 0 & n \neq m \end{cases} \quad \text{and}$$

$$\sum_{n=n_a}^{n_b} \delta[n-m] = \begin{cases} 1 & n_a \leq m \leq n_b \\ 0 & m < n_a \text{ or } m > n_b \end{cases}.$$

Because Kronecker deltas are discrete, sums are used in place of integrals in these expressions. The relationship between Kronecker and Dirac deltas is that $\delta[n]$ is analogous to $dx\, \delta(x)$.

4.3.3   Continuous and Discrete Acquisitions

Equation (4.1) indicates that continuous-time signals $g(t)$ acquired from continuous-time objects $f(t)$ using a linear measurement device may be represented by a noisy integral transformation. The equivalent discrete-time representation is

$$[\mathbf{g}]_m = g[m] = \mathcal{H}f[n] = \sum_{n=1}^{N} h[m,n]\, f[n] + e[m], \tag{4.11}$$

where now $\mathcal{H} \triangleq e[m] + \sum_{n=1}^{N} h[m,n]$.

A big advantage of $\mathcal{H}$ notation is that it simply states data are acquired, then leaves the important but often messy details to be defined elsewhere. Barrett and Myers [10] label $\mathcal{H}$ in (4.1) as an example of a *continuous-to-continuous linear transformation*, which is useful for design. $\mathcal{H}$ in (4.11) exemplifies a *discrete-to-discrete transformation*, which is useful for numerical computations. Mixed forms are available, too. For example, a *noise-free continuous-space to discrete-time transformation* may be written as

$$g[m] = \mathcal{H}\, f(x) = \int_{-\infty}^{\infty} dx\, h_m(x)\, f(x) = \int_{-\infty}^{\infty} dx\, h(x, mT)\, f(x). \tag{4.12}$$

This measurement maps objects in space into data in time. In your work, you should choose the level of equation detail necessary to accurately tell the story you need to tell with a model. Some form of (4.12) is applied to describe many realistic digital recordings. We will learn much more about these equations as we apply them.

4.3.4   1-D Correlation

Equation (3.41) in Chapter 3 described time-averaged autocorrelation for stochastic functions. We found this quantity useful for estimating the ensemble autocorrelation of a random process under the ergodic assumption and when we have only a limited number of waveforms. We can use the same concept to quantify the similarity of two deterministic functions as one is shifted relative to the other. The time-averaged *cross-correlation* for f_1 and f_2, each of duration T_t [s], is

$$\varphi_{12}(t) \triangleq [f_1 \star f_2](t) \triangleq \lim_{T_t \to \infty} \frac{1}{T_t} \int_0^{T_t} dt'\, f_1^*(t')\, f_2(t' - t), \qquad (4.13)$$

where $\varphi_{12} < \infty$ and $f_1^*(t')$ is the complex conjugate of $f_1(t')$. $\varphi_{12}(t)$ is proportional to the *mutual energy* between the two functions. For φ_{12} to exist, f_1 and f_2 must be square integrable, meaning $\int_{\Omega \in S} dt\, |f_j(t)|^2 < \infty$.

Figure 4.7a shows real f_1 and f_2 (top) and the corresponding cross-correlation (bottom) as a function of the time lag. These two functions are most similar when f_2 is delayed by three to four units of time. Also note that the correlation functional is mapped according to the vertical edge of f_2 in Figure 4.7. Selecting another reference would shift φ.

We can ignore the subscripts of φ when there is no ambiguity about the functions used or the order of functions being correlated. If $f_1(t) = f_2(t)$, then (4.13) is the time-averaged *autocorrelation function*, φ_{11}, which is symmetric about $t = 0$. The code used to generate the plots in Figure 4.7a is given here.

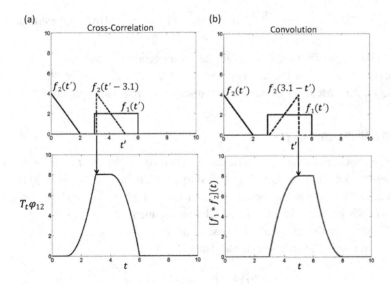

Figure 4.7 (a) The process of cross-correlation is illustrated with real 1-D f_1 and f_2. (b) Convolution of the same two functions is illustrated. Functions shown as dashed lines are shifted and/or reversed copies of f_2. The arrows show the value of t where each function initially peaks. The location of peaks in time depend on where you decide to map the output of the functional.

```
tp=0:0.01:10;t=tp;                              % set up t and t' axes
f1=zeros(size(t));f2=f1;                        % initialize functions
f1(301:600)=2;f2(1:200)=4-2*tp(1:200);          % create f1 and f2
plot(tp,f1,tp,f2);hold on;plot(tp+3.1,f2);hold off   % view them
pp=xcorr(f1,f2)*0.01;p=pp(1001:2001);           % correlate f1 and f2
figure;plot(t,p)                                % display the result
```

Because the sampling increment is $T = 0.01$, we multiplied the `xcorr` result by 0.01 to include dt' in the numerical approximation of the integral. This leaves us with $T_t \varphi_{12}(t)$. I would have multiplied by $dt'/T_t = 1/N$ if f_1 and f_2 did not include so many zero points, making normalization by the total waveform duration unnecessary. Make your decisions carefully when scaling is important. If `xcorr` is applied to two N-point functions, the result is $2N - 1$ points long, necessitating extraction of the central N points.

In practice, all functions have finite length, and therefore all time-averaged correlation functions are estimates that depend on T_t. We can estimate correlation $\varphi(t; T_t) = \hat{\varphi}(t)$ at least four ways in MATLAB, where two estimators are for continuous real functions and two estimators for discrete real functions:

$$\hat{\varphi}^{(b)}(t) = \frac{1}{T_t} \int_0^{T_t} dt'\, f_1(t')\, f_2(t' - t) \qquad \text{Biased estimate}$$

$$\hat{\varphi}^{(u)}(t) = \frac{1}{T_t - |t|} \int_0^{T_t - t} dt'\, f_1(t')\, f_2(t' - t) \qquad \text{Unbiased estimate}$$

$$\hat{\varphi}^{(b)}[\ell] = \frac{1}{N} \sum_{n=1}^{N} f_1[n]\, f_2[n - \ell] \qquad \text{Biased estimate} \qquad (4.14)$$

$$\hat{\varphi}^{(u)}[\ell] = \frac{1}{N - |\ell|} \sum_{n=1}^{N-\ell} f_1[n]\, f_2[n - \ell] \qquad \text{Unbiased estimate}$$

The continuous forms are most helpful for analytical discussions. In switching from integrals to sums for the last two forms, we use $dt'/T_t = 1/N$. Providing that lag $t \ll T_t$, the biased and unbiased estimators are approximately equal.

4.3.5   Correlation Symmetry

In some texts, time-averaged correlation is defined as $\int dt'\, f_1^*(t')\, f_2(t'+t)$, and sometimes the conjugate operation is applied to f_2. Also the definition (4.13) fixes f_1^* and scans f_2. Does the result change if f_2 is fixed and f_1^* is scanned? The answer to this question is given by the rules of correlation symmetry.

To simplify the following discussion, assume $T_t \to \infty$ for f_1 and f_2 while each function remains square-integrable. Then (4.13) becomes

$$\varphi_{12}(t) = \int_{-\infty}^{\infty} dt'\, f_1^*(t')\, f_2(t' - t). \qquad (4.15)$$

The subscripts in φ_{12} tell us the order of functions in the integral, where the second function has been shifted. Using a change of variables $w = t' - t$, we find $dw = dt'$ and the integration limits are unchanged. Consequently,

$$\varphi_{12}(t) = \int_{-\infty}^{\infty} dw \, f_1^*(w+t) \, f_2(w) = \int_{-\infty}^{\infty} dw \, f_2(w) \, f_1^*(w-(-t)) = \varphi_{21}(-t).$$

$$(4.16)$$

Comparing (4.15) and (4.16), we find that scanning f_2 in one direction is equivalent to scanning f_1^* in the other direction. The second form of (4.16) shows that reversing the order of functions in the integral results in a time-reversed version of the original correlation function. Further, if one of the functions is symmetric, e.g., if $f_1(t) = f_1(-t)$, then $\varphi_{12}(t) = \phi_{21}(t)$.

Correlation functions have several applications in the management of noisy data from instruments. One such application is *matched filtering*, which is appropriate when known deterministic signals are hidden in random noise or background signals.

Example 4.3.1. *Autocorrelation: Calculate the autocorrelation of $g(t)$ where $g(t) = A \cos(\Omega_0 t)$, Ω_0 [rad/s] $= 2\pi u_0$, and u_0 is temporal frequency [Hz]. Note that the period of the wave is $T_0 = 2\pi/\Omega_0 = 1/u_0$. Since $g(t)$ is infinitely long, i.e., $T_t = \infty$, but periodic over time T_0, it is sufficient to compute $\hat{\phi}$ over T_0 (see Section 5.3). From line 1 of (4.14),[4]*

$$\hat{\varphi}(t) = \frac{1}{T_0} \int_0^{T_0} dt' \, g(t') \, g(t'-t)$$

$$= \frac{\Omega_0 A^2}{2\pi} \int_0^{2\pi/\Omega_0} dt' \, \cos(\Omega_0 t') \cos(\Omega_0(t'-t))$$

$$= \frac{\Omega_0 A^2}{4\pi} \int_0^{2\pi/\Omega_0} dt' \left[\cos(2\Omega_0 t' - \Omega_0 t) + \cos(\Omega_0 t) \right]$$

$$= \left[\frac{\Omega_0 A^2}{4\pi} \int_0^{2\pi/\Omega_0} \!\!\! dt' \, \cos(2\Omega_0 t' - \Omega_0 t) \right] + \left[\frac{\Omega_0 A^2 \cos(\Omega_0 t)}{4\pi} \int_0^{2\pi/\Omega_0} dt' \right]$$

$$= \frac{A^2}{2} \cos(\Omega_0 t).$$

The autocorrelation of a cosine function with constant amplitude, frequency, and phase is another cosine at the same frequency and phase but with a different amplitude.

4.3.6 Convolution and Invariance

The mechanics of the *convolution* operation are similar to those of correlation, except that one of the two functions is time-reversed, as shown in Figure 4.7b, and integration is explicitly over the entire range of the integration variable;

---

[4] $\cos a \cos b = (\cos(a+b) + \cos(a-b))/2$.

$$[f_1 * f_2](t) = \int_{-\infty}^{\infty} dt' \, f_1(t - t') \, f_2(t').$$ (Convolution equation)   (4.17)

No complex conjugate is taken, and the output is not assigned a special character, unlike with correlation. The symmetry relations are described next.

The convolution integral of (4.17) is a cornerstone of linear system analysis. It describes acquisition operator $\mathcal{H}$ when the linear system is invariant in the independent variable:

$$g(t) = [h * f](t) = \int_{-\infty}^{\infty} dt' \, h(t - t') f(t').$$ (LTI acquisition)   (4.18)

We can estimate the impulse response by presenting to the measurement device an impulse at object time t_0, $f(t') = \delta(t' - t_0)$, while measuring output $g(t)$:

$$g(t) = \int_{-\infty}^{\infty} dt' \, h(t - t')\delta(t' - t_0) = h(t - t_0).$$

Such an experiment can be difficult to pull off, because realistic "impulses" within objects are weak signals. Thus, successful measurement of h might depend on achieving a very large SNR. If an experiment can be cleanly designed, then you can determine if a system is time invariant by just moving the impulse to other values of t_0 and observing whether $h(t - t_0)$ remains unchanged.

4.3.7   Multidimensional Convolution

For functions of N variables, $f(\mathbf{x})$ may be convolved in up to N dimensions. An N-D convolution is specified using

$$[f_1 \overset{N}{*} f_2](\mathbf{x}) = \int_{-\infty}^{\infty} dx'_N \cdots \int_{-\infty}^{\infty} dx'_1 \, h(x_1 - x'_1, \cdots, x_N - x'_N) \, f(x'_1, \cdots, x'_N)$$

$$= \int_{-\infty}^{\infty} d\mathbf{x}' \, h(\mathbf{x} - \mathbf{x}') \, f(\mathbf{x}').$$ (4.19)

The second line also includes N integrals via shorthand vector notation.

Impulse response functions can be composed of a series of convolutions when the measurement process involves a series of linear measurements. Moreover, convolution does not need to be applied over every variable in a function. For example, the pulse-echo impulse response of a linear shift-invariant (LSI) ultrasonic imaging system is given by the convolution of the transmitted and received impulse responses along the axis of pulse propagation. This is written as a 1-D integral over 4-D impulse responses:

$$h(\mathbf{x},t) = [h_t \overset{x_1}{*} h_r](\mathbf{x},t) = \int_{-\infty}^{\infty} dx_1' \, h_t(x_1 - x_1', x_2, x_3, t) \, h_r(x_1', x_2, x_3, t).$$

4.3.8   Symmetry Relations

Time-averaged correlation and convolution are similar mathematical operations. Here, we summarize the definitions and relationships for continuous and discrete, infinite, square integrable time series.

$$\varphi_{12}(t) = \int_{-\infty}^{\infty} dt' \, f_1^*(t') \, f_2(t' - t) = \int_{-\infty}^{\infty} dw \, f_2(w) \, f_1^*(w - (-t))$$

$$= \varphi_{21}(-t), \quad \text{where } w = t' - t$$

$$\varphi_{12}[\ell] = \lim_{N \to \infty} \frac{1}{N} \sum_{n=-N/2}^{N/2} f_1^*[n] \, f_2[n - \ell]$$

$$[f_1 * f_2](t) = \int_{-\infty}^{\infty} dt' \, f_1(t - t') \, f_2(t') = -\int_{\infty}^{-\infty} dw \, f_1(w) \, f_2(t - w)$$

$$= \int_{-\infty}^{\infty} dw \, f_2(t - w) \, f_1(w) = [f_2 * f_1](t), \quad \text{where } w = t - t' \quad (4.20)$$

$$(f_1 * f_2)[\ell] = \sum_{n=-\infty}^{\infty} f_1[\ell - n] \, f_2[n]$$

$$\phi_{12}(t) = \int_{-\infty}^{\infty} dt' \, f_1^*(t') \, f_2(t' - t) = \int_{-\infty}^{\infty} dt' \, f_1^*(t - (-t')) \, f_2(t')$$

$$= [f_1^*(-t') * f_2](t) = [f_1^* * f_2(-t')](t). \quad (4.21)$$

The last line relates correlation and convolution operations. It shows that shorthand notations can be confusing. To ensure clarity, it is best to write out the integrals.

Example 4.3.2. *Display: Although ECG traces are not sinusoidal waveforms, they are nearly periodic. Let's oversimplify the ECG measurement problem by supposing we wish to measure and display the electrical ECG waveform from a patient f that we assume oscillates sinusoidally in time. Specifically, let $f(t') = B\cos(2\pi u_0 t')$ at temporal frequency u_0 (see Figure 4.8). Assume the response of an LTI device for measuring ECG waveforms is the rectangular function,*

$$h(t') = A \, \text{rect}\left(\frac{t'}{c}\right).$$

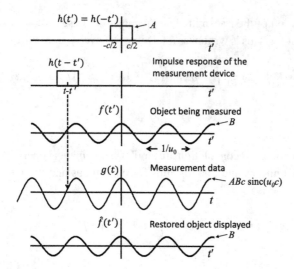

Figure 4.8 Illustration of an LTI measurement using the convolution of Example 4.3.2. The top two plots illustrate the time reversal and shift steps by which $h(t') \rightarrow h(t - t')$. In the numerical example, the width of the rectangular function for h is c and the frequency of the sinusoidal function f is u_0. We set $c = 1/2u_0$ so that $g(0)$ is a maximum. The bottom plot displays the data $\hat{f}$, which in this case is simply found by scaling g.

From f and h, predict the measurement data function $g(t)$ produced by this device and devise a display operator that displays the measured data $\hat{f}(t')$ as accurately as possible. Also find the width of h that maximizes g. That value of c will help maximize the point SNR for display.

Solution

Applying (4.18) and the symmetry relationship of (4.20), the acquisition is[5]

$$g(t) = \int_{-\infty}^{\infty} dt' \, h(t') \, f(t - t')$$

$$= AB \int_{-\infty}^{\infty} dt' \, \mathrm{rect}\left(\frac{t'}{c}\right) \cos(2\pi u_0 (t - t'))$$

$$= AB \int_{-c/2}^{c/2} dt' \left[\cos(2\pi u_0 t) \cos(2\pi u_0 t') + \underbrace{\sin(2\pi u_0 t) \sin(2\pi u_0 t')}_{0} \right]$$

$$= AB \cos(2\pi u_0 t) \int_{-c/2}^{c/2} dt' \, \cos(2\pi u_0 t') = AB \cos(2\pi u_0 t) \left. \frac{\sin(2\pi u_0 t')}{2\pi u_0} \right|_{-c/2}^{c/2}$$

$$= \left[AB c \frac{\sin(\pi u_0 c)}{\pi u_0 c} \right] \cos(2\pi u_0 t) = [AB c \, \mathrm{sinc}(u_0 c)] \cos(2\pi u_0 t). \qquad (4.22)$$

---

[5] Note the trig identity $\cos(a - b) = \cos(a) \cos(b) + \sin(a) \sin(b)$. A time integral centered at zero over the anti-symmetric sine-wave term equals zero. Also, integration variables do not survive definite integration and must not appear on the left-hand side of the expression.

The bracketed factor in the last line is a constant amplitude, where $\text{sinc}(x) \triangleq$
$\sin(\pi x)/(\pi x)$. *Like* f, g *is a cosine wave with the same frequency and phase.*
The amplitude of the data is not equal to that of the object, demonstrating the
need for display-stage processing. The amplitude distortion depends on the length
of the rectangular function c *relative to the period of the cosine wave,* $T_0 =$
$1/u_0$. *Specifically, when* $c = 1/2u_0$, *the amplitude is at a maximum and equal to*
$ABc \sin(\pi/2)/(\pi/2) = 2ABc/\pi$. *When* $c = 1/u_0$, *the amplitude is at a minimum*
and equal to zero, which we want to avoid. There are many minima and maxima and
even phase reversals, but we need consider only the maximum where SNR peaks.

With this problem, we set $c = 1/2u_0$ *and scale* g *by* $\pi/2Ac$ *to restore the object*
amplitude. Consequently, $\hat{f}(t') = \mathcal{O}g(t) = (\pi/2Ac)g(t' = t) \simeq f(t')$. *We will also*
need to have characterized the measurement device to know parameters A *and* c. *The*
approximation acknowledges our ignorance of the noise, which we accept provided
that the point SNR is large. You might worry that because you don't know frequency
u_0, *this design requires too much inside information to work. I would respond that you*
can find u_0 *by adjusting* c *until the amplitude of* $g(t)$ *is at a maximum; at that point,*
this display-stage operator is valid provided $f(t')$ *is sinusoidal.*

The sinusoidal simplification of an ECG waveform is not so farfetched if you
consider a Fourier decomposition of these patient signals. In that case, we consider
each sinusoidal frequency channel separately. Integral transformations are often much
easier and faster to compute in the Fourier domain as long as we can make the LTI/LSI
assumptions. This point will become clear in Chapters 5 and 6. Nevertheless, we can
do some design work in the spatiotemporal domain even with our limited understanding, as we now show.

4.4   An Ultrasonic Measurement

Consider the pulse-echo ranging experiment illustrated in Figure 4.9. The task is to
measure x_0, the distance from a transducer surface to the wall of a plastic tank filled
with degassed water, by acquiring an ultrasonic echo-voltage time series $g(t)$. We can
simulate this experiment with a model [122].

Placing a short-duration electrical pulse across the transducer's piezoelectric element generates a pulse of compressional sound waves that travel in water toward the
wall on the right in Figure 4.9. At the wall surface, a portion of the pulse energy
is reflected because of the change in acoustic impedance between water and plastic.
The reflected waves travel back to the transducer, where some of the mechanical
energy is transduced back into electrical energy that we sense as a voltage signal. The
transmitted and reflected pulsed-wave voltages are both displayed on an oscilloscope
time trace as $g(t)$ in Figure 4.9.

The measurement clock is initialized by auto-triggering of the oscilloscope, which
occurs with the first appearance of the transmitted pulse. The first appearance of the
reflected pulse appears a time t_0 later. Since the pulse must travel to the wall and back,

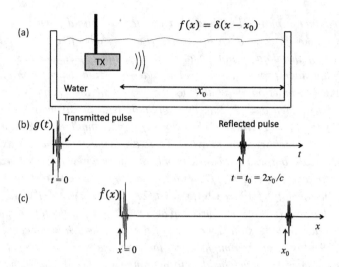

Figure 4.9 (a) An ultrasonic transducer (TX) transmits a short-duration sound pulse into water toward the right wall of the water tank a distance x_0 away. The task is to estimate x_0 from the pulse-echo acquisition recorded during an experiment (b). (c) The data displayed in object space.

we estimate the distance x_0 from the pulse-echo equation $2x_0 = ct_0$, where c is the speed of sound in water at room temperature and pressure (1.5 mm/μs). This equation is important because it relates measurement time and distance for display purposes.

Modeling f(x).
Acoustic impedance is the object property that interacts with the pulse. It is defined as $Z = \rho c$, where ρ [kg m^{-3}] is mass density and c [m s^{-1}] is sound speed. The units of Z are therefore [kg m^{-2} s^{-1}]. However, the units of $f(x)$ are [kg m^{-4} s^{-1}] because $f(x) = \partial^2 Z(\mathbf{x},t)/\partial x_1^2$, the second spatial derivative of impedance in the direction of wave propagation, x_1, which we simplify to x because the waves travel along one spatial axis. This simple experimental geometry allows us to simplify f from a 4-D function of space and time, $f(\mathbf{x},t')$, to $f(x)$.

The simplest object model assumes an impulse at the origin $x = 0$ that represents the transmitted pulse, and another impulse at $x = x_0$ with reflection coefficient $-1 \leq R \leq 1$ that represents the wall reflection,

$$f(x) = \delta(x) + R\,\delta(x - x_o). \tag{4.23}$$

Modeling g(t).
The echo voltage measured by $g(t)$ has units of volts [V]. Assume the object and data functions are continuous in space and time. If we further assume the impulse response represents a linear, space-time–invariant device and ignore acquisition noise, we can estimate the voltage waveform using a convolution,

$$g(t) = \int_{-\infty}^{\infty} dt'\, h'(t - t')\, f(ct'/2) = \int_{-\infty}^{\infty} dx\, h(t - 2x/c)\, f(x). \tag{4.24}$$

Either form works, although the latter is preferred. In addition, $h(t - 2x/c) = (c/2)h'(t - t')$. The Rosetta Stone role of the impulse response is to transform objects from object space into measurements in measurement space, while also accounting for the units. For this task, the pulse-echo equation $x = ct'/2$ is key.

Modeling h(t,x).
A realistic 1-D expression for a broadband ultrasound pulse is a *Gabor function,*[6] given by $A \exp(-t^2/2\sigma^2)\, \sin(\Omega_0 t)$. Replacing t with $t - 2x/c$ as instructed by (4.24), we find the LTI impulse response for pulse amplitude A is

$$h(t - 2x/c) = A\, e^{-(t-2x/c)^2/2\sigma^2}\, \sin(\Omega_0(t - 2x/c)), \tag{4.25}$$

where $\Omega_0 = 2\pi u_0$ [radian/μs] and u_0 is the temporal transmission frequency of the pulse [MHz]. Also, σ [μs] is the 1-D temporal pulse-duration parameter. The carrier wave, $\sin(\Omega_0(t - 2x/c))$, enables the sound pulse to propagate in space-time. A short pulse length, i.e., $\sigma c/2 \ll x_0$, provides the high spatial resolution needed for precise echo ranging.

Computing g(t).
Applying (4.23) and (4.25) to (4.24), and noting the sifting property described by (4.7), the measurement model predicts

$$g(t) = \int_{-\infty}^{\infty} dx\, h(t - 2x/c)\, f(x) = \int_{-\infty}^{\infty} dx\, h(t - 2x/c)[\delta(x) + R\delta(x - x_0)]$$

$$= h(t) + R\, h(t - 2x_0/c) \tag{4.26}$$

$$= A\left[e^{-t^2/2\sigma^2}\, \sin(\Omega_0 t) + R\, e^{-(t-2x_0/c)^2/2\sigma^2}\, \sin(\Omega_0(t - 2x_0/c)) \right].$$

Of course! If the object function is composed of impulses, the time-series measurement must be scaled copies of the pulse at those locations. That is why h is called the impulse response. If we were to plot this result, it would be close to the result for $g(t)$ plotted in Figure 4.9, albeit with some differences that we explore in Problem 4.8.

Computing $\hat{f}(x)$.
The display operator is fairly simple, $\hat{f}(x) = g(t = 2x/c)$. As long as $\sigma c/2 \ll x_0$, we will find that $\hat{f}(x)$ represents a 1-D profile of the impedance fluctuation along the axis of pulse propagation. We could scale g by $1/R$, but the height of the reflected pulse doesn't influence the task for our noise-free simulation, so we didn't bother. An observer can read off the value of x_0 from the display (see the bottom trace in Figure 4.9.)

---

[6] A Gabor pulse is a Gaussian-modulated sinusoid. With proper consideration of phase, it is the real or imaginary part of the Gabor wavelet, $e^{-(t-t')^2/2\sigma^2}\, e^{-i2\pi u_0 t}$, used as the kernel of a short-time Gabor transform [38].

4.5   1-D Convolution by Matrix Multiplication

Numerical computations of convolution can be slow for large files. The simplest implementations of convolution involve Fourier techniques, as described in Chapters 5 and 6. As with convolution, Fourier methods are limited to LTI/LSI systems. For a more general treatment that is computationally fast, we turn to matrix methods for implementing convolution.

The goal is to arrange the data so that a single matrix multiplication completes the convolution. We will demonstrate this process using the following example. Assume time series $h[n]$ and $f[n]$ are expressed as column vectors $\mathbf{h}, \mathbf{f} \in \mathbb{R}^{N \times 1}$, where $N = 19$, as shown in Figure 4.10. This impulse-response vector has "edge detection" features, meaning the device being represented acts on object functions to enhance edges and suppress constant regions; compare $\mathbf{f}$ and $\mathbf{g}$ in Figure 4.10. This computational problem is small, and we quickly compute the convolution using g=conv(h,f,'same').

If vector lengths were significantly longer, we might benefit from pre-applying the time reversal to $h[n]$ and then shifting that result N times, placing each single-shifted result into subsequent rows of an $N \times N$ circulant matrix $\mathbf{H}$ (see Figure 4.10 and Appendix A). Then all we need to do is multiply vector $\mathbf{f}$ times matrix $\mathbf{H}$ to compute the convolution.[7] In this situation, the acquisition operation $\mathcal{H}f(x)$ is given by multiplying object vector $\mathbf{f}$ and convolution matrix $\mathbf{H}$ and then adding noise vector $\mathbf{e}$.

If the impulse response is linear but shift varying (LSV), then $h[m - n]$ is not accurate for modeling the acquisition process. Instead, the mth row of $\mathbf{H}$ must contain $h[m,n]$, which is not just a shifted replica of other rows. For LSV systems, matrix $\mathbf{H}$ is not circulant.

The impulse response in Figure 4.10 is symmetric, so the time-reversal step is unnecessary. As row vector $h[m - n]$ is shifted in either direction by one element, the extra element wraps around to replace the missing element on the other side of the row. The code for implementing such a *circular convolution* is shown at the end of this section.

This specific impulse response is a positive spike surrounded by negative values – that is, it is a form of *high-pass filtering*. Since convolution is a linear filtering process, LTI measurements are essentially applying a device filter to an object function. The filter in Figure 4.10 emphasizes changes in $\mathbf{f}$, which explains why we refer to it as *edge detection*. As $h[n]$ is circularly shifted to form rows of $\mathbf{H}$, we can decide to start the cycle anywhere we wish. However, beginning the cycle as we did (see the top row of $\mathbf{H}$ in Figure 4.10) places $\mathbf{g}$ in phase with $\mathbf{f}$.

The symmetric structure of $\mathbf{H}$ makes it a *Toeplitz matrix* – all elements in each diagonal have the same value. Furthermore, because rows of $\mathbf{H}$ are shifted copies

---

[7] The matrix acquisition operator $\mathbf{H}$ does not contain a noise term even though the more general operator $\mathcal{H}$ does.

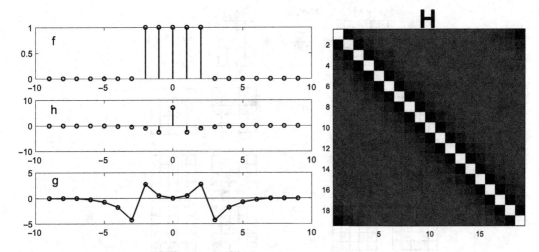

Figure 4.10 Example of 1-D circular convolution $g[m] = (h * f)[m]$ that was executed by first building a circulant matrix $\mathbf{H}$, whose rows are shifted and wrapped copies of $h[m - n]$. Note that row 10 of $\mathbf{H}$ is plotted as $\mathbf{h}$. The convolution results from the matrix-vector product $\mathbf{g} = \mathbf{Hf}$. Notice the changing vertical axis scales.

of $\mathbf{h}^\top$ that wrap around in a circular fashion, $\mathbf{H}$ is a circulant approximation of a Toeplitz matrix. (See Appendix A.)

```
%%%%%%%%%%%%%%%%%%%%%%%  Code to generate Fig. 4.10  %%%%%%%%%%%%%%%%%%%%%%%%%
%generate edge detector impulse response; implement as circulant matrix H
h=[7 -2.5 -1 -0.5 -0.25 0 0 0 0 0 0 0 0 0 -0.25 -0.5 -1 -2.5];
N=length(h);
H(1,:)=h; mid=floor(N/2);t=[-mid:mid];
for j=2:N                              % create circulant matrix from 1-D h
    H(j,1:j-1)=h(N-j+2:N);
    H(j,j:end)=h(1:N-j+1);
    end
imagesc(H);colormap gray,axis square   % image of H
f=[0 0 0 0 0 0 1 1 1 1 0 0 0 0 0 0];  % create object
g=H*f';                                % measurement of rect function object
figure;subplot(3,1,1);stem(t,f,'k')   % plot object
subplot(3,1,2);stem(t,H(mid+1,:),'k') % plot impulse response
subplot(3,1,3);plot(t,g, 'k-o')       % plot measurement data
```

4.6   2-D Direct Numerical Convolution

In this section, we seek to achieve graphical understanding of the direct process of 2-D circular convolution, while limiting ourselves to toy problems that are easily visualized. Here, we model object properties that interact with the measurement device as discretely sampled functions in two spatial dimensions, $f(\mathbf{x}) = f(x_1, x_2) \rightarrow f(m'X, n'X) = f[m', n']$. We chose to sample this continuous function on the same interval X in both the x_1 and x_2 dimensions, although that is not, strictly speaking, a requirement.

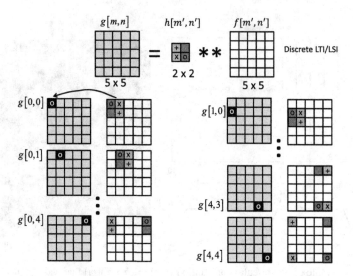

Figure 4.11 Graphical depiction of a direct 2-D circular convolution to model data acquisition. When h is applied to f for the convolution, it is first reversed along both axes. h wraps circularly at boundaries. You decide which pixel in measurement **g** receives the convolution output.

The impulse response of the measurement device must have two dimensions in the object space (x_1,x_2) and two dimensions in the data space (t_1,t_2). We write this as the discrete 4-D function $h[m,m',n,n']$, where the primed indices must be sampled on the same interval as the object, X, because they map from the same object space. An LSI/LTI measurement is required to employ convolution, so the four dimensions of h reduce to two, $h[m-m',n-n']$. Notice that both dimensions are reversed with respect to the object coordinates.

To simplify, assume $(x_1,x_2) \to (t_1,t_2)$, so the 2-D discrete circular convolution is

$$g(t_1,t_2) = [h \overset{2}{*} f](t_1,t_2)$$

$$g[m,n] = \sum_{n'=-\infty}^{\infty} \sum_{m'=-\infty}^{\infty} h[m-m',n-n']\, f[m',n']. \tag{4.27}$$

This equation is illustrated in Figure 4.11 for a 5×5 object function f convolved with a 2×2 impulse response h to generate a 5×5 data function g. By following the shading and markings on h pixels, you can follow the sequence by which elements of h are reversed and matched to f before multiplication to generate each pixel in g. We label this a 2-D circular convolution because h is wrapped around f near the boundaries in Figure 4.11. We can also pad the periphery of f with extra zero-valued elements. This approach negates the need for wrapping h elements near boundaries, giving us a 2-D *linear convolution*. When scaled up to practical dimensions, the direct 2-D numerical process can be slow to compute.

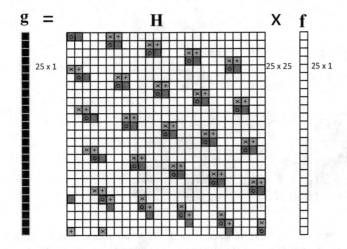

Figure 4.12 To execute the 2-D convolution $g(t) = [h \overset{2}{*} f](t)$ as a matrix multiplication $\mathbf{g} = \mathbf{Hf}$, we must sample $f(x_1, x_2) \rightarrow f[m', n']$ and reshape the 5×5 matrix into the 25×1 vector $\mathbf{f}$. The hard part is reshaping 2-D h to become rows of $H[m' - m, n' - n]$, the step illustrated here. h is the same 2×2 function shown in Figure 4.11. Examples using MATLAB to form $\mathbf{H}$ matrices from linear shift-invariant (LSI) and linear shift-variant (LSV) 1-D impulse responses are given in Section 4.5 and Section 7.4.1, respectively.

4.7   2-D Convolution by Matrix Multiplication

For LSI/LTI systems, computational speed may be improved at the cost of memory allocation by building circulant $\mathbf{H}$ for the 2-D impulse response in Figure 4.12, similar to what we did for the 1-D convolution in Figure 4.10. First we restructure the $N \times N$ object function $f[m', n']$ into a column vector. To do so, we simply "cut" the object into column strips and attach the top element of the second column $f[1, 2]$ below the bottom element of the first column $f[N, 1]$. Then we attach the third column below the second column, etc., converting the $N \times N$ object function into an $N^2 \times 1$ column vector $\mathbf{f}$. This action is achieved in MATLAB with reshape.

Reshaping from 2D to 1D can occur any way you like, provided the ordering scheme is consistently applied. It is a little more challenging to convert a 2-D impulse response function $h[m, n]$ to become rows of system matrix $\mathbf{H}$. The Toeplitz function in MATLAB does not generate a circulant matrix, so we adapted the code for 1-D circulant matrices from Section 4.5. Next, we offer a MATLAB example that applies the just described methods to simulate a ultrasonic B-mode image.

Example 4.7.1. *B-mode imaging:* *The following code may be used to simulate a 1-mm² ultrasonic B-mode image using a 20-MHz pulse from an LSI imaging system. The results are displayed in Figure 4.13.*

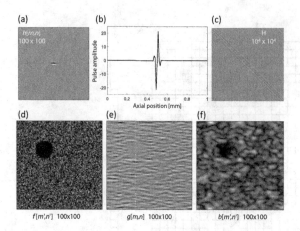

Figure 4.13 Elements of the 2-D ultrasonic B-mode image simulation from Example 4.7.1. (a) An image of the $N \times N$ LSI $h[m,n]$ ultrasound pulse, where $N = 100$. (b) The central vertical line through the pulse profile. (c) **H**, which is $N^2 \times N^2$; it is difficult to see the details without significant magnification. (d) The scattering object $f[m',n']$. (e) The acquisition $g[m,n]$. (f) $b[m'n'] \triangleq \hat{f}[m',n']$. This program ran in about 3 s on a laptop.

```
%%%%%%%%%%%%%%%%%%%%%%%%%%%%  Code to generate Fig. 4.13 in Example 4.7.1 %%%%%%%%%%%%%%%%%%%%%%%%%%%%
tic;dx=0.01; X=1;                                      % sampling interval, field size
x=0:dx:X;N=length(x);h=zeros(N);Ns=N^2;N2=floor(N/2);H=zeros(Ns); % region of interest and parameters
b=zeros(N);u0=20/0.77;sx=0.015;sy=0.065;               % u0=temp freq -> sp freq, sx,sy pulse widths
hx=exp(-(x-X/2).^2/(2*sx^2))/(sqrt(2*pi)*sx).*sin(2*pi*u0*x);  % h(x) along the pulse propagation axis
plot(x,hx)
for j=1:N                                              % generate full 2D pulse, h(x,y)
    h(:,j)=hx*exp(-(x(j)-X/2)^2/(2*sy^2)/(sqrt(2*pi)*sy));
end
figure;imagesc(h);colormap gray;axis square;          % image the 2-D pulse
%
%%%%%%%%%%%%%%%%% Convert 2-D pulse to circulant matrix H %%%%%%%%%%%%%%%%%%%%%%%%%%%%%%%%%%%%%%%%%%%%%
for k=1:Ns                                             % create circulant H from h(x,y)
    h1=fftshift(reshape(h,1,Ns));                      % center of pulse at first element
    hq=h1(Ns:-1:1);                                    % time-reverse for convolution
    H(k,k:Ns)=hq(1:Ns-k+1);
    if k-1>0;H(k,1:k-1)=hq(Ns-k+2:Ns);end;            % other rows of H
end
%%%%%%%%%%%%%%%%%%%%%%%%%%%%%%%%%%%%%%%%%%%%%%%%%%%%%%%%%%%%%%%%%%%%%%%%%%%%%%%%%%%%%%%%%%%%%%%%%%%%%%%%
figure;imagesc(H);colormap gray;axis square;          %image the huge H matrix
%
%%%%%%%%%%%%%%%%%%%%%% random scattering field with circular inclusion %%%%%%%%%%%%%%%%%%%%%%%%%%%%%%%%
ff=randn(N);                                           %normally distributed stationary scattering...
for j=1:N;                                             %... except for circle reduced to 30% of bkg
    for i=1:N;                                         % make the circle
        if(j<floor(N/3)+sqrt(floor(N/10)^2-(i-floor(N/3))^2)...
        &&j>floor(N/3)-sqrt(floor(N/10)^2-(i-floor(N/3))^2));
        ff(j,i)=ff(j,i)*0.3; end;
    end;
end;
f=reshape(ff,Ns,1);                                    % make N^2 x 1 vector from NxN object
figure;imagesc(abs(ff));colormap gray;axis square      % visualize object function
%
%%%%%%%%%%%%%%%%%%%%%%%% B-mode image of echo signal g %%%%%%%%%%%%%%%%%%%%%%%%%%%%%%%%%%%%%%%%%%%%%%%%
g=H*f;figure;imagesc(reshape(g,N,N));colormap gray;axis square %vf=vn=1(var's)
n=randn(Ns,1);vg=var(g);scale=vg/20;vn=scale*var(n);eSNR=10*log10(vg/vn);
gn=g+sqrt(scale)*n;%%%%%%%%%%denom of scale is eSNR(dB)
figure;imagesc(abs(hilbert(reshape(g,N,N))));colormap gray;axis square;toc
%
% %%%%%%%%%%%% See Appendix C for details on demodulation used to make a B-mode image. %%%%%%%%%%%%%%%
```

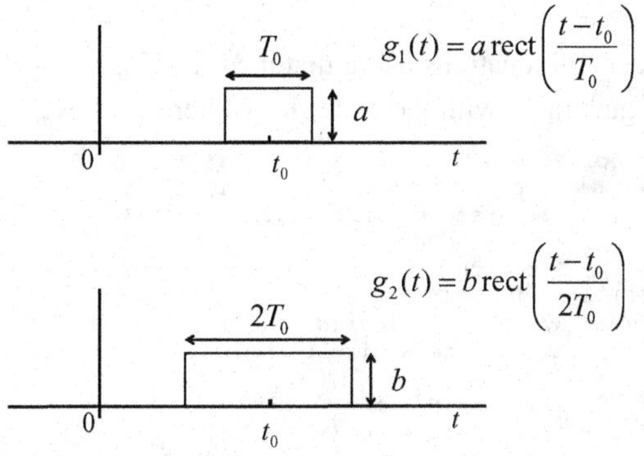

Figure 4.14 Illustration of rectangular function from Problem 4.3.

4.8   Problems

4.1   (a)   Solve:

$$\int_{-\infty}^{\infty} dt \ y(t) \left(\frac{d}{dt} \text{step}(t - t_0) \right)$$

(b)   Find $\psi(t)$ where

$$\int_{-\infty}^{\infty} dt \ y(t) \ \psi(t) = \int_{0}^{\infty} dt \ y(t) - \int_{-\infty}^{0} dt \ y(t)$$

4.2   (a)   Compute the autocorrelation function $\varphi_{gg}(t)$ mathematically in the time domain for $g(t) = \exp(-t^2/2\sigma^2)$ via (4.15). Note that in general

$$\int_{-\infty}^{\infty} dx \ \exp(-a^2 x^2 \pm bx) = \sqrt{\pi/a^2} \exp(b^2/4a^2).$$

(b)   Plot $g(t)$ and $\varphi(\tau)$ on top of each other.

(c)   Compute analytically the full-width-at-half-maximum (FWHM) values for both $g(t)$ and $\varphi(\tau)$, and find their ratio.

4.3   (a)   Figure 4.14 illustrates two rectangular functions $g_1(t)$ and $g_2(t)$. Compute analytically the cross-correlation function $\varphi_{12}(t)$. Plot it along with the MATLAB solution via `phi=xcorr(g1,g2)`. To do this numerically, select $t_0 = 3$, $T_0 = 1$, $a = 1$, and $b = 2$.

(b)   Explain in words how the result from (a) changes if you convolve g_1 and g_2.

4.4   The following code generates a crude representation of a noisy microscopy image showing eight structures that vaguely resemble cells. The model $f(\mathbf{x}) = s(\mathbf{x}) + e(\mathbf{x})$ considers a stochastic signal (cell walls in elliptical patterns) and a uniform (stationary) noise source. Assume the impulse response is compact enough to be ignored for our purposes. Cell walls fluorescing give the optical signal, and the surrounding areas give off-target photons.

(a) Compute the SNR.

(b) Devise a plan to detect cell locations using matched filtering. The field for analysis is generated with the code shown here.

```
%%%%%%%%%%%%%%%%%%%%%%%%%%%%%%%%%%%  Script for Problem 4.4  %%%%%%%%%%%%%%%%%%%%%%%%%%%%%%%%%%%%%%%%%%%%
N=101;f=zeros(N);x0=floor(2*N/3);y0=x0;    % parameter initialization
b=1;c=2;d=0.8;e=1.8;                       % parameter initialization
for j=1:N;
    for i=1:N;                             % search x,y for locations within circle or ellipse
        if(j<floor(N/3)+sqrt(floor(N/10)^2-(i-floor(N/3))^2) ...
            && j>floor(N/3)-sqrt(floor(N/10)^2-(i-floor(N/3))^2));
            f(j,i)=1;                      % indicate a circle region
        end;
        if(j<floor(N/3)+sqrt(floor(N/12)^2-(i-floor(N/3))^2) ...
            && j>floor(N/3)-sqrt(floor(N/12)^2-(i-floor(N/3))^2));
            f(j,i)=0;                      % subtract interior to simulate circular cell wall
        end;
        if(j<y0+sqrt((c*10)^2-(c*(i-x0)/b)^2) ...
            && j>y0-sqrt((c*10)^2-(c*(i-x0)/b)^2));
            f(j,i)=1;                      % indicate elliptical region
        end;
        if(j<y0+sqrt((e*10)^2-(e*(i-x0)/d)^2) ...
            && j>y0-sqrt((e*10)^2-(e*(i-x0)/d)^2));
            f(j,i)=0;                      % subtract interior to simulate elliptical cell wall
        end;
    end;
end;
g=[f f';f' f];                             % expand field to 201 x 201 pixels with 8 structures
rng('default');                            % fix the random number seed for the random fields
sig=poissrnd(10,202,202);                  % simulate photons from the cell walls
nois=poissrnd(20,202,202);                 % simulate photons from off-target sources (noise)
gg=g.*sig+nois;                            % this is the signal to be analyzed
figure;imagesc(gg);colormap gray
```

4.5   Let

$$h(t') = \frac{1}{\tau}\exp\left(-\frac{t'}{\tau}\right) \times \text{step}(t') \quad \text{and} \quad f(t') = \text{rect}\left(\frac{t'}{2T_0}\right).$$

Find $g(t) = [h * f](t)$ analytically.

4.6   The system matrix $\mathbf{H}$ that generates the results found in Figure 4.3b is linear but shift varying, such that each row has a length of constant, nonzero values given by the row number. However, the area under the elements in each row of $\mathbf{H}$ is 1. There is no need to add noise. Find $\mathbf{H}$ that gives $\mathbf{g}$ in Figure 4.3, where $f[n] = \cos(2\pi t)$ for $0 \le t \le 10$, and show your results.

4.7   Let's examine the inclusion of transients (end effects) in modeling measurements. You have a device for measuring respiration that was designed for large veterinary patients, such as dogs. You want to see if you can adapt this device for tracking respiration in a rat. Ignoring the obvious practical difficulties associated with fastening a harness to a rat that was designed for a dog, you are concerned that the most fundamental instrumentation problem is that the designer did not allow the system to respond fast enough to record the four-fold higher respiration rate in rats without distortion effects. The respiration rate for a dog is 16–20 min^{-1}, while the respiration rate for a rat is 66–114 min^{-1}. Use respiration rates of 20 (dog) and 80 (rat) in this problem. Assume the input function is sinusoidal at frequency u_0 in Hz,

$$f(t) = (1 - \cos(2\pi u_0 t))\,\text{step}(t).$$

The object function is the volume of air in the lungs as a function of time. At time $t < 0$, the instrument is at rest, $f(t) = 0$. Then suddenly at $t = 0$, a sinusoidal input (breathing) begins. The manufacturer tells you that the device is an LTI system with impulse response

$$h(t) = \frac{1}{\tau} \exp(-t/\tau)\, \text{step}(t), \quad \text{where} \quad \tau = 0.5s.$$

(a) Can the respiration rate u_0 of a rat be measured accurately with this device? (b) Can the tidal volume be measured accurately? Tidal volume is the amount of air in the lungs inhaled and exhaled with each normal breath. We measure it indirectly from the amplitude of the sensor. To address these questions, model the outputs for both animals.

4.8 Looking closely at the voltage trace in Figure 4.9, we see the reflected pulse at $t = t_0$ is not identical to the transmitted pulse at $t = 0$, as the simple model in Section 4.4 suggests. Clearly, our simple model is missing something, but should we be concerned?

The transmitted pulse is an odd function because of the sine function in the Gabor pulse of (4.25), while the reflected pulse is an even function.[8] A better model of the reflected pulse might present it as a scaled derivative of the transmitted pulse.

We can do that by assuming

$$f(x) = \delta(x) + \frac{R}{x_0}\, \text{step}(x - x_0), \tag{4.28}$$

which models the reflection portion as a scaled step function. R/x_0 is the reflection coefficient. Since the derivative of a step function equals a Dirac delta, this could be a good guess even if we need to work a bit harder to solve the integral equation. Model this process.

---

[8] If you fold an even pulse at its center, the two sides will symmetrically overlap.

5   Basis Decomposition I

The intrinsic properties of a system can be uncovered by decomposing the model equations describing that system into bases. The linear-algebra concept of *basis vectors* or functions describes fundamental modes of operation for many linear (or linearized) systems. Knowing the system's properties is essential for predicting output responses to an input stimulus. For example, the motion of a bridge suspended over a river in windy conditions is described by its eigenstates that specify which movements are possible [40]. Materials are selected and assembled in designs that produce long-lasting bridges that are able to justify the significant investment made in them. Similarly, infectious diseases can be modeled to predict the spread of a virus through populations, so that we can then develop strategies to mitigate the devastating effects on public health and the economy. Bioengineers are becoming increasingly comfortable decomposing models of living organisms, especially with respect to cell metabolism and signaling, to help define states of health and disease [78].

The functional modes of a measurement process are also revealed in this manner. Properties of measurement acquisition and display may be found from a basis decomposition of the corresponding operators that map an object into data. However, device engineering alone cannot completely predict clinical performance. That is, the basis necessary to define overall measurement performance extends beyond the device, to include both properties of the object under investigation and properties of the decision maker.

This chapter describes the essential elements of basis decomposition that will help us analyze everything from measurement devices to biological systems. Furthermore, it discusses a means for looking into data sets to find and extract the information that first prompted the measurement. Before getting into the details, we begin with an example.

5.1   Introduction to Principal Components Analysis

Example 5.1.1. *Principal components analysis (PCA): Consider the simple harmonic oscillator experiment diagrammed in Figure 5.1. We use three linear time-invariant (LTI) sensors to measure the back-and-forth motion of a sphere along the x_1 axis. The sensors are positioned far enough away from the sphere to track its*

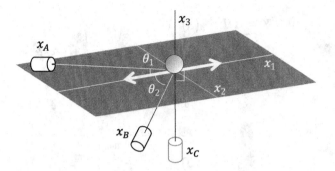

Figure 5.1 Geometry of a sphere moving as a harmonic oscillator ± 1 unit along x_1 within rectilinear coordinates (x_1, x_2, x_3). Three sensors acquire data (x_A, x_B, x_C) along the measurement axes shown. Sensor C axis is oriented along x_3, while the sensor axes for A and B are in the (x_1, x_2) plane at angles θ_1 and θ_2 with respect to the linear motion along x_1. $N = 1,001$ points are recorded for each of $M = 3$ measurements.

movement, but each is only sensitive to motion along the sensor axis. Sensors A and B have axes in the (x_1, x_2) plane at angles $\theta_1 = 50°$ and $\theta_2 = 60°$ relative to x_1, and sensor C is oriented parallel to x_3. We have no prior knowledge of the sphere's path, but we have measured sensor noise properties and hope to use these data to measure the amplitude of sphere motion. Which measurement features are fundamental to our ability to describe the sphere's movement from these data? How might we decide if these measurements are sufficient for achieving the task?

The true sphere position is given by vector $\mathbf{x} = (x(t)\boldsymbol{\mu}_1, 0\boldsymbol{\mu}_2, 0\boldsymbol{\mu}_3)$, where $x(t) = A_0 \cos(2\pi u_0 t + \phi)$, $\boldsymbol{\mu}_j$ are unit directional vectors for the Cartesian coordinates, A_0 is oscillation amplitude, u_0 is oscillation frequency, and $\phi = \cos^{-1}(x(0)/A_0)$ is the initial phase value. These are the objective facts of the experiment, but experimenters do not know these details; they only have access to the sensor data. From a privileged viewpoint of knowing the motion parameters, we could estimate sphere movement with just one sensor by aligning it along x_1 or at a known angle relative to x_1. Naively, we measure motion on the three noisy sensors over 10 seconds at a rate of 100 Hz to give 1,001 data points per sensor. Measurements are simulated from the geometry of Figure 5.1, where $x_A(t) = x(t)\cos(\theta_1) + e_A(t)$, $x_B(t) = x(t)\cos(\theta_2) + e_B(t)$, and $x_C(t) = x(t)\cos(90°) + e_C(t) = e_C(t)$ (sensor C is all noise).

Keep the following questions in mind as you read through the solution. (1) How can we determine whether all sensors offer task-relevant information? (2) Do we generally need three sensors? (3) How should we analyze sensor data to determine the sphere motion?

Solution

The simulation and processing code is given here. The simulated time series from each sensor yields $M = 3$ measurement variables represented by $N \times 1$ vectors $\mathbf{x}_A$, $\mathbf{x}_B$, and $\mathbf{x}_C$, with $N = 1,001$ in this example. Data matrix $\mathbf{X} = (\mathbf{x}_A, \mathbf{x}_B, \mathbf{x}_C) \in \mathbb{R}^{N \times M}$ is assembled to calculate the sample temporal covariance via (3.37). The simulation

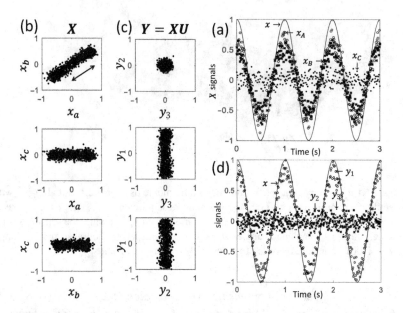

Figure 5.2 Measurements associated with sensing harmonic oscillator motion. $A_0 = 1$, $\phi = 0$, $\theta_1 = 50°$, $\theta_2 = 60°$, $\sigma_A = \sigma_B = \sigma_C = 0.1$, where $e_j \sim \mathcal{N}(0, \sigma_j^2)$. Data in (a) and (b) are raw sensor data, while those in (c) and (d) are the same data after PCA. Plotted lines in (a) and arrows in (b) indicate true sphere motion.

parameters are listed in the caption of Figure 5.2. We expect $\mathbf{K}_x \in \mathbb{R}^{M \times M}$ in (5.1) to show us that $\mathbf{x}_A$ and $\mathbf{x}_B$ offer task information while $\mathbf{x}_C$ does not:[1]

$$\mathbf{K}_x = \frac{1}{N-1}\mathbf{X}^\top \mathbf{X} = \begin{pmatrix} \text{var}(\mathbf{x}_A) & \text{cov}(\mathbf{x}_A, \mathbf{x}_B) & \text{cov}(\mathbf{x}_A, \mathbf{x}_C) \\ \text{cov}(\mathbf{x}_B, \mathbf{x}_A) & \text{var}(\mathbf{x}_B) & \text{cov}(\mathbf{x}_B, \mathbf{x}_C) \\ \text{cov}(\mathbf{x}_C, \mathbf{x}_A) & \text{cov}(\mathbf{x}_C, \mathbf{x}_B) & \text{var}(\mathbf{x}_C) \end{pmatrix}$$

$$= \begin{pmatrix} 0.2168 & 0.1616 & -0.0008 \\ 0.1616 & 0.1359 & -0.0003 \\ -0.0008 & -0.0003 & 0.0110 \end{pmatrix}. \tag{5.1}$$

The time series for the three variables are shown in Figure 5.2a. The variances for $\mathbf{x}_A$ and $\mathbf{x}_B$ in (5.1) are indeed the largest given their respective signal amplitudes, $A_0 \cos(\theta_1) = 0.64$ and $A_0 \cos(\theta_2) = 0.50$. Each includes additive noise with parameter $\sigma_j = 0.1$, but $\text{var}(\mathbf{x}_C) = 0.011 \simeq \sigma_j^2$ indicates that only noise is present in that recording. Sensors A and B record projections of true sphere motion $x(t)$ onto those measurement axes. If the sensor axes were all perpendicular to each other, the measurements would be independent and $\mathbf{K}_x$ would be diagonal. However, we are

---

[1] In this example, the mean of each measurement is zero. If the means are not zero, then $\mathbf{K}_x$ is computed from $\mathbf{X}'[n] = (x_A[n] - \bar{x}_A, x_B[n] - \bar{x}_B, x_C[n] - \bar{x}_C)$.

naive experimenters who know nothing about the direction of motion except what we can learn from the data.

```
%%%%%%%%%%%%%%%%%%% Code for Example 5.1.1 that generates data in Figure 5.2 %%%%%%%%%%%%%%%%%%%
t=0:0.01:10;A0=1;u0=1;w0=2*pi*u0;phi=0;N=length(t);        % initialize
theta1=50;theta2=60;e0a=0.1;e0b=0.1;e0c=0.1;               % initialize
x=A0*cos(w0*t+phi);plot(t,x,'k');hold on;                  % oscillator motion
ea=e0a*randn(size(t));eb=e0b*randn(size(t));               % additive noise
xa=x*cosd(theta1)+ea;xb=x*cosd(theta2)+eb;                 % sensor A & B measurements
ec=e0c*randn(size(t));xc=ec;                               % sensor C measurement
plot(t,xa,'.r');plot(t,xb,'.b');plot(t,xc,'.k');hold off
X=[(xa-mean(xa))' (xb-mean(xb))' (xc-mean(xc))'];          % subtract means
Kx=X'*X/(N-1);                                             % Kx = sensor signal covariance
figure;subplot(3,1,1);plot(xa,xb,'.k');title('x_A, x_B')
subplot(3,1,2);plot(xa,xc,'.k');title('x_A, x_C')
subplot(3,1,3);plot(xb,xc,'.k');title('x_B, x_C')
[U,S]=eig(Kx);Y=U'*X';Ky=Y*Y'/(N-1);                       % transformed covariance Ky
figure;subplot(3,1,1);plot(Y(1,:),Y(2,:),'.k');title('y_2, y_3')
subplot(3,1,2);plot(Y(1,:),Y(3,:),'.k');title('y_1, y_3')
subplot(3,1,3);plot(Y(2,:),Y(3,:),'.k');title('y_1, y_2')
figure;plot(t,Y(3,:),'ko');hold on;plot(t,x,'k')           % careful about order!!
plot(t,Y(2,:),'kx');plot(t,Y(1,:),'k.');hold off
```

The four covariances in (5.1) involving $\mathbf{x}_C$ are all less than var($\mathbf{x}_C$), as expected for signal independent iid noise from a normal process. In contrast, cov($\mathbf{x}_A, \mathbf{x}_B$) = cov($\mathbf{x}_B, \mathbf{x}_A$) are on the order of var($\mathbf{x}_A$) and var($\mathbf{x}_B$), so the variables are highly correlated: The correlation coefficient (3.35)

$$\rho = \frac{\text{cov}(\mathbf{x}_A, \mathbf{x}_B)}{(\text{var}(\mathbf{x}_A)\text{var}(\mathbf{x}_B))^{1/2}} = 0.94.$$

Since we know the noise properties, we see immediately that $\mathbf{x}_C$ can be discarded because its variance equals the noise variance. We also see that $\mathbf{x}_A$ and $\mathbf{x}_B$ have variances greater than that contributed by noise, indicating both contain signal from the moving sphere, but the high correlation coefficient suggests the information is redundant.

Plotting one set of sensor data against another to eliminate the time axis in Figure 5.2b, we find the diagonal pattern of points in $\mathbf{x}_A$ versus $\mathbf{x}_B$. The $\mathbf{x}_A$ versus $\mathbf{x}_C$ and $\mathbf{x}_B$ versus $\mathbf{x}_C$ plots show that variances for the three measurements are quite different from each other, but $\mathbf{x}_C$ appears uncorrelated to the other two variables. Essential features in the plots of Figure 5.2b are summarized by $\mathbf{K}_x$ in (5.1). Because motion is a major source of variance in the data, we want to determine the direction of greatest variance.

$\mathbf{K}_x$ is a well-conditioned matrix due to acquisition noise where, from (A.11) in Appendix A, we find $\kappa(\mathbf{X}) = 36$. If noise could be reduced, the condition number would increase as var($\mathbf{x}_C$) $\rightarrow$ 0. A signal rank of 2 tells us that two measurement variables contribute signal information. Yet it is also possible that the intrinsic dimensionality of this problem is less than 2 because of the redundancy indicated by large cov($\mathbf{x}_A, \mathbf{x}_B$). Because these sensors are linear systems with normally distributed noise, uncorrelating the measurements can reveal the intrinsic dimensionality of the measurement.

5.1.1   PCA via Matrix Diagonalization

All covariance matrices computed from measurement data are symmetric and positive definite. As such, they are diagonalized by eigenanalysis, as shown in Appendix A and relation (6.2) in Chapter 6. The eigenvalues of $\mathbf{K}_x$ are real and the corresponding eigenvectors are orthogonal if the eigenvalues are distinct.

As $N \to \infty$, the sample covariance for ergodic process $\mathbf{X}$ approaches ensemble covariance,

$$\mathbf{K}_x = \lim_{N \to \infty} \frac{1}{N-1} \mathbf{X}^\top \mathbf{X} = \mathcal{E}\{\mathbf{X}^\top \mathbf{X}\}.$$

Applying eigenanalysis,

$$\mathbf{U}^\dagger \mathbf{K}_x \mathbf{U} = \mathbf{U}^\dagger \mathcal{E}\{\mathbf{X}^\top \mathbf{X}\} \mathbf{U} \tag{5.2}$$

$$\mathbf{\Lambda} = \mathcal{E}\{(\mathbf{X}\mathbf{U})^\dagger (\mathbf{X}\mathbf{U})\},$$

where $\mathbf{U} \in \mathbb{R}^{M \times M}$ is the eigenvector matrix for the M measurements.[2] The columns of $\mathbf{U}$ are the eigenvectors of $\mathbf{K}_x$ and the diagonal elements of $\mathbf{\Lambda}$ are the corresponding eigenvalues. The *principal components* are

$$\mathbf{Y} \triangleq \mathbf{X}\mathbf{U} \in \mathbb{R}^{N \times M}. \tag{5.3}$$

$\mathbf{K}_y = \mathbf{\Lambda}$ is diagonal and composed of the eigenvalues of $\mathbf{K}_x$. Hence $\mathbf{K}_y$ has the same properties as $\mathbf{K}_x$. Computing $\mathbf{K}_y$ numerically from (5.1), we find

$$\mathbf{K}_y = \begin{pmatrix} \lambda_1 & 0 & 0 \\ 0 & \lambda_2 & 0 \\ 0 & 0 & \lambda_3 \end{pmatrix} = \begin{pmatrix} 0.343 & -0.000 & -0.000 \\ -0.000 & 0.011 & -0.000 \\ 0.000 & -0.000 & 0.010 \end{pmatrix}.$$

The elements of diagonal $\mathbf{K}_y$ are the eigenvalues of $\mathbf{K}_x$, as expected. Note that MATLAB chooses to orient eigenvalues in increasing order instead of the usual descending order, so I flipped results in the preceding array.

These eigenvalues can be interpreted in terms of signal and noise energies. The second and third diagonal elements approximate the input noise variance $\sigma^2 = 0.01$. In contrast, $\mathbf{K}_y(1,1) = 0.343 \simeq 0.51 = (1/T_0) \int_0^{T_0} dt \, \cos^2(2\pi u_0 t) = 0.50$ plus noise variance. The three signals of the diagonalized matrix $\mathbf{Y} = (\mathbf{y}_1, \mathbf{y}_2, \mathbf{y}_3)$ are plotted in Figure 5.2d. The approximately 17 dB SNR is too low to faithfully reproduce the sphere motion by examining $\mathbf{y}_1$, so $\text{var}(\mathbf{y}_1) = 0.34$ is lower than the value of 0.51 found analytically. The first eigenvalue tells us the restored amplitude is 0.82 instead of the true value 1.0 – an improvement over the signal amplitudes from either $\mathbf{x}_1$, which is 0.64, or $\mathbf{x}_2$, which is 0.50. We are ultimately limited by measurement noise. Plotting $\mathbf{y}_2$ versus $\mathbf{y}_3$ in Figure 5.2c, we see a circular cluster of points centered at zero, where the radius of the circle is between 0.1 and 0.2 as expected for

---

[2] $\mathbf{U}$ is real here. In general, $\mathbf{U} \in \mathbb{C}^{M \times M}$ for M measurements. So let's replace transpose $\top$ with conjugate transpose $\dagger$. The latter always works because when $\mathbf{U}$ is real, $\mathbf{U}^\dagger = \mathbf{U}^\top$.

zero-mean iid noise with $\sigma = 0.1$. Only plots including the $\mathbf{y}_1$ data extend toward an amplitude of 1.[3]

Most importantly, the measurement in Example 5.1.1 is *intrinsically one-dimensional*. Let's look at this statement carefully. We had reason to believe our sensors could be described as linear systems, and we knew the noise properties of those sensors. With that information, PCA was helpful. The number of sensors required for a successful measurement depended on our prior knowledge about sphere motion. We knew nothing in this example, so we naively chose three sensors. If we knew motion would be confined to a plane, then just two sensors would be required, e.g., x_A and x_B. However, PCA indicated only one sensor would be needed if we knew exactly where to position it. This sensing problem is intrinsically 1-D because only one eigenvalue of $\mathbf{K}_y$ was greater than the noise variance. With no prior knowledge of sphere movement, we would need three sensors oriented orthogonally to each other if possible. Then, if the sphere moved in a planar arc, we would find a noise-free $\mathbf{K}_y$ to be rank 2, indicating an intrinsically 2-D problem. If the sphere moved along a spiral path, PCA would tell us this is an intrinsically 3-D problem.

The process of diagonalizing $\mathbf{K}_x$ determined the eigenvector matrix $\mathbf{U}$ applied in (5.3) that rotated the measurement axes into the principal axes of the measurement data. The first principal component is the *direction of maximum variance*, which is assumed to have the most signal amplitude for high-SNR data. In this example, it was a linear combination of $\mathbf{x}_A$ and $\mathbf{x}_B$. The linear combination resulting in $\mathbf{y}_1$ removed the redundancy in data variables, allowing us to reduce the three measurements to one without loss of information within the noise limitations.

By sampling the data as frequently as we did (100 Hz over 10 s), we obtained better estimates of the population covariance. If there are costs associated with sampling, we should consider reducing the rate, but of course there are limits to doing so (as discussed in Section 5.7).

The importance of PCA is seen when we determine the *intrinsic dimensionality of a measurement*. Since biomedical measurements all incur costs and risks, knowing the smallest number of measurements necessary is essential for optimal engineering design.

5.2   Data Spaces

This section begins to formalize the approach taken in Example 5.1.1 by defining essential concepts related to mathematical descriptions of data. Appendix A explains which collections of vectors qualify to form *vector space* $\mathbb{V}$. Simply put, for $\mathbb{V}$ to constitute a vector space, any vector formed by scaling and summing members of $\mathbb{V}$ must also be found in $\mathbb{V}$. By establishing $\mathbb{V}$ as existing in a D-dimensional *field* of

---

[3] There are an infinite number of eigenvectors that satisfy the eigendecomposition. Running the code in MATLAB shows the eigenvectors can change sign, which inverts the phase of $\mathbf{y}_1$. Rerun the code to see this effect.

real numbers, i.e., $\mathbb{V} \in \mathbb{R}^D$, or a field of complex numbers, $\mathbb{V} \in \mathbb{C}^D$, we validate applications of foundational analyses such as algebra and calculus to the data. We leave that important stuff to other texts.

A simple example of a 3-D vector space that you have likely encountered many times since high school is one composed of spatial vectors from Euclidean geometry, $\{\mathbf{x}\} \in \mathbb{R}^3$. There are an infinite number of vectors in this space, yet each one can be synthesized by linearly combining just three members that form a *basis* for $\mathbb{V}$. While basis vectors are not unique, the number of basis vectors – the intrinsic dimensionality of the space – is unique and a fundamental property of the data.

5.2.1   Dimension

Dimension is defined mathematically as the cardinality of the basis for the vector space of the data. Dimension specifies the size of the largest set of *linearly inde-pendent*[4] *functions or vectors* in that vector space. For example, spatial vectors $\mathbf{x} = (x_1, x_2, x_3) \in \mathbb{R}^3$ with cardinality 3 cannot have subsets larger than three that are linearly independent.

Consider the standard Euclidean basis, $(1,0,0),(0,1,0),(0,0,1)$, and a nonstandard basis, $(\frac{\sqrt{3}}{2},\frac{1}{2},0),(-\frac{1}{2},\frac{\sqrt{3}}{2},0),(0,0,1)$, found by rotating the x_1, x_2 plane about x_3 by $30°$ (Figure 5.3a). Both are *orthonormal bases*[5] for this Euclidean vector space. Both bases span the vector space and, therefore, can be used to generate any vector in that space, including all possible solutions to a problem you may be trying to solve.

It is straightforward to demonstrate these properties in numerical examples. How-ever, these concepts are quite general. For example, consider this binary-response question: Does patient X have disease Y? If there is a basis in biomedical sensing for the disease in that population and we know its dimensionality, we can select a range of measurements that span the space and be sure that an accurate diagnosis is possible if we have sufficient SNR and degrees of freedom (defined later in this section). Statistics cannot identify the most sensitive measurement variables; that requires a deep understanding of the medical problem. Nevertheless, we can apply statistical concepts to data to find the smallest set of decision variables formed from linear combinations of the measurement variables, as we saw in Example 5.1.1

5.2.2   Big Picture

Unfortunately, there is no systemic procedure for determining the dimensionality of systems as large and complex as the human body. We do know quite a bit about par-ticular subspaces associated with specific diseases. Medicine often tracks observable

---

[4] Vectors μ and v are linearly independent if, for scalars c_1, c_2, the only solution to $c_1\mu + c_2 v = \mathbf{0}$ is $c_1 = c_2 = 0$.
[5] When any pair of vectors in a set has an inner product of zero, e.g.,
$\langle \mu, v \rangle = \mu^\dagger v = (\mu_1^* v_1 + \mu_2^* v_2 + \mu_3^* v_3) = 0$, those vectors are orthogonal. If, in addition, each vector has ℓ_2 norm $\|\mu\|_2 = \sqrt{\langle \mu, \mu \rangle} = \sqrt{\mu_1^2 + \mu_2^2 + \mu_3^2} = 1$, the set is orthonormal. Vector products and norms are discussed in Appendix A.

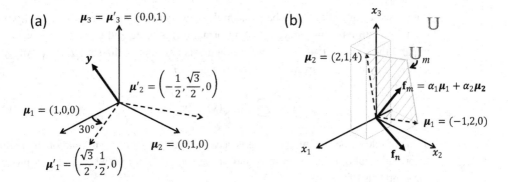

Figure 5.3 (a) The standard Euclidean basis (solid axes, μ_1, μ_2, μ_3) defining measurement **y** is rotated 30° about the μ_3 axis to define another orthonormal basis (dashed axes, μ'_1, μ'_2, μ'_3). (b) Illustration of a 2-D *ambient subspace*, $\mathbb{U}_m \in \mathbb{R}^2$ (shaded plane), within a 3-D object space, $\mathbb{U} \in \mathbb{R}^3$. The subspace must contain the origin. $\mathbb{U}_m$ describes the subspace accessible by measurement $\mathcal{H}$. It contains vector $\mathbf{f}_m$, which leads to measurement $\mathbf{g}_m$ (not shown). $\mathbf{f}_m(\mathbf{x})$ may be defined by orthogonal (not orthonormal) basis μ_1, μ_2. In contrast, $\mathbf{f}_n(\mathbf{x})$ lies in the null space of the measurement because it is not within the ambient subspace $\mathbb{U}_m$.

features such as the presence of fever or changes in blood chemistry or radiological anatomy, which could be combinations of a larger and unknown basis describing health. Imagine the possibilities when we are able to identify the basis of health! Observable features, such as fever, can tell us that something is wrong with a patient, but that information is nonspecific because many variables contribute to this observable feature.

Fortunately, measurement instruments are considerably less complex than organisms are. Expressing the acquisition and display processes for patient data using linear algebra provides a rigorous basis for analyzing and designing measurements that span health states as best we know them. The ultimate goal for vector-space analysis of biomedical data is to arrive at a first-principles approach to measurement-systems engineering.

5.2.3   Data Subspaces

Engineering mathematics provides a means to represent the functions associated with measurements. Assume that task-related properties of objects can be represented by functions in vector space $\mathbb{U}$. Further assume that data generated from measurements of these objects exist in vector space $\mathbb{V}$. If data acquisition is facilitated by $\mathcal{H}$, a linear transformation mapping object functions f from $\mathbb{U}$ into data functions g in $\mathbb{V}$, then the ability of that measurement to capture task information can be defined using the rules of linear algebra and the accessibility of $\mathcal{H}$ to elements of $\mathbb{U}$.

Let a set of object functions $f(\mathbf{x}) \in \mathbb{U}$ be square integrable, which means their *inner product* is finite; i.e., $\langle f, f \rangle = \int d\mathbf{x}\, f^*(\mathbf{x})\, f(\mathbf{x}) < \infty$. Although continuous $f(\mathbf{x})$ is composed of an infinite number of samples, let's say its intrinsic dimensionality (the number of dimensions carrying task information) is finite and equal to N. Hence, we expand the $f(\mathbf{x})$ in $\boldsymbol{\mu} \in \mathbb{C}^{\infty \times 1}$ where the basis consists of the set $\{\boldsymbol{\mu}_k\}_{k=0}^{N}$. Any f in $\mathbb{U}$ can be synthesized from the *synthesis equation*,

$$f(\mathbf{x}) = \sum_{k=0}^{N-1} \alpha_k \, \boldsymbol{\mu}_k(\mathbf{x}), \quad \text{for } k = 0, 1, 2 \dots. \tag{5.4}$$

The complex coefficient α_k assigns a weight to $\boldsymbol{\mu}_k$ before summation to include information about the magnitude and phase (how much and where) of that mode in the synthesis of f.

In the simple geometric example of Figure 5.3b, functions $\boldsymbol{\mu}_1$ and $\boldsymbol{\mu}_2$ form the orthogonal basis of a 2-D *subspace* given by a plane through the origin. If we scale $\boldsymbol{\mu}_1(\mathbf{x})$ by $1/\sqrt{2}$ and $\boldsymbol{\mu}_2(\mathbf{x})$ by $1/3\sqrt{2}$, then the basis is orthonormal. Because $\boldsymbol{\mu}_1, \boldsymbol{\mu}_2$ span $\mathbb{U}_m$ and are *linearly independent* (Appendix G), the rules of linear algebra ensure that their linear combinations describe any function within that subspace. It may be that there is only a small subset in the *ambient subspace*[6] $\mathbb{U}_m$ that qualifies as physically realizable. In Section 6.8, we discuss the application of *low-dimensional signal models* that approximately represent measurement data.

Let $f(\mathbf{x})$ be fully represented by vector $\mathbf{f} \in \mathbb{R}^{N \times 1}$. Also let the basis vectors of (5.4) be arranged into columns of matrix $\mathbf{U} = (\boldsymbol{\mu}_1, \dots, \boldsymbol{\mu}_k, \dots, \boldsymbol{\mu}_N) \in \mathbb{C}^{N \times N}$. The synthesis equation of (5.4) becomes

$$\mathbf{f} = \sum_{k=0}^{N-1} \alpha_k \, \boldsymbol{\mu}_k = \mathbf{U}\boldsymbol{\alpha}, \quad \text{where the } n\text{th element of } \mathbf{f} \text{ is} \quad f_n = \sum_{k=0}^{N-1} \alpha_k [\mathbf{U}]_{nk}, \tag{5.5}$$

and $\boldsymbol{\alpha} = (\alpha_1, \dots, \alpha_k, \dots, \alpha_N)^\top \in \mathbb{C}^{N \times 1}$. The discrete synthesis equation of (5.5) is graphically illustrated in Figure 5.4a.

Finding coefficients once we select a basis is easy. The orthogonal decomposition of functions and vectors into coefficients follows the *analysis equation*,

$$\alpha_k = \int_{-\infty}^{\infty} d\mathbf{x} \, f(\mathbf{x}) \, \mu_k^*(\mathbf{x}) \quad \text{Continuous } f(\mathbf{x})$$

$$= \sum_{n=0}^{\infty} f_n \, [\boldsymbol{\mu}_k]_n^* = \boldsymbol{\mu}_k^\dagger \mathbf{f} \quad \text{or} \quad \boldsymbol{\alpha} = \mathbf{U}^\dagger \mathbf{f} \quad \text{Discrete } \mathbf{f}, \tag{5.6}$$

where $[\boldsymbol{\mu}_k]_n^*$ is the conjugate transpose of the kth basis vector, $\boldsymbol{\mu}_k$, as illustrated in Figure 5.4b. As discussed in Appendix A, $\mathbf{U}$ is a unitary matrix where $\mathbf{U}^{-1} = \mathbf{U}^\dagger$.

The basis decomposition in (5.6) breaks down *observable features* from a patient/object, where the dimensions are often coupled, into *uncoupled properties* because $\{\boldsymbol{\mu}_k\}$ are orthogonal. If there are N possible coefficients α_k representing patient features, and we find that only $R < N$ are nonzero with respect to a specific task, the

[6] An ambient space is the space encompassing the object function under investigation.

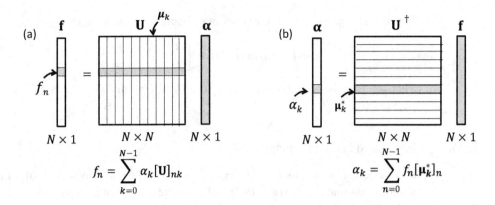

Figure 5.4 Geometry of (a) the discrete synthesis equation (5.5) and (b) the analysis equation (5.6).

feature space for measurement has a task dimension no larger than R. In such a case, we can say that R is the *bandwidth* of the task regardless of the basis selected.

Finally, note that the measurement of object vectors $\mathbf{f}_m \in \mathbb{U}_m$ (the object subspace accessible by acquisition operator $\mathcal{H}$ in Figure 5.3b) results in data vectors $\mathcal{H}\mathbf{f}_m = \mathbf{g}_m \in \mathbb{V}$. However, those vectors in $\mathbb{U}$ that are outside $\mathbb{U}_m$ will fall into the *null space* of the measurement, i.e., $\mathcal{H}\mathbf{f}_n = \mathbf{0}$. The following sections discuss standard bases.

5.3   Orthonormal Basis

Let's expand real continuous function $f(x)$ of length X_0 in a *Fourier basis* given by $\mu_k(x) = \sqrt{u[1]} \exp(i2\pi kx/X_0)$.[7] $u[1] = u_0 = 1/X_0$ is the fundamental spatial frequency set by *object support*, X_0. Harmonic frequencies are integer multiples of the fundamental, $u[k] = ku_0$ for $k = 0, \pm 1, \pm 2 \ldots$, giving the spatial-frequency axis in units of length^{-1}. Before examining this point further, we want to demonstrate the properties of the basis.

The continuous Fourier basis is orthonormal because, for integers k and ℓ,

$$\int_{-\infty}^{\infty} dx\, \mu_k(x)\mu_\ell^*(x) = \frac{1}{X_0}\int_{-\infty}^{\infty} dx\, e^{i2\pi(k-\ell)x/X_0} = \delta(ku_0 - \ell u_0). \qquad (5.7)$$

Expressing the discrete Fourier basis for $N \times 1$ object vector $\mathbf{f}$ as $\mu_k = \exp(i2\pi kn/N)/\sqrt{N}$, where x is sampled on the interval X and $N = X_0/X$, we find the discrete equivalent of (5.7),

$$\sum_{k=-\infty}^{\infty} \mu_k^\dagger \mu_\ell = \delta[\ell - k] = \delta[k - \ell]. \qquad (5.8)$$

---

[7] μ_k is a basis vector, $\mu_k(x)$ is a basis function, and $u[k] = u_k = ku_0 = k/X_0$ are spatial frequencies.

In general, if $\{\mu_k(\mathbf{x})\}$ constitutes a set of orthogonal functions, then

$$\int d\mathbf{x}\,\mu_k^*(\mathbf{x})\,\mu_\ell(\mathbf{x}) = \begin{cases} C \neq 0 & k = \ell \\ 0 & k \neq \ell \end{cases}.$$

This set is orthonormal when $C = 1$ or when normalizing basis functions via $\mu_k(\mathbf{x})/\sqrt{C}$.

5.3.1   Integrating Periodic Functions

Euler's equation tells us that $\exp(ix) = \cos(x) + i\sin(x)$. So why do the infinite duration sinusoids in (5.7) and (5.8) yield finite results when integrated?

For real constants a, b, integer constant N, and function g that is periodic over interval a, we have $\frac{1}{a}\int_0^a dt\, g(t) = \frac{1}{a}\int_b^{b+a} dt\, g(t) = \frac{1}{Na}\int_b^{b+Na} dt\, g(t)$. Once normalized, integrating over N cycles is the same as integrating over one cycle:

$$\frac{1}{Na}\int_b^{b+Na} dt\, g(t) = \frac{N}{Na}\int_0^a dt\, g(t) = \frac{1}{a}\int_0^a dt\, g(t).$$

As a consequence, for temporal sinusoids at radial frequency $\Omega_0 = 2\pi/T_0$, period T_0, and total duration $T_t = NT_0$,

$$\lim_{T_t \to \infty} \frac{1}{T_t}\int_0^{T_t} dt\, \sin(\Omega_0 t)\cos(\Omega_0 t) = \frac{1}{T_0}\int_0^{T_0} dt\, \sin(\Omega_0 t)\cos(\Omega_0 t) = 0. \quad (5.9)$$

More generally, for integers n and m and $\cos\alpha\,\cos\beta = [\cos(\alpha - \beta) + \cos(\alpha + \beta)]/2$, we find

$$\frac{1}{T_0}\int_0^{T_0} dt\, \cos(n\Omega_0 t)\cos(m\Omega_0 t) = \begin{cases} 1 & \text{for } m = n = 0 \\ 1/2 & \text{for } m = n \neq 0 \\ 0 & \text{for } m \neq n \end{cases}. \quad (5.10)$$

You can see that this result holds for sine waves once you apply the identity $\sin\alpha\,\sin\beta = [\cos(\alpha - \beta) - \cos(\alpha + \beta)]/2$. This useful result tells us that sines and cosines *at harmonic frequencies* form an orthonormal basis (if we are careful with the scaling). All discrete frequencies $k\Omega_0$ that compose the basis are harmonics of the fundamental Ω_0.

Example 5.3.1. *Orthogonality:* *To demonstrate orthogonality numerically, consider an example from [94]. Five harmonically related sine and cosine waves and one nonharmonic wave are generated. The inner products of any two waveforms can be found using correlation coefficient ρ.*

```
N=256;Tt=1;T=Tt/N;t=T:T:Tt;      % initialize parameters and independent variables
G(:,1) = 1.0*sin(2*pi*1*t)';     % time-harmonic functions
G(:,2) = 2.0*cos(2*pi*1*t)';     % tick marks make column vectors
G(:,3) = 1.5*sin(2*pi*2*t)';     % cos(2pi ut)=sin(2pi ut + pi/2)
G(:,4) = 3.0*cos(2*pi*2*t)';     % sines & cosines at u=1,2,3.5 Hz
G(:,5) = 2.5*sin(2*pi*3*t)';
G(:,6) = 1.75*cos(2*pi*3.5*t);   % nonharmonic waveform
r = corrcoef(G) =                % correlation coeff matrix
```

1.0000	-0.0000	-0.0000	-0.0000	-0.0000	-0.0566
-0.0000	1.0000	-0.0000	-0.0000	0.0000	-0.0078
-0.0000	-0.0000	1.0000	0.0000	0.0000	-0.1544
-0.0000	-0.0000	0.0000	1.0000	-0.0000	-0.0078
-0.0000	0.0000	0.0000	-0.0000	1.0000	-0.5878
-0.0566	-0.0078	-0.1544	-0.0078	-0.5878	1.0000

Only the sixth waveform is nonharmonic, and therefore has nonzero off-diagonal terms.

5.4   Fourier Series

The Fourier series (FS) expands a continuous function of space or time into a discrete spectrum of harmonic frequencies at integer k multiples of the fundamental, i.e., $ku_0 = k/T_0$. Let's expand $g(t)$ of duration T_0 using (5.4) with Fourier basis $e^{i2\pi kt/T_0}$,

$$g(t) = \sum_{k=-\infty}^{\infty} G[k]\, e^{i2\pi kt/T_0}. \qquad \text{(Inverse FS)} \qquad (5.11)$$

We relabel complex constants α_k as complex constants $G[k]$ when applying the Fourier basis. Equation (5.11) is referred to as the *Fourier-series synthesis equation* or *inverse Fourier series* because it assembles the original function from spectral coefficients.

We find the unique set of coefficients $\{G[k]\}$ from the general analysis equation of (5.6), which we implement by multiplying both sides of (5.11) by the conjugate basis and integrating:

$$\int_{-T_0/2}^{T_0/2} dt\, g(t)\, e^{-i2\pi \ell t/T_0} = \sum_{k=-\infty}^{\infty} G[k] \int_{-T_0/2}^{T_0/2} dt\, e^{i2\pi(k-\ell)t/T_0}. \qquad (5.12)$$

The integral on the right-hand side is found by applying Euler's equation,

$$\int_{-T_0/2}^{T_0/2} dt\, e^{i2\pi(k-\ell)t/T_0} = \int_{-T_0/2}^{T_0/2} dt\, \cos(2\pi(k-\ell)t/T_0)$$
$$+ i \int_{-T_0/2}^{T_0/2} dt\, \sin(2\pi(k-\ell)t/T_0),$$

and noting the imaginary term is zero because the integrand is antisymmetric and the integration is over one period. Applying the identity $\cos(\alpha - \beta) = \cos\alpha\cos\beta + \sin\alpha\sin\beta$ and (5.10), we have

$$\int_{-T_0/2}^{T_0/2} dt\, e^{i2\pi(k-\ell)t/T_0} = \int_{-T_0/2}^{T_0/2} dt\, \cos(2\pi kt/T_0)\cos(2\pi \ell t/T_0)$$
$$+ \int_{-T_0/2}^{T_0/2} dt\, \sin(2\pi kt/T_0)\sin(2\pi \ell t/T_0)$$

$$= \begin{cases} T_0\delta[\ell] & k = 0 \\ \frac{T_0}{2}\delta[k - \ell] + \frac{T_0}{2}\delta[k - \ell] & k \neq 0 \end{cases}$$

$$= T_0\delta[k - \ell] \qquad \forall \; k. \tag{5.13}$$

Substituting (5.13) into (5.12) to find $G[\ell]$ and then renaming dummy integer index ℓ as k, we find the *forward Fourier series* or *Fourier-series analysis equation* of (5.6),

$$G[k] = \frac{1}{T_0} \int_{-T_0/2}^{T_0/2} dt \; g(t) \, e^{-i2\pi kt/T_0}. \qquad \text{(Forward FS)} \tag{5.14}$$

Thus, we went through the details to show the origin of scale factor $1/T_0$.

Equations (5.11) and (5.14) are the usual forms of the Fourier-series pair for $g(t)$. Equation (5.14) is an example of a *continuous-to-discrete transformation* that is represented using shorthand operator notation via $G[k] = \mathcal{F}g(t)$. The corresponding *discrete-to-continuous transformation* from (5.11) is $g(t) = \mathcal{F}^{-1}G[k]$. $\mathcal{F}$ is the Fourier operator, while $\mathcal{F}^{-1}$ is the inverse Fourier operator. Of course, $g(t) = \mathcal{F}^{-1}\mathcal{F}g(t)$. $\mathcal{F}$ is a transformation within one vector space, a topic we will return to later.

How can we know a Fourier series operation is implied by $\mathcal{F}$ and not a discrete Fourier transform (see Section 5.8) for which the same symbol is used? The answer depends on the context – that is, we must know that g is a continuous function and G is a discrete function. Finally, note that functions that are discrete in one domain are periodic in the other. In the Fourier-series pair, since $G[k]$ is discrete, we must assume $g(t)$ is periodic on the interval T_0. More on this later.

The basis functions given in (5.11) are orthogonal but not orthonormal. Fortunately, each is normalized with a simple change. From (5.13), we see that

$$\mu_k(t) = \frac{1}{\sqrt{T_0}} e^{i2\pi kt/T_0}$$

is an orthonormal basis. If we use this form, we have a more symmetric and yet entirely equivalent version of the Fourier series equations:

$$g(t) = \frac{1}{\sqrt{T_0}} \sum_{-\infty}^{\infty} G[k] \, e^{i2\pi kt/T_0} \quad \text{and} \quad G[k] = \frac{1}{\sqrt{T_0}} \int_{-T_0/2}^{T_0/2} dt \; g(t) \, e^{-i2\pi kt/T_0}.$$

$$\tag{5.15}$$

With the orthonormal basis, if the units of $g(t)$ are [V] and time is in units of [s], the units of $G[k]$ are [V s$^{1/2}$]. While this latter version is a true orthonormal basis representation, (5.11) and (5.14) are more commonly used. So why are multiple forms of Fourier analysis found in texts and throughout the literature? The short answer is that authors are free to adapt the form to suit their specific problems because all of these versions are technically correct provided $\mathcal{F}^{-1}\mathcal{F}g(t) = g(t)$, including units. The trick is to be consistent with whatever representation is selected.

The complex exponential basis is the most convenient form of the Fourier-series expansion. Applying Euler's equation again, the result may be also expressed entirely in terms of positive frequencies and real functions using three terms:

$$g(t) = \frac{A[0]}{2} + \sum_{k=1}^{\infty} A[k] \cos(2\pi kt/T_0) + \sum_{k=1}^{\infty} B[k] \sin(2\pi kt/T_0), \qquad (5.16)$$

where, for $k > 0$,

$$A[0] = \frac{2}{T_0} \int_{-T_0/2}^{T_0/2} dt\ g(t)$$

$$A[k] = \frac{2}{T_0} \int_{-T_0/2}^{T_0/2} dt\ g(t)\ \cos(2\pi kt/T_0) \qquad (5.17)$$

$$B[k] = \frac{2}{T_0} \int_{-T_0/2}^{T_0/2} dt\ g(t)\ \sin(2\pi kt/T_0).$$

$A[k]$ and $B[k]$ are the coefficients for the even and odd parts of $g(t)$, and they have the same units. Again, to have an orthonormal basis, we would need to split the coefficients between the basis functions and the coefficients, though that is rarely done. The problem with this form is that you need to remember that the scale factor for the zero-frequency term, $A[0]$, is different from the coefficients at other frequencies, $A[k]$ and $B[k]$ for $k > 0$.

If $g(t)$ is an even function of t, then only $A[k]$ coefficients are nonzero and (5.15) and (5.17) are the *cosine transform pair*. Similarly, if $g(t)$ is an odd function, then only $B[k]$ coefficients are nonzero, and the equations form a *sine transform pair*.

I have used the term "frequency" when referring to integer k. To be precise, $u_0 = 1/T_0$ is the temporal frequency increment, and the temporal frequency variable [Hz] is $u = ku_0 = k/T_0$. The radial temporal frequency variable [radians/s] is designated as $\Omega = k\Omega_0 = 2\pi k/T_0$. We say that t and u are *conjugate Fourier variables* because transformations from one domain to the other result in a function of t or u as the independent variable. Similarly, we can use n and k as *conjugate Fourier indices* for 1-D transforms. $g(t)$ and $G[k]$ convey the same information about the data, albeit in different domains as indicated by their independent variables t and k.

Fourier basis functions are applied to transport functions between these domains. Functions of time and temporal frequency lie in the *same vector space*; i.e., transformations $\mathcal{F}$ and $\mathcal{F}^{-1}$ are from $\mathbb{U}$ to $\mathbb{U}$ or from $\mathbb{V}$ to $\mathbb{V}$ (Figure 5.5). Conversely, acquisition $\mathcal{H}$ and display $\mathcal{O}$ operators transform information between the two vector spaces. In both cases, we transform functions of continuous or discrete variables, which alters the forms of the equations but not the essence of the operations.

We can apply the Fourier-series expansion to a function of a continuous spatial variable, $f(x)$, which is defined over the spatial interval X_0. The orthonormal bases are

$$\mu_k(x) = e^{i2\pi kx/X_0},$$

and the inverse and forward transformations are, respectively,

$$f(x) = \sum_{-\infty}^{\infty} F[k]\ e^{i2\pi kx/X_0} \quad \text{and} \quad F[k] = \frac{1}{X_0} \int_{-X_0/2}^{X_0/2} dx\ f(x)\ e^{-i2\pi kx/X_0}.$$

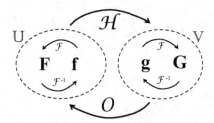

Figure 5.5 Illustration of acquisition and display operators $(\mathcal{H}, \mathcal{O})$ applied between vector spaces $(\mathbb{U}, \mathbb{V})$ and the Fourier series operators $(\mathcal{F}, \mathcal{F}^{-1})$ applied within each vector space.

The quantity u can denote temporal or spatial frequency. With spatial frequency, $u_0 = 1/X_0$, $u = k/X_0$. There should be no ambiguity because the reader need only look for the conjugate variable t or x to understand the context. Finally, for multidimensional functions $\mathbf{x} = (x_1, x_2, x_3)^\top$, we have $\mathbf{u} = (u_1, u_2, u_3)^\top$.

Let's apply these ideas to the computation of a Fourier series for a periodic square wave *analytically* and compare it with the MATLAB results.

Example 5.4.1. *Fourier series of rectangular function: Find the Fourier series coefficients for the rect function of Figure 5.6 that is continuous in time and periodic over duration T_0.*

Since the function is even, we only need to compute the even coefficients $A[k]$ from (5.17). To do this, we explicitly express the rectangular function of width c and height h on the interval T_0,

$$g(t) = h \operatorname{rect}(t/c) \text{ for } -T_0/2 \le t \le T_0/2,$$

where h and c are defined in Figure 5.6. Analyzing one period in a cosine transform,

$$
\begin{aligned}
G[k] &= \frac{1}{T_0} \int_{-T_0/2}^{T_0/2} dt\, g(t)\, e^{-i2\pi kt/T_0} \\[2mm]
&= \frac{2}{T_0} \int_{-T_0/2}^{T_0/2} dt\, g(t)\, \cos(2\pi kt/T_0) = A[k] \quad \textit{Forward cosine transform} \\[2mm]
&= \frac{2h}{T_0} \int_{-T_0/2}^{T_0/2} dt\, \operatorname{rect}\left(\frac{t}{c}\right) \cos(2\pi kt/T_0) = \frac{2h}{T_0} \int_{-c/2}^{c/2} dt\, \cos(2\pi kt/T_0) \\[2mm]
&= \frac{2h T_0}{2\pi k T_0} [\sin(2\pi kt/T_0)]_{-c/2}^{c/2} \\[2mm]
&= \frac{2h}{\pi k} \sin(\pi kc/T_0) = \left(\frac{2hc}{T_0}\right) \frac{\sin(\pi kc/T_0)}{\pi kc/T_0} \\[2mm]
&= 2hcu_0 \operatorname{sinc}(cku_0), \quad \text{where } \operatorname{sinc}(x) \triangleq \frac{\sin(\pi x)}{\pi x} \text{ and } u_0 = 1/T_0. \quad (5.18)
\end{aligned}
$$

$G[k]$ *has units of $g(t)$ as indicated by units of h.*

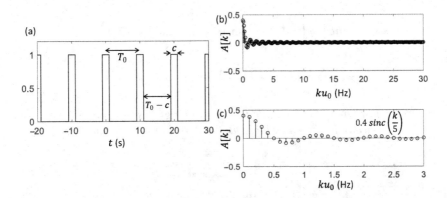

Figure 5.6 (a) The rect function from Example 5.4.1 has amplitude $h = 1$ and width $c = 2$ s. It is periodic on the interval $T_0 = 10$ s. Frames (b) and (c) show its Fourier series at two time scales.

Figure 5.6 displays the result computed numerically from the following code:

```
%%%%%%%%%%%%%%%%%%%%%%%%%%%%%% Script generating Fig. 5.6 %%%%%%%%%%%%%%%%%%%%%%%%%%%%%%%%%%%%%%
t=0:0.01:10;g=zeros(size(t));        % 10 s time axis is one period of g(t)
g(1:101)=1;g(901:1001)=1;            % create square wave g(t) for c=2s
for k=1:301;                         % 301 bases for 301 coefficients
   C(:,k)=cos(2*pi*(k-1)*t/10)';     % basis functions
end
A=2/10*g*C*0.01;                     % Fourier coefficients
u=0:0.1:30;                          % prepare a limited-frequency axis
subplot(2,1,1);stem(u,A,'k')  % plot Fourier series coefficients
subplot(2,1,2);stem(u(1:31),A(1:31),'k','linewidth',1) % focus on low frequencies
```

Although $g(t)$ is periodic, $G[k]$ is not. The amplitude of a sinc function is 1, which is found by applying l'Hôpital's rule. Equation (5.18) predicts that the peak value for $A[k]$, occurring at $k = 0$, is $2hc/T_0 = 2(1)(2)/10 = 0.4$, and that's what we find in Figure 5.6. The zeros of the sinc function predict $A[k]$ will be zero at $\pi kc/T_0 = n\pi$, which occurs at frequencies $u = k/T_0 = n/c = \frac{1}{2}, 1, \frac{3}{2}, 2, \ldots$ for nonzero integer n (a different integer from k). Consequently, narrow rect functions produce broad sinc functions, where zero crossings move further from the origin. It is generally true that narrow functions in one domain imply broad functions in the other.

Example 5.4.2. *Synthesize rectangular function from Fourier coefficients:* Synthesize $g(t)$ from the coefficients found in Example 5.4.1 in two ways.
(1) Combine the analytical result of (5.18) with (5.16).

We must decide how many terms in the infinite summation to include.

$$g(t) = \sum_{k=-\infty}^{\infty} G[k] \, e^{i2\pi kt/T_0} \quad \text{Real-even functions can use cosine transforms.}$$

$$= \frac{A[0]}{2} + \sum_{k=1}^{\infty} A[k] \cos(2\pi kt/T_0) \quad \text{Inverse cosine transform}$$

$$= hc/T_0 \left(1 + 2 \sum_{k=1}^{\infty} \text{sinc}(kc/T_0) \cos(2\pi kt/T_0) \right) \qquad (5.19)$$

$$= h \sum_{n=-\infty}^{\infty} \text{rect}\left(\frac{t - nT_0}{c} \right).$$

The last line of (5.19) explicitly shows that g(t) is periodic. We implemented (5.19) numerically in the following code for 10^4 terms in the sum over four rect-function cycles:

```
%%%%%%%%%% Implementing (5.19) to synthesize g(t) %%%%%%%%%%%%
t=-20:0.01:20;T0=10;c=2;h=1;q=zeros(size(t));
for k=1:10000
    q=q+sin(pi*k*c/T0)*cos(2*pi*k*t/T0)/(pi*k*c/T0);
end
g=(h*c/T0)*(1+2*q);plot(t,g)
```

(2) Numerically reconstruct g(t) from different numbers of Fourier series coefficients, with and without additive noise.

    See the code presented here for the solution. We may apply up to $N/2$ coefficients (positive frequencies) to reconstruct g(t). The top row of Figure 5.7 shows the original signals without (a) and with (b) noise. The middle row uses all 500 Fourier series coefficients, and the bottom row uses just the first 50 coefficients. Applying 50 coefficients provides too few high frequencies to fully recover sharp edges. The noise-free reconstructed signal (b, bottom plot) oscillates near sharp transitions (Gibb's ringing). However, including 500 coefficients provides an excellent reconstruction except for a spike at g(0).

    With noise and 500 coefficients, both the signal and noise are recovered. With 50 coefficients, however, noise is reduced much more than the signal. Spectrally white

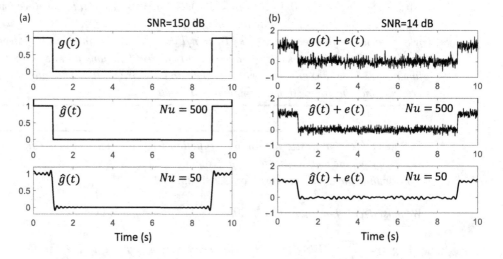

Figure 5.7 Numerical solutions from the calculation in Example 5.4. The top row shows the original signals, while the middle and bottom rows reconstruct signals using 500 and 50 FS coefficients. Results in (a) are noise free while results in (b) have additive Gaussian noise at 14 dB SNR. Note that 150 dB is "noise free" because that is the roundoff error for the simulation (one part in 10^{15}).

noise adds equal amounts of noise energy across the frequency range. However, signal bandwidth is confined mostly to low frequencies, so by zeroing A[k] for k > 50 we end up filtering significant noise energy.

```
%%%%%%%%%%%%%%%%%%%%% From Example 5.4.2 resulting in parts of Fig. 5.7 %%%%%%%%%%%%%%%%%%%%%%%%
T0=10;c=2;h=1;dt=T0/1000;t=0:dt:T0;N=length(t);g1=zeros(size(t));   % initialize
g1(1:floor(N/10))=h;g1(ceil(0.9*N):N)=h;Nu=floor(N/20);            % square wave g(t) for c=2s
e=0.2*randn(size(t));C=zeros(N,Nu);g1=g1+e;                        % add noise
for k=1:Nu;
    C(:,k)=cos(2*pi*(k-1)*t/T0)';                                 % basis functions for FS
end
A=(2/T0)*g1*C*dt;                                                 % Fourier coefficients
subplot(3,1,1);plot(t,g1)                                         % plot time series
u=0:1/T0:(Nu-1)/T0;                                              % prepare frequency axis
subplot(3,1,2);stem(u,A)                                         % plot Fourier series coefficients
A(1)=A(1)/2;g2=C*A';                                             % synthesize g1
subplot(3,1,3);plot(t,g2)
```

A Fourier-series expansion is a decomposition of continuous periodic functions. Sharp readers will notice that we assume $g(t)$ is continuous, yet we use digital data to compute Fourier series coefficients in Figure 5.6. The value of the Fourier series, in my view, is to uncouple the sampling parameters in the time domain from those in the frequency domain, with the exception that $u_0 = 1/T_0$. Uncoupling means the signals can be approximately continuous with fine sampling of the frequency axis. Uncoupling also means the Fourier coefficients are not periodic functions of frequency. Unfortunately, FS analysis is a slow computation compared with other tools, which we will examine.

5.5 Continuous-Time Fourier Transform

Fourier analysis can be extended to nonperiodic continuous-time functions. For this task, we must include the complex Fourier coefficients from an infinite number of frequencies when synthesizing the time-domain data $g(t)$. Therefore continuous-time transforms are something we compute analytically, not numerically. Continuous-time Fourier transform (CT-FT) is a tool for design and discussion, rather than being appropriate for numerical modeling or data processing.

Beginning with the Fourier series results, set $T_0 \to \infty$ so that the fundamental frequency u_0 becomes infinitesimally small. The result is a continuum of Fourier coefficients $G[k] \to du\, G(u)$ as the discrete frequency increment $1/T_0$ becomes a continuous differential du. The infinite-duration time axis implies $g(t)$ has an infinite period, which is a nonperiodic function. Importantly, if the units of $g(t)$ and $G[k]$ are volts [V], the units of $G(u)$ become [V/Hz] or [V s]. Units are an important difference between FS and CT-FT coefficients.

The sum of (5.11) becomes an integral:

$$g(t) = \int_{-\infty}^{\infty} du\, G(u)\, e^{i2\pi ut}. \qquad \text{(Inverse CT-FT)} \qquad (5.20)$$

Equation (5.20) is the *CT Fourier synthesis equation* or the *inverse Fourier transform*, since it recovers the original function from complex coefficients. The *CT Fourier analysis equation* or the *forward Fourier transform* decomposes $g(t)$ into complex coefficients via

$$G(u) = \int_{-\infty}^{\infty} dt\, g(t)\, e^{-i2\pi ut}. \qquad \text{(Forward CT-FT)} \qquad (5.21)$$

Notice the symmetry between the forward and inverse forms. Both are continuous-to-continuous transformations generating nonperiodic functions because t and u are both continuous. The Fourier integrals exist if the integrand is absolutely integrable,[8] which does not naturally apply to random processes such as white noise. Several notations may be used to represent the transform pair:

$$g(t) \overset{FT}{\longleftrightarrow} G(u) \qquad\qquad G(u) = \mathcal{F}g(t) \qquad\qquad g(t) = \mathcal{F}^{-1}G(u).$$

CT-FTs are a great source of closed-form solutions in part because of the many theorems that simplify these calculations (Appendix D). The downside is that you cannot exactly implement CT-FT numerically, except using symbolic computational methods in MATLAB as available (see the end of Appendix D).

I prefer to use a Fourier kernel of $\exp(i2\pi ut)$, while others use $\exp(i\Omega t)$, where $\Omega = 2\pi u$. As you can see from (5.20) and (5.21), the first form has no scaling constants. If you use the latter form, the inverse transform needs to be scaled: $g(t) = \frac{1}{2\pi}\int_{-\infty}^{\infty} d\Omega\, G(\Omega)\, e^{i\Omega t}$.

Let's revisit Example 5.4.1 using CT-FT and compare results. In this example, there is just one square in the waveform (nonperiodic) and we'll position its center at $t = b$, which was zero in the previous example.

Example 5.5.1. *CT-FT: Compute the CT-FT for a rect function analytically and then approximate that result in* MATLAB

$$g(t) = h\,\text{rect}\left(\frac{t-b}{c}\right) = \begin{cases} h; & b - c/2 \le t \le b + c/2 \\ 0; & \text{otherwise} \end{cases}.$$

$$G(u) = \int_{-\infty}^{\infty} dt\, g(t)\, e^{-i2\pi ut} = \int_{-\infty}^{\infty} dt\left[h\,\text{rect}\left(\frac{t-b}{c}\right)\right] e^{-i2\pi ut}.$$

$$= h \int_{b-c/2}^{b+c/2} dt\, e^{-i2\pi ut}$$

---

[8] If $\|f\|_1 = \int_{-\infty}^{\infty} dt\, |f(t)| < \infty$, then $\mathcal{F}f(t)$ exists. It is sufficient for f to be square integrable, $\|f\|_2^2 = \int_{-\infty}^{\infty} dt\, |f(t)|^2 < \infty$, i.e., to have a finite squared ℓ_2 norm. All experimental data satisfy these criteria because of conservation laws and finite-duration acquisitions. The concern generally lies with models that include Dirac deltas and other noncontinuously differentiable "functions."

Letting $t' = t - b$,

$$= h \int_{-c/2}^{c/2} dt' \, e^{-i2\pi u(t'+b)} = he^{-i2\pi ub} \int_{-c/2}^{c/2} dt' \, e^{-i2\pi ut'}$$

$$= he^{-i2\pi ub} \left[\frac{-\left(e^{-i2\pi uc/2} - e^{i2\pi uc/2}\right)}{i2\pi u} \right]$$

$$= hc \, e^{-i2\pi ub} \frac{\sin(\pi uc)}{\pi uc} = hc \, e^{-i2\pi ub} \, \mathrm{sinc}(uc). \tag{5.22}$$

Examining units helps us compare the CT-FT result of (5.22) with the FS result of (5.18). While coefficients $A[k]$, $B[k]$, and $G[k]$ have the same units as $g(t)$, $G(u)$ has the units of $g(t)$ divided by frequency, which are the units of hc in (5.22). $G(u)$ has the form of a continuous probability density. Also, shifting $g(t)$ to a later time by the amount b results in multiplication of the Fourier transform of $g(t)$ by the complex exponential $\exp(-i2\pi ub)$. Think about what this factor does to the transform result (and then see Example 5.8.2). MATLAB provides some insights into the meaning and effects of phase and other transform properties, but first we will discuss sampling, describe transformations of sampled signals, and introduce the discrete Fourier transform (DFT) and its cousin the fast Fourier transform (FFT).

Example 5.5.2. *Dirac delta: What is the Fourier transform of $\delta(t)$?*

$$\mathcal{F}\{\delta(t)\} = \int_{-\infty}^{\infty} dt \, \delta(t) \, e^{-i2\pi ut} = e^{-i2\pi u \times 0} = \cos(0) - i\sin(0) = 1$$

The sifting property of deltas is important for this result. In addition, note that $\mathcal{F}^{-1}\{\delta(u)\} = 1$ using the same reasoning. The ultimate in narrow functions, $\delta(t)$, transforms into the ultimate in broad functions, a constant over all frequencies! It must be true, then, that either transform of constant C results in

$$\mathcal{F}\{C\} = C \int_{-\infty}^{\infty} dt \, e^{-i2\pi ut} = C\delta(u) \quad \text{and} \quad \mathcal{F}^{-1}\{C\} = C \int_{-\infty}^{\infty} du \, e^{i2\pi ut} = C\delta(t).$$

Example 5.5.3. *Shifted Dirac delta: What is the Fourier transform of $\delta(t - t_0)$?*
We know this answer from the discussion in Examples 5.5.1 and 5.5.2.

$$\mathcal{F}\{\delta(t - t_0)\} = \int_{-\infty}^{\infty} dt \, \delta(t - t_0) \, e^{-i2\pi ut}$$

Letting $t' = t - t_0$,

$$= \int_{-\infty}^{\infty} dt' \, \delta(t') \, e^{-i2\pi u(t'+t_0)} = e^{-i2\pi ut_0} \int_{-\infty}^{\infty} dt' \, \delta(t') \, e^{-i2\pi ut'} = e^{-i2\pi ut_0}.$$

This result proves the shift theorem in Appendix D: $\mathcal{F}f(t - t_0) = \exp(-i2\pi ut_0)F(u)$.

5.6   Fourier Convolution Theorem

Of the many useful theorems in Fourier analysis, it can be argued that the *Fourier convolution theorem* is among the most important and useful. Simply stated, the convolution of two functions in one domain (time or frequency) is equivalent to the multiplication of their transforms in the other domain (frequency or time). That is,

$$[h * f]t \overset{FT}{\longleftrightarrow} H(u)\,F(u) \quad \text{and} \quad h(t)\,f(t) \overset{FT}{\longleftrightarrow} [H * F](u).$$

To show this is true, assume continuous-time functions $g(t), h(t), f(t)$ and continuous-frequency functions $G(u), H(u), F(u)$. Then

$$g(t) = [h * f](t) = \int_{-\infty}^{\infty} dt'\, h(t - t')\, f(t')$$

$$= \int_{-\infty}^{\infty} dt' \left[\int_{-\infty}^{\infty} du\, H(u) e^{i2\pi u(t - t')} \right] \left[\int_{-\infty}^{\infty} du'\, F(u') e^{i2\pi u't'} \right]$$

$$= \int_{-\infty}^{\infty} du' \int_{-\infty}^{\infty} du\, H(u) F(u') e^{i2\pi ut} \int_{-\infty}^{\infty} dt'\, e^{-i2\pi(u - u')t'}$$

$$= \int_{-\infty}^{\infty} du' \left[\int_{-\infty}^{\infty} du\, H(u) F(u') e^{i2\pi ut} \delta(u - u') \right]$$

$$= \int_{-\infty}^{\infty} du'\, H(u') F(u') e^{i2\pi u't}.$$

Taking the forward FT of both sides,

$$\int_{-\infty}^{\infty} dt\, g(t) e^{-i2\pi ut} = \int_{-\infty}^{\infty} dt \left[\int_{-\infty}^{\infty} du'\, H(u') F(u') e^{i2\pi u't} \right] e^{-i2\pi ut}$$

$$G(u) = \int_{-\infty}^{\infty} du'\, H(u') F(u') \int_{-\infty}^{\infty} dt\, e^{-i2\pi(u - u')t}$$

$$= \int_{-\infty}^{\infty} du'\, H(u')\, F(u')\, \delta(u - u') = H(u)\, F(u).$$

By following this calculation, you can compute many CT-FT problems by hand.

5.7   Discrete-Time Fourier Transform

In previous sections, we discussed the decomposition of continuous-variable functions of time or space into either discrete-frequency (FS) or continuous-frequency (CT-FT) domains. However, most devices sample signals before any processing occurs. So we now examine the effects of sampling on Fourier analysis to discuss the *discrete-time Fourier transform* (DT-FT), a discrete-to-continuous transformation.

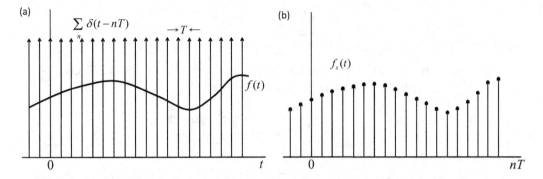

Figure 5.8 (a) Illustration for Section 5.7.1 of applying compound sampling operator $\mathcal{S}^\dagger \mathcal{S}$ to continuous function $f(t)$ to obtain (b) a continuously sampled function $f_s(t)$.

5.7.1 Sampling Operator

Let's compute the Fourier transform of $f(t)$ from sampled function $f_s(t)$. First, define the uniform sampling operator $\mathcal{S}$ [10] using the comb function from (4.9). $\mathcal{S}$ is a continuous-to-discrete convolution operator that, when applied to $f(t)$, generates $f[n]$:

$$f[n] = \mathcal{S}f(t) = \int_{-\infty}^{\infty} dt\, f(t)\, \delta(nT - t).$$

This operator takes f from a continuous-time space to a discrete-time space. We can return to continuous time by applying the adjoint operator $\mathcal{S}^\dagger$ (see Appendix D) to $f[n]$ [10]. That is,

$$f_s(t) \triangleq \mathcal{S}^\dagger f[n] = \mathcal{S}^\dagger \mathcal{S} f(t) = \sum_{n=-\infty}^{\infty} f[n]\, \delta(t - nT)$$

$$= f(t) \sum_{n=-\infty}^{\infty} \delta(t - nT) = \left[\frac{1}{T}\mathrm{comb}\left(\frac{t}{T}\right)\right] \times f(t). \tag{5.23}$$

These functions are illustrated in Figure 5.8. We used (4.9) and the peculiar properties of deltas that give $\sum f[n]\delta(t - nT) = \sum f(t)\delta(t - nT) = f(t)\sum \delta(t - nT)$ so we could separate $f(t)$ from the sum in (5.23). The result of this *compound sampling operator* is effectively a product of $f(t)$ and $\mathcal{S}^\dagger \mathcal{S}$. More on this in a minute.

5.7.2 Expressing the Sampling Operator

Before computing the Fourier transform of $f_s(t)$, let's examine the *Dirichlet kernel* as a means of finding other useful forms of the comb function. We begin with the general equation for a geometric series:

$$\sum_{n=-M}^{M} as^n = a\frac{s^{-M} - s^{M+1}}{1 - s},$$

where M and n are integers, a is a constant that we set to 1, and s is a function to be specified. Multiply the righthand side by $s^{-1/2}/s^{-1/2}$,

$$\sum_{n=-M}^{M} s^n = \frac{s^{-M-1/2} - s^{M+1/2}}{s^{-1/2} - s^{1/2}},$$

and let $s = \exp(i2\pi t/T)$ and $N = 2M + 1$. This gives an expression for the Dirichlet kernel:

$$\sum_{n=-(N-1)/2}^{(N-1)/2} e^{i2\pi n t/T} = \frac{e^{-i2\pi t(M+1/2)/T} - e^{i2\pi t(M+1/2)/T}}{e^{-i\pi t/T} - e^{i\pi t/T}}$$

$$= \frac{\sin(2\pi(M + 1/2)t/T)}{\sin(\pi t/T)} = \frac{\sin(\pi N t/T)}{\sin(\pi t/T)}.$$

To observe the Dirichlet kernel graphically as a function of time, we use the `diric` function in MATLAB, which plots $\sin(\pi N t/T)/(N\sin(\pi t/T))$ (note the $1/N$ factor!).

```
x=-2*pi:0.01:2*pi;                                          % plot 2 cycles about 0...
subplot(3,1,1);plot(x,diric(x,7));title('Dirichlet, N=7');   % ...for N = 7 ...
subplot(3,1,2);plot(x,diric(x,21));title('Dirichlet, N=21'); % ...for N = 21 ...
subplot(3,1,3);plot(x,diric(x,211));title('Dirichlet, N=211'); % ...for N = 211
```

If you run this code, you can watch how the ratio of sinc functions converges toward a Kronecker delta as N becomes large because `diric` scales by $1/N$. We want the Dirac delta version for our analytical problem (remove the $1/N$ factor), which is found asymptotically using

$$\lim_{N\to\infty} \sum_{n=-(N-1)/2}^{(N-1)/2} e^{i2\pi n t/T} = \lim_{N\to\infty} \frac{\sin(\pi N t/T)}{\sin(\pi t/T)}$$

$$= \sum_{n=-\infty}^{\infty} \delta\left(\frac{t}{T} - n\right) \triangleq \mathrm{comb}\left(\frac{t}{T}\right). \qquad (5.24)$$

In summary, essential sampling operator forms include

$$\boxed{\mathrm{comb}\left(\frac{t}{T}\right) = \sum_{n=-\infty}^{\infty} \delta\left(\frac{t}{T} - n\right) = T\sum_{n=-\infty}^{\infty} \delta(t - nT) = \sum_{n=-\infty}^{\infty} e^{i2\pi n t/T}. \quad (5.25)}$$

The middle forms use scaling and shifting properties of deltas. Each form is unitless.

5.7.3   DT-FT Equation

Applying (5.23) and the information from Section 5.6, we can derive the DT-FT equation as follows:

$$\mathcal{F} f_s(t) = \left[\mathcal{F}\{f(t)\} * \mathcal{F}\left\{ \frac{1}{T}\mathrm{comb}\left(\frac{t}{T}\right)\right\}\right](u). \qquad (5.26)$$

We know that $\mathcal{F} f(t) = F(u)$, and from (5.25),

$$\mathcal{F}\left\{ \frac{1}{T}\mathrm{comb}\left(\frac{t}{T}\right)\right\} = \frac{1}{T}\int_{-\infty}^{\infty} dt \left[\sum_{n=-\infty}^{\infty} e^{i2\pi nt/T}\right] e^{-i2\pi ut}$$

$$= \frac{1}{T} \sum_{n=-\infty}^{\infty} \int_{-\infty}^{\infty} dt \, e^{-i2\pi(u-n/T)t}$$

$$= \frac{1}{T}\sum_{n=-\infty}^{\infty} \delta\left(u - \frac{n}{T}\right) = \sum_{k=-\infty}^{\infty} \delta\left(Tu - k\right) = \mathrm{comb}(Tu).$$

The Fourier transform of a comb spaced with interval T is another comb on the interval $1/T$. The last line changed the dummy time index n to frequency index k to emphasize that we are now in the frequency domain. Assembling parts, (5.26) becomes

$$\mathcal{F} f_s(t) = \left[\mathcal{F}\{f(t)\} * \mathcal{F}\left\{ \frac{1}{T}\mathrm{comb}\left(\frac{t}{T}\right)\right\}\right](u)$$

$$= F(u) * \mathrm{comb}(Tu) = \int_{-\infty}^{\infty} du' \, F(u') \,\mathrm{comb}(T(u - u'))$$

$$= \int_{-\infty}^{\infty} du' \, F(u') \left[\sum_{k=-\infty}^{\infty} \delta\left(Tu - Tu' - k\right)\right]$$

$$= \int_{-\infty}^{\infty} du' \, F(u') \left[\frac{1}{T}\sum_{k=-\infty}^{\infty} \delta\left(u - u' - \frac{k}{T}\right)\right]$$

$$= \frac{1}{T}\sum_{k=-\infty}^{\infty} \int_{-\infty}^{\infty} du' \, F(u') \delta\left(u - \frac{k}{T} - u'\right)$$

$$\boxed{\; F_s(u) = \mathcal{F} f_s(t) = \frac{1}{T}\sum_{k=-\infty}^{\infty} F\left(u - \frac{k}{T}\right). \qquad \text{Forward DT-FT} \qquad (5.27) \;}$$

We refer to (5.27) as a DT-FT because the time domain is sampled. However, because of the compound sampling operator, the transformation is from continuous time $f_s(t)$ to continuous frequency $F_s(u)$. The CT-FT result gave $\mathcal{F} f(t) = F(u)$. Now we find in (5.27) that the sampled functions are transformed into an infinite number of scaled copies of the CT-FT result, $F(u)$. Each is separated by sampling frequency $u_s = 1/T$ before being summed. Copies centered at $u = 0, \pm 1/T, \pm 2/T \dots$ are each scaled by the factor $1/T$, so the units of $F_s(u)$ are the same as those of $f(t)$.

Functions that are sampled in one domain are periodic in the other. This periodicity stems from summed copies of $F(u)$ that are separated along the frequency axis by integer multiples of temporal sampling frequency u_s. Let's examine the conditions under which $F(u)$ can be recovered from $F_s(u)$.

5.7.4   Band-Limited Functions and the Sampling Theorem

Figure 5.9a shows $f(t) = a^{-2} \text{sinc}^2(t/a) = a^{-1} \text{sinc}(t/a) \times a^{-1} \text{sinc}(t/a)$. Expressing $f(t)$ as a product allows us to apply the convolution and scaling theorems (Appendix D) that immediately give the CT-FT result, $F(u) = [\text{rect} * \text{rect}](au) = \text{tri}(au)$. Ideally, this *triangle function* is exactly *band-limited* because values outside the support region $|u| < 1/a$ are exactly zero. The $|u| = 1/a$ frequency bound is the *Nyquist frequency* u_n.

Applying sampling operator $\mathcal{S}^\dagger \mathcal{S}$ in Figure 5.9b, we find $f_s(t)$ in Figure 5.9c. The DT-FT expression of (5.27) reminds us there will be summed copies of $F(u)$. If copies overlap, the spectrum will be distorted by *aliasing*. In our example, we can clearly define the bandwidth of $f(t)$ as $2u_n$. The *Nyquist sampling theorem* for real signals states that aliasing is avoided provided the sampling frequency u_s exceeds twice the Nyquist frequency u_n,

$$u_s \geq 2u_n. \qquad \text{Nyquist sampling criterion} \qquad (5.28)$$

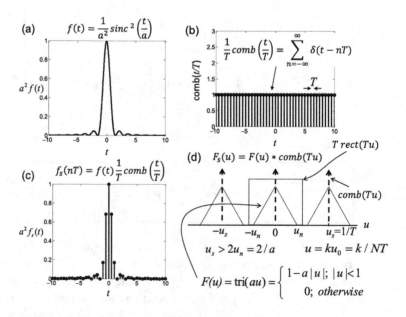

Figure 5.9 Continuous-time $f(t)$ (a) is multiplied by sampling operator $\mathcal{S}^\dagger \mathcal{S} = (1/T)\text{comb}(t/T)$ (b) to give $f_s(t)$ (c). (d) DT-FT $F_s(u) = \frac{1}{T}\sum_{k=-\infty}^{\infty}$ tri $(a(u - k/T))$. Recovery operator $\mathcal{R}$ restores the CT-FT result $F(u)$. In practice, imperfect band limits and acquisition noise prevent full recovery.

In this example, the Nyquist criterion is met when $u_s \geq 2/a$, and since $u_s = 1/T$ the sampling interval must be set at $T \leq a/2$. We can avoid aliasing if we sample the Fourier basis with the highest-frequency nonzero coefficient at least twice per wavelength.

Measurement data are not perfectly band-limited as in this example, so a clean result like $u_n = 1/a$ is rare. Realistic measurements are always finite in duration, and data with compact support in time extend to infinity in frequency. An *effective signal bandwidth* can be established that *sufficiently minimizes* sampling-related signal distortions. It makes sense to first select the sampling rate and then choose a low-pass filter at or below u_n.

In situations where (5.28) is strictly true, we can recover $F(u)$ from $F_s(u)$ by applying *recovery operator* $\mathcal{R}$. This operator simply multiplies $F_s(u)$ by a scaled rectangular function in the frequency domain:

$$F(u) = \mathcal{R} F_s(u) = T \operatorname{rect}(Tu) F_s(u). \tag{5.29}$$

The result is a frequency-domain version of an ideal *finite-impulse-response* (FIR) low-pass filter with amplitude T and corner frequency $1/2T$ (see Figure 5.9d). The units of $F(u)$ are the units of $f(t)$ times time, as expected for the CT-FT result.

If the sampling frequency greatly exceeds the Nyquist rate, $u_s \gg 2u_n$, the extra data do not reveal any new information about the object and performance may be diminished when noise is present. Performing some degree of oversampling or pre-sampling filtering is often the wise course to avoid noise aliasing. Sampling considerations are important in digital measurements ranging from music and video recordings to all types of medical imaging. Recent breakthroughs have shown that sub-Nyquist sampling can be recovered, as described in Section 6.8.

Protection against aliasing is found by low-pass filtering the signal *before sampling* the function. Realistic *anti-aliasing (AA) filters* distort acquisition $g(t)$ at frequencies near the band limits, but that distortion is often preferable to aliasing. The full DT-FT procedure, given in operator notation, applies the AA filter via $\mathcal{W}$, samples the result via $\mathcal{S}^\dagger \mathcal{S}$, applies DT-FT via $\mathcal{F}$, and recovers the CT-FT result via $\mathcal{R}$. That is, $\hat{G}(u) = \mathcal{R}\mathcal{F}\mathcal{S}^\dagger \mathcal{S}\mathcal{W} g(t)$. The result is a cascade of operators.

Example 5.7.4. *Digital anti-aliasing filters: Butterworth filters are examples of simple low-pass anti-aliasing filters. AA filters are generally applied in the time domain, but we can show their spectral responses using* MATLAB *as follows.*

```
%%%%%%%%%%%%%%%%%%%%%%%%%%%%%%%%%%%%%%%  Code for results in Fig. 5.10  %%%%%%%%%%%%%%%%%%%%%%%%%%%%%%%%%%%
[b,a]=butter(9,0.8);N=1001;                    % 9th-order LP Butterworth coeff w/cutoff at 0.8u_n
x=zeros(1,N);x(1)=1;                           % poor man's delta
w=filter(b,a,x);                               % filter x=delta to obtain w(t)
W=abs(fft(w));WdB=20*log10(W);                 % display |W(u)| in dB
u=1:N;                     % frequency axis extends from 0 to u_s.  We will plot from 0 to u_n=u_s/2
e=10^(-30/20)*randn(1001,10);E=mean(abs(fft(e))/sqrt(N),2);EdB=20*log10(E); % noise and frequency spectrum
HF=exp(-(u-200).^2/(2*65^2))+exp(-(u-800).^2/(2*65^2));       % FT(h*f)=HF for SNR=30 dB
HFdB=20*log10(HF);                             % signal spectrum G in dB
HFpEdB=20*log10(HF'+E);                        % signal + noise spectrum
plot(u(1:501),EdB(1:501));hold on;plot(u(1:501),HFdB(1:501))
plot(u(1:501),HFpEdB(1:501));hold off
WHFpEdB=20*log10(W'.*(HF'+E));                 % LP filter the noisy signal
figure;plot(u(1:501),WdB(1:501));hold on;plot(u(1:501),HFpEdB(1:501))
plot(u(1:501),WHFpEdB(1:501));hold off
```

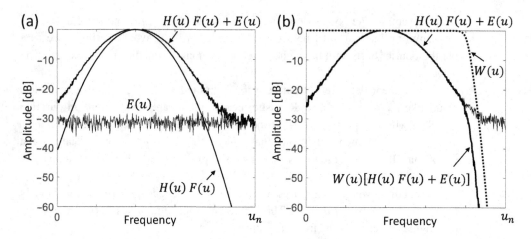

Figure 5.10 (a) We see a Gaussian-shaped signal spectrum $H(u)F(u)$, a white noise spectrum $E(u)$ at the peak-signal SNR of 30 dB, and the sum giving $G(u) = H(u)F(u) + E(u)$. The frequency axis is specified by 0 and Nyquist frequency u_n. (b) The frequency response of a ninth-order low-pass Butterworth filter $W(u)$ is shown with corner at $0.8u_n$ to avoid noise aliasing. This preserves the signal spectrum up to about $0.8u_n$, above which everything is severely attenuated.

Figure 5.10a shows a signal and noise spectra, and the sum of the two on a dB scale over positive frequencies up to the Nyquist frequency u_n. Figure 5.10b gives the frequency response of a ninth-order Butterworth low-pass anti-aliasing filter (dotted line), where every 6 dB is a factor of 2 loss of amplitude. By reducing any spectral amplitude below 60 dB (factor of 1,000) at frequencies as close to u_n as we can manage, we can feel confident that the white noise spectrum is not aliasing the measured spectrum. Selecting a lower-order filter is faster and less costly, but the frequency response is not as sharp.

5.8   Discrete Fourier Transform

Although DT-FTs are rarely used in practice, they are a great way to explain the effects of sampling. The *discrete Fourier transform* (DFT) is most frequently used and similar to DT-FT, so most of the work describing DFT is already done. DFT applies to data for which both the time and frequency axes are discretely sampled. Consequently, functions in both domains are periodic and the conjugate axes are entirely coupled. Sampled data in both domains are encountered whenever numerical methods are applied. FFT is a family of algorithms used to quickly compute DFTs.

We begin with the Fourier-series decomposition expression of (5.14),

$$G[k] = \frac{1}{T_0} \int_{-T_0/2}^{T_0/2} dt \, g(t) \, e^{-i2\pi kt/T_0}.$$

However, we have sampled data $g_s(t)$ and not $g(t)$. From (5.23),

$$G[k] = \frac{1}{T_0} \int_{-T_0/2}^{T_0/2} dt\, g_s(t)\, e^{-i2\pi k u_0 t}$$

$$= \frac{T}{T_0} \sum_{-\infty}^{\infty} \int_{-T_0/2}^{T_0/2} dt\, g(t)\, e^{-i2\pi k u_0 t}\, \delta(t - nT)$$

$$= \frac{1}{N} \sum_{-\infty}^{\infty} g(nT)\, e^{-i2\pi k u_0 nT} = \frac{1}{N} \sum_{-\infty}^{\infty} g[n]\, e^{-i2\pi kn/N}.$$

We used the relation that the period T_0 over which $g(t)$ is repeated is related to the sampling interval T and the number of samples N by $T_0 = NT$. Also, $u_0 T = T/T_0 = 1/N$. Hence, the DFT analysis/synthesis equation pair is[9]

$$G[k] = \frac{1}{N} \sum_{n=0}^{N-1} g[n]\, e^{-i2\pi kn/N} \qquad \text{Forward DFT}$$

$$g[n] = \sum_{k=0}^{N-1} G[k]\, e^{i2\pi kn/N}. \qquad \text{Inverse DFT} \qquad (5.30)$$

Yes, with DFT, we are stuck with remembering the constant $1/N$ for the 1-D forward transformation. Now, however, the units of $g[n]$ and $G[k]$ are the same.

Example 5.8.1. *Comparing FS and FFT: Compute the Fourier series of $g(t)$ and compare it to the DFT (via FFT) result, where $g(t)$ is a mixture of cosines at frequencies 1, 10, and 25 Hz.*

```
%%%%%%%%%%%%%%%%%%%%%%%%%%%%%% Script associated with Fig. 5.11 %%%%%%%%%%%%%%%%%%%%%%%%%%%%%%%%%%
T=0.01;T0=10;t=0:0.01:10;                          % 10 s time axis, generate three sinusoids
g1=1.5*cos(2*pi*1*t);g2=3.0*cos(2*pi*10*t);g3=4.5*cos(2*pi*25*t);
g=g1+g2+g3;                                        % noise-free waveform to be analyzed
for k=0:300;                                       % basis set up to 30Hz
  C(:,k+1)=cos(2*pi*k*t/10)';
end
A=2/10*g*C*0.01;                                   % cosine transform coefficients
u=0:0.1:30;                                        % frequency axis
subplot(3,1,1);plot(t(1:501),g(1:501));subplot(3,1,2);stem(u,A);
G=fft(g);                                          % DFT result via fft
subplot(3,1,3);stem(u,2.0/10*abs(G(1:301))*0.01)   % plot magnitude
```

The Fourier series coefficients in Figure 5.11 precisely represent the correct amplitudes at the exact frequencies of the three cosine components. This solution is able to provide an accurate decomposition because the function $g(t)$ is exactly periodic. As we vary the length of $g(t)$ via T_0 in Figure 5.11, we change the frequencies of the basis functions. The FFT algorithm requires time series $g[n]$ to have N equal to a power of 2. Since $g[n]$ has only $N = 1,001$ points, the FFT algorithm appends $g[n]$ with 23 zeros, making it $N = 1,024 = 2^{10}$ points. The padded function is an approximation of

---

[9] Several different notations are used for discrete transforms. For example, consider Section 2.7 in [69]. The important thing is to adopt one notation and use it consistently.

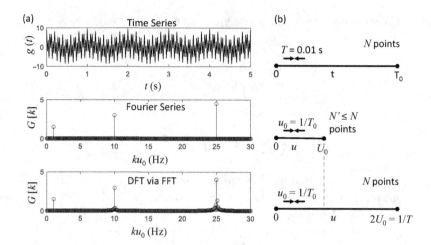

Figure 5.11 (a) The Fourier series and DFT (via FFT) of waveform $g(t)$ are compared. $g(t)$ is the sum of sinusoids at 1, 10, and 25 Hz. Here $T = 10$ ms, $T_0 = 10$ s, and $N = T_0/T + 1 = 1{,}001$, but only the first 501 time points are plotted. (b) A diagram of the time-axis parameters that shows how they determine the frequency-axis parameters for the FS and DFT analyses.

the original, so the FFT result in Figure 5.11 differs slightly from the FS result. Instead of generating only deltas at 1, 10, and 25 Hz as FS analsyis achieves, the deltas from the FFT are convolved with sinc functions.

5.8.1   Time Axis, Frequency Axis, and the Frequency Spectrum

MATLAB knows nothing of units, so it is your responsibility at the beginning of the code to establish units and then be consistent in their use. For time series, the master clock is set by selecting sampling interval T and the total duration T_0. Then $t = nT$, the number of points is $N = T_0/T$,[10] and frequency-axis parameters are

$$u[1] - u[0] = u_0 = 1/T_0 \qquad 2U_0 = u[N] = 1/T.$$

The parameter relationships are illustrated in Figure 5.11, where $u_0 = \Delta u$ or du is the numerical frequency increment; $2U_0$ is the maximum frequency, also equal to the sampling frequency u_s; and U_0 is the Nyquist frequency u_n. Everything is determined by your selection of T and T_0 and the units you assign to those parameters.

For example, to set up a 10-unit time axis sampled at 0.01 units, the following line of code determines the frequency axis, here selected to be $-U_0 \le u \le U_0$ instead of $0 \le u \le 2U_0$:

```
T=0.01;T0=10;t=0:T:T0;N=length(t);du=1/T0;U0=1/(2*T);u=-U0:du:U0;
```

[10] Often $N = T_0/T + 1$, depending on how points are counted.

Since the transformed values are generally complex values, it is common practice to plot the absolute value of the transform, $|G[k]| = \sqrt{G^*[k]\,G[k]}$. $|G[k]|$ is the *frequency spectrum*, while $|G[k]|^2$ is proportional to the *power spectral density* (see Section 5.12). Either may be presented as the "spectrum."

The advantage of the fft algorithm is its reduction of computational redundancies as compared with a straightforward implementation of (5.30). Specifically, FFT reduces the number of operations from N^2 to $N \ln N$, which provides substantial time savings for vectors with large N. For an $N = 1,024$ point FFT, the reduction in computation time is nearly a factor of 150. Standard texts on this topic provide details [42]. Let's apply the code in an example.

Example 5.8.2. *Where is zero? The computational example in Figure 5.12 shows that the periodic nature of signals in both domains requires a careful adjustment of signal origin and phase. In Figure 5.12b, the signal is symmetric about $t = 0$ s. Consequently, $g(t) = \text{rect}(t/2)$ transforms to $G[k] = 2\text{sinc}(2u)$. In Figure 5.12a, however, the signal is centered at $t = 5$ s. Now, $g(t) = \text{rect}((t-5)/2)$ transforms to $G[k] = 2\text{sinc}(2u)e^{-i2\pi 5u}$ as given by the shift theorem. However, the frequency and power spectra of both functions are equal because the phase factor is eliminated.*

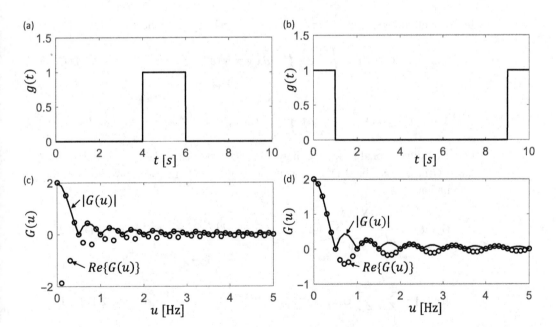

Figure 5.12 (a) Plot of $g(t) = \text{rect}[(t-5)/2]$. (c) The real part of fft (g) ∗dt, which is analytically $2\exp(-i2\pi5u)\text{sinc}(2u) = 2\cos(2\pi5u)\text{sinc}(2u)$ (plotted as points) and the frequency spectrum $2|\text{sinc}(2u)|$ (line). (b) Plot of $\text{rect}(t/2)$. (d) The real part, $2\text{sinc}(2u)$, and the magnitude of its Fourier transform, $2|\text{sinc}(2u)|$ (line). Note that $G(u)$ is now real. The comparisons demonstrate how delaying the signal in time produces a complex exponential factor in the Fourier transform.

```
%%%%%%%%%%%%%%%%%%%%%%%%%%%%%%%% Script to generate Fig. 5.12 %%%%%%%%%%%%%%%%%%%%%%%%%%%%%%%%
dt=0.01;T0=10;t=0:dt:T0;                        % set time axis...
du=1/T0;U0=1/(2*dt);u=0:du:U0;                  % ...to find frequency axis
g=zeros(size(t));g(401:601)=1;                  % 2 s long rect function centered at 5 s
G=fft(g)*dt;                                    % transform the result and scale with dt
subplot(2,1,1);plot(t,g)                        % plot time function
subplot(2,1,2);plot(u(1:51),abs(G(1:51)));      % plot first 51 pts of frequency spectrum
hold on;plot(u(1:51),real(G(1:51)),'o');        % plot first 51 pts of real part of transform
g=zeros(size(t));                               % reset the variable space
g(1:101)=1;g(901:1001)=1;                        % same rect function but now centered at 0 s
G=fft(g)*dt;                                    % transform this result
figure;subplot(2,1,1);plot(t,g)                 % plot the time function
subplot(2,1,2);plot(u(1:51),abs(G(1:51)));      % plot frequency spectrum
hold on;plot(u(1:51),real(G(1:51)),'o');        % plot real part of transform
```

As you compute CT-FTs analytically (on paper) and DFTs computationally (via FFT in MATLAB), you will notice that each requires careful attention to get the details correct, such as the scaling factors. Figure 5.13 summarizes the four types of Fourier analysis methods. Each is also discussed in classic texts such as [77]. The form of the equations used depends on the form of the data to be analyzed: The two considerations are discrete versus continuous and periodic versus nonperiodic.

5.9   Fourier Analysis of 2-D LTI/LSI Measurement Systems

In two continuous spatial dimensions (x, y used in imaging applications), the FT pair is

$$\mathcal{F} g(x,y) = G(u,v) = \int_{-\infty}^{\infty} dy \int_{-\infty}^{\infty} dx\, g(x,y)\, e^{-i2\pi(ux+vy)} \quad \text{Forward 2-D CT-FT}$$

(5.31)

$$\mathcal{F}^{-1} G(u,v) = g(x,y) = \int_{-\infty}^{\infty} dv \int_{-\infty}^{\infty} du\, G(u,v)\, e^{i2\pi(ux+vy)}. \quad \text{Inverse 2-D CT-FT}$$

For multidimensional data, we may use $g(\mathbf{x})$ to mean $g(x, y, \ldots)$ and $G(\mathbf{u})$ to mean $G(u, v, \ldots)$. The 2-D–DFT pair yielding $N \times N$ matrices involves separable summations:

$$G[k,\ell] = \frac{1}{MN} \sum_{m=0}^{N-1} \sum_{n=0}^{N-1} g[n,m]\, e^{-i2\pi(kn+\ell m)/N} \quad \text{Forward 2-D DFT}$$

(5.32)

$$g[n,m] = \sum_{\ell=0}^{N-1} \sum_{k=0}^{N-1} G[k,\ell]\, e^{i2\pi(kn+\ell m)/N}. \quad \text{Inverse 2-D DFT}$$

The conjugate variables to the continuous spatial variable x, y are the continuous spatial-frequency variables u, v. Conjugate to the discrete spatial indices n, m are the discrete frequency indices k, ℓ. As in the 1-D cases, continuous-to-continuous transforms are nonperiodic functions in both domains, while discrete-to-discrete transforms are assumed to be periodic in both domains.

Fourier Series:

Continuous time (periodic) ⟷ Discrete frequency (nonperiodic)

$$g(t) = \sum_{k=-\infty}^{\infty} G[k]\exp(i2\pi\, kt/T_0) \qquad G[k] = \frac{1}{T_0}\int_0^{T_0} dt\; g(t)\exp(-i2\pi\, kt/T_0)$$

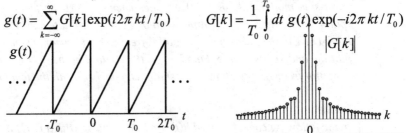

Discrete-Time Fourier Transform:

Discrete time (nonperiodic) ⟷ Continuous frequency (periodic)

$$g_s(nT) = \int_{-\infty}^{\infty} du \sum_{k=-\infty}^{\infty} G(u-k/T)\, e^{i2\pi unT} \qquad G_s(u) = \frac{1}{T}\sum_{k=-\infty}^{\infty} G(u-k/T)$$

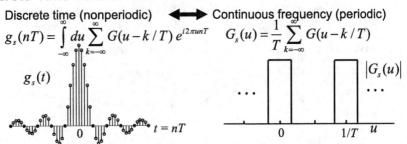

Continuous-Time Fourier Transform:

Continuous time (nonperiodic) ⟷ Continuous frequency (nonperiodic)

$$g(t) = \int_{-\infty}^{\infty} du\, G(u)\exp(i2\pi ut) \qquad G(u) = \int_{-\infty}^{\infty} dt\; g(t)\exp(-i2\pi ut)$$

$$g[n] = \sum_{k=0}^{N-1} G[k]\exp(i2\pi kn/N)$$

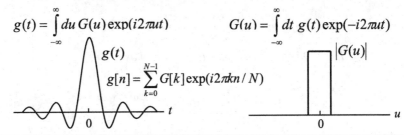

Discrete Fourier Transform:

Discrete time (periodic) ⟷ Discrete frequency (periodic)

$$g[n] = \sum_{k=0}^{N-1} G[k]\exp(i2\pi kn/N) \qquad G[k] = \frac{1}{N}\sum_{n=0}^{N-1} g[n]\exp(-i2\pi kn/N)$$

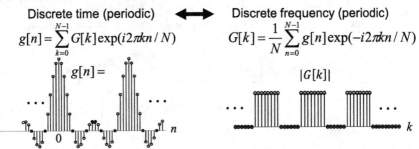

Figure 5.13 Fourier pairs resulting from four different data types.

Example 5.9.1. *Spectrum of 2-D Gabor pulse: Generate an 8-MHz Gabor pulse (Gaussian-modulated sinusoid) in time. Then convert the temporal pulse to a 2-D spatial pulse where waves propagate along one coordinate axis. Plot the magnitude of the 2-D FT of the pulse in a 2-D spatial-frequency spectrum. Do not use FFT; instead, plot the analytical results. Finally, find the column in the 2-D FT and convert it to a temporal frequency to show the result in comparison with the temporal pulse profile.*

Solution

In the following code, I formed h(t), an 8-MHz pulse with a FWHM bandwidth of 3.62 MHz (3.62/8.0 ≃ 45% fractional bandwidth). See Figure 5.14. Next, we find the 2-D spatial pulse h(x, y) that is Gaussian in both directions, but twice as wide as tall; i.e., the y-axis bandwidth is about 45%/2 = 23% of the pulse frequency (about 1.8 MHz).

```
%%%%%%%%%%%%%%%    Script to generate results in Fig. 5.14    %%%%%%%%%%%%%%%
T0=2;T=0.01;t=0:T:T0;f0=8;s1=0.1;                                 % initialize in time
ht=exp(-(t-T0/2).^2/(2*s1^2))/(s1*sqrt(2*pi)).*sin(2*pi*f0*t);    % temporal pulse
subplot(1,4,1);plot(t,ht,'k','linewidth',2);
x=t*0.77;u0=f0/0.77;sx1=s1*0.77;sx2=2*sx1;h=zeros(length(x));     % initialize space
h1=exp(-(x-T0*0.77/2).^2/(2*sx1^2))/(sx1*sqrt(2*pi)).*sin(2*pi*u0*x);
for n=1:length(x)                                                 % 2-D spatial pulse
    h(:,n)=h1*exp(-((x(n)-T0*0.77/2)^2)/(2*sx2^2))/(sx2*sqrt(2*pi));
end
subplot(1,4,2);imagesc(h);colormap gray;axis off
u=-1/(4*T):1/T0:1/(4*T);H=zeros(length(u));                       % initialize frequency axes
H1=0.5*(exp(-2*pi^2.*(u-u0).^2*sx1^2) - exp(-2*pi^2.*(u+u0).^2*sx1^2));  % analytic transform 1D
for k=1:length(u)
    H(:,k)=H1*exp(-2*pi^2*u(k)^2*sx2^2);                          % again in 2D
end
subplot(1,4,4);imagesc(abs(H));colormap gray;axis off
subplot(1,4,3);plot(u*0.77,abs(H(:,51)),'k','linewidth',2)
```

The 2-D spatial pulse may be expressed as

$$h(x, y) = \frac{1}{2\pi\sigma_1\sigma_2} e^{((x-x_0)^2/2\sigma_1^2 + (y-x_0)^2/2\sigma_2^2)} \sin(2\pi u_0 x_1).$$

It is shifted in x and y to the center of the square field using x_0, which is why x_0 is the shift for both axes. Applying theorems, we can quickly compute the 2-D transform of h(x, y). Because a 2-D Gaussian function is separable in x and y,

$$|H(u, v)| = \left| \mathcal{F}\left\{ \frac{1}{2\pi\sigma_1\sigma_2} e^{((x-x_0)^2/2\sigma_1^2 + (y-x_0)^2/2\sigma_2^2)} \right\} * \mathcal{F}\{\sin(2\pi u_0 x_1)\} \right|$$

$$= \left| \left\{ e^{-i2\pi x_0(u+v)} e^{2\pi^2(u^2\sigma_1^2 + v^2\sigma_2^2)} \right\} * \left\{ \frac{i}{2}(\delta(u+u_0) - \delta(u-u_0)) \right\} \right|$$

$$= \frac{1}{2} \left[e^{2\pi^2((u+u_0)^2\sigma_1^2)} - e^{2\pi^2((u-u_0)^2\sigma_1^2)} \right] e^{2\pi^2 v^2\sigma_2^2}. \qquad (5.33)$$

The 2-D frequency spectrum is illustrated in Figure 5.14d. The center column at |H(u, 0)| is plotted in Figure 5.14c after scaling the spatial-frequency axis to become

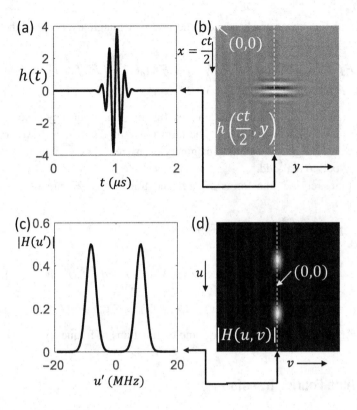

Figure 5.14 An 8-MHz/45% fractional-BW Gabor pulse is plotted in time in (a). The full 2-D pulse is imaged in spatial coordinates x, y in (b). The space-time relation in the direction of pulse propagation is $x = ct/2$. The magnitude of the 1-D FT of (a) is plotted versus temporal frequency u' in (c). The magnitude of the 2-D FT of (b) is imaged in (d), where the axes are spatial frequencies, u, v. The transforms are computed analytically.

temporal frequency. This example is intended to show how theorems offer much intuition about the properties of 1-D and 2-D functions in time, space, and frequency. Being able to visualize equations is very important for understanding signals and their spectra.

5.10 Even–Odd Real–Imaginary Symmetries

Let $f_e(x)$ and $f_o(x)$ each be complex functions that are, respectively, the even and odd parts of a general function,

$$f(x) = f_e(x) + f_o(x) = \Re f_e(x) + i\Im f_e(x) + \Re f_o(x) + i\Im f_o(x),$$

where $\Re$ and $\Im$, respectively, denote the real and imaginary parts of the complex functions. Then

$$\mathcal{F}f(x) = \mathcal{F}\Re f_e(x) + i\mathcal{F}\Im f_e(x) + \mathcal{F}\Re f_o(x) + i\mathcal{F}\Im f_o(x)$$
$$F(u) = \Re F_e(u) + i\Im F_e(u) + i\Im F_o(u) + \Re F_o(u) = F_e(u) + F_o(u),$$

where $F_e(u)$ and $F_o(u)$ are both complex functions of frequency. The second line matches the transforms in the first line term for term. That is, real-even transforms to real-even; imaginary-even to imaginary-even; real-odd to imaginary-odd; and imaginary-odd to real-odd.

Sometimes cosine transforms $\mathcal{F}_c$ or sine transforms $\mathcal{F}_s$ are used. These are defined as

$$F(u) = \int_{-\infty}^{\infty} dx \ f(x) e^{-2i\pi ux}$$

$$= 2\int_{0}^{\infty} dx \ f_e(x)\cos(2\pi ux) - i2\int_{0}^{\infty} dx \ f_o(x)\sin(2\pi ux)$$

$$= \mathcal{F}f(x) = \mathcal{F}_c f_e(x) - i\mathcal{F}_s f_o(x) = F_e(u) + F_o(u).$$

Again, $F_e(u)$ and $F_o(u)$ can both be complex functions of frequency.

5.11   Short-Time Fourier Transform

Functions adapted to model stochastic objects and noisy recorded data regularly violate assumptions required of Fourier analysis. For example, measured signals all have finite duration, are nonperiodic, and often result from time/shift-varying and nonlinear device responses. We ignore these and other assumptions at some peril, since the spectra generated may not be physically representative, leading to bad medical decisions. We must use Fourier analysis with full knowledge of the approximations being made.

If nonstationary and shift-variant signals can be considered stationary and shift-invariant *locally*, over a small region of space-time, one remedy is to simply segment the data to study local properties despite the effects of segmentation on the Fourier coefficients. This section examines the effects of transforming a waveform segment.

Consider Fourier analysis of a time-series segment with duration $2T_0$,[11]

$$\hat{F}(u; T_0) = \int_{-\infty}^{\infty} dt \ f(t) \, \text{rect}\left(\frac{t}{2T_0}\right) e^{-i2\pi ut} = \int_{-T_0}^{T_0} dt \ f(t) e^{-i2\pi ut}. \quad \text{(ST-FT)}$$

---

[11] We write $\hat{F}(u; T_0)$ to show that T_0 parameterizes ST-FT, even though $2T_0$ is the segment length.

This nonstandard form is called the *short-time Fourier transform (ST-FT)*. The estimate changes as T_0 increases, which we indicate using the parameterized approximate symbol, $\hat{F}(u; T_0)$. To find $F(u) = \lim_{T_0 \to \infty} A(T_0) \hat{F}(u; T_0)$, we seek scaling factor $A(T_0)$ that enables convergence.

First, note that signal energy, not amplitude, is the quantity averaged in stochastic phenomena. This is the reason that variances – and not standard deviations – are averaged. Since $\hat{F}(u; T_0)$ are amplitude coefficients, selecting $A(T_0) = 1/\sqrt{2T_0}$ achieves the goal of scaling the energy by $1/2T_0$, yielding

$$F(u) = \lim_{T_0 \to \infty} \frac{1}{\sqrt{2T_0}} \hat{F}(u; T_0) = \lim_{T_0 \to \infty} \frac{1}{\sqrt{2T_0}} \int_{-T_0}^{T_0} dt\ f(t) e^{-i2\pi u t}. \qquad (5.34)$$

This discussion assumes use of CT-FT notation, although these ideas generalize to other FT forms.

5.12 Power Spectral Density

5.12.1 The Sample Spectrum

We are rarely interested in the *phase* of the stochastic coefficients $F(u)$. However, we are frequently interested in estimates of the *signal energy or power* contained within each frequency channel. The reason is that variance is a fundamental property of stochastic data, related to signal energy, and the power spectral density function is the frequency spectrum of sample variance for zero-mean samples.

We can estimate the *power spectral density* of stochastic function f as follows:

$$\hat{S}_f(u; T_0) = \left[\frac{1}{\sqrt{2T_0}} \hat{F}(u; T_0) \right] \left[\frac{1}{\sqrt{2T_0}} \hat{F}(u; T_0) \right]^* = \frac{1}{2T_0} |\hat{F}(u; T_0)|^2.$$

(PSD estimate)   (5.35)

In the limit, but only for stationary f,

$$\hat{S}_f(u) = \lim_{T_0 \to \infty} \hat{S}_f(u; T_0) = \lim_{T_0 \to \infty} \frac{1}{2T_0} |\hat{F}(u; T_0)|^2,$$

and we are able to eliminate the parameterization. The term "power" comes from considering voltage signals, whose square normalized by impedance is [power] or [energy/time]. Even when the units of $f^2(t)$ are not energy, the concept can be generalized to those cases. $\hat{S}_f$ has units of [power/frequency], or signal power within each frequency channel (i.e., power density). Since [power/frequency] = [energy], $\hat{S}_f$ is referred to as the *sample power spectral density* or the *sample energy spectrum*. Integrating both sides of (5.35) over frequency before taking the limit gives

$$\int_{-\infty}^{\infty} du \, \hat{S}_f(u; T_0) = \frac{1}{2T_0} \int_{-\infty}^{\infty} du \, \hat{F}(u; T_0) \hat{F}^*(u; T_0)$$

$$= \frac{1}{2T_0} \int_{-\infty}^{\infty} du \, \hat{F}(u; T_0) \int_{-T_0}^{T_0} dt \, f(t) e^{i2\pi ut}$$

$$= \frac{1}{2T_0} \int_{-T_0}^{T_0} dt \, f(t) \int_{-\infty}^{\infty} du \, \hat{F}(u; T_0) e^{i2\pi ut}$$

$$= \frac{1}{2T_0} \int_{-T_0}^{T_0} dt \, f^2(t) = \langle f^2(t) \rangle. \tag{5.36}$$

The last form is the temporal sample mean-squared value of real f. If f is *zero mean*, we find from (5.36) that

$$\mathrm{v\hat{a}r}(f) = \langle (f(t) - \mathcal{E}f(t))^2 \rangle = \langle f^2(t) \rangle = \int_{-\infty}^{\infty} du \, \hat{S}_f(u; T_0). \tag{5.37}$$

The "hat" in the sample spectrum $\hat{S}(u; T_0)$ indicates a temporally or spatially averaged quantity, distinguishing it from the ensemble statistic discussed in the next subsection.

5.12.2 The Ensemble Spectrum (Wiener–Khintchine Theorem)

The preceding discussion combined Fourier analysis with sample statistics. We now examine a population statistic for wide-sense stationary (WSS) processes that may not be ergodic, resulting in the *Wiener–Khintchine theorem*. The W-K theorem states that the *ensemble spectrum* $S(u)$ is the Fourier transform of the *ensemble autocorrelation function*, $\phi_f(t)$:

$$S_f(u) = \int_{-\infty}^{\infty} dt \, \phi_f(t) e^{-i2\pi ut}. \tag{5.38}$$

To verify this result, first take the ensemble average of the sample spectrum. From (5.35) for the linear expectation operator $\mathcal{E}$ and real WSS f, we have

$$S_f(u) = \mathcal{E}\hat{S}_f(u) = \lim_{T_0 \to \infty} \frac{1}{2T_0} \mathcal{E}\left\{ |\hat{F}(u; T_0)|^2 \right\}$$

$$= \lim_{T_0 \to \infty} \frac{1}{2T_0} \mathcal{E}\left\{ \int_{-T_0}^{T} dt' \int_{-T_0}^{T_0} dt \, f(t) f(t') e^{-i2\pi u(t-t')} \right\}$$

$$= \lim_{T \to \infty} \frac{1}{2T_0} \int_{-T_0}^{T_0} dt' \int_{-T_0}^{T_0} dt \, \mathcal{E}\left\{ f(t) f(t') \right\} e^{-i2\pi u(t-t')}$$

$$= \lim_{T \to \infty} \frac{1}{2T_0} \int_{-T_0}^{T_0} dt' \int_{-T_0}^{T_0} dt \, \phi_f(t - t') e^{-i2\pi u(t-t')}. \tag{5.39}$$

To move from the third to the fourth line of (5.39), we assumed f is WSS. Following a straightforward geometric argument, it can be shown [13] that

$$\frac{1}{2T_0} \int_{-T_0}^{T_0} dt' \int_{-T_0}^{T_0} dt \, \phi_f(t - t') e^{-i2\pi u(t-t')} = \frac{1}{2} \int_{-T_0}^{T_0} dt' \left(1 - \frac{|t|}{T_0}\right) \phi_f(t') e^{-i2\pi u t'}$$

$$= \int_{-\infty}^{\infty} dt' \phi_f(t'; T_0) e^{-i2\pi u t'}. \quad (5.40)$$

Since $|e^{-i2\pi u t'}| = 1$, we can see that $\phi_f(t'; T_0) e^{-i2\pi u t'} \leq \phi_f(t')$, where $\phi_f(t')$ is defined over all time. Given this condition, we can deploy the Lebesgue dominated convergence theorem [11], which states that for the n elements in f_n,

$$\lim_{n \to \infty} \left(\int_B dx f_n \right) = \int_B dx \left(\lim_{n \to \infty} f_n \right) = \int_B dx f$$

provided $|f_n| \leq |f|$. Combining this theorem with (5.39) and (5.40), we find

$$S_f(u) = \lim_{T_0 \to \infty} \frac{1}{2T_0} \int_{-T_0}^{T_0} dt' \int_{-T_0}^{T_0} dt \, \phi_f(t - t') e^{-i2\pi u(t-t')}$$

$$= \lim_{T_0 \to \infty} \int_{-\infty}^{\infty} dt' \, \phi_f(t'; T_0) e^{-i2\pi u t'}$$

$$= \int_{-\infty}^{\infty} dt' \lim_{T_0 \to \infty} \phi_f(t'; T_0) e^{-i2\pi u t'}$$

$$= \int_{-\infty}^{\infty} dt' \, \phi_f(t') e^{-i2\pi u t'}, \quad (5.41)$$

which proves (5.38) and yields the Fourier transform pair

$$S_f(u) = \int_{-\infty}^{\infty} dt \, \phi_f(t) e^{-i2\pi u t} \qquad \text{(W-K theorem)}$$

$$\phi_f(t) = \int_{-\infty}^{\infty} du \, S_f(u) e^{i2\pi u t}. \qquad (5.42)$$

Similar to the results for sample statistics that we saw in (5.37), the result for population statistics is

$$\text{var}(f) = \phi_f(0) = \int_{-\infty}^{\infty} du \, S_f(u) \qquad (5.43)$$

for zero-mean f.

Note on Units. If $g(t)$ has units of [data], then the autocorrelation of $g(t)$, ϕ_g, has units of $[\text{data}^2]$ and [via the W-K theorem, (5.38)] the power spectral density S_g has units of $[\text{data}^2\text{-time}]$. The units are the same if we use the short-time Fourier transform from (5.35). However, if $g(x, y)$ [data], then S_g $[\text{data}^2\text{-area}]$. In summary, as the independent variable and its dimensionality change, so do the units of $S(u)$.

5.13   Passing Random Processes through Linear Systems

In this section, we revisit the effects of LTI transformations on statistical properties from Chapter 3, including spectral densities. For example, if we make measurements of a known WSS function using a linear measurement device with known properties, what are the statistical properties of the data?

Beginning with the acquisition equation, $g(t) = \mathcal{H}f(t') = [h * f](t) + e(t)$, we assume a deterministic, continuous-to-continuous LTI transformation of a WSS object from object space to data space via h. Furthermore, we assume a stochastic WSS object function f and additive WSS iid noise e.

5.13.1   An LTI Transformation of a WSS Process Is WSS

To show that a WSS process remains WSS after passing through a linear-time invariant system, we need to examine the effects of a deterministic LTI system transformation on first- and second-order statistics of the WSS process $f(t')$. Initially, we assume $\mathcal{H}$ contributes no acquisition noise to the measurement; i.e., $g(t) = \mathcal{H}\{f(t')\} = [h * f](t)$.

As an example of first-order statistics, the ensemble mean of the data is

$$\mathcal{E}g(t) = \mathcal{E}\mathcal{H}f(t') = \mathcal{E}\left\{\int_{-\infty}^{\infty} dt' \, h(t') \, f(t - t')\right\}.$$

Since $\mathcal{E}$ is a linear operator,

$$= \int_{-\infty}^{\infty} dt' \, h(t') \, \mathcal{E}f(t - t').$$

If f is WSS, $\mathcal{E}f(t - t') = \mathcal{E}f(t') = \mathcal{E}f$ is time invariant, so

$$= \mathcal{E}f \int_{-\infty}^{\infty} dt' \, h(t') = H(0) \, \mathcal{E}f(t')$$

$$= \mathcal{H}\mathcal{E}f(t'). \tag{5.44}$$

We see that *the acquisition and ensemble operators commute for LTI h and WSS f*, and that $\mathcal{E}g(t)$ is time invariant. Note that $H(0)$ is the Fourier coefficient at zero frequency.

As an example of second-order statistics, consider the autocorrelation function $\phi_{ff}(t_1', t_2') = \mathcal{E}\{f^*(t_1') \, f(t_2')\}$ from (3.40). For WSS f, the autocorrelation does not depend on absolute times t_1' or t_2', but only on relative time $t_1' - t_2' = \tau'$. Therefore, $\phi_{ff}(t_1', t_2') = \phi_{ff}(\tau')$. Applying the acquisition operator to $f(t_1')$ in $\phi_{ff}(t_1', t_2')$, we find [80],

$$\phi_{gf}(t_1, t_2') = \int_{-\infty}^{\infty} dt_1' \, \phi_{ff}(t_1', t_2') \, h(t_1 - t_1') = \mathcal{H}_1 \, \phi_{ff}(t_1', t_2').$$

Notice how acquisition transforms the data from the t_1' to the t_1 axis. Repeating the process for t_2' gives

$$\phi_{gg}(t_1, t_2) = \int_{-\infty}^{\infty} dt_2' \int_{-\infty}^{\infty} dt_1' \, \phi_{ff}(t_1', t_2') \, h(t_1 - t_2') \, h(t_2 - t_2')$$

$$= \mathcal{H}_2 \mathcal{H}_1 \, \phi_{ff}(t_1', t_2') = \mathcal{H}_2 \mathcal{H}_1 \, \phi_{ff}(\tau') = \phi_{gg}(\tau),$$

where $\tau = t_1 - t_2$. Because first- and second-order statistics are time invariant, applying an LTI operator to a WSS random process generates another WSS random process.

5.13.2   Averaging over Different Sources of Variability

Let's revisit (5.44), and now assume the acquisition operator includes *signal-independent additive noise*; $g(t) = \mathcal{H} f(t') = \int_{-\infty}^{\infty} dt' \, h(t - t') f(t') + e(t)$. In this case, computing $\mathcal{E} g(t)$ asks us to average over both f and e. We do this by taking the mean over noise e for a fixed realization of f; i.e., we apply the *conditional expectation operator* $\mathcal{E}_{e|f}$ [9]. Only then do we average over f via $\mathcal{E}_f$:

$$\mathcal{E}\{g(t)\} \triangleq \mathcal{E}_f \left\{ \mathcal{E}_{e|f} \left\{ \int_{-\infty}^{\infty} dt' \, h(t') \, f(t - t') + e(t) \right\} \right\}$$

$$= \mathcal{E}_f \left\{ \int_{-\infty}^{\infty} dt' \, h(t') \, f(t - t') + \mathcal{E}_{e|f} e(t) \right\}$$

$$= \int_{\infty}^{\infty} dt' \, h(t') \mathcal{E}_f\{f(t - t')\} + \mathcal{E}_e\{e(t)\},$$

and if f is WSS,

$$\mathcal{E}\{g(t)\} = \mathcal{E}_f\{f(t)\} \int_0^{\infty} dt' \, h(t') + \mathcal{E}_e\{e(t)\} = H(0) \, \mathcal{E}_f\{f(t)\} + \mathcal{E}_e\{e(t)\}. \quad (5.45)$$

Equation (5.45) for signal-independent noise offers experimental opportunities. Say noise process e is zero-mean normal and the signal process f is Poisson and therefore not zero mean. If the signal process can be fixed while acquiring many realizations of noise, the two random processes can be separated experimentally.

5.13.3   Autocorrelation and Power Spectral Density

For real WSS measurements $g(t)$ with zero-mean *signal-independent* noise $e(t)$,

$$\phi_g(t_1, t_2) = \phi_g(t', t' - t) = \mathcal{E}\{g(t') \, g(t' - t)\}$$

$$= \mathcal{E}\left\{ \left[\int_0^{\infty} da \, h(a) \, f(t' - a) + e(t') \right] \right.$$

$$\left. \times \left[\int_0^{\infty} db \, h(b) \, f(t' - t - b) + e(t' - t) \right] \right\}$$

$$= \int_0^\infty da \int_0^\infty db\, h(a)\, h(b)\, \mathcal{E}_f\{f(t'-a)\, f(t'-t-b)\}$$

$$+ \int_0^\infty da\, h(a)\, \mathcal{E}_{e|f}\{f(t'-a)\, e(t'-t)\}$$

$$+ \int_0^\infty db\, h(b)\, \mathcal{E}_{e|f}\{f(t'-t-b)\, e(t)\} + \mathcal{E}_e\{e(t')\, e(t'-t)\}$$

$$= \int_0^\infty da \int_0^\infty db\, h(a)\, h(b)\, \phi_f(t+b-a) + \phi_e(t)$$

$$= [h * \phi_f * h(-b)](t) + \phi_e(t). \tag{5.46}$$

The cross terms are zero because the noise and signal are statistically independent processes. The autocorrelation of the measurement equals the autocorrelation of the object, which is then twice convolved[12] with the impulse response of the measurement plus the autocorrelation of the noise. This isn't very transparent yet, so let's take the Fourier transform of the second-to-last line of (5.46) to find the power spectral density of measurement $g(t)$:

$$S_g(u) = \int_{-\infty}^\infty dt \int_0^\infty da \int_0^\infty db\, h(a)\, h(b)\, \phi_f(t+b-a)\, e^{-i2\pi ut}$$

$$+ \int_{-\infty}^\infty dt\, \phi_e(t)\, e^{-i2\pi ut}$$

$$= \int_0^\infty da\, h(a)\, e^{-i2\pi ua} \int_0^\infty db\, h(b)\, e^{i2\pi ub}$$

$$\times \int_{-\infty}^\infty dt\, \phi_f(t+b-a)\, e^{-i2\pi u(t+b-a)} + S_e(u).$$

$S_e(u)$ is the power spectral density of the noise process from (5.42). Note that changing the sign of the Fourier kernel on the integral over b is the same as taking the complex conjugate of the integral for real impulse response functions. So the components of the measured power spectrum are (if we avoid end effects)

$$S_g(u) = H(u)\, H^*(u)\, S_f(u) + S_e(u) = |H(u)|^2 S_f(u) + S_e(u). \tag{5.47}$$

Applying the Wiener–Khintchine theorem one more time, we find what might be a more transparent expression for the measured autocorrelation function than (5.46):

$$\phi_g(t') = \int_{-\infty}^\infty du\, |H(u)|^2\, S_f(u)\, e^{i2\pi ut'} + \phi_e(t'). \tag{5.48}$$

12 Equivalently, $\phi_f(t)$ is first convolved by $h(t)$ and then cross-correlated with $h(t)$.

Example 5.13.1. *Noise filtering: The following example is intended more for those readers with electrical engineering backgrounds. Nevertheless, most readers should be able to use the code provided to gain some intuition.*

Consider an additive signal-in-noise example, where $g(t) = Q(\cos(\Omega_0 t) + e(t))$. Here, the signal and noise amplitudes are the same and are given by scalar value Q. The goal is to filter the noise and then observe how the filter influences the statistical properties of the measurement signal.

We are unsure of the signal frequency at the time of measurement (the simulation actually applies $\Omega_0 = 2\pi \times 100$ Hz), but we know that $e \sim \mathcal{N}(0, 1)$ and the population noise power spectrum is "white," having equal values at frequencies over the measurement bandwidth. The device samples measured signals at 6 kHz and we're pretty sure the signal we are looking for has $\Omega_0/2\pi < 1$ kHz. Clearly, we have oversampled this process and the data are suffering from all the noise above 100 Hz, although we are unsure where to set the corner frequency for the low-pass filter since we don't know the signal frequency.

Let's examine both first- and second-order resistive-capacitive (RC) low-pass (LP) filters, which might be applied before sampling to reduce the broadband noise. The system response functions (Fourier transforms of the impulse responses) for these anti-aliasing and noise-reducing filters are

$$W_1(u) = \frac{1}{1 + i(\Omega/\Omega_1)} \quad \textit{First-order filter}$$

$$W_2(u) = \frac{1}{1 - A + iB} \quad \textit{Second-order filter,}$$

where A and B are functions of the resistor and capacitor values as well as frequency, as shown in Figure 5.15 in the circuit diagrams. Note that we use temporal frequency u rather radial frequency Ω in the plot. $\Omega_1 = 1/R_1 C_1$ is the corner frequency for

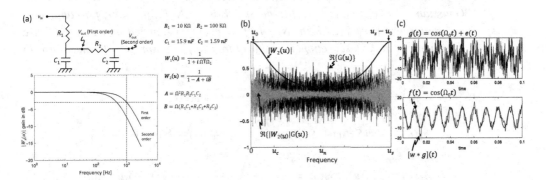

Figure 5.15 (a) Circuit diagrams for simple first- and second-order RC filters and their system response functions, $H_n(u)$ for $n = 1, 2$ where n indicates the order. (b) $|H_2(u)|$ is replotted on a linear scale with the real parts of transform $G(u)$ and the real part of the product $H_2(u)G(u)$ that is the filtered spectrum. Transforming back, we see (c) the cosine signal with and without noise and the filtered signal. Only a tenth of each time series is shown here.

the first-order filter at which signals are attenuated −3 dB. We show Bode plots of the filter gains; Bode plots are the magnitudes of the complex filter spectra plotted in dB, i.e., $20 \log |W_n(u)|$. Frequencies with 0 dB gain impose no change in amplitude; negative values denote attenuation. Selecting R_1 and C_1 so that $\Omega_1/2\pi = 1$ kHz, we set the filter gain at that frequency to −3 dB. The second-order filter adds another RC stage, where $R_1 C_1 = R_2 C_2$, such that the −3 dB point moves down to 0.5 kHz.

The real part of $G(u) = \mathcal{F}g(t)$ shows a white noise spectrum as well as a spike at $u_0 = 100$ Hz and another spike at $u_s − u_0 = 5{,}900$ Hz (note that the cyclic nature of the transforms means this latter frequency is equivalent to $−u_0 = −100$ Hz). Applying a second-order LP filter to this signal spectrum shows that high-frequency noise is suppressed, but the transition from the pass band to the stop band is very gradual, as shown by the filtered spectrum in gray. Transforming back, we can see the effect of filtering on the time series. The code to generate the data in Figure 5.15 follows:

```
%%%%%%%%%%%%%%%%%%  Code to generate Fig. 5.15 from Example 5.13.1  %%%%%%%%%%%%%%%%%%
R1=10e3;C1=15.9e-9;u=0:3000;Om=2*pi*u;Om1=1/(R1*C1);          % initiate parameters
W=1./(1+1i*Om/Om1);
W1=1./(1+(Om/Om1).^2);semilogx(u,10*log10(W1));hold on        % first-order filter
R2=100e3;C2=1.59e-9;A=Om*R1*R2*C1*C2;B=Om*(R1*C1+R1*C2+R2*C2);
W2=1./(-A+1i*B+1);W12=W2.*conj(W2);                          % second-order filter
semilogx(u,10*log10(W12));hold off
dt=1/6000;t=0:dt:1;
f=10*cos(2*pi*100*t);g=f+10*randn(size(t));G=fft(g)*dt;      % noisy signals
WW(1:3001)=W1;WW(3001:6001)=W1(end:-1:1);
WWW(1:3001)=W2;WWW(3001:6001)=W2(end:-1:1);
uu=0:6000;figure;plot(uu,WW,uu,2*real(G),uu,2*real(G.*WWW))
gw=ifft(G.*WWW)*6000;figure;subplot(2,1,1);plot(t(1:600),g(1:600))  % view filtered response
subplot(2,1,2);plot(t(1:600),real(gw(1:600)),t(1:600),f(1:600))
```

Frankly, we would not use these simple filters in practice because the transition zone from the pass band to the stop band is too gradual. They are useful here because we can analyze them mathematically to gain some intuition about the effects of filtering on data. For example, for $g(t) = Q(\cos(\Omega_0 t) + e(t))$ where we have white Gaussian noise with power spectral density $S_e(u) = N_0/2$, $|u| < \infty$ for continuous time series or $|u| < u_n$ (u_n is the Nyquist frequency) for sampled signals. We can use the equations in the last section, the table of transforms in Appendix D, and $W_1(u)$ to show that

$$g(t) = Q(f(t) + e(t)) = Q(\cos(2\pi u_0 t) + e(t))$$

$$S_g(u) = Q^2(S_f(u) + S_e(u)) = \frac{Q^2}{2}\left[\delta(u + u_0) + \delta(u - u_0) + N_0\right]$$

$$S_{w_1 * g}(u) = |W_1(u)|^2 S_g(u) = \frac{Q^2}{4(1 + (u/u_1)^2)}\left[\delta(u + u_0) + \delta(u - u_0) + N_0\right].$$

How did we obtain $S_f(u)$? Recall from Example 4.3.1 that if $f(t) = A\cos(2\pi u_0 t)$, then $\phi_f(t) = (A^2/2)\cos(2\pi u_0 t)$, a scaled version of the same sinusoid. Then, apply the Wiener–Khintchine theorem.

The most instructive part of this exercise is what happens to the white Gaussian (standard normal) noise. Since $S_e(u) = N_0/2$ (a constant), we can immediately write $\phi_e(t) = (N_0/2)\delta(t)$. Originally, the noise is completely uncorrelated in time. However, filtering the waveform via $[w_1 * g](t)$ reduces the high-frequency noise components, where there is no signal, and correlates the remaining noise samples, thereby creating "red" noise. To see this, we can compute the autocorrelation of the convolution $[w_1 * e](t)$, i.e., $\phi_{w_1*e}(t)$:

$$\phi_{w_1*e}(t) = \int_{-\infty}^{\infty} du \, |W_1(u)|^2 \, S_e(u) \, e^{i2\pi ut}$$

$$= \frac{N_0}{2} \int_{-\infty}^{\infty} du \, \frac{e^{i2\pi ut}}{1 + (u/u_1)^2}$$

$$= \frac{N_0 u_1}{4} e^{-u_1|t|}.$$

Since $2\pi u_1 = 1/(R_1 C_1)$, the amount of correlation among noise samples depends on the corner frequency of the filter, u_1. More closely correlated noise samples provide fewer degrees of freedom for time averaging and further noise suppression.

5.14 Examples from Research

This section explores two problems from research that apply statistical and Fourier analyses.

5.14.1 Estimating Structure in Liver Parenchyma

The first example is a research problem I studied while working with Bob Wagner and David Brown at the Food and Drug Administration and Brian Garra at the National Institutes of Health in 1985. Bob found a paper by Graham Sommer et al. [102] that suggested the average spacing between dominant acoustic scatterers in liver could be a strong diagnostic indicator of inflammation. We know that liver structure is fractal over a large scale range (the log-log histogram of structure sizes has a slope of approximately approximately 2.35 – the fractal dimension of liver tissue). Nevertheless, the *measurable* acoustic-scattering intensity of liver has a characteristic spacing. The hepatic portal triads scatter sound more than other structures because they are rich in collagen. The interaction between the ultrasound beam and tissue structures is illustrated in the insert to Figure 5.16. These structures present to the ultrasound beam as strong "periodic" reflectors spaced about 1 mm from each other in healthy adults [39]. Inflammation is a first reaction of liver to nonalcoholic steatohepatitis (NASH) and other infections that swell liver tissue and may lead to cancers if left untreated. We hoped a simple ultrasonic test could reliably and safely indicate liver inflammation from changes in portal-triad spacing. The solution was to apply the autocorrelation and power spectra techniques described earlier in this chapter.

Example 5.14.1. *Liver analysis:* *We begin by simulating idealized 1-D periodic struc-*
tures embedded in random structures for which echo acquisitions could be formed
and analyzed. The object function in Figure 5.16a includes 1-mm-spaced reflectors
surrounded by weakly scattering random structures over a length of 10 mm. An object
model allows us to model the object autocorrelation function $\hat{\varphi}_f$, which was not avail-
able experimentally. Figure 5.16b shows that $\hat{\varphi}_f$ reinforces the periodic refections
while suppressing echoes from randomly positioned structures. Object autocorrela-
tion, averaged over $L = 10$ lines of sight, is expressed analytically as a convolution
via (4.21),

$$\hat{\varphi}_f[n] = \frac{1}{L} \sum_{\ell=1}^{L} \left[(f * f[-n'])[n] \right]_\ell, \tag{5.49}$$

where

$$(f * f[-n'])[n] = \sum_{n=0}^{N-1} f[n'] f[n' + n].$$

Can you see the small peaks at 1 and 2 mm in Figure 5.16b where the periodic
reflectors accumulate reflected signal energy? The larger peak at the origin is where
both signal and noise contribute.

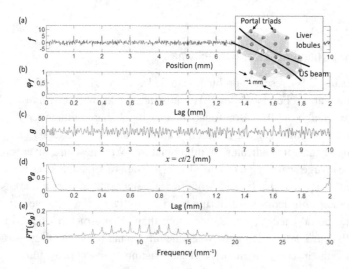

Figure 5.16 Simulation results showing ultrasonic measurements of scattering structures from
one line-of-sight in liver parenchyma (inset). (a) One realization of object function f is
generated by placing low-contrast spikes separated by 1 mm in a stochastic scattering field. (b)
Temporal autocorrelation for the object $\hat{\varphi}_f$ is shown. (c) The echo signal g is found by
convolving (a) with the impulse response of the measurement. The echo-signal autocorrelation
$\hat{\varphi}_g$ is in (d). Finally, $S_g[k] = \mathcal{F}\varphi_g[m]$ is plotted in (e).

Now pass the object through an acquisition operator, $g[m] = (h * f)[m] + e[m]$ at SNR = 32 dB, to generate the echo data in Figure 5.16c. The reflector contrast is too weak to observe the periodic structure unless we know where to look. However, the periodic structure becomes visible once the echo data are autocorrelated, as shown in Figure 5.16d. We now express the spatially averaged temporal correlation function as a convolution:

$$\hat{\varphi}_g[m] = \frac{1}{L} \sum_{\ell}^{L} \left[\left(g * g[-n] \right)[m] \right]_\ell.$$

Substituting the acquisition expression for g, we find

$$\hat{\varphi}_g[m] = \frac{1}{L} \sum_{\ell}^{L} \left[\left((f * h) * (f[-n] * h[-n]) \right)[m] \right]_\ell.$$

The parameters selected for this simulation suggest the measurement could be sensitive to 0.1-mm changes in periodicity. Based on this prediction, we began a clinical investigation.

Applying our knowledge of Fourier transforms, we can see how the autocorrelation of the object is related to that of the echo data:

$$\hat{\varphi}_g[m] = \mathcal{F}^{-1} \left\{ H(u) \left[\frac{1}{L} \sum_{\ell}^{L} \left(F(u)F^*(u) \right)_\ell \right] H^*(u) \right\}$$

$$= \mathcal{F}^{-1} \left\{ |H(u)|^2 \mathcal{F}\{\varphi_f[n]\} \right\}$$

$$= \left(h * \hat{\varphi}_f * h[-n] \right)[m]. \tag{5.50}$$

We wrote the result compactly, using the shorthand notation discussed in (4.21). The bottom line of (5.50) tells us that the echo autocorrelation of the object is blurred twice by convolutions with the ultrasound pulse (one is time-reversed). We needed to keep the pulse length relatively short to remain sensitive to small changes in average spacing.

After much clinical testing, the method was determined to be unreliable for diagnosis because the intrapatient variability among hepatic scatterer spacings was too large to give a detectable correlation peak for every patient. Oh, well, that happens in research. The points for analysis are as follows: (a) We can intermix frequency and spatiotemporal representations of data when doing so helps clarify the analysis; (b) operators such as ensemble averages must be interpreted in terms of experimentally realizable methods, and (c) the CT-FT notation is nice for understanding the approach, but eventually we must rewrite the equation in discrete form to implement methods experimentally.

The frequency domain offers a different view of the same information. Applying the Wiener–Khintchine theorem, we can view the echo-signal power spectrum,

$$\hat{S}_g(u; T_0) = \mathcal{F}\{\hat{\varphi}_g(t)\} = |H(u)|^2 \hat{S}_f(u; T_0) = \frac{1}{2T_0} |H(u)|^2 |F(u; T_0)|^2.$$

The result plotted in Figure 5.16e reveals a broad pulse spectrum centered at 10 mm^{-1} from $|H(u)|^2$. Embedded in the pulse spectrum are spikes at integer frequency intervals. The spikes are from $S_f(u)$, which we approximate using the short-time Fourier transform expression,

$$\hat{S}_f(u; T_0) = \frac{1}{2T_0} |F(u; T_0)|^2. \tag{5.51}$$

The power spectral density shows how the spectral response of the instrument provides a frequency-domain window into a periodic object spectrum.

One last point: It is straightforward to show that $\mathcal{F}f(-t) = F(-u)$ using the change of variable $w = -t$. Also, since h and f are real functions, each potentially with both even and odd components, we must assume F and H both have real-even and imaginary-odd transforms (see Section 5.10). For example, $F(u) = \Re F_e(u) + i\Im F_o(u)$ because $\Re F_o(u) = \Im F_e(u) = 0$. Consequently, F and H are Hermitian; viz., $F(-u) = F^*(u)$. That is the reason we could use $\mathcal{F}f(-t) = F^*(u)$ in the preceding analysis. Since physically realizable functions are all real, this is a common assumption.

5.14.2 Array Beamforming and Fourier Optics

All detectors have *apertures* that accept incoming radiation fields. The shape of the aperture partially determines the *directivity* response of that detector. Directivity describes the sensitivity of a detector to input energy from different directions. Similarly, the aperture of a radiation source determines the shape of the outgoing beam at a distance far from the source – the *far field* of the source. See Figure 5.17. Radiation wave fields that diffract have far-field beam pattern shapes that are determined by the Fourier transform of the aperture shape. This curious relationship is central to the design of many source-sensor devices and is referred to as *Fourier optics*. A wonderful text by Joseph Goodman [43] is entirely dedicated to the principles of Fourier optics.

Since the eighteenth century, during the time of Jean-Baptiste Joseph Fourier and his contemporaries Augustin-Jean Fresnel and Joseph Ritter von Fraunhofer, we have known how waves diffract through apertures. Therefore, much is known about sensor beam formation, focusing, and steering of emitted and received diffracting wave energy. We refer readers to standard references on optics and beam formation [10, 18, 43, 67, 106, 111] that describe lens theory and electronic focusing using arrays. Applications of *beamforming* principles are wide ranging, including those in medical imaging, radar, sonar, teleconferencing, radio astronomy, seismology, and wireless communications. Here, we concentrate on the Fourier optics of beam formation from array apertures.

Example 5.14.2. *Single-element aperture: Consider the rectangular aperture on the left side of Figure 5.17. The far-field diffraction pattern of wave energy transmitted through the aperture is given by the 2-D Fourier transform of the aperture at $x_1 = 0$.*

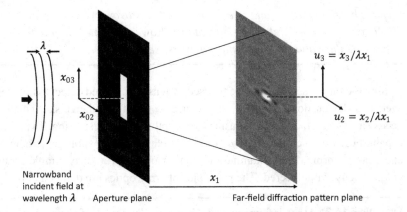

$u_3 = x_3/\lambda x_1$

$u_2 = x_2/\lambda x_1$

Narrowband incident field at wavelength λ

Aperture plane

x_1

Far-field diffraction pattern plane

Figure 5.17 The geometry of Example 5.14.2 showing a rectangular aperture at $x_1 = 0$ and its narrow-band far-field diffraction pattern at $x_1 \gg x_{2\,max}, x_{3\,max}$ from irradiation at wavelength λ. Note that spatial frequencies u_2 and u_3 that are conjugate to aperture axes x_{02} and x_{03}, respectively, can be interpreted spatially as x_2 and x_3 in the diffraction pattern. The far-field diffraction pattern along the ith axis, for $i = 2, 3$, expands in proportion to $\lambda x_1/c_i$.

The aperture is the white rectangular region on the left side of Figure 5.17. The dark regions absorb all incident light. This aperture is expressed by $A(x_{02}, x_{03}) = \mathrm{rect}(x_{02}/c_2)\,\mathrm{rect}(x_{03}/c_3)$, where $c_3 = 4c_2$, $(0, x_{02}, x_{03})$ are coordinates defining the aperture plane, and the center of the aperture is the coordinate origin. To find the far-field paraxial pattern of narrowband light located a distance x_1 from the aperture, we apply CT-FT methods because the 2-D spatial domain of the aperture and its conjugate 2-D spatial-frequency domain of the far-field pattern are both continuous and nonperiodic.

Applying the methods of Section 5.5 in two dimensions, we find the following amplitude pattern from the incident radiation field at axial distance x_1 in far-field plane (x_2, x_3):

$$P(x_2, x_3|x_1) = \mathcal{F}_{2D}\,A(x_{02}, x_{03}) = c_1c_2\mathrm{sinc}(c_2u_2)\,\mathrm{sinc}(c_3u_3)$$
$$= c_1c_2\mathrm{sinc}(c_2x_2/\lambda x_1)\,\mathrm{sinc}(c_3x_3/\lambda x_1).$$

As expected, the narrow aspect of the aperture c_2 results in a broader sinc pattern along x_2. It is curious that a Fourier decomposition transfers a function of space $A(x_{02}, x_{03})$ into another function of space $P(x_2, x_3|x_1)$, until you realize that spatial frequencies u_2 and u_3 can be interpreted geometrically in terms of the far-field plane x_2, x_3, as detailed in [43, 67]. This relationship between the spatial and spatial-frequency domains works because Fourier analysis applies transformations from one vector space to the same vector space. The arguments of P show that it is a function of x_2, x_3 and is parameterized by x_1, which along with wavelength λ and aperture size c_2, c_3 acts to scale the diffraction pattern, P.

The relationship between A and P can guide sensor design. Ideally, we would select a large aperture, short wavelength, and short focal length x_1 if our goal is to *keep the beam as narrow and intense as possible. Narrow beams provide high spatial resolution as well as high SNR acquisitions. However, numerous practical concerns complicate this valid, but simplistic parameter selection.*

Single-element sensors may be focused by lenses or shaped detector surfaces. However, this design does not allow the device to adapt to other tasks, which usually needs variable foci, dynamic focusing over depth, and beam steering. To provide these capabilities, we assemble an *array* of sensor elements and apply electronic focusing and beam steering. Let's examine the Fourier optics of a fairly simple array that is neither focused nor steered. The principles of array optics are detailed in [67, 111].

Example 5.14.3. *Array of apertures: Consider the array of sensors illustrated in Figure 5.18. The aperture of this array is defined by seven identical rectangles that are uniformly sensitive to incoming energy. The method of Fourier optics predicting the far-field sensitivity pattern of this array of elements is simply the Fourier transform of the 2-D aperture. Each array element is $w \times h$ in area and separated by distance d with a total aperture length of $D = Nd$ and a sensitive detector area of Nwh. The aperture lies in the $(0, x_{02}, x_{03})$ plane, and we want to estimate the paraxial sensitivity pattern in a parallel plane at (x_1, x_2, x_3). Assume we satisfy the Fraunhofer criteria, where focal length $x_1 \gg \sqrt{2\pi(x_{2max}^2 + x_{3max}^2)}$. (See chapters 4 and 5 in [43].) For odd N, the aperture function can have two equivalent forms [67]:*

$$A(0, x_{02}, x_{03}) = \left\{ \left[\text{rect}\left(\frac{x_{02}}{D}\right) \frac{1}{d}\text{comb}\left(\frac{x_{02}}{d}\right) \right] * \text{rect}\left(\frac{x_{02}}{w}\right) \right\} \text{rect}\left(\frac{x_{03}}{h}\right)$$

$$= \text{rect}\left(\frac{x_{03}}{h}\right) \sum_{n=-(N-1)/2}^{(N-1)/2} \text{rect}\left(\frac{x_{02} - nd}{w}\right). \tag{5.52}$$

(a) Find the 2-D far-field sensitivity pattern at axial distance x_1, which is given by $P(x_2, x_3|x_1) = \mathcal{F}_{2D}\{A(0, x_{02}, x_{03})\}$ at monochromatic wavelength λ. The

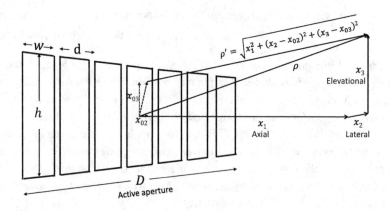

Figure 5.18 The geometry of a linear array aperture and axes describing its far-field geometry.

simplest solution is found using the first form of (5.52), which is easier to transform analytically.

(b) Plot the lateral profile from (a) for the range -30 mm $\leq x_2 \leq 30$ mm. Assume $\lambda = c/u_0 = [1.54 \text{ mm}/\mu s]/10 \text{ MHz} = 0.154 \text{ mm}$, $x_1 = 50$ mm, $d = 0.2$ mm, $w = 0.18$ mm, $N = 7$, $D = Nd = 1.4$ mm, and $h = 1$ mm. (These values are not paraxial but are otherwise practical.)

Solution

(a) Applying theorems in 2-D from Appendix D, we have the far-field approximation

$$P(v_2, v_3 | x_1) = \mathcal{F}A(x_{02}, x_{03}) = \{[D \operatorname{sinc}(Dv_2) * \operatorname{comb}(v_2 d)] \, w \operatorname{sinc}(wv_2)\} \, h \operatorname{sinc}(hv_3),$$

where

$$\mathcal{F}\operatorname{rect}\left(\frac{x_{02}}{D}\right) = D \operatorname{sinc}(Dv_2)$$

$$\mathcal{F}\operatorname{comb}\left(\frac{x_{02}}{d}\right) = d \operatorname{comb}(v_2 d) = d \sum_{n=-\infty}^{\infty} \delta(v_2 d - n) = \sum_{n=-\infty}^{\infty} \delta(v_2 - n/d)$$

$$\mathcal{F}\left\{\operatorname{rect}\left(\frac{x_{02}}{w}\right) \operatorname{rect}\left(\frac{x_{03}}{h}\right)\right\} = hw \operatorname{sinc}(wv_2) \operatorname{sinc}(hv_3). \quad \textit{(Separable functions)}.$$

We could stop here, but this form is difficult to plot. It can be helpful to apply identities and simplify the convolution resulting from the transformation as follows:

$$D \operatorname{sinc}(Dv_2) * \operatorname{comb}(v_2 d) = D \int_{-\infty}^{\infty} dv \operatorname{sinc}(Dv) \operatorname{comb}(d(v_2 - v))$$

$$= \frac{D}{d} \int_{-\infty}^{\infty} dv \operatorname{sinc}(Dv) \sum_{n=-\infty}^{\infty} \delta\left(v_2 - v - \frac{n}{d}\right)$$

$$= N \sum_{n=-\infty}^{\infty} \int_{-\infty}^{\infty} dv \operatorname{sinc}(Dv)\delta\left(v_2 - v - \frac{n}{d}\right)$$

$$= N \sum_{n=-\infty}^{\infty} \operatorname{sinc}(D(v_2 - n/d)).$$

Assembling parts and substituting for v_2 and v_3,

$$P(x_2, x_3 | x_1) = \mathcal{F}A(0, x_{02}, x_{03}) \hspace{3cm} \textit{(Exact)} \quad (5.53)$$

$$= Nwh\left[\operatorname{sinc}\left(\frac{wx_2}{\lambda x_1}\right) \sum_{n=-\infty}^{\infty} \operatorname{sinc}\left(D\left(\frac{x_2}{\lambda x_1} - \frac{n}{d}\right)\right)\right] \operatorname{sinc}\left(\frac{hx_3}{\lambda x_1}\right).$$

The peak response is given by the active area of the array, Nwh. This result can be plotted, which we do in the answer for (b), but we prefer a closed-form solution to avoid convergence concerns regarding the summation. Either way, we need to make an approximation.

Returning to (5.24) and (5.25), we remind ourselves of the following:

$$\text{comb}\left(\frac{t}{T}\right) \triangleq \sum_{n=-\infty}^{\infty} \delta\left(\frac{t}{T} - n\right) = \lim_{N\to\infty} \sum_{n=-(N-1)/2}^{(N-1)/2} e^{i2\pi nt/T}$$

$$\simeq \sum_{n=-(N-1)/2}^{(N-1)/2} e^{i2\pi nt/T} = \frac{\sin(\pi Nt/T)}{\sin(\pi t/T)}.$$

$D = Nd$ and we want N to be a large number, so $D \gg d$. In part (b), we set $N = 7$, which doesn't sound like a great approximation. Practical arrays use $N = 128$ or larger. Nevertheless, let's go with this approximation, even though $N = 7$ in this example. From Section 5.7 we know that

$$\mathcal{F}\left\{\frac{1}{d}\text{comb}\left(\frac{x_{02}}{d}\right)\right\} = \frac{1}{d}\int_{-\infty}^{\infty} dx_{02}\left[\sum_{-\infty}^{\infty} e^{i2\pi nx_{02}/d}\right]e^{-i2\pi v_2 x_{02}}.$$

Adopting the approximation,

$$\mathcal{F}\left\{\text{rect}\left(\frac{x_{02}}{Nd}\right)\frac{1}{d}\text{comb}\left(\frac{x_{02}}{d}\right)\right\} \simeq \frac{1}{d}\int_{-\infty}^{\infty} dx_{02}\left[\sum_{-(N-1)/2}^{(N-1)/2} e^{i2\pi nx_{02}/d}\right]e^{-i2\pi v_2 x_{02}}$$

$$= \frac{1}{d}\sum_{-(N-1)/2}^{(N-1)/2}\int_{-\infty}^{\infty} dx_{02}\, e^{-i2\pi(v_2 - n/d)x_{02}}$$

$$= \frac{1}{d}\sum_{-(N-1)/2}^{(N-1)/2} \delta(v_2 - n/d)$$

$$= \frac{1}{d}\sum_{-(N-1)/2}^{(N-1)/2} \delta\left(\frac{x_2}{\lambda x_1} - n/d\right)$$

$$= \sum_{-(N-1)/2}^{(N-1)/2} \delta\left(\frac{x_2 d}{\lambda x_1} - n\right) = \frac{\sin\left(\frac{\pi N x_2 d}{\lambda x_1}\right)}{\sin\left(\frac{\pi x_2 d}{\lambda x_1}\right)}.$$

The approximate beam pattern in closed form is

$$P(x_2, x_3 | x_1) = wh\,\text{sinc}\left(\frac{wx_2}{\lambda x_1}\right)\frac{\sin\left(\frac{\pi N x_2 d}{\lambda x_1}\right)}{\sin\left(\frac{\pi x_2 d}{\lambda x_1}\right)}\text{sinc}\left(\frac{hx_3}{\lambda x_1}\right). \qquad \text{(Closed form)}.$$

$$(5.54)$$

Note the Dirichlet kernel, $\sin(Nx)/\sin(x)$, scales the plot by a factor of N, so the exact and approximate expressions are equal in scale.

(b) Figure 5.19 depicts the 1-D and 2-D far-field sensitivity profiles. The plot is the sensitivity along just the x_2 axis. (See Figure 5.18 for geometric details.) We plotted the exact and closed-form solutions and find they agree well. The result is the continuous-wave 10-MHz ultrasonic pressure profile from a linear array. To obtain a 10-MHz

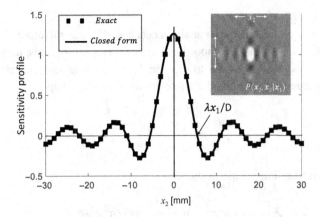

Figure 5.19 The image (inset) is the 2-D sensitivity profile $P(x_2, x_3 | x_1 = 50)$ from aperture $A(0, x_{02}, x_{03})$, where the center of the beam at peak amplitude is at the center of the image. The plots are exact and closed-form estimates of $P(x_2 | x_1 = 50, x_3 = 0)$ (the central segment of the middle horizontal line in the grayscale image). Spatial resolution of this beam can be defined by the first zeros of the beam profile, which appear at $x_2 = \pm \lambda x_1 / D$, where we find $x_2 = \pm 5.5$ mm. Low resolution is improved by increasing $D = Nd$.

pulsed profile, we just weight and sum the solutions at other frequencies in the pulse, assuming the system is linear in terms of wave propagation and instrumentation. Note that zero crossings in the lateral profile occur at $m\lambda x_1 / D$ for $m = 0, \pm 1, \pm 2 \ldots$. The code used to generate Figure 5.19 is also given here.

The 2-D diffraction pattern for the array in Figure 5.19 resembles that of one element in Figure 5.17, albeit with a few differences. The array is wider than the element, so the sinc pattern is narrower along x_2 in Figure 5.19 than in Figure 5.17. Also, the ratio of sine functions in (5.54) shows that the sinc pattern is periodic, repeating along the x_2 axis. This is a consequence of array elements behaving as a sampled rectangular aperture, so the Fourier transform is periodic along the sampled axis. These repeated patterns are called grating lobes, *whereas oscillations near the axial main lobe seen in Figure 5.19 are called* side lobes. *Grating lobes are located more distal from the beam axis compared to side lobes; both sense off-axis energy that is attributed to energy sensed along the beam axis, and therefore contrast resolution is reduced. Grating lobes become prominent when the beam is electronically steered, as in phase array technologies [67].*

```
%%%%%%%%%%%%%  MATLAB script for Example 5.14.3 that generates Fig. 5.19  %%%%%%%%%%%%%
%10 MHz, single frequency beam. Lengths are all in mm.
x2=-30:0.01:30;x1=length(x2);lam=1.54/10;x1=50;d=0.2;w=0.18;N=7;  %initialize
D=N*d;h=1;q=pi/(lam*x1);P2=zeros(81,length(x2));                  %initialize
P1=w*h*(sin(q*w*x2)./(q*w*x2)).*(sin(q*N*x2*d)./sin(q*x2*d));     %closed form, (5.54)
for n=1:81
    P2(n,:)=sin(D*pi*(x2/(lam*x1)-(n-41)/d))./(D*pi*(x2/(lam*x1)-(n-41)/d));
end
P=N*w*h*sin(q*w*x2)./(q*w*x2).*sum(P2);                           %exact solution, (5.53)
plot(x2,P1,'k',x2(1:100:end),P(1:100:end),'k*')
```

Summary.
This chapter concentrated on Fourier analysis because of its central importance to data analysis. To understand basis decompositions at a deeper level, we must turn to matrix methods and extend the discussion to eigenanalysis and singular-value decompositions that generalize Fourier methods. These are the topics of Chapter 6.

5.15   Problems

5.1   Use closed form integration to show

$$\mathcal{F}\left\{\exp(-a^2x^2)\right\} = \frac{\sqrt{\pi}}{a}\exp(-\pi^2u^2/a^2).$$

5.2   Use the result of Problem 5.1 to find the following without performing an integral:

$$\mathcal{F}\left\{h(x)\right\},$$

where

$$h(x) = \frac{1}{\sqrt{2\pi}\sigma}\exp(-(x-x_0)^2/2\sigma^2).$$

5.3   Prove the modulation theorem shown here without using any theorems except the convolution theorem:

$$\mathcal{F}\left\{h(x)\cos(2\pi u_0 x)\right\} = \frac{1}{2}\left[H(u-u_0) + H(u+u_0)\right].$$

5.4   Compute

$$\mathcal{F}\left\{\frac{d}{dt}\text{rect}\left(\frac{t}{2T_0}\right)\right\}$$

in two different ways.

(a)   Transform the derivative of the rect function and sketch the derivative.
(b)   Apply the derivative theorem and then compute the transform. The results of (a) and (b) must be equal. Identify the theorems applied in your solution.

5.5   Use closed-form integration to show that the 2-D Fourier transform of a circular aperture is given by

$$\mathcal{F}_{2D}\text{circ}(r/a) = \pi a^2\text{jinc}(a\rho).$$

The shorthand $\text{circ}(r/a)$ indicates a circle of radius a centered at the origin, where all values inside the circle at $r < a$ equal 1 and values outside equal zero. Also $r^2 = x^2 + y^2$, $\rho^2 = u^2 + v^2$ and $\tan\theta = y/x$, $\tan\varphi = v/u$, where (x,y) are conjugate variables to (u,v). Finally, $\text{jinc}(\rho) \triangleq 2J_1(2\pi\rho)/(2\pi\rho)$.

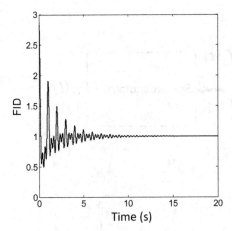

Figure 5.20 Illustration of the free-induction decay signal in Problem 5.6.

The following relations will be helpful:

$$J_0(a) = \frac{1}{2\pi} \int_0^{2\pi} d\theta \, e^{-ia \cos(\theta - \varphi)} \quad \text{and} \quad \int_0^a d\beta \, \beta \, J_0(\beta) = a \, J_1(a),$$

where J_n is a Bessel function of the first kind, nth order.

5.6 The magnetic resonance (MR) free-induction decay (FID) signal in Figure 5.20 may be expressed as the sum of three exponentially decaying sinusoids plus a constant:

$$\text{FID}(t) = \left[M_0 + \sum_{n=1}^{3} M_n \, e^{-t/\tau_n} \, \cos(\Omega_n t) \right] \text{step}(t),$$

where $\Omega_n = 2\pi u_n$

and u_n is a constant representing temporal frequency in hertz for the nth signal.
(a) Find the Fourier transform of $\text{FID}(t)$ to compute the spatial distribution of magnetization from the three sources.
(b) Find the real part of the result in part (a).
(c) Use MATLAB and assume $M_0 = 1$, $M_1 = 1$, $\tau_1 = 1$ s, $u_1 = 1$ Hz; $M_2 = 0.5$, $\tau_2 = 2$ s, $u_2 = 2$ Hz; $M_3 = 0.333$, $\tau_3 = 3$ s, $u_3 = 3$ Hz. Plot the result from part (b), i.e., $\text{Re}\{\mathcal{F}\,\text{FID}(t)\}$, from $0 \le u \le 5$ Hz using MATLAB, being sure to label the axes.

5.7 You borrowed a spectrophotometer (a photometer that measures optical radiation intensity as a function of frequency or wavelength) from the lab next door because it is 10 times faster than the one in your lab. To calibrate it before use, you illuminate its sensor with a 600-nm wavelength (5×10^{14} Hz) source known to give a Gaussian-shaped spectrum. Expecting to see the smooth spectrum shown at the left side of Figure 5.21, you are surprised to observe the scalloped spectrum on the right side of the figure. Let's see if we can figure out what is wrong with this system.

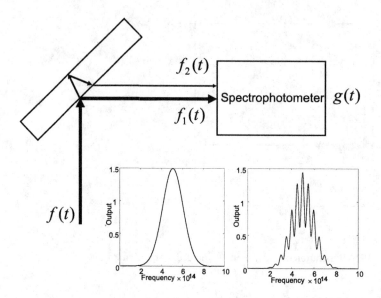

Figure 5.21 Spectrophotometer experiment from Problem 5.7. Note that power spectral densities are plotted and labeled as "output."

The user's manual indicates that the system uses a mirror to reflect photons $f(t)$ into the sensor. To make it oscillate faster, the manufacturer had to make the mirror lighter, which also allowed some of the incident light to penetrate the mirror, reflect off of the back surface, and then add back into the main signal but with a $t_0 = 2$ fs (femtosecond = 10^{-15} s) time delay. Therefore f_2 is a time-shifted copy of f_1 except for the lower magnitude, $|f_2| = a|f_1|$, where the reflection coefficient is $a = 0.2$.

(a) Ignoring system effects, i.e., $g_1(t) = f_1(t)$, write an expression for net output $g(t)$ in terms of $f_1(t)$.

(b) Calculate analytically the power spectrum of g using the information given. Plot your final answer and show that you obtain the scalloped spectrum shown in Figure 5.21.

5.8 A new flow cytometer was purchased to size cells suspended in fluid. While calibrating, this device you observe the calibration signal $c(t)$ and a negligible amount of random noise. However, you also observe a cosine wave that you suspect the system is receiving from EM radiation emitted by fluorescent lights ($u_0 = 60$ Hz). Figure out if the cosine signal has contaminated your experiment to the point that the results are meaningless, or if the data can be used after filtering the signal. You will need the following to make a decision: $g(t) = c(t) + 0.5\cos(2\pi u_0 t)$, where $c(t) = \exp(-t^2/(2\sigma^2))/(\sigma\sqrt{2\pi})$ and $\sigma = 3/u_0$.

5.9 A common shorthand notation for convolution reads

$$g(t) = \int_{-\infty}^{\infty} dt'\, h(t - t')\, f(t') = h(t) * f(t).$$

We adopted a different shorthand notation, $[h * f](t)$, because of the following situation.

You are using an LTI optical system to estimate the concentration of a substance in blood as a function of time. A light beam is positioned over the patient's vein and the backscattered energy is recorded to give the measurement $g(t)$. The system is defined by its impulse response $h(t)$ and the substance in the blood as $f(t)$. The substance is injected as a bolus so that $f(t)$ looks like a rectangular function immediately after injection, but it spreads out in the bloodstream according to $f(at - b)$, where a and b are constants. This model suggests that the substance is shifted in time b and scaled by the factor a. Therefore our measurements can be expressed as

$$g(t) = \int_{-\infty}^{\infty} dt' \, h(t - t') \, f(at' - b)$$

and the notation $h(t) * f(t)$ doesn't tell the story. (Note that a must have no units and b has units of time if the function is dimensionally consistent.) Beginning with the equation just given, compute the Fourier transform $G(u)$.

5.10 If X and Y are two independent Poisson processes with parameters λ_x and λ_y, let $Z = X + Y$. Find $p_Z(z)$ and an expression for λ_z.

5.11 We know that $g[m] = \mathcal{F}^{-1}\mathcal{F}g[m]$ for $1 \leq m \leq M$. We also know from Section 5.10 that measurable signals from experiments are real valued but generally have both even and odd components. Consequently, $G[k] \in \mathbb{C}^{M \times 1}$ is complex. Thus, if we computed the inverse DFT of just the real part of $G[k]$, i.e., $\mathcal{F}^{-1}\{\Re G[k]\} = \mathcal{F}^{-1}\{\Re \mathcal{F}g[m]\}$, then we would have no reason to expect to fully recover the original signal $g[m]$. After all, some of the information is missing.

A paper by So and Paliwal [101] pointed out that since $\frac{1}{2}(g[m] + g^*[-m]) \xleftrightarrow{\mathcal{F}} \Re G[k]$, we can zero pad $g[m]$ by a factor of 2 using

$$g'[m] = \begin{cases} g[m] & 1 \leq m \leq M \\ 0 & M + 1 \leq m \leq 2M \end{cases},$$

and recover $g[m]$ exactly using $g[m] = \mathcal{F}^{-1}\{2\Re G'[k]\}$. Zero padding in time is sinc interpolation in frequency. However, the effect of zero padding by a factor of 2 here is to provide space for $g^*[-m]$ to sum into the original data $g[m]$ without aliasing.

Use MATLAB to show $g[m] \neq g''[m] = \mathcal{F}^{-1}\{\Re \mathcal{F}g[m]\}$ and yet $g[m] = \mathcal{F}^{-1}\{2\Re \mathcal{F}g'[m]\}$ for $1 \leq m \leq M$.

5.12 The relationship between stress σ and strain ϵ is determined by the relaxation modulus $G(t)$ for a viscoelastic medium according to the Boltzmann superposition principle,

$$\sigma(t) = \int_{-\infty}^{t} dt' \, G(t - t') \frac{d\epsilon(t')}{dt'}.$$

This linear relationship holds if the applied strain is "small," defined as $\epsilon \ll 1$. For $\Omega = 2\pi u$ and applying sinusoidal strain $\epsilon = \epsilon_0 \sin \Omega t$, show that the stress is

$$\sigma(t) = \epsilon_0 \left(G'(\Omega) \sin \Omega t + G''(\Omega) \cos \Omega t \right)$$

where

$$G'(\Omega) = \Omega \int_0^\infty d\tau \, G(\tau) \sin \Omega t$$

$$G''(\Omega) = \Omega \int_0^\infty d\tau \, G(\tau) \cos \Omega t.$$

G' is the storage modulus given by the sine transform of the relaxation modulus and G'' is the loss modulus given by the cosine transform.

5.13 Consider the code provided in Appendix C that was used to generate Figure C.1, and modify the third line to read

```
g=cos(2*pi*u0*t)+0.5*cos(2*pi*(u0+us)*t);
```

(It will help to read the material in that appendix.) This code change modifies the ultrasonic echo spectrum to have a strong component at the carrier frequency at $u_0 = 1$ and a weaker component at $u_0 + u_s = 1.2$, where, of course, $u_s = 0.2$. I'm avoiding specifying the units, which will unnecessarily complicate the solution if all the experimental details are attended to. Consider the strong spectral component at u_0 caused by tissue scattering, which we refer to as the "clutter" spectrum and which needs to be eliminated. The weaker signal at $u_0 + u_s$ is the blood-flow echo component, and we want to estimate u_s. If you make the code change specified, you will obtain a plot similar to Figure C.1, where each baseband component has two nonzero elements.

(a) Compute the centroid of the power spectrum via $\bar{u} = \sum u \, S(u) / \sum S(u)$, where $S(u)$ is the power spectral density of $\bar{g}_a(t)$, which is the complex envelope of $g(t)$. Use the result to estimate u_s. It will not be accurate.

(b) You will find that you must first eliminate the clutter signal from the complex envelope before you can accurately estimate u_s. Propose a simple clutter filter and show how well you can estimate u_s.

6 Basis Decomposition II

This chapter begins by revisiting matrix forms of data acquisition equations from Chapter 4 and the DFT discussion of Section 5.8. While (5.30) provides transformation details and intuition about the analysis, these equations can be less practical. This section introduces the DFT matrix $\mathbf{Q}^\dagger$ that, when multiplied by vector $\mathbf{g}$, executes the forward transformation $\mathcal{F}\mathbf{g}$. The original vector is restored by a second matrix multiplication via $\mathbf{g} = \mathbf{Q}\mathbf{Q}^\dagger\mathbf{g} = \mathcal{F}^{-1}\mathcal{F}\mathbf{g}$ reverses the effects of the first. Matrix forms greatly simplify notation and speed numerical computation through array processing. Aside from notational convenience and implementation speed, the matrix view guides applications of (5.4) and (5.6) when developing generalizations of Fourier analysis.

6.0.1 Acquisition Equation

In Section 4.5 and Section 4.6, we showed that the operator for LTI/LSI acquisitions is a circular convolution implemented by matrix product $\mathbf{g} = \mathbf{H}\mathbf{f}$, where system matrix $\mathbf{H} \in \mathbb{R}^{M \times N}$ is a circulant approximation to Toeplitz (see Appendix A and Section 4.5). Data vector $\mathbf{g}$ is periodic with an M-sample period, while object vector $\mathbf{f}$ has an N-sample period. Let's examine the structure of $\mathbf{H}$ with a specific numerical example that illustrates the process.

$$\mathbf{g} = \mathbf{H}\mathbf{f} \quad \text{becomes} \quad \begin{pmatrix} g_1 \\ g_2 \\ g_3 \\ g_4 \\ g_5 \\ g_6 \\ g_7 \\ g_8 \end{pmatrix} = \begin{pmatrix} h_1 & h_0 & 0 & 0 & 0 & 0 & 0 & h_2 \\ h_2 & h_1 & h_0 & 0 & 0 & 0 & 0 & 0 \\ 0 & h_2 & h_1 & h_0 & 0 & 0 & 0 & 0 \\ 0 & 0 & h_2 & h_1 & h_0 & 0 & 0 & 0 \\ 0 & 0 & 0 & h_2 & h_1 & h_0 & 0 & 0 \\ 0 & 0 & 0 & 0 & h_2 & h_1 & h_0 & 0 \\ 0 & 0 & 0 & 0 & 0 & h_2 & h_1 & h_0 \\ h_0 & 0 & 0 & 0 & 0 & 0 & h_2 & h_1 \end{pmatrix} \begin{pmatrix} f_1 \\ f_2 \\ f_3 \\ f_4 \\ f_5 \\ f_6 \\ f_7 \\ f_8 \end{pmatrix},$$

(6.1)

when $M = N = 8$ and $\mathbf{h}^\top = [h_0 \ h_1 \ h_2 \ h_3 \ h_4 \ h_5 \ h_6 \ h_7] = [0.25 \ 0.50 \ 0.25 \ 0 \ 0 \ 0 \ 0 \ 0]$ has three unequal nonzero elements that sum to 1. The zero elements pad the length of $\mathbf{h}$ so it equals the length of $\mathbf{f}$. Time-reversed and shifted copies of $\mathbf{h}$ become the rows of $\mathbf{H}$. It is somewhat arbitrary where we begin the cycle, although that decision will shift the phase of output vector $\mathbf{g}$.

6.1   Eigenanalysis with a Fourier Basis

We reviewed the fundamentals of eigenanalysis in Appendix A. Here we apply those methods to analyze square system matrix $\mathbf{H}$ in (6.1). Specifically, we seek nonsingular $N \times N$ matrix $\mathbf{Q}$ that *diagonalizes* $\mathbf{H}$,

$$\mathbf{Q}^{-1}\mathbf{H}\mathbf{Q} = \Lambda \quad \text{or} \quad \mathbf{H}\mathbf{Q} = \mathbf{Q}\Lambda, \tag{6.2}$$

where $N \times N$ Λ is both diagonal and *similar* to $\mathbf{H}$. The application of $\mathbf{Q}$ is a *similarity transformation* that preserves the properties (eigenvalues) of $\mathbf{H}$; this preservation is essential if we hope to analyze the measurement device in this manner. Because $\mathbf{Q}$ has complex elements and $\mathbf{Q}^{\dagger} = \mathbf{Q}^{-1}$, $\mathbf{Q}$ is a *unitary matrix* that has wonderful properties. Let's look at it more closely.

Equation (6.2) is an eigenvalue decomposition of $N \times N$ system matrix $\mathbf{H}$ that we implement using $\mathbf{Q}$. Fourier basis vectors compose the N columns of $\mathbf{Q}$, and the associated N eigenvalues are the diagonal elements of Λ, viz., $(\lambda_1, \ldots, \lambda_N)^{\top} = \text{diag}(\Lambda)$.

More specifically,

$$\mathbf{Q}[n,k] = \frac{1}{\sqrt{N}} \begin{pmatrix} 1 & 1 & \cdots & 1 & \cdots & 1 \\ 1 & e^{i2\pi/N} & \cdots & e^{i2\pi k/N} & \cdots & e^{i2\pi(N-1)/N} \\ \vdots & \vdots & & \vdots & & \vdots \\ 1 & e^{i2\pi n/N} & \cdots & e^{i2\pi nk/N} & \cdots & e^{i2\pi n(N-1)/N} \\ \vdots & \vdots & & \vdots & & \vdots \\ 1 & e^{i2\pi(N-1)/N} & \cdots & e^{i2\pi(N-1)k/N} & \cdots & e^{i2\pi(N-1)^2/N} \end{pmatrix}. \tag{6.3}$$

Note that each complex exponent includes the index product nk that steps from 0 to $N - 1$ as follows:

$$\begin{pmatrix} & & & \text{fixed } k & & \\ & & & \downarrow & & \\ & 0(0) & \cdots & 0(k) & \cdots & 0(N-1) \\ & 1(0) & \cdots & 1(k) & \cdots & 1(N-1) \\ & \vdots & & \vdots & & \vdots \\ \text{fixed } n \rightarrow & n(0) & \cdots & n(k) & \cdots & n(N-1) \\ & \vdots & & \vdots & & \vdots \\ & (N-1)(0) & \cdots & (N-1)(k) & \cdots & (N-1)(N-1) \end{pmatrix}.$$

Let's examine the eighth column of $\mathbf{Q}$ in our $N = 8$ example:

$$\mathbf{q}[7] = \frac{1}{\sqrt{8}} \begin{pmatrix} 1 \\ e^{i2\pi(1)7/8} \\ e^{i2\pi(2)7/8} \\ e^{i2\pi(3)7/8} \\ e^{i2\pi(4)7/8} \\ e^{i2\pi(5)7/8} \\ e^{i2\pi(6)7/8} \\ e^{i2\pi(7)7/8} \end{pmatrix} = \frac{1}{\sqrt{16}} \begin{pmatrix} \sqrt{2} \\ 1+i \\ i\sqrt{2} \\ -1+i \\ -\sqrt{2} \\ -1-i \\ -i\sqrt{2} \\ 1-i \end{pmatrix}.$$

Using the numerical values for $\mathbf{h}^{\top}$ provided after (6.1), we multiply the system matrix by this vector and find $\mathbf{H}\,\mathbf{q}[7] = 0.8536\,\mathbf{q}[7]$. *Anytime we multiply a matrix by a vector and obtain a scaled copy of the same vector, we know that vector is an eigenvector of that matrix. Its scaling constant is the corresponding eigenvalue.* In fact, if you enter the numerical example of (6.1) for $\mathbf{H}$ into MATLAB and execute `[V D] = eig(H)`, you will find 0.8536 as the $D[5,5]$ and $D[6,6]$ elements of D. The columns of output matrix V are the *eigenvectors* of $\mathbf{H}$, $\mathbf{v}[k]$, and the diagonal elements of D are the *eigenvalues*, $\lambda[k]$. In general, the eigenvectors that MATLAB finds may not be the same as those you find working out the problem by hand – that's expected since *eigenvectors are not unique.*

Fourier transformation of a circulant matrix is equivalent to an eigenvalue decomposition of that matrix using a Fourier basis. If $\mathbf{H}$ is a real symmetric matrix and a Fourier basis is applied, the eigenvalues are real Fourier coefficients. Circulant $\mathbf{H}$ matrices are generally asymmetric, so the Fourier coefficients are complex valued.[1]

6.2 What Do Eigenstates Describe?

The paired combination of eigenvector and eigenvalue is often referred to as an *eigenstate* of a system. We interpret eigenvalues of $\mathbf{H}$ as the uncoupled properties of the measurement system represented by that matrix. Each eigenvector is an uncoupled *functional mode* of activity through the device that connects input stimulus to output response. A functional mode is not a linear path through the interconnected system network describing a measurement process. Instead, eigenvectors are more like all possible parallel paths through the network but each with a distinct weighting pattern that is orthogonal to the other modes. For example, one vibrating guitar string has fundamental and harmonic overtones that are each modes (sinusoidal eigenvectors) of that string occurring simultaneously on all elements of that string. Eigenvalues linearly

---

[1] This is a consequence of the shift theorem that we saw at work in Figure 5.12.

weight each modal response in space-time to give the system properties responsible for the overall measurement features we observe – features including frequency
and phase.[2]

Example 6.2.1. *Symmetric matrix: If we apply an eigenvector to the input of a system,
we will observe the same eigenvector at the output but weighted by its eigenvalue that
is determined by the properties of that system. To offer a very simple example, consider
real symmetric* **A** *and real eigenvector* **f**:

$$\mathbf{A} = \begin{pmatrix} 1 & 2 \\ 2 & 1 \end{pmatrix} \text{ and } \mathbf{f} = \begin{pmatrix} -\cos \pi + i \sin \pi \\ \cos 2\pi - i \sin 2\pi \end{pmatrix} = \begin{pmatrix} 1 \\ 1 \end{pmatrix}, \quad \mathbf{g} = \mathbf{Af} = \begin{pmatrix} 3 \\ 3 \end{pmatrix} = 3\mathbf{f}.$$

The statement **Af** $= 3\mathbf{f}$ *shows that* **f** *is an eigenvector or* **A**. *The symmetry of* **A**
means it has real eigenvalues, one of which is $\lambda = 3$. *The corresponding orthonormal
eigenvector is* $\frac{\mathbf{f}}{\|\mathbf{f}\|} = \frac{1}{\sqrt{2}} \begin{pmatrix} 1 \\ 1 \end{pmatrix}$. *Try it in MATLAB or by hand.*

Example 6.2.2. *Asymmetric matrix: Consider real asymmetric matrix* **B** *and complex
eigenvector* **f**:

$$\mathbf{B} = \begin{pmatrix} 3 & -2 \\ 4 & -1 \end{pmatrix} \text{ and } \mathbf{f} = \begin{pmatrix} \cos 0 + i \sin \pi \\ \cos 2\pi + i \sin \pi/2 \end{pmatrix} = \begin{pmatrix} 1 \\ 1+i \end{pmatrix},$$

$$\mathbf{g} = \mathbf{Bf} = \begin{pmatrix} 1 - 2i \\ 3 - i \end{pmatrix} = (1 - 2i)\mathbf{f}.$$

This eigenvalue is complex, $\lambda = 1-2i$. *Every possible output from these and all linear
systems is a weighted linear superposition of that system's eigenstates. The number
of eigenstates is given by the rank of the system matrix, which is 2 for both examples.
Eigenstates are fundamental because they describe properties of linear systems that
determine how they respond to input stimuli.*

A Fourier basis is a generic collection of harmonically related sines and cosines that
depend on a fundamental frequency determined by the spatial or temporal duration
of the input/output sequence. The ability of a device to capture task information is
described by the Fourier coefficients for LTI/LSI systems. Because **A** in Example
6.2.1 is circulant, the eigenvalues are also the Fourier coefficients of **A**. However, **B** in
Example 6.2.2 is not circulant, so eigenanalysis applies but Fourier analysis does not
provide a system spectrum. Clearly, Fourier analysis is a *special case* of eigenanalysis. The importance of basis coefficients for assessing measurement instruments is
apparent from the fact that instrument "quality" metrics in common use, such as SNR,
dynamic range, and spatial and temporal resolutions, can each be defined in terms of
basis coefficients (discussed later).

---

[2] Tone and timbre describe combinations of fundamental frequencies weighted by overtones to give
different instruments a unique sound even when they play the same notes.

Given this background, it should be clear why eigenstates are important in measurement science. We will show in this chapter that eigenanalysis is also a special case of a more general decomposition, known as singular-value decomposition, which is needed to fully exploit the fact that measurement operators live in two vector spaces. Nevertheless, within one vector space, eigenanalysis is the decomposition of choice, revealing uncoupled properties of objects, patients, and data sets. Before looking further into these points, we first connect (6.2) to DFT notation given that the object and data are identically sampled.

6.3   Eigenanalysis Connections to DFT

Elements of $N \times N$ matrix $\mathbf{Q}$ in (6.3) form the orthonormal basis,

$$Q[n,k] = \frac{1}{\sqrt{N}} e^{i2\pi nk/N}, \tag{6.4}$$

where integers n and k are the indices of conjugate variables t and u, respectively, representing time and temporal frequency variables (or space and spatial frequency variables, x and u). Since $T/T_0 = 1/N$ is the frequency increment, $k/N \sim u$ is a unitless version of the discrete-frequency variable. Equation (6.4) is the *Fourier kernel* used in DFT synthesis calculations, except here we use the orthonormal basis by applying factor $1/\sqrt{N}$ to both the analysis and synthesis kernels. With that in mind, let's examine some properties of $\mathbf{Q}$.

Any two columns of $\mathbf{Q} = (\mathbf{q}_0, \dots, \mathbf{q}_k, \dots, \mathbf{q}_\ell, \dots, \mathbf{q}_{N-1})$, are orthonormal, e.g.,

$$\mathbf{q}_k^\dagger[n]\mathbf{q}_\ell[n] = \frac{1}{N} \sum_{k=0}^{N-1} e^{-i2\pi nk/N} \, e^{i2\pi n\ell/N} = \delta_{k\ell}. \tag{6.5}$$

Consequently, elements of the inverse matrix can be written by inspection as equal to the conjugate transpose via

$$Q^{-1}[n,k] = \frac{1}{\sqrt{N}} e^{-i2\pi nk/N} = Q^*[k,n] = [\mathbf{q}_k^\dagger][n], \tag{6.6}$$

a row vector, and showing $\mathbf{Q}$ is a unitary matrix, $\mathbf{Q}^\dagger = \mathbf{Q}^{-1}$. We also explicitly define $[\mathbf{Q}^\dagger]_{nk} \triangleq Q^*[k,n]$. The statement $Q^{-1}[n,k] = Q^*[k,n]$ can be puzzling unless you account for basis periodicity, $\exp(i2\pi(N-n)k/N) = \exp(-i2\pi nk/N)$.

6.3.1   The Acquisition Equation

From (6.2), the unitary property $\mathbf{Q}^\dagger\mathbf{Q} = \mathbf{Q}\mathbf{Q}^\dagger = \mathbf{I}$, and the rules of matrix algebra, we apply the forward Fourier operator to the acquisition equation for a noiseless discrete-to-discrete LTI device as follows:

$$\mathbf{g} = \mathbf{Hf}$$
$$\text{and} \quad \mathbf{Q}^\dagger\mathbf{g} = \mathbf{Q}^\dagger\mathbf{HQQ}^\dagger\mathbf{f} = \Lambda\mathbf{Q}^\dagger\mathbf{f}. \tag{6.7}$$

(For the moment, assume $\mathbb{U} \subset \mathbb{V}$.) Interpreting each part of (6.7) as a separate equation, we find the following DFT relationships:

$$\mathbf{Q}^{\dagger}\mathbf{f} = \frac{1}{\sqrt{N}} \sum_{n=0}^{N-1} f[n]\,e^{-i2\pi nk/N} = F[k] \qquad\qquad \text{(Analysis)}$$

$$\mathbf{QQ}^{\dagger}\mathbf{f} = \frac{1}{\sqrt{N}} \sum_{k=0}^{N-1} \left[\frac{1}{\sqrt{N}} \sum_{n=0}^{N-1} f[n]\,e^{-i2\pi nk/N} \right] e^{i2\pi nk'/N} = f[n] \quad \text{(Synthesis)}$$

$$\mathbf{Q}^{\dagger}\mathbf{g} = \frac{1}{\sqrt{N}} \sum_{n=0}^{N-1} g[n]\,e^{-i2\pi nk/N} = G[k] \qquad\qquad \text{(Analysis)}$$

$$\text{(6.8)}$$

$$\mathbf{QQ}^{\dagger}\mathbf{g} = \frac{1}{\sqrt{N}} \sum_{k=0}^{N-1} \left[\frac{1}{\sqrt{N}} \sum_{n=0}^{N-1} g[n]\,e^{-i2\pi nk/N} \right] e^{i2\pi nk'/N} = g[n] \quad \text{(Synthesis)}$$

$$\text{diag}\,\mathbf{Q}^{\dagger}\mathbf{HQ} = \Lambda_{kk} = \lambda[k] = \frac{1}{\sqrt{N}} \sum_{n=0}^{N-1} h[n]\,e^{-i2\pi nk/N} = H[k]. \qquad \text{(Analysis)}$$

The quantities on the far left in the first four equations are column vectors that equal the DFT functions from Section 5.8 shown on the far right. Uppercase $F[k]$ and $G[k]$ denote Fourier-domain variables and not matrices; they correspond to $f[n]$ and $g[n]$ in the time domain. The fifth line describes diagonal elements of Λ and its relationship to the system response function $H[k]$.[3]

Assembling the parts, the 1-D DFT of the acquisition equation is

$$\mathbf{Q}^{\dagger}\mathbf{g} = \Lambda\mathbf{Q}^{\dagger}\mathbf{f} \qquad \text{and} \qquad G[k] = H[k]F[k], \qquad \text{Analysis}$$

$$\mathbf{QQ}^{\dagger}\mathbf{g} = \mathbf{Q}\Lambda\mathbf{Q}^{\dagger}\mathbf{f} \qquad \text{and} \qquad \mathcal{F}^{-1}\{G[k]\} = \mathcal{F}^{-1}\{H[k]F[k]\}, \qquad \text{(6.9)}$$

$$\mathbf{g} = \mathbf{Hf} \qquad\qquad \text{and} \qquad g[n] = (h * f)[n]. \qquad \text{Synthesis}$$

This is a matrix expression for the Fourier convolution theorem.

Example 6.3.1. *Rank of Fourier basis: To illustrate numerically, consider the 8×8 system matrix $\mathbf{H}$ in (6.1), which is symmetric, real, and positive semi-definite. Using* `rank(H)` *in MATLAB, we see it has rank 7 (try it). To compute the Fourier transform of $\mathbf{H}$, we form an 8×8 unitary matrix $\mathbf{Q}$ via (6.3).[4] The properties of $\mathbf{H}$ ensure the Fourier coefficients are real, nonnegative, and given by the diagonal values of $\Lambda = \mathbf{Q}^{\dagger}\mathbf{HQ}$,*

$$1.00,\ 0.8536,\ 0.8536,\ 0.50,\ 0.50,\ 0.1464,\ 0.1464,\ 0.00,$$

---

[3] Significantly, the rows of $\mathbf{H}$ are time reversed so that $\sum h[-n]\,e^{i2\pi nk/N} = \sum h[n]\,e^{-i2\pi nk/N}$.

[4] All eigenvalues of $\mathbf{Q}$ are nonzero. Despite non-unique eigenvalues, the algebraic multiplicity (multiplicity of eigenvalues) equals the geometric multiplicity (dimension of corresponding eigenspaces), so $\mathbf{Q}$ is full rank.

when placed in descending order. Since **H** *has one zero eigenvalue, it is not full rank. The same values are obtained in MATLAB using* [U D] = eig(H). *There are only* $N/2$ *unique, nonzero Fourier coefficients.*

6.3.2   Information Channels, Degrees of Freedom, Sensitivity, and Noise

Generally, eigenvectors are not frequency channels, but rather *functional modes* by which an instrument responds to input. The *eigenvectors* specifically for LTI/LSI measurement systems are uncoupled frequency channels through which task information flows from patient to decision maker. These modes are uncoupled in the sense that given an arbitrary input signal, each eigenvector responds to only those signal components tuned to that mode or eigenchannel. For LTI/LSI systems, Fourier coefficients of **H** are the system eigenvalues. In essence, *eigenvalues* quantify the information *sensitivity* of that eigenchannel.

The performance of an LTI/LSI device is ultimately limited by its *bandwidth* as well as by the sensitivity and SNR of each frequency channel in that bandwidth. The acquisition-stage bandwidth includes the width and number of independent frequency channels that an instrument makes available for transferring information from object-function inputs into data-function outputs. Bandwidth[5] is a measure of the *degrees of freedom* for a measurement; it is the number of independent pieces of information from which the state of the object can be determined. Instrument *sensitivity* combines the dynamic range and resolution (spatial, contrast, temporal) of a device for expressing Fourier coefficients/eigenvalues. (See Chapter 7.)

Measurement *noise* may be defined as any contribution to the acquisition that does not provide information related to the task. This is why task-based analysis is fundamental to device design and evaluation: You must know the task before you can define the signal and the noise. Put simply, signals for one task can be noise for another. The ideal measurement would respond to all input frequencies for the task with unit gain while adding no acquisition noise. Of course, that never happens in practice, so the job of measurement engineering is to make the most of the available task information for each situation.

6.3.3   Estimating Degrees of Freedom

With Fourier/eigenanalysis tools, we can begin investigating how the degrees of freedom for a system combine with the task to determine *measurement performance*. Consider a method suggested 40 years ago by Andrews and Hunt in Section 4.5 of [5]. Suppose we have a continuous-to-discrete LTI measurement device where impulse response $h(t)$ is adjustable. Although $h(t)$ is a continuous function of time, the device records data at discrete time intervals that we set at $T = 1$. The acquisition equation

---

[5] A sensor array has a spatial bandwidth and a temporal bandwidth, each of which contributes degrees of freedom to measurements.

for a continuous-to-discrete LTI measurement is $g[m] = \int dt'\, h(mT - t')\, f(t')$. The top three plots in Figure 6.1 illustrate the measurement process for Gaussian impulse-response functions of varying widths specified by parameter σ. Each Gaussian plot describes the impulse response being applied to the continuous object function at various values of mT. Large values of σ/T correlate with measurement samples more closely than do small values.

The number of degrees of freedom (dofs) of a measurement device may also be described as the minimum number of independent data samples required to accomplish a measurement task in the presence of noise. A total of 21 data samples are measured in each of the three plots at the top of Figure 6.1, so all we know at this point is that dofs ≤ 21 depending on the correlation among measurement samples. Among the three devices, only $\sigma/T = 0.1$ yields nearly uncorrelated measurements. In the other two cases, the measurement correlations make it difficult to count degrees of freedom in the spatiotemporal domain. Basis decomposition can help us here.

Overlap is quantified by examining the deterministic temporal autocorrelation matrix, φ_h, with elements[6]

$$\varphi_h = \int_{-\infty}^{\infty} dt\, h_i(t)\, h_j(t) \quad i, j = -21, \ldots, 21. \tag{6.10}$$

The 41×41 matrix in this example represents a maximum of 21 possible degrees of freedom, one for each recorded measurement sample. φ_h matrices for the three devices represented by σ/T values are also given in Figure 6.1. However, only for $\sigma/T = 0.1$ does φ_h appear diagonal.

Eigenanalysis specifies the number of linearly independent rows in φ_h, which is the rank of the matrix and an estimate of the number of degrees of freedom. Eigenvalues[7] are the diagonal elements of

$$\Lambda_h = \mathbf{U}^\dagger \varphi_h \mathbf{U} \tag{6.11}$$

that we plot for the three examples at the bottom of Figure 6.1. Adjacent pairs of eigenvalues in the plots are equal because the symmetry $\varphi_h[i, j] = \varphi_h[j, i]$ introduces information redundancy; i.e., only half of the eigenvalues for φ_h carry independent information.[8] We define each *unique and nonzero* eigenvalue of φ_h as one *degree of freedom*. When there is additive acquisition noise, we may set an eigenvalue threshold slightly above zero (depending on the SNR) when counting signal degrees of freedom.

At one extreme in Figure 6.1, we have a device where $\sigma/T = 0.1$. Adjacent data samples are essentially uncorrelated. Since the eigenvalues are equal, the sensitivity

---

[6] A cleaner way to express φ_h in (6.10) is as $\varphi_h = \mathbf{H}\mathbf{H}^\mathsf{T}$, where $\mathbf{H} \in \mathbb{R}^{21 \times \infty}$. Stated this way, $\varphi_h \in \mathbb{R}^{21 \times 21}$ eliminates redundancies.

[7] You can find $\mathbf{U}$ and Λ from φ in MATLAB using [U,D]=eig(P), where D = Λ and P = φ.

[8] The Fourier basis matrix is full rank despite two-fold multiplicity because the eigenvalue pairs are linearly independent (e.g., positive and negative frequencies). When we consider correlation/spectral density functions, however, the functions are squared so the distinction between pairs is lost. The rank N Fourier basis matrix has a spectral density with only $N/2$ degrees of freedom.

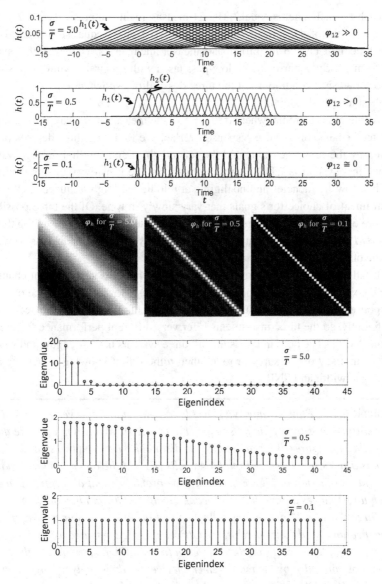

Figure 6.1 Each of the three top plots shows Gaussian impulse responses $h(t)$ separated by one unit of time over 20 units. The ratio of the σ width parameter for the three $h(t)$ functions divided by the one-unit sampling interval T is given. Smaller values of σ/T have less overlap and therefore provide more temporal resolution. These functions are used to compute 41-pt temporal autocorrelation function matrices φ_h via (6.10) as rendered in the grayscale images. The bottom three plots give the eigenspectra (diagonal elements of Λ) of the three φ_h matrices. There are 41 eigenvalues but only half, 21, are potential degrees of freedom. As the impulse response broadens across sampling intervals, the eigenvalue distribution changes, even though they all sum to the value 41. All of the potential degrees of freedom do not participate equally in generating measurements. The eigenvalues for the $\sigma/T = 5.0$ system show that it concentrates most of its sensitivity in the first few (low-frequency) channels, whereas the $\sigma/T = 0.1$ system spreads its sensitivity equally across the 41-channel bandwidth. Because we are not considering a task here, we are showing the potential for the three devices to record information.

of each eigenchannel is equal.[9] The 21 measurement samples generate 21 degrees of freedom (out of 41 autocorrelation channels), each of equal sensitivity. This is the instrument of choice for broadband measurement tasks. However, this system can perform poorly at narrowband, low-pass tasks with spectrally white acquisition noise under low-SNR conditions. Of the three possibilities shown in Figure 6.1, it offers the highest temporal resolution.

At the other extreme, where $\sigma/T = 5.0$, the measurement samples are highly correlated because of impulse-response overlap. We find only three degrees of freedom given by the first eigenvalue and the second–third and fourth–fifth eigenvalue pairs. Since this system is LTI/LSI, we can also say the sensitivity of this device is very high at low frequencies and negligible at middle and high frequencies. It will be the instrument of choice for signals that vary slowly in time. Of the three possibilities in Figure 6.1, it offers the greatest immunity to white acquisition noise. The third device has $\sigma/T = 0.5$. Its sensitivity distribution across the eigenchannels is between those to the other two systems.

In all three cases, the sum of eigenvalues is 41. This sum will not change unless the energy within each impulse response changes. However, the shape of an impulse response generates eigenvalues/Fourier coefficients that activate device modes very differently, so the three instruments offer very different performance features. Which is the "best"? The bottom line is that all three systems have strengths and weaknesses. We cannot say one is superior to another without first knowing something about the task bandwidth and SNR.

Example 6.3.2. *Flow cytometry: Consider a flow cytometer given the task of counting and sorting cells from a blood sample (Figure 6.2a). We need to evaluate the devices described in Figure 6.1 for their ability to distinguish between two similar-size targets: One is a stack of aggregated red blood cells (R) and the other is a large macrophage cell (M). At the top of Figure 6.2b, the time profile of light absorption is modeled as each target moves across the laser beam, giving the object function $f(t)$.*

Targets R and M are individually sampled according to our measurement-device specifications, and their correlation matrices φ_f are computed using $f(t)$ and (3.41). In this example, we focus on the degrees of freedom (dofs) required by the task, rather than those provided by the measurement device. We then apply the eigenvector matrix found in (6.11) to estimate eigenvalues for these object functions as sensed by the measurement device. Specifically, we use $\Lambda_f = U^\dagger \varphi_f U$. The eigenspectra for the two objects (diagonal elements of Λ_f) are plotted in Figure 6.2b. They show that R requires sensitivity of the system over a broader range of eigenvalues compared to M. We arranged the experiment such that the eigenvectors are equivalent to frequency channels, which means R requires more frequency bandwidth than M.

Comparing the object and device spectra, we find the low-bandwidth device ($\sigma/T = 5.0$ from Figure 6.1) will be about three times more sensitive to either large

---

[9] Here we set the energy contained in an impulse response to unity, i.e., $\int dt\, |h(t)|^2 = 1$. Consequently, the eigenvalues for the uncorrelated samples equal 1.

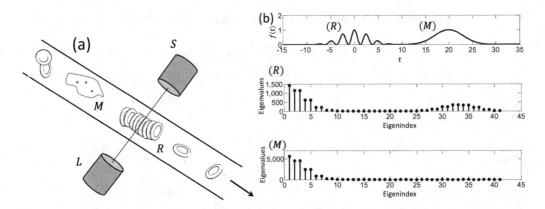

Figure 6.2 (a) Measurement of two target particles in a flow cytometer is illustrated. Fluid in a tube carries a stack of aggregated red blood cells (rouleau R) followed by a large macrophage M cell. To detect R, the task is to count the aggregate as a whole and also to recognize it is composed of individual cells. The task for M is to count the single large cell as such. L is the laser source and S is the transmitted light sensor. (b) The two targets are modeled as $f(t)$ and the eigenspectra for each task is given below.

target than the high-bandwidth device. However, only the high-bandwidth device
($\sigma/T = 0.1$) has the temporal resolution necessary to reliably distinguish between
the two targets. Its performance at detecting individual RBCs in rouleau also depends
on the SNR in the subspace defined by eigenvectors 28–39, over which there are 6
degrees of freedom.

6.3.4   Summary

Objects are continuous functions of space and time, and it is difficult to find a countable number of degrees of freedom in phenomena described by continuous functions. A digital measurement of a continuous object imposes a continuous-to-discrete transformation as the measurement device acquires samples, and this offers us an opportunity to determine degrees of freedom for a specific measurement task. The number of acquired samples fixes the upper bound on the number of degrees of freedom available to the decision maker. Correlations among samples will reduce the degrees of freedom to be less than the number of samples. Eigen/Fourier analysis then helps us find uncoupled properties of LTI measurement devices. When we apply the eigenvector matrix to a model of the object, we can determine the degrees of freedom required for the task and those provided by the measurement, enabling us to predict the system's performance. Figure 6.2b illustrates that the low-bandwidth system is far superior at detecting cells of size on the order of a macrophage. In contrast, if the task is to discriminate macrophages from rouleau, the high-bandwidth system is far superior. In summary, performance evaluation is task dependent.

Some of the results of Figures 6.1 and 6.2 were generated by the following script.

```
%%%%%%%%%%%%%%%%%%%%%%%%%% Code for results in Figs. 6.1 and 6.2 %%%%%%%%%%%%%%%%%%%%%%%%%%%%%
dt=0.01;t=-100:dt:100;                       % initialize parameters
x=1/sqrt(2*pi);s=10*dt;a=100;                % s is sigma of the impulse response (IR)
T=a*dt;                                      % and T = a*dt is sampling rate in time
h0=x/s*exp(-t.^2/(2*s^2));                   % IR at zero shift (s set to 10, 50, or 500)
lp=21;lp2=2*lp-1;
plot(t,h0);hold on;
hh=zeros(1,lp);
for m=1:lp                                   % generate IR sampled @ T to find ACF
    h1 = x/s*exp(-(t-(m-1)*T).^2/(2*s^2));plot(t,h1,'k')
    hh(m)=h1*h0'/(h0*h0');
end;hold off
h(1:lp)=hh;h(lp+1:lp2)=hh(lp:-1:2);          % first line of the ACF
H=zeros(lp2);H(1,:)=h;                        % make circulant ACF matrix
for j=2:lp2
    H(j,1:j-1)=h(lp2-j+2:lp2);
    H(j,j:lp2)=h(1:lp2-j+1);
end;figure;imagesc(H);colormap(gray);axis square % image the ACF matrix
[U,D]=eig(H);d=diag(D);                       % find eigenstates
figure;stem(d(lp2:-1:1),'k')
                                             % generate object function with two features
b=(cos(2*pi*t/5).^2).*exp(-t.^2/(2*3^2))+exp(-(t-20).^2/(2*3^2));
figure;plot(t,b)                              % use this & line above only to visual obj functs
% Now generate the two objects separately and form ACF sampled at rate aT
f=exp(-t.^2/(2*3^2)).*cos(2*pi*t/5).^2;
pff=xcorr(f,f);ppf=pff(10001:100:30001);P=zeros(lp2);
P(1,1:21)=ppf(101:121);P(1,22:41)=ppf(81:100);plot(P(1,:))
% pf is the autocorrelation of f sampled @ aT, like above,
% and is the same length as h so we can apply  U matrix and find
% the eigenvalues of f for this sampling rate (aT where a=100)
%
for j=2:lp2
    P(j,1:j-1)=P(1,lp2-j+2:lp2);
    P(j,j:lp2)=P(1,1:lp2-j+1);
end;figure;imagesc(P);colormap(gray);axis square & image the ACF matrix
DD=U'*P*U;dd=diag(DD);                         % find eigenstates using
figure;stem(dd(lp2:-1:1),'k')                  % same eigenvectors
diag(DD)'*diag(D)
```

6.4   Measurements as Vector-Space Transformations

Thus far in this chapter, we have analyzed functions and vectors in one vector space. By contrast, acquisition and display are mappings between different vector spaces, which is the topic of the next two sections. The analytical development in this section borrows from that of Barrett and Myers [10].

The measurement process can be modeled in two stages, as illustrated in Figure 5.5. An acquisition stage with additive noise transforms object vectors from vector space $\mathbb{U}$ into data vectors in vector space $\mathbb{V}$ via the linear operation $\mathbf{g} = \mathcal{H}\mathbf{f}$. A display stage prepares the acquired data for presentation to an observer by formatting the object under study via operator $\hat{\mathbf{f}} = \mathcal{O}\mathbf{g}$. In doing so, the display-stage operators remap data from data space $\mathbb{V}$ back into object space $\mathbb{U}$. When observers view the acquired data before the display stage is applied, they must mentally perform display-stage processes. The details implied by the general operators $\mathcal{H}$ and $\mathcal{O}$ become clear once you consider the nature of input $\mathbf{f}$ and output $\mathbf{g}$. This section discusses how these staged processes fundamentally limit task performance.

Consider the diagram shown in Figure 6.3. Objects $\mathbf{f}$ in vector space $\mathbb{U}$ that are in the range of acquisition operator $\mathcal{H}$ (let's call them $\mathbf{f}_m$) appear as $\mathbf{g}$ in vector space $\mathbb{V}$ as a result of the transformation. In the figure, a 3-D object vector is mapped into a 2-D data vector. Consistent with the results of Example 5.1.1, a reduced-dimension acquisition is not only acceptable but actually preferred, provided sufficient task information is captured to achieve decision-performance specifications.

6.4.1   One-to-One Operators

If each $\mathbf{g}$ in the range of $\mathcal{H}$ results from one and only one $\mathbf{f}$ in $\mathbb{U}$, then $\mathcal{H}$ is a *one-to-one operator*. For a one-to-one operator, the left-inverse operator exists; i.e.,

$$\mathcal{H}^{-1}\mathcal{H}\mathbf{f} = \mathbf{f}. \tag{6.12}$$

This equation states that compound operator $\mathcal{H}^{-1}\mathcal{H}$ gathers and then completely restores all object vectors in the range of $\mathcal{H}$ without distortion. A simple example is found by letting $f = a$ and operator $\mathcal{H} = 2\times$ be a two-fold amplifier of scalar a. Then, $\mathcal{H}^{-1} = 0.5\times$ (an attenuator) exists and (6.12) holds exactly. However, if $\mathcal{H}$ forms $\mathbf{g}$ by taking a self-inner product, e.g., $\mathbf{g} = \mathbf{f}^\dagger\mathbf{f}$, then $\mathcal{H}$ is not one-to-one. For example, the two complex scalar inputs $f_1 = a + ib$ and $f_2 = a - ib$ both give the same output, $g = a^2 + b^2$. With two input vectors in the range of $\mathcal{H}$ that map into the same output vector, the inverse operator does not strictly exist. Other examples of operators that are not one-to-one include measurement instruments with limited spatial or temporal resolution that are unable to uniquely map fine-scale spatiotemporal objects from $\mathbb{U}$ into data in $\mathbb{V}$. This was the case for two of the three flow cytometers described in Example 6.3.2.

6.4.2   Onto Operators

If all vectors in $\mathbb{V}$ are in the range of $\mathcal{H}$, then $\mathcal{H}$ is an *onto operator*. Figure 6.3 shows that some vectors in $\mathbb{V}$ are not in the range of $\mathcal{H}$ (outside of the shaded region in $\mathbb{V}$), which indicates $\mathcal{H}$ is not onto. If $\mathcal{H}$ is onto, the right-inverse operator exists; i.e.,

$$\mathcal{H}\mathcal{H}^{-1}\mathbf{g} = \mathbf{g}. \tag{6.13}$$

Existence of the right-inverse operator ensures that application of $\mathcal{H}^{-1}$ to $\mathbf{g}$, which maps from $\mathbb{V}$ to $\mathbb{U}$, is completely restored by then applying $\mathcal{H}$, which returns all vectors back to $\mathbb{V}$. Linear algebra shows that two operators, or two matrices, do not generally commute. Part of the reason is that operation order determines the space in which the resulting vector resides. An example of a non-onto operator is additive noise in $\mathbf{g}$ that does not pass through $\mathcal{H}$.

6.4.3   Invertible Measurements

A measurement that is strictly invertible is called a *perfect measurement*. Such a situation occurs when $\mathcal{H}$ is nonsingular so its inverse exists and is both one-to-one

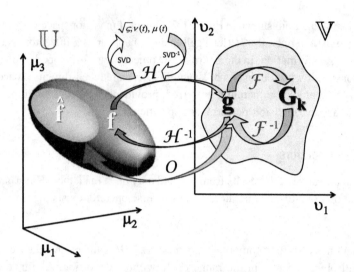

Figure 6.3 A vector-space representation of the measurement process and associated analyses. Acquisition is a mapping of $\mathbf{f}_m$ from $\mathbb{U}$ into data vectors $\mathbf{g}$ in $\mathbb{V}$ via $\mathbf{g} = \mathcal{H}\mathbf{f}$, where $\mathbf{f} = \mathbf{f}_m + \mathbf{f}_n$ and $\mathcal{H}\mathbf{f}_n = \mathbf{0}$. The shaded region in $\mathbb{V}$ indicates all $\mathbf{g}$ in the range of $\mathcal{H}$. If $\mathcal{H}$ is nonsingular, its inverse exists such that $\mathbf{f}_m = \mathcal{H}^{-1}\mathbf{g}$ can be found. In practice, $\mathcal{H}$ is singular, so we deploy display operator $\mathcal{O}$ to give estimates $\hat{\mathbf{f}} = \mathcal{O}\mathbf{g}$ designed to address the task. Also shown are Fourier transformations of data vectors within $\mathbb{V}$ and SVD of the acquisition operator between $\mathbb{U}$ and $\mathbb{V}$.

and onto. The procedure for displaying measurement data using an invertible acquisition operator is straightforward: Acquire $\mathbf{g}$ and apply $\mathcal{H}^{-1}$ to recover an undistorted description of $\mathbf{f}$. Unfortunately, no practical measurement devices have realistic operator representations that are strictly invertible. Thus, we are stuck with devising the suboptimal display operator $\mathcal{O}$ that approximates $\mathcal{H}^{-1}$ for the task; see Figure 6.3.

6.4.4   Null Spaces

If the rank of $\mathcal{H}$ does not equal the dimension of vector space $\mathbb{U}$ for a task, then there is a subspace of $\mathbb{U}$ that is not in the range of $\mathcal{H}$. That operator is singular, and there is a subspace in $\mathbb{U}$, called the *null space*, that contains vectors $\mathbf{f}_n$. The implication for measurements is that $\mathbf{f}_n$ cannot be recorded by $\mathcal{H}$. Invertible acquisition operators have only trivial null spaces, where $\mathcal{H}\mathbf{0} = \mathbf{0}$. *Singular operators* have nontrivial null spaces, where $\mathcal{H}\mathbf{f}_n = \mathbf{0}$ even though $\mathbf{f}_n \neq \mathbf{0}$. A consequence of $\mathcal{H}$ having a null space is

$$\mathbf{g} = \mathcal{H}\mathbf{f} = \mathcal{H}\{\mathbf{f}_m + \mathbf{f}_n\} = \mathcal{H}\mathbf{f}_m,$$

which is the commonsense result that the only object vectors appearing in the data are those in the range of $\mathcal{H}$. $\mathbf{f}_m$ are vectors in the *measurement subspace of* $\mathbb{U}$. Barrett and Myers point out that all vectors in the measurement subspace are orthogonal to those in the null subspace, so together the two subspaces span $\mathbb{U}$.

6.4.5   Adjoint Operators

If a perfect measurement does not exist outside of a model, then all realistic measurement operators are singular. Although inverse operator $\mathcal{H}^{-1}$ may not exist experimentally, *adjoint operator* $\mathcal{H}^{\dagger}$ always exists. Adjoint operators are discussed in Appendix A; they equal the inverse operators when $\mathcal{H}$ is unitary. Although we can write $\hat{\mathbf{f}} = \mathcal{H}^{\dagger}\mathcal{H}\mathbf{f}$, which returns us to the object space for display, we will rarely be happy with the results. Chapter 7 explores this topic with simulated images. The next section shows that adjoint operators have other useful functions.

6.5   Singular-Value Decomposition

Fourier analysis and eigenanalysis describe basis decompositions within a single vector space, as with the forward and inverse Fourier transforms illustrated in Figure 6.3. Generally, measurements are transformations between different vector spaces. We made this point explicit in the ultrasonic imaging example of Section 4.4, where in (4.24) acquisition was modeled as a transformation between a spatially varying object $f(x)$ and temporally varying data $g(t)$.

This section describes *singular-value decomposition (SVD)* as a general linear analysis and synthesis of measurement operators (see Figure 6.3). You may have run into SVD in linear algebra as a way to compute eigenvalues for rectangular matrices. Our focus now shifts to the matrix operator $\mathbf{H}$.

In the noise-free, discrete-to-discrete mapping, $\mathbf{g} = \mathbf{H}\mathbf{f}$, object vector $\mathbf{f}$ in $\mathbb{U}$ has size $N \times 1$,[10] data vector $\mathbf{g}$ in $\mathbb{V}$ has size $M \times 1$, and the linear matrix operator $\mathbf{H}$ that transforms between these vector spaces has size $M \times N$, where often $M \neq N$. Equation (5.4) decomposes $\mathbf{f}$ into an orthonormal basis $\boldsymbol{\mu}_k$ that spans $\mathbb{U}$:

$$\mathbf{f} = \sum_{k=0}^{N-1} \alpha_k \boldsymbol{\mu}_k. \tag{6.14}$$

Similarly, we can expand data vector $\mathbf{g}$,

$$\mathbf{g} = \sum_{k=0}^{N-1} \beta_k \boldsymbol{v}_k, \tag{6.15}$$

by choosing basis $\boldsymbol{v}_k$ that spans $\mathbb{V}$. Since $\mathbf{H}$ is a transformation between these two vector spaces, its decomposition must include both bases. Our goal in the following sections is to obtain (6.15) from (6.14) and in doing so reveal the inner workings of SVD.

---

[10] As we saw in Section 4.6 functions of two or more dimensions can be reshaped to form vector $\mathbf{f}$, which enables this approach to be generally applied to any object that can be represented as a function or data array.

6.5.1   Matrix Operators in $\mathbb{U}$

Rectangular $M \times N$ matrix $\mathbf{H}$ cannot be Hermitian. However, pre-multiplying by its adjoint matrix results in $\mathbf{H}^\dagger \mathbf{H} \in \mathbb{R}^{N \times N}$, which is Hermitian since $(\mathbf{H}^\dagger \mathbf{H})^\dagger = \mathbf{H}^\dagger \mathbf{H}$ and it is positive semi-definite. Therefore, $\mathbf{H}^\dagger \mathbf{H}$ has real nonnegative eigenvalues that describe the ability of the measurement process to transfer object information into data. We note that $\text{rank}(\mathbf{H}) = \text{rank}(\mathbf{H}^\dagger \mathbf{H}) = R$ and is bounded by $R \leq \min(M, N)$.[11] Equality indicates that appropriate acquisition bandwidth is possible.

$\mathbf{H}^\dagger \mathbf{H}$ is a transformation from $\mathbb{U}$ to $\mathbb{V}$, and then back to $\mathbb{U}$, so we can use the same basis found in the expansion of the object vector $\mathbf{f}$ in (6.14). The eigenstates of the $N \times N$ matrix $\mathbf{H}^\dagger \mathbf{H}$ are found from the characteristic equation,

$$\mathbf{H}^\dagger \mathbf{H} \boldsymbol{\mu}_k = \varsigma_k \boldsymbol{\mu}_k \quad 0 \leq k \leq N - 1. \tag{6.16}$$

Here, $\boldsymbol{\mu}_k$ are the eigenvectors and ς_k are the corresponding eigenvalues in decreasing order,

$$\varsigma_0 > \varsigma_1 > \cdots > \varsigma_{R-1} > \varsigma_R > \cdots \varsigma_k \cdots > \varsigma_{N-1}.$$

We are using a variation on the sigma symbol ς_k to represent eigenvalues in this section. For $\text{rank}(\mathbf{H}^\dagger \mathbf{H}) = R$, real $\varsigma_k > 0$ for $0 \leq k \leq R - 1$ and $\varsigma_k \simeq 0$ for $R \leq k \leq N - 1$.

6.5.2   Matrix Operators in $\mathbb{V}$

Left-multiplying $\mathbf{H}$ on both sides of (6.16) gives [10]

$$\mathbf{H} \mathbf{H}^\dagger (\mathbf{H} \boldsymbol{\mu}_k) = \varsigma_k (\mathbf{H} \boldsymbol{\mu}_k), \tag{6.17}$$

which is a mapping from $\mathbb{U}$ to $\mathbb{V}$. This expression shows that $\mathbf{H} \boldsymbol{\mu}_k$ is an eigenvector of the complementary compound matrix operator $\mathbf{H} \mathbf{H}^\dagger$ having size $M \times M$. (Parentheses identify the new unscaled eigenvector.) However, $\mathbf{H}^\dagger \mathbf{H}$ and $\mathbf{H} \mathbf{H}^\dagger$ are not equal. Although they share the same eigenvalues, ς_k, they have different eigenvectors, $\{\boldsymbol{\mu}_k\}$ and $\{\mathbf{H} \boldsymbol{\mu}_k\}$, respectively.

As seen from (6.16), eigenvectors $\mathbf{H} \boldsymbol{\mu}_k$ are orthogonal but not orthonormal since

$$(\mathbf{H} \boldsymbol{\mu}_k)^\dagger \mathbf{H} \boldsymbol{\mu}_\ell = \boldsymbol{\mu}_k^\dagger \mathbf{H}^\dagger \mathbf{H} \boldsymbol{\mu}_\ell = \varsigma_k \delta_{k\ell} = \begin{cases} \varsigma_k & k = \ell \\ 0 & \text{otherwise} \end{cases},$$

and there is no reason for all the eigenvalues ς_k to be 1. However,

$$\frac{1}{\sqrt{\varsigma_k}} \boldsymbol{\mu}_k^\dagger \mathbf{H}^\dagger \mathbf{H} \boldsymbol{\mu}_k \frac{1}{\sqrt{\varsigma_k}} = 1$$

---

[11] This statement is evidence of the fundamental role of object properties in assessments of measurement capabilities and performance. M depends on the acquisition bandwidth and sampling rate. Interpreting N as an object subspace rank encompassing the task, there is insufficient resolution when $M < N$.

ensures orthonormality. The new orthonormal basis is designated as

$$v_k = \frac{1}{\sqrt{\varsigma_k}} H\mu_k \quad \text{for } 0 \leq k \leq R-1, \text{ where } \varsigma_k > 0. \tag{6.18}$$

It makes sense that the basis for the data vectors is given by the basis for the object vectors after passing them through the acquisition operator and scaling. Because all data vectors $\mathbf{g}$ must fall in the range of $\mathbf{H}$, the size of basis set ν required to span $\mathbb{V}$ is limited to the rank of $\mathbf{H}$.

$\mathbf{H}$ and $\mathbf{H}^\dagger$ can be synthesized [10] from singular components $(\varsigma_k, \mu_k, \nu_k)$ by combining (6.16) and (6.18) for each eigenvector and summing the resulting nonzero matrices $0 \leq k \leq R-1$:

$$\mathbf{H}^\dagger \mathbf{H} \mu_k = \varsigma_k \mu_k$$
$$\mathbf{H}^\dagger(\sqrt{\varsigma_k} \nu_k) = \varsigma_k \mu_k$$
$$\mathbf{H}^\dagger \nu_k \nu_k^\dagger = \mathbf{H}^\dagger \delta_{kk} = \sqrt{\varsigma_k} \mu_k \nu_k^\dagger.$$

Summing over the R nonzero elements, we have

$$\mathbf{H}^\dagger = \sum_{k=0}^{R-1} \sqrt{\varsigma_k} \mu_k \nu_k^\dagger, \tag{6.19}$$

$$\mathbf{H} = (\mathbf{H}^\dagger)^\dagger = \left(\sum_{k=0}^{R-1} \sqrt{\varsigma_k} \mu_k \nu_k^\dagger \right)^\dagger = \sum_{k=0}^{R-1} \sqrt{\varsigma_k} \nu_k \mu_k^\dagger.$$

To check for consistency, substitute both expressions from (6.19) into (6.16):

$$\mathbf{H}^\dagger \mathbf{H} \mu_k = \left[\sum_{k'=0}^{R-1} \sqrt{\varsigma_{k'}} \sum_{\ell=0}^{R-1} \sqrt{\varsigma_\ell} \, \mu_{k'} \nu_{k'}^\dagger \nu_\ell \mu_\ell^\dagger \right] \mu_k$$

$$= \left[\sum_{k'=0}^{R-1} \sqrt{\varsigma_{k'}} \sum_{\ell=0}^{R-1} \sqrt{\varsigma_\ell} \, \mu_{k'} \delta_{k'\ell} \mu_\ell^\dagger \right] \mu_k$$

$$= \left[\sum_{k'=0}^{R-1} \varsigma_{k'} \, \mu_{k'} \mu_{k'}^\dagger \right] \mu_k = \sum_{k'=0}^{R-1} \varsigma_{k'} \, \mu_{k'} \delta_{k'k} = \varsigma_k \, \mu_k.$$

Equation (6.19) shows that the decomposition of $\mathbf{H}$ involves the bases of both the object and the data, as it must. μ_k spans object space $\mathbb{U}$, which is composed of the measurement $\{\mathbf{f}_m\}$ and null $\{\mathbf{f}_n\}$ subspaces, and ν_k spans the data space $\mathbb{V}$ with elements $\{\mathbf{g}\}$, which are limited to the range of $\mathbf{H}$. We found that $\{\varsigma_k\}$ are the eigenvalues of both $\mathbf{H}^\dagger \mathbf{H}$ and $\mathbf{H}\mathbf{H}^\dagger$. Since they act as the interface between the two vector spaces, it makes sense that $\{\sqrt{\varsigma_k}\}$ are the *singular values* for both $\mathbf{H}$ and $\mathbf{H}^\dagger$.

---

[12] The square root of a diagonal matrix is just the square root of each element.
[13] MATLAB may not give the same eigenvectors found by hand calculations.

6.5.3   Matrix Forms

We may compose matrix $\mathbf{U}$ to have columns made up of μ_k and matrix $\mathbf{V}$ to have columns composed of v_k. If the diagonal $M \times N$ matrix[12] is $\mathbf{\Sigma}^{1/2}$, where $\left[\mathbf{\Sigma}^{1/2}\right]_{kk} = \sqrt{\varsigma_k}$, the SVD equations of (6.19) become matrix expressions:[13]

$$\mathbf{H} = \mathbf{V}\,\mathbf{\Sigma}^{1/2}\,\mathbf{U}^\dagger, \qquad \mathbf{H}^\dagger = \mathbf{U}\,\mathbf{\Sigma}^{1/2}\,\mathbf{V}^\dagger \quad \text{and} \quad \mathbf{\Sigma}^{1/2} = \mathbf{V}^\dagger\,\mathbf{H}\,\mathbf{U}. \quad \text{(SVD)} \qquad \text{(6.20)}$$

The columns of $\mathbf{U}$ are *singular vectors* describing object modes accessible during acquisition, while those of $\mathbf{V}$ describe data modes resulting from acquisition. Consequently, eigenvalues ς_k summarize properties of the connection between the object task and the measurement device. Let's try a simple numerical example to get the mechanics of SVD down.

Example 6.5.1. *Decompose real 2×3 matrix $\mathbf{A}$ using SVD, where* $\mathbf{A} = \begin{pmatrix} 1 & 0 & 2 \\ 0 & 1 & 0 \end{pmatrix}$.

Solution
To find $\mathbf{V} \in \mathbb{C}^{2 \times 2}$:

$$\mathbf{A}\mathbf{A}^\top = \begin{pmatrix} 1 & 0 & 2 \\ 0 & 1 & 0 \end{pmatrix} \begin{pmatrix} 1 & 0 \\ 0 & 1 \\ 2 & 0 \end{pmatrix} = \begin{pmatrix} 5 & 0 \\ 0 & 1 \end{pmatrix}.$$

The result is diagonal (convenient!), so the eigenvalues are $\varsigma = 5, 1$.
   For eigenvalue $\varsigma = 5$, we solve $(\mathbf{A}\mathbf{A}^\top - \varsigma\mathbf{I})\mathbf{v} = \mathbf{0}$ for $\mathbf{v}$:

$$\begin{pmatrix} 5-5 & 0 \\ 0 & 1-5 \end{pmatrix} \begin{pmatrix} v_1 \\ v_2 \end{pmatrix} = \begin{pmatrix} 0 & 0 \\ 0 & -4 \end{pmatrix} \begin{pmatrix} v_1 \\ v_2 \end{pmatrix} = \begin{pmatrix} 0 \\ 0 \end{pmatrix} \quad \rightarrow \quad \begin{matrix} v_1 = r \\ v_2 = 0 \end{matrix} \quad v = r \begin{pmatrix} 1 \\ 0 \end{pmatrix}.$$

Since r is any constant, we see there are an infinite number of vectors that satisfy the equation. However, we are most interested in orthonormal vectors, which narrows the possibilities to that satisfying $\|\mathbf{v}\| = r\sqrt{1^1 + 0^2} = 1$. Solving for r, we find $r = 1$.
   For eigenvalue $\varsigma = 1$:

$$\begin{pmatrix} 5-1 & 0 \\ 0 & 1-1 \end{pmatrix} \begin{pmatrix} v_1 \\ v_2 \end{pmatrix} = \begin{pmatrix} 0 \\ 0 \end{pmatrix} \quad \rightarrow \quad \begin{matrix} 4v_1 = 0 \\ v_2 = s \end{matrix} \quad v = s \begin{pmatrix} 0 \\ 1 \end{pmatrix}.$$

Of course, $s = 1$ normalizes this eigenvector.
   The singular vector and singular value matrices are $\mathbf{V} = \begin{pmatrix} 1 & 0 \\ 0 & 1 \end{pmatrix}$, $\mathbf{\Sigma}^{1/2} = \begin{pmatrix} \sqrt{5} & 0 & 0 \\ 0 & 1 & 0 \end{pmatrix}$.
   To find $\mathbf{U} \in \mathbb{C}^{3 \times 3}$:

$$\mathbf{A}^\top\mathbf{A} = \begin{pmatrix} 1 & 0 \\ 0 & 1 \\ 2 & 0 \end{pmatrix} \begin{pmatrix} 1 & 0 & 2 \\ 0 & 1 & 0 \end{pmatrix} = \begin{pmatrix} 1 & 0 & 2 \\ 0 & 1 & 0 \\ 2 & 0 & 4 \end{pmatrix}.$$

The eigenvalues are

$$\begin{vmatrix} 1-\varsigma & 0 & 2 \\ 0 & 1-\varsigma & 0 \\ 2 & 0 & 4-\varsigma \end{vmatrix} = 0 \qquad \begin{aligned} (1-\varsigma)(1-\varsigma)(4-\varsigma) - 4(1-\varsigma) = 0 \\ (1-\varsigma)[\varsigma^2 - 5\varsigma + 4 - 4] = 0 \\ (1-\varsigma)(\varsigma - 5)\varsigma = 0 \end{aligned}$$

or $\varsigma = 5, 1, 0$ but we already knew that from computing the eigenvalues for **V**. *Recall that eigenvector matrices* **U** *and* **V** *share the same eigenvalues. Two of them were found earlier to be 5 and 1, so the remaining one must be 0.*
   For $\varsigma = 5$:

$$\begin{pmatrix} -4 & 0 & 2 \\ 0 & -4 & 0 \\ 2 & 0 & -1 \end{pmatrix} \begin{pmatrix} u_1 \\ u_2 \\ u_3 \end{pmatrix} = \mathbf{0} \ \rightarrow \ \begin{aligned} -4u_1 + 2u_3 = 0 \\ -4u_2 = 0 \\ 2u_1 - u_3 = 0 \end{aligned} \quad \begin{aligned} u_2 = 0 \\ u_3 = 2u_1 \end{aligned}$$

$$\mathbf{u} = r \begin{pmatrix} 1 \\ 0 \\ 2 \end{pmatrix} \Rightarrow \frac{1}{\sqrt{5}} \begin{pmatrix} 1 \\ 0 \\ 2 \end{pmatrix},$$

because $\|\mathbf{u}\| = r\sqrt{1^2 + 2^2} = 1$ gives $r = 1/\sqrt{5}$ as the orthonormal eigenvector.
   For $\varsigma = 1$:

$$\begin{pmatrix} 0 & 0 & 2 \\ 0 & 0 & 0 \\ 2 & 0 & 3 \end{pmatrix} \begin{pmatrix} u_1 \\ u_2 \\ u_3 \end{pmatrix} = \mathbf{0} \ \rightarrow \ \begin{aligned} 2u_3 = 0 \\ u_2 = s \\ 2u_1 + 3u_3 = 0 \end{aligned}$$

$$\begin{aligned} u_1 = u_3 = 0 \\ u_2 = s \end{aligned} \quad \mathbf{u} = s \begin{pmatrix} 0 \\ 1 \\ 0 \end{pmatrix} \Rightarrow \begin{pmatrix} 0 \\ 1 \\ 0 \end{pmatrix}.$$

For $\varsigma = 0$:

$$\begin{pmatrix} 1 & 0 & 2 \\ 0 & 1 & 0 \\ 2 & 0 & 4 \end{pmatrix} \begin{pmatrix} u_1 \\ u_2 \\ u_3 \end{pmatrix} = \mathbf{0} \ \rightarrow \ \begin{aligned} u_1 + 2u_3 = 0 \\ u_2 = 0 \end{aligned} \quad \begin{aligned} u_3 = t \\ u_2 = 0 \\ u_1 = -2t \end{aligned}$$

$$\mathbf{u} = t \begin{pmatrix} -2 \\ 0 \\ 1 \end{pmatrix} \Rightarrow \frac{1}{\sqrt{5}} \begin{pmatrix} -2 \\ 0 \\ 1 \end{pmatrix}.$$

Assembling the singular vectors as columns, $\mathbf{U} = \begin{pmatrix} \frac{1}{\sqrt{5}} & 0 & -\frac{2}{\sqrt{5}} \\ 0 & 1 & 0 \\ \frac{2}{\sqrt{5}} & 0 & \frac{1}{\sqrt{5}} \end{pmatrix}$. *The order that these vectors are placed in* **U** *must be consistent with $\Sigma^{1/2}$ elements.*
   Finally, (and notice the transpose!)

$$\mathbf{A} = \mathbf{V}\Sigma^{1/2}\mathbf{U}^\top = \begin{pmatrix} 1 & 0 \\ 0 & 1 \end{pmatrix} \begin{pmatrix} \sqrt{5} & 0 & 0 \\ 0 & 1 & 0 \end{pmatrix} \begin{pmatrix} \frac{1}{\sqrt{5}} & 0 & -\frac{2}{\sqrt{5}} \\ 0 & 1 & 0 \\ \frac{2}{\sqrt{5}} & 0 & \frac{1}{\sqrt{5}} \end{pmatrix}^\top = \begin{pmatrix} 1 & 0 & 2 \\ 0 & 1 & 0 \end{pmatrix}.$$

We use transpose here because all matrix elements are real. Consequently, $\mathbf{U}$ and $\mathbf{V}$ are orthogonal in this case; e.g., $\mathbf{U}^{-1} = \mathbf{U}^{\top}$. Also, using (6.19), we find

$$\mathbf{H} = \sum_R \sqrt{\varsigma_k} \boldsymbol{v}_k \boldsymbol{\mu}_k^{\top} = \sqrt{5} \begin{pmatrix} 1 \\ 0 \end{pmatrix} \begin{pmatrix} 1/\sqrt{5} & 0 & 2/\sqrt{5} \end{pmatrix} + \sqrt{1} \begin{pmatrix} 0 \\ 1 \end{pmatrix} \begin{pmatrix} 0 & 1 & 0 \end{pmatrix}$$

$$= \begin{pmatrix} 1 & 0 & 2 \\ 0 & 1 & 0 \end{pmatrix}.$$

It works!

6.5.4   Pseudoinverse

The Moore–Penrose pseudoinverse from Appendix A is accurately computed using SVD [41]. Selecting $\mathbf{A} \in \mathbb{R}^{M \times N}$ from Example 6.5.1, where $M = 2$ and $N = 3$, and applying $\mathbf{x} = \begin{pmatrix} 1 \\ 2 \\ 3 \end{pmatrix}$, we find $\mathbf{y} = \mathbf{Ax} = \begin{pmatrix} 7 \\ 2 \end{pmatrix}$. Let's assume we know only $\mathbf{y}$ and $\mathbf{A}$ and we want to estimate unknown $\mathbf{x}$. Since $\mathbf{A}$ is singular, its inverse does not exist. However, you may compute the pseudoinverse $\mathbf{A}^+$ and then use $\hat{\mathbf{x}} = \mathbf{A}^+\mathbf{y}$. Let's try.

Example 6.5.2. *Inverting a linear system of equations: Find $\hat{\mathbf{x}}$.*

Solution
If $\mathbf{A}$ was nonsingular, then for

$$\mathbf{A} = \mathbf{V}\boldsymbol{\Sigma}^{1/2}\mathbf{U}^{\dagger}$$

we have

$$\mathbf{A}^{-1} = \left(\mathbf{V}\boldsymbol{\Sigma}^{1/2}\mathbf{U}^{\dagger}\right)^{-1} = (\mathbf{U}^{\dagger})^{-1}\boldsymbol{\Sigma}^{-1/2}\mathbf{V}^{-1} = \mathbf{U}\boldsymbol{\Sigma}^{-1/2}\mathbf{V}^{\dagger},$$

recalling that $\mathbf{U}$ and $\mathbf{V}$ are unitary. However, $\mathbf{A}$ is not square, so the pseudoinverse may be applied:

$$\mathbf{A}^+ = \mathbf{U}[\boldsymbol{\Sigma}^{1/2}]^+\mathbf{V}^{\dagger}. \tag{6.21}$$

We know $\mathbf{U}$ and $\mathbf{V}$ from Example 6.5.1. Also, since $\boldsymbol{\Sigma}^{1/2} = \begin{pmatrix} \sqrt{5} & 0 & 0 \\ 0 & 1 & 0 \end{pmatrix}$, then $[\boldsymbol{\Sigma}^{1/2}]^+$ is the transpose of the matrix where all the nonzero elements are inverted:

$$\begin{pmatrix} \frac{1}{\sqrt{5}} & 0 \\ 0 & 1 \\ 0 & 0 \end{pmatrix}.$$

Consequently,

$$\hat{\mathbf{x}} = \mathbf{A}^+\mathbf{y} = \mathbf{U}[\boldsymbol{\Sigma}^{1/2}]^+\mathbf{V}^\dagger\mathbf{y}$$

$$= \begin{pmatrix} \frac{1}{\sqrt{5}} & 0 & -\frac{2}{\sqrt{5}} \\ 0 & 1 & 0 \\ \frac{2}{\sqrt{5}} & 0 & \frac{1}{\sqrt{5}} \end{pmatrix} \begin{pmatrix} \frac{1}{\sqrt{5}} & 0 \\ 0 & 1 \\ 0 & 0 \end{pmatrix} \begin{pmatrix} 1 & 0 \\ 0 & 1 \end{pmatrix} \begin{pmatrix} 7 \\ 2 \end{pmatrix} = \begin{pmatrix} \frac{7}{5} \\ 2 \\ \frac{14}{5} \end{pmatrix} = \begin{pmatrix} 1.4 \\ 2 \\ 2.8 \end{pmatrix} \simeq \begin{pmatrix} 1 \\ 2 \\ 3 \end{pmatrix}.$$

As we can see, this example does not give a great approximate inverse. However, a functional inverse is nevertheless provided even when the matrix does not strictly have one.

In MATLAB computations, matrix elements larger than ϵ are considered nonzero, while all others are replaced by zeros. Tolerance ϵ is computed using $\epsilon = \Delta \times \max(M,N) \times \|\mathbf{A}\|$. $\Delta \triangleq$ eps is the *machine precision*. To obtain double-precision values in 64-bit MATLAB, specify eps=2.2204e-16. In this specific example, $\max(M,N) = $ max(size(A))=3 and $\|\mathbf{A}\| = \sqrt{5} = $ norm(A)=2.2361. Tolerance was not a factor in this toy problem. However, in large matrices with stochastic elements, tolerance ϵ does become an important consideration.

6.5.5   Visualizing Singular Components

To illustrate SVD analysis in a 2-D imaging format, we created the small 3×3 plus-sign–shaped 2-D impulse-response function, $h(x_1,x_2)$, shown in Figure 6.4a. It is reshaped to form a 9×9 noncirculant system matrix $\mathbf{H}$ that is decomposed using

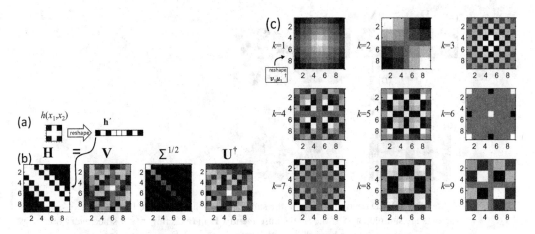

Figure 6.4 (a) Illustration of an SVD decomposition of 2-D impulse response $h(x_1,x_2)$. Assuming an LSI system, and reshaping $h(x_1,x_2)$, we obtain the Toeplitz (but noncirculant) system matrix $\mathbf{H}$. (b) It is decomposed into the singular system $\mathbf{V}$, $\boldsymbol{\Sigma}^{1/2}$ and $\mathbf{U}^\dagger$. (c) The first nine response modes are illustrated from reshaped outer products, $v_k\mu_k^\dagger$, $1 \leq k \leq 9$.

(6.20). Recall that **H** encodes shifted copies of impulse response **h** in rows that execute a convolution when **H** is multiplied by **f**.

We set up this 2-D convolution to index one sample at a time. Here, **f** is 9×1, so we know **g** will also be 9×1 and **H** is 9×9, i.e, $M = N = 9$. The three 9×9 matrices resulting from the SVD MATLAB function $[\texttt{V},\texttt{D},\texttt{U}] = \texttt{svd(H)}$; are not recognizable except for $\texttt{D} \triangleq \mathbf{\Sigma}^{1/2}$, which is diagonal and in this case full rank as each $\varsigma_k > 0$. In Figure 6.4(c), we formed the nine *outer products* $\boldsymbol{v}_k \boldsymbol{\mu}_k^\dagger$ that are the function modes of the impulse response – the SVD equivalent of the sinusoidal basis in Fourier analysis. Just as $\sqrt{\varsigma_k}$ in SVD analysis is analogous to $H[k]$ in Fourier analysis, SVD modes $\boldsymbol{v}_k \boldsymbol{\mu}_k^\dagger$ are analogous to the orthonormal Fourier basis $\exp(-2\pi nk/N)/\sqrt{N}$. When the modes are scaled by $\sqrt{\varsigma_k}$ and summed, the result is exactly **H**, as (6.19) predicts.

Downsampling.

In Figure 6.5, the sampling rate was reduced by a factor of 2. With the same impulse response as that in Figure 6.4 (notice the rows of **H** have not changed), the lower sampling rate effectively discards even rows of the **H** in Figure 6.4. Consequently, the 9×9 matrix **H** in Figure 6.5 is now rectangular and 5×9. The reduced-size **H** increases the size of the null space, making it more difficult to successfully record fine-scale object information. At the same time, it speeds computation as needed when the problem is scaled up.

The diagonal elements of the $\mathbf{\Sigma}^{1/2}$ matrices in Figures 6.4 and 6.5 are the SVD spectrum similar to the Fourier frequency spectrum. The SVD spectrum is monotonically decreasing only because we order the real, nonnegative singular values

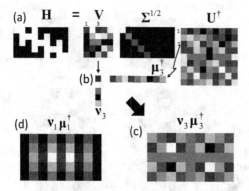

Figure 6.5 An SVD decomposition of the same 2-D impulse response from Figure 6.4, but now the data are sampled at half the rate. Consequently, **H** in (a) is now 5×9 since every other row of **H** in Figure 6.4 (a) has been discarded. The basis vectors have changed to accommodate the change in sampling rate. We now have $N = 9$, $M = 5$, and rank $= 5$. (b) shows how the third column of **V** and the third row of $\mathbf{U}^\dagger$ are isolated; from their outer product, we obtain the third mode in (c). (d) shows the first mode. Note that basis-vector elements can be positive or negative and that the matrix images were auto-scaled.

that way. The number of spectrum values and the spectrum shape gives a sense of measurement bandwidth.

These examples graphically illustrate SVD analysis on simplistic toy problems. Of course, the methods examined in these small-scale examples can be scaled up to address more realistic measurement analyses.

6.6 SVD Indicates Dimensionality of Motion

Fourier analysis is fairly straightforward to understand as a weighted linear combination of sinusoidal bases. SVD analysis is less intuitive because there are two bases $\mathbf{U}$ and $\mathbf{V}$ from the object and data vector spaces, whose columns combine via $v_k \mu_k^\dagger$ to form functional modes. This section illustrates some of the information provided by SVD analysis by considering the three data sets summarized in Figure 6.6b. Each 3-D array $\mathbf{X}(x_1, x_2, t) \in \mathbb{R}^{N \times N \times 30}$ views a $N \times N$ ($N = 100$) stationary, spatially random field move over 30 frames.

After moving, the random frame is reordered using reshape (X,10000,1) into a $N^2 \times 1$ spatial vector $(x_1, x_2) \to x$. Thirty column vectors are assembled into 2-D array $\mathbf{X}(x, t) \in \mathbb{R}^{10000 \times 30}$, with one vector for each time frame. Decomposing that matrix via SVD,[14] $\mathbf{X} = \mathbf{V}\mathbf{\Sigma}^{1/2}\mathbf{U}^\dagger$, we plot the singular values in decreasing order for each set in Figure 6.6c. Also shown in Figure 6.6 are the singular-vector matrices $\mathbf{U}$ and $\mathbf{V}$ for the constant motion (d) and gradient motion (e) arrays.

There will be a singular value for each of the $\min(N^2, 30) = 30$ modes describing variance Figure 6.6c. For the *stationary* array that is just 30 exact copies of one random field, there is just one nonzero singular value. We can rewrite the general form of the function as $\mathbf{X}(x_1, x_2, t) = X(x)$ for this data array. Only one spatial dimension carries variance, $(x_1, x_2) \to x$, which is (trivially) independent of temporal variance. The one singular value suggests this stationary process has just one degree of freedom.

For the *constant* velocity array, the first singular value is significant, but the next 29 singular values are approximately equal and nonzero. The Casorati form of the data matrix[15] has rank 30, but the near-equality of 29 singular values tells us they are not providing independent information. We can write $\mathbf{X}(x_1, x_2, t) = X(x_1 - v_1 t)X(x_2 - v_2 t)$ for this array, where $v_1 = v_2$ are the constant velocity components. Two variables (degrees of freedom) are required to describe the separable function, but they are linear combinations of the original three: $(x_1, x_2, t) \to (x_1 - v_1 t, x_2 - v_2 t)$.

For the *gradient* velocity array, pixels move down and to the right, but mostly down with $v_2 t \propto x_2$. These singular values resemble the constant-motion data except that singular values decrease, indicating that new information is provided with each time

[14] The code in this section uses "economy" SVD, as indicated by the "0" in the argument.
[15] This Casorati matrix of spatiotemporal data orients spatial variations along one axis and temporal variations along the other axis. Like the Wronskian matrix of two differential functions, a principal application of the Casorati matrix is to test for linear independence.

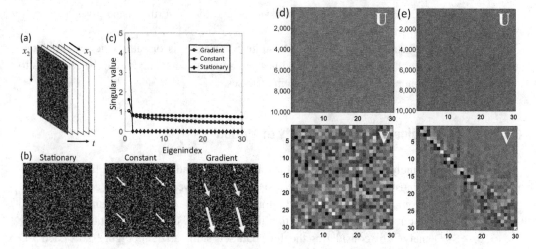

Figure 6.6 (a) $\mathbf{X}(x_1, x_2, t) \in \mathbb{R}^{N_1 \times N_2 \times 30}$ is a time series of 2-D spatially random fields that can vary over time. (b) The array marked *stationary* is 30 copies of one stationary random field. The *constant* array has the random field moving uniformly down and to the right an equal amount between frames. The *gradient* array imposes a gradient of motion between frames as indicated by arrows. (c) The singular values for the three data sets are also plotted. Singular-vector matrices $\mathbf{U}$ and $\mathbf{V}$ for the constant motion (d) and gradient motion (e) arrays are also shown.

frame. I'm not sure how to write $\mathbf{X}(x_1, x_2, t)$ in detail for this data set, but we can say the problem is 3-D and that the spatial and temporal features are statistically dependent.

The spatial singular matrices in parts (d) and (e) of Figure 6.6, designated as $\mathbf{U}$, are similar for constant and gradient motions. Both reflect a stationary random field. However, the temporal singular matrix $\mathbf{V}$ for the gradient array shows a linear pattern reflecting the velocity gradient, whereas this pattern is not seen in $\mathbf{V}$ for constant motion.

The following script generates the gradient array from Figure 6.6. Note how every particle in the random field is tracked so it can be moved continuously between frames with minimum discretization error. We return to *list-mode* data representations in Chapter 9.

```
%%%%%%%%%%%%%%%%%%%%%%%%%%%%%% Code for data in Fig. 6.6 %%%%%%%%%%%%%%%%%%%%%%%%%%%%%%%%%%%%%%%
%fist generate one random frame
x=0:0.01:10;N=length(x);s=0.01;xp=x(451:551);Np=length(xp);       % distance=mm
h1=exp(-(xp-5).^2/(2*s^2))/(s*sqrt(2*pi));h=zeros(Np);             % impulse response, s controls blur
for j=1:Np                                                        % 2D Gaussian blur
    h(j,:)=h1.*exp(-(xp(j)-5)^2/(2*(0.7*s)^2))/(0.7*s*sqrt(2*pi));
end
Nx=10^5;f=zeros(N,N,30);g=zeros(N,N,900);                         %10^5 pt targets in 10x10 mm random field
p=1000*rand(1,Nx);q=1000*rand(1,Nx);
pp=ceil(p);qq=ceil(q);
for j=1:Nx                                                        % position randomly
    f(qq(j),pp(j),1)=f(qq(j),pp(j),1)+1;
end
ff=f(:,:,1);ff(1:Np,1:Np)=ff(1:Np,1:Np)+h/10;
gp=conv2(f(:,:,1),h,'valid');                                     % measure the random field
```

```
pq=p;qp=q;shift=100*sqrt(2)/30;g=zeros(length(gp),length(gp),30);
g(:,:,1)=gp;
shiftq=zeros(size(qp));SS=zeros(1,30);
for j=1:Nx                                       % move elements to form 30-frame sequence
   shiftq(j)=16*shift*qp(j)/max(qp);             % set shift gradient
end
for k=2:30
   pq=pq+shift;qp=qp+shiftq;                     % use shift in place of shiftq for constant motion
   for l=1:Nx
      if pq(l) > 1001;pq(l)=1001;end             % stay in range
      if qp(l) > 1001;qp(l)=1001;end             % stay in range
   end
   pp=ceil(pq);qq=ceil(qp);
   for j=1:Nx
      f(qq(j),pp(j),k)=f(qq(j),pp(j),k)+1;
   end
   g(:,:,k)=conv2(f(:,:,k),h,'valid');           %only use when blurring data
end
%z=uint8(100*f); %unblurred data
z=uint8(g/7);                                    % blurred data
implay(z)                                        % play movie of motion
%%
% Reshape central 100x100 sections and run svd analysis
F=zeros(10000,30);
for j=1:30
   F(:,j)=reshape(g(451:550,451:550,j),10000,1); %change to either f or g!!!
end
[U,S,V]=svd(F,0);                                % use economical svd (set by "0")
for j=1:30
   SS(j)=S(j,j);
end
plot(SS)                                         % plot the singular values
%%
figure
Q1=V(:,1)*U(:,1)';subplot(3,1,1);imagesc(Q1,[0 0.005]);colormap gray;axis square  % modal images
Q2=V(:,5)*U(:,5)';subplot(3,1,2);imagesc(Q2,[-0.01 0.01]);colormap gray;axis square
Q3=V(:,15)*U(:,15)';subplot(3,1,3);imagesc(Q3,[-0.01 0.01]);colormap gray;axis square
```

6.7   Linear Shift-Varying Measurement Systems

Recall that Fourier analysis describes acquisitions in frequency space as transform products, e.g., $G[k] = H[k] F[k]$, as long as the acquisitions can be modeled as LTI/LSI. The intuition is straightforward: If the response of a linear-system instrument does not vary significantly with time or spatial range, we can characterize the instrument response function using a single spatial/temporal function composed of one variable that includes both the object the x' and data x domains, e.g., $h(x - x')$. The response of a system at one spatial frequency applies to all positions in the analysis range, and the response at one location is reflected by a weighted linear combination of all frequency responses. This is the principle of invariance that makes Fourier analysis so intuitive and valuable.

However, if the system response changes significantly within the positional range of data analyzed, then that instrument's response depends on two spatial variables, one from the object domain and one from the data domain, which we write as $h(x, x')$. We may also write this LSV impulse response as $h_x(x')$ to remind ourselves that each acquisition occurring at position x requires that the measurement be represented by a different impulse response function. Frequency analysis under these conditions should depend on two frequencies u, u' [120, 121] that are conjugate variables to x, x'. Each conjugate pair x, u or x', u' is associated with its own set of basis vectors. SVD analysis includes one basis spanning the vector space of the object and another basis from the data, which can analyze linear system responses having shift-varying transformation operators. This dual-basis feature of SVD makes the analysis results a little less intuitive than the results of Fourier analysis, yet familiar concepts such as bandwidth and spectrum still apply.

Combining (6.14) and (6.19), the elements of the acquisition equation for a shift-varying system are decomposed as follows [10]:

$$
\begin{aligned}
\mathbf{g} = \mathbf{Hf} &= \sum_{k=0}^{R-1} \sqrt{\varsigma_k}\, \mathbf{v}_k \boldsymbol{\mu}_k^\dagger \sum_{n=0}^{R-1} \alpha_n \boldsymbol{\mu}_n \\
&= \sum_{k=0}^{R-1} \sum_{n=0}^{R-1} \alpha_n \sqrt{\varsigma_k}\, \mathbf{v}_k\, \delta[k-n] \qquad \text{(Shift-varying acquisition)} \\
&= \sum_{k=0}^{R-1} \alpha_k \sqrt{\varsigma_k}\, \mathbf{v}_k = \sum_{k=0}^{R-1} \beta_k \mathbf{v}_k, \qquad\qquad (6.22)
\end{aligned}
$$

where coefficients $\beta_k \triangleq \alpha_k \sqrt{\varsigma_k}$. This closes the loop between (6.14) and (6.15). Also note that all object information contained within the null space terms, $\mathbf{f}_n = \sum_{n=R}^{N-1} \alpha_n \boldsymbol{\mu}_n$, is lost. Therefore, if the bandwidth of the measurement is more limited than the bandwidth of the task, information will be lost during acquisition. If the lost information results in lower diagnostic task performance, we can be sure that the α_k coefficients between $R \leq k \leq N - 1$ contained vital diagnostic information.

6.7.1   Comparisons with Fourier Analysis

Generally, object functions/vectors can be decomposed using the linear basis we labeled $\{\boldsymbol{\mu}_k\}$, which spans the vector space of the object, $\mathbb{U}$. Properties of the object are represented by the associated coefficients $\{\alpha_k\}$. Measurements of the object yield data vectors that can also be decomposed, albeit using a related basis $\{\mathbf{v}_k\}$ in the vector space of the data $\mathbb{V}$. Data properties are represented by coefficients $\beta_k = \alpha_k \sqrt{\varsigma_k}$, which are composed of the data properties of the object and measurement. To analyze a general linear measurement system, we include the bases from both the object decomposition and the data decomposition via the SVD expression $\mathbf{H} = \mathbf{V}\boldsymbol{\Sigma}^{1/2}\mathbf{U}^\dagger$.

In the special case of LTI/LSI measurement systems, where the system matrix is square and Hermitian, and $\mathbb{U} = \mathbb{V}$, the same Fourier basis may be applied to both object and data, giving the system matrix described as $\mathbf{H} = \mathbf{Q}\boldsymbol{\Lambda}\mathbf{Q}^\dagger$. The coefficients from the SVD analysis α_k, β_k, and $\sqrt{\varsigma_k}$ pair with the Fourier coefficients F_k, G_k, and H_k, respectively, but for different bases. The Fourier basis is a fixed orthonormal set of complex exponentials that adjusts only to input function/vector length. The index k for Fourier coefficients tracks easily interpreted harmonic frequencies. SVD basis vectors are orthonormal but, in contrast with the situation in Fourier analysis, the kth basis vector does not assume a prescribed functional form. Although the eigenvalues/singular values are fundamental to the process they describe, we are free to select any basis consistent with the eigenvalues/singular values that span the vector space. The generality of the SVD basis means the coefficient index is just an index; it does not track a physical quantity, such as frequency, as we will see in the application sections in later chapters.

Finally, eigenanalysis generates orthogonal eigenvectors only if the matrix being decomposed is Hermitian. SVD generates eigenvectors that are always orthogonal because compound operators $\mathbf{H}^\dagger\mathbf{H}$ and $\mathbf{H}\mathbf{H}^\dagger$ are always Hermitian.

6.8 Compressed Sensing/Compressive Sampling Methods

Compressed sensing applies and extends many of the topics from this chapter by (a) introducing twenty-first-century data-sampling methods and (b) generalizing vector-basis concepts, leading to an introduction of *frames* (see Appendix A). We introduce compressive sampling through a simple example. Readers are referred to Baraniuk [8] for a signal-processing introduction and to Eldar and Kutyniok [32], especially chapter 1, where concepts are developed and explored in much detail.

In Section 5.7.4, we saw that the Nyquist sampling theorem is a safe and reliable guide to measurement sampling. The rule is simple: Real functions can be recovered exactly from a set of equally spaced samples acquired at sampling rate u_s that is at least twice the highest frequency in the signal bandwidth u_n; i.e., $u_s \geq 2u_n$. This minimum sampling rate is *sufficient* to recover the original function, given appropriate application of anti-aliasing filters when noise is present, but it is not always *necessary* to sample at that rate. Measurement efficiency encourages acquisition of only those data necessary to achieve the task. Efficient sampling minimizes operating cost and processing time. Systems using biomedical measurements, like most modern technologies, are now overwhelmed by the volume of data being generated. Developments in medical imaging, optical spectroscopy, classification using machine learning, and genomic analysis increasingly include sparse acquisitions as a key performance element [49, 50, 85, 123].

Assume that acquisitions satisfying the Nyquist rate result in data vectors of length N. Smaller acquisitions are possible without performance consequences only if the object function has a sparse or compressible representation [25, 29]. *Sparse data* are *fully represented* by $K \ll N$ nonzero coefficients when decomposed by basis or frame vectors or dictionary atoms.[16] *Compressible data* are *well represented* by $K \ll N$ nonzero coefficients, which may be fewer in number than the total number of nonzero coefficients. For example, realistic measurements often contain broad-spectrum noise

---

[16] A *frame* is an overcomplete spanning set; i.e., it is similar to a basis but has linearly dependent vectors. Frame properties are discussed in Appendix A. A *dictionary* further loosens the constraints by allowing nonorthogonal vectors (called *atoms*) in the overcomplete spanning set. Frames and dictionaries can be highly redundant, offering a rich set of descriptors that can provide a sparser set than a basis.

such that no basis coefficient is zero. Yet, most of the signal energy is limited to just K coefficients. These data are labeled *K-compressible*, indicating that data of ambient dimension N have intrinsic dimension K when applied to appropriate tasks. Compressive sampling is a method for directly acquiring data at rates lower than that dictated by the Nyquist theorem, such that when the information is displayed there is no reduction in the performance of the task that motivated the measurement. Consequently, signal models are integral components of compressive sampling design.

Example 6.8.1. *Sampling a sine wave: Consider the simple object function $f(t) = \sin(2\pi u_0 t)$. We first apply the LTI sampling operator from Section 5.7.1, where the sampling rate is set to oversample, $u_s \simeq 3u_0 = 3u_n$ yielding vector $\mathbf{f} = \mathcal{S}f(t) \in \mathbb{R}^N$. Next, we sinc interpolate the sampled data by zero padding the Fourier coefficients to exactly recover (within computational error) the original function at any temporal resolution. $\mathbf{f}$ is generated over $t = 1{,}024$ s, where $u_s = 1$ sample/s and $u_0 = 300/1{,}024 \simeq 0.3$ Hz. It is sinc interpolated a factor of 10 to obtain the smooth function plotted as a solid line from 200 s to 300 s in Figure 6.7d.*

$\mathbf{f}$ is not sparse in time, but its coefficient vector $\boldsymbol{\alpha}$, found from basis decomposition (5.5) $\mathbf{f} = \mathbf{U}\boldsymbol{\alpha}$, is sparse. In this problem, we select a Fourier basis, i.e., $\mathbf{U} = \mathbf{Q}$, where $\mathbf{f}$ is an eigenvector of $\mathbf{Q}$. There are only $K = 2$ nonzero coefficients out of $N = 1{,}024$ because $\mathcal{F}f(t) = (1/2i)(\delta(u - u_0) - \delta(u + u_0))$. The two are $-0.5i$ at $u_0 = 300/1{,}024 = 0.293u_s$ and $0.5i$ at $u_0 = 1 - 300/1{,}024 = 0.707u_s$ (equivalent to $u_0 = -0.293u_s$).

Acquisition.
In this simple example, the intrinsic dimension $K = 2$ and the ambient dimension $N = 1{,}024$. To efficiently sample $f(t)$ without significant loss, we seek to determine acquisition matrix $\boldsymbol{\Phi} \in \mathbb{R}^{M \times N}$, where N is set and M is to be determined within the bounds $K < M \ll N$. Neglecting noise, the acquisition equation becomes

$$\mathbf{g} = \boldsymbol{\Phi}\mathbf{f} = \boldsymbol{\Phi}\mathbf{U}\boldsymbol{\alpha} = \mathbf{A}\boldsymbol{\alpha}. \tag{6.23}$$

The different forms of (6.23) indicate we may consider measurement vector $\mathbf{g}$ as resulting from the transformation of object $\mathbf{f}$ via $\boldsymbol{\Phi}$, analogous to LTI acquisitions via $\mathbf{H}$. More fundamentally, $\mathbf{g}$ is a transformation of $\boldsymbol{\alpha}$, the sparse basis coefficients for the object, via $\mathbf{A} = \boldsymbol{\Phi}\mathbf{U}$. Both views are illustrated in Figure 6.7a. In this problem, we set $\mathbf{U} = \mathbf{Q}$, the DFT basis via (6.3) given the nature of $\mathbf{f}$. It is smart to know something about $\mathbf{f}$ so that we can select an appropriate decomposition matrix $\mathbf{U}$.

We don't know which coefficients are significant, so we seek a method capable of selecting the K significant coefficients for any data set. This is achieved by finding $\boldsymbol{\Phi}$ that satisfies three conditions. First, $f(t)$ must be K-compressible (in this case, $f(t)$ is K-sparse). Second, $\boldsymbol{\Phi}$ is stable and preserves the length of $\mathbf{f}$, which for some $\epsilon \in (0,1)$ is

$$1 - \epsilon \le \frac{\|\boldsymbol{\Phi}\mathbf{f}\|_2^2}{\|\mathbf{f}\|_2^2} \le 1 + \epsilon. \tag{6.24}$$

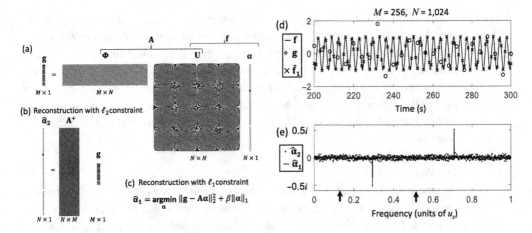

Figure 6.7 An illustration of compressive sampling of **f** to record **g** in (a), and reconstruction methods in (b) and (c). $\mathbf{U} = \mathbf{Q}$, the DFT basis. (d) Plot of original object values **f** between $200 \leq t \leq 300$ s sampled at $u_s = 1$ sample/s (approximately 3 samples/cycle) and sinc interpolated $\times 10$. Also plotted are **g**, which is compressively sampled via the methods of (a) at $u_s = 4$ samples/s, and reconstructed signal $\hat{\mathbf{f}}_1$, found using an IST algorithm. Looking closely at (e), you will see there are $\hat{\alpha}_2$ data points near the peaks of the lined spikes at $\pm u_0$ that describe the IST $\hat{\alpha}_1$ solution. $\hat{\mathbf{f}}_1$ in (d) is found from $\hat{\alpha}_1$ in (e) via $\hat{\mathbf{f}}_1 = \mathbf{Q}\hat{\alpha}_1$. Notice that the negative frequencies in (e) appear from $u_s/2$ to u_s to clarify points about sampling and aliasing. Arrows in (e) indicate Nyquist frequencies at two sampling rates as discussed in the text.

Equation (6.24) is the restricted isometry property *[25]. It holds for any subset* $\mathbf{\Phi} \in \mathbb{C}^{M \times P}$, *where* $0 < P < N$, *and corresponding-size* **f**. *You are asked to verify this property in the end-of-chapter problems. Third, the* incoherence property *states that columns of* **U** *cannot sparsely represent rows of* $\mathbf{\Phi}$.

A random matrix achieves the second and third conditions with high probability [8, 25, 29]. Specifically, the elements of $\mathbf{\Phi}$ *are drawn from a normal pdf with zero mean and variance equal to* $1/N$. *Finally, we select* M *that satisfies the restricted isometry property through a theorem [8, 32] stating that* $M \geq 0.28K \ln(N/K)$, *which for our problem gives* $M > 2K$. *A more conservative practical limit might be* $M \geq 3K$.[17]

The measurement matrix $\mathbf{\Phi}$ *shown in Figure 6.7a has* $M = 256 = N/4 = 128K$. *We selected a very conservative four-fold reduction to keep the problem simple, yet*

---

[17] Actually, the restricted isometry property in (6.24) needs to hold for $1 - \epsilon \leq \frac{\|\mathbf{A}\alpha\|_2^2}{\|\alpha\|_2^2} \leq 1 + \epsilon$, where $\mathbf{A} = \mathbf{\Phi}\mathbf{U}$. That is, the "sparsifying" transform matrix [86] **U** should be included to truly satisfy the property. As you can see from Figure 6.7, the DFT matrix used for **U** adds structure that increases coherence to **A** (despite it being difficult to see in $\mathbf{A}^+$). The consequence of failing to satisfy RIP and the incoherence property is a failure to achieve high rates of compression with low error. Problems 6.14 and 6.15 illustrate these consequences.

even a factor of 4 is quite aggressive with respect to the Nyquist theorem. Computing $\mathbf{g} = \mathbf{A}\boldsymbol{\alpha}$ via (6.23), we obtain the sparse random sampling indicated in Figure 6.7d by circles. The spectrum in Figure 6.7e shows that the Nyquist limit for the original sampling is $u_n = 0.5u_s = 0.5$ samples/s (rightmost arrow on the frequency axis). Indeed, $u_0 \simeq 0.3 < u_n = 0.5$. However, reducing u_s by a factor of 4 moves u_n to $0.125u_s$ (leftmost arrow on the frequency axis), where $u_0 \simeq 0.3 > u_n = 0.125$, violating the Nyquist theorem. If all goes well, the sine wave will not be distorted by aliasing.

Display.

Recovery of the object function involves estimating coefficients $\hat{\boldsymbol{\alpha}} \in \mathbb{R}^N$ from the sparsely sampled $\mathbf{g} \in \mathbb{R}^M$, and then transforming the coefficients to arrive at the time series via $\hat{\mathbf{f}} = \mathbf{U}\hat{\boldsymbol{\alpha}}$. First, note that the result of compressive sampling $\mathbf{g} = \mathbf{A}\boldsymbol{\alpha}$ is not unique, meaning there are many null-space vectors $\boldsymbol{\alpha}_n$ for which $\mathbf{g} = \mathbf{A}(\boldsymbol{\alpha} + \boldsymbol{\alpha}_n) = \mathbf{A}\boldsymbol{\alpha}'$ is also possible. Although none of the linear methods we have discussed so far can save the day, the restricted isometry and incoherence properties considered when we proposed $\boldsymbol{\Phi}$ ensure with high probability that $\boldsymbol{\alpha}$ can be restored from $\mathbf{g}$. Our goal now is to show that is true.

The general approach to restoration is to estimate via constrained least-squares optimization (see Appendix E). That is, we seek solutions $\boldsymbol{\alpha}'$ that minimize the squared norm of our model $\|\mathbf{g} - \mathbf{A}\boldsymbol{\alpha}'\|$. First, we subject solutions to the constraint that $\|\boldsymbol{\alpha}'\|_2^2 < \beta$, where β is an acceptable error bound. The objective function for this problem is

$$\hat{\boldsymbol{\alpha}}_2 = \underset{\boldsymbol{\alpha}' \in \mathbb{R}^N}{\operatorname{argmin}} \|\mathbf{g} - \mathbf{A}\boldsymbol{\alpha}'\|_2^2 + \beta\|\boldsymbol{\alpha}'\|_2^2. \tag{6.25}$$

Equation (6.25) has a closed-form solution given by $\hat{\boldsymbol{\alpha}}_2 = \mathbf{A}^\dagger(\mathbf{A}\mathbf{A}^\dagger)^{-1}\mathbf{g} = \mathbf{A}^+\mathbf{g}$, where $\mathbf{A}^+$ is the Moore–Penrose pseuoinverse[18] of $\mathbf{A}$.

Unfortunately, the ℓ_2 norm constraint has a low probability of finding a sparse solution, because it constrains solutions based on signal energy rather than signal sparsity. The results of restoration using (6.25) are shown in Figure 6.7e as $\hat{\boldsymbol{\alpha}}_2$ (noisy). We see $\hat{\boldsymbol{\alpha}}_2 = -0.5i$ at $u = u_0$ and $\hat{\boldsymbol{\alpha}}_2 = 0.5i$ at $u = u_s - u_0$ as expected for a sine wave spectrum (look closely to see the dots). We also see low-level values over the bandwidth, so the solution is not sparse at all! This noise-like signal where we expect zeros is actually the Gibbs ringing from sampling a finite segment of the sine wave. However, the randomness of $\boldsymbol{\Phi}$ breaks up the coherence of the ringing effect so it appears random.

A better approach than the energy constraint is to constrain solutions to only those with the expected number of nonzero coefficients. An ℓ_0 "norm" is a count of nonzero

---

[18] The Moore–Penrose (MP) pseudoinverse is introduced in Appendix A. Note that rows of $\mathbf{A}$ must be linearly independent for $\mathbf{A}\mathbf{A}^\dagger$ to be invertible. Given that $\mathbf{A}$ is generally a frame and not a basis, linear independence is not always true. In that case, the general MP pseudoinverse expression is $\mathbf{A}^+ = \mathbf{A}^\dagger(\mathbf{A}\mathbf{A}^\dagger)^+$.

elements; i.e., $\|\hat{\boldsymbol{\alpha}}'\|_0 = K$. See Appendix A for a discussion of norms. Unfortunately, optimization with an ℓ_0 constraint is a very difficult problem.[19]

Donoho [29] showed that constrained optimization (see Appendix E) with an ℓ_1 norm,

$$\hat{\boldsymbol{\alpha}}_1 = \underset{\boldsymbol{\alpha}' \in \mathbb{R}^N}{\operatorname{argmin}} \|\mathbf{g} - \mathbf{A}\boldsymbol{\alpha}'\|_2^2 + \beta \|\boldsymbol{\alpha}'\|_1, \qquad (6.26)$$

has a high probability of exactly recovering $\boldsymbol{\alpha}$ from M samples. Equation (6.26) is a convex function with a single minimum, unlike the ℓ_0 problem. The convex ℓ_1 norm is the complex envelope of the nonconvex ℓ_0 norm [87].

Optimization constrained by the ℓ_1 norm is a problem that has been widely pursued [20, 24, 32, 107], and code for it can be found online. An iterative soft-thresholding (IST) approach was applied to solve (6.26). The code provided in this section generated the results shown in Figure 6.7. This approach iteratively computes derivatives of (6.26) and then applies a threshold that truncates the low-amplitude values. These ideas are expressed mathematically as follows [12]:

$$\hat{\boldsymbol{\alpha}}_{1,j+1} = \mathcal{T}_{\beta\tau}\left(\hat{\boldsymbol{\alpha}}_{1,j} - 2\tau\mathbf{A}^{\dagger}(\mathbf{g} - \mathbf{A}\hat{\boldsymbol{\alpha}}_{1,j})\right). \qquad (6.27)$$

The term in the parentheses is the difference between the current solution and its derivative, $[d\|\mathbf{g} - \mathbf{A}\boldsymbol{\alpha}'\|_2^2/d\boldsymbol{\alpha}']_{\boldsymbol{\alpha}'=\hat{\boldsymbol{\alpha}}_1}$, which becomes the next estimate in the iteration. First, however, that result becomes the argument of a soft-thresholding operator $\mathcal{T}$, which is defined for parameters β and τ as

$$\mathcal{T}_{\beta\tau}(x)_j = \operatorname{sgn}(x_j)(|x_j| - \beta\tau),$$

and enforces the constraint. The function sgn is defined in Appendix D. Examine the code we provide on the next page, and you will find that MATLAB *offers a soft-thresholding function* wthresh *that works well.*

The results are plotted as $\hat{\boldsymbol{\alpha}}_1$ in Figure 6.7e, and $\hat{\mathbf{f}}_1 = \mathbf{U}\hat{\boldsymbol{\alpha}}_1$ is plotted in Figure 6.7d. I requested 10 iterations for $\beta = 0.1$ in the code, but the solution converged in just 2 iterations. The variables cost_iter_IST *and* err_iter_IST *give measurements of the cost and relative error versus iteration number (see the end-of-chapter problems).*

Thresholding is a highly nonlinear operation *that works well in this situation, but it is nothing like the linear operations we spent so much time developing. I encourage readers to play with the code provided here by adding noise to the sparse acquisitions, generating other sparse signals that are not as simple as a sine wave in some other frame or basis, and addressing 2-D objects. I suggest that you work through the two end-of-chapter problems first. Note that compressive sampling methods are becoming increasingly mainstream approaches.*

---

[19] This optimization problem is highly nonconvex (multiple local minima), NP-complete [numerically complex, requiring a comprehensive search of all $\binom{N}{K}$ locations], and unstable [25, 32].

```
%%%%%%%%%%%%%%%%%%%  Compressive sensing demo from, Example 6.8.1 %%%%%%%%%%%%%%%%%%%%%%
%
N=1024;t=0:N-1;u0=300/N;Q=zeros(N);          % initialize parameters
for k=1:N                                     % frequency axis of basis matrix Q
    for j=1:N                                 % data axis of basis matrix Q
        Q(k,j)=exp(i*2*pi*(j-1)*(k-1)/N);     % DFT basis
    end
end
f=sin(2*pi*u0*t)';                            % true object function f
u=t/N;                                        % frequency axis
F=Q'*f/N;                          % apply adjoint of DFT basis to forward transform f
M=N/4;                             % M is number of samples in sparse acquisition
psi=randn(M,N)/sqrt(N);g=psi*f;    % random sensing matrix
A=psi*Q;a=(N/M)*(pinv(A)*g);       % a = \hat{\alpha}_2 reconstructed via pseudoinverse
subplot(2,1,1);plot(t,f);hold on;  % original sampled f at u_s (not upsampled)
plot(t(1:N/M:N),g','ro');          % sparsely sampled object at u_s/4
subplot(2,1,2);plot(u,imag(F));hold on; % original alpha spectrum (it's imaginary)
plot(u,imag(a),'k.');hold off      % display \hat{\alpha}_2
title(['M = ', num2str(M), ', N = ',num2str(N)]) % display sampling parameters
%
% Fan Lam's IST algorithm to reconstruct alpha => \hat{\alpha}_1
%
[a_hat,nIter,cost_iter_IST,err_iter_IST] = SolveIST_Fourier(g+randn(size(g)),A,F,0.1,10);
figure;plot(u,imag(a_hat))  % display a_hat = \hat{\alpha}_1. For M/N=4, I set beta=0.1, maxit=10
f1_hat=Q*a_hat;             % the restored object function at the original u_s sampling rate.
figure(1);subplot(2,1,1);plot(t,real(f1_hat),'kx');hold off
%
%
% Function to solve argmin 0.5*||y-A*x||_2^2 + beta*||x||_1 by Fan Lam.
function [x_hat,nIter,cost_iter_IST,err_iter_IST] = SolveIST_Fourier(y,Mask,xt,beta,maxit)
% y is sampled data g, Mask is A=phi*U, xt is the true object function f for error estimation
% beta is the scaling constant applied to the L1 constraint, and maxit is max # of iterations
cost_iter_IST = [];                                      % initialize the while loop
err_iter_IST  = [];
tau = 2;
N   = size(Mask',1);
xbp = zeros(N,1);                                        % initialize the solution vector
nIter = 1;                                               % initialize the iteration counter
while nIter < maxit
    cost_iter_IST(nIter) = 0.5*norm(y-Mask*xbp)^2 + beta*norm(xbp,1);  % L1 norm selected
    err_iter_IST(nIter)  = norm(xbp - xt)/norm(xt);      % relative error
    xbp = xbp + 2*tau*Mask'*(y-Mask*xbp)/N;              % update estimates
    xbp = sign(xbp).*wthresh(abs(xbp), 'h', tau*beta);   % apply soft threshold
    nIter = nIter+1;
end
x_hat = xbp;                                    % result returned after maxit iterations
end
```

6.9   Analyzing Systems of Chemical Reactions

We have been studying measurement systems and the data they generate. The ultimate goal is to answer questions about object properties that are encoded in eigenstates. Organisms are also complex systems that may be modeled and analyzed using the same analytical and computational techniques. These models can guide measurement design and data interpretation. We now describe models of simple systems whose properties are illuminated by the application of SVD methods. Note that systems of ordinary differential equations are the topic of Chapter 10; the discussion here is a preview of that more expansive coverage.

Many chemical reactions can be modeled as linear systems [78, 92] and consequently analyzed via SVD. Systems composed of interacting compounds are modeled with input stimuli specified by reaction fluxes[20] that produce chemical compounds. This approach has become a powerful tool for discovering functional "pathways" through large biochemical networks in *systems biology*.

We illustrate the approach using a simple example from [78] that involves two chemical compounds, AB and BA, each composed of moieties[21] A and B. The stoichiometric equation

$$AB \underset{f_2}{\overset{f_1}{\rightleftharpoons}} BA \tag{6.28}$$

shows how flux vector $\mathbf{f} = \begin{pmatrix} f_1 \\ f_2 \end{pmatrix}$, describing the rate and direction of reactions, determines compound concentrations $\mathbf{x} = \begin{pmatrix} x_1 \\ x_2 \end{pmatrix} \triangleq \begin{pmatrix} [AB] \\ [BA] \end{pmatrix}$. These reactions are described by a noise-free linear transformation

$$\dot{\mathbf{x}} \triangleq \frac{d\mathbf{x}}{dt} = \mathbf{Sf}, \tag{6.29}$$

where $\dot{\mathbf{x}} = \begin{pmatrix} \dot{x}_1 \\ \dot{x}_2 \end{pmatrix} = \begin{pmatrix} dx_1/dt \\ dx_2/dt \end{pmatrix}$ describes temporal changes in compound concentrations.

Column vectors of *stoichiometric matrix* $\mathbf{S} = (\mathbf{s}_1 \ \mathbf{s}_2)$ describe how each *flux vector* component connects to all compounds that are present. In (6.28), f_1 describes how a reduction in the concentration $x_1 = [AB]$ leads to an increase in $x_2 = [BA]$. That is, $\mathbf{s}_1 = \begin{pmatrix} -1 \\ 1 \end{pmatrix}$. Fluxes pointing away from compounds are negative, those pointing toward compounds are positive, and those that do not connect have a value of zero. Since f_2 describes the opposite reaction, we have $\mathbf{s}_2 = \begin{pmatrix} 1 \\ -1 \end{pmatrix}$. Equivalent forms of the same system of equations describing (6.28) are given by

$$\dot{\mathbf{x}} = \mathbf{Sf}$$

$$\begin{pmatrix} \dot{x}_1 \\ \dot{x}_2 \end{pmatrix} = \begin{pmatrix} -1 & 1 \\ 1 & -1 \end{pmatrix} \begin{pmatrix} f_1 \\ f_2 \end{pmatrix}$$

$$dx_1/dt = -f_1 + f_2 \tag{6.30}$$

$$dx_2/dt = f_1 - f_2.$$

$\mathbf{S}$ is functioning as a map that connects all reacting compounds in the sample through all possible fluxes. Thus, the modes, i.e., eigenvectors, of $\mathbf{S}$ describe the dynamic and steady-state pathways on which this chemical system operates. Eigenvalues that are zero identify steady-state modes, while nonzero eigenvalues identify dynamic modes. Let's look at this issue more closely.

---

[20] Flux is a vector quantity describing the magnitude and direction of the rate of movement of molar concentrations of compounds through a surface over time.
[21] Characteristic components of molecular structures.

6.9.1 Rank Range and Null Space of the Transformation

This simple problem is quickly characterized so we can exercise our linear algebra skills. Since $s_1 = -s_2$, it is pretty clear that the columns are linearly dependent and the rank(S) $= 1$. This yields two vector spaces[22] that offer much physical significance.

The *kernel of the transformation* between the concentrations $[AB]$ and $[BA]$ consists of all flux vectors that result in no change in compound concentrations. This is the *null space of matrix* S:

$$\mathbf{0} = S\mathbf{f} = \begin{pmatrix} -1 & 1 \\ 1 & -1 \end{pmatrix} \begin{pmatrix} f_1 \\ f_2 \end{pmatrix} = \begin{pmatrix} 0 \\ 0 \end{pmatrix}$$

$$\longmapsto \quad \begin{matrix} -f_1 + f_2 = 0 \\ f_1 - f_2 = 0 \end{matrix}, \quad \text{and so null}(S) = \left\{ r \begin{pmatrix} 1 \\ 1 \end{pmatrix} \right\}.$$

The presence of arbitrary constant r shows that the set of potential null space vectors is infinite. These compose the *steady-state solutions* found when the system is in *equilibrium*; i.e., $f_1 = f_2$. An orthonormal basis for the 1-D kernel subspace in the $\mathbb{R}^2$ domain is $\frac{1}{\sqrt{2}} \begin{pmatrix} 1 \\ 1 \end{pmatrix}$.

The *range of the transformation* is spanned by the column vectors of S, with dimension rank(S). We can find the range from the reduced echelon form of $S^\top$ (this is simple to do by hand, where row 2 becomes R2 = R2 + R1 and R1 = $-$R1. In MATLAB, use `rref(S')`):

$$S^\top = \begin{pmatrix} -1 & 1 \\ 1 & -1 \end{pmatrix} \underset{\text{rref}}{\longrightarrow} \begin{pmatrix} 1 & -1 \\ 0 & 0 \end{pmatrix}, \quad \text{and so rank}(S) = \left\{ s \begin{pmatrix} 1 \\ -1 \end{pmatrix} \right\},$$

where s is an arbitrary constant. An orthonormal basis for the 1-D range subspace in the $\mathbb{R}^2$ domain is $\frac{1}{\sqrt{2}} \begin{pmatrix} 1 \\ -1 \end{pmatrix}$ and, alternatively, $\frac{1}{\sqrt{2}} \begin{pmatrix} -1 \\ 1 \end{pmatrix}$. Range vectors compose the solution for the system under dynamic or nonequilibrium conditions; e.g., $f_1 \neq f_2$.

The domain of this reaction is the sum

$$\text{domain} = \text{rank}(S) + \text{null}(S).$$

6.9.2 SVD Approaches

The previous section defined states of the system as subspaces of the transformation. SVD provides the same information and more. Analyzing the system matrix in (6.30),

$$S = V\Sigma^{1/2}U^\dagger = \frac{1}{\sqrt{2}} \begin{pmatrix} -1 & 1 \\ 1 & 1 \end{pmatrix} \begin{pmatrix} 2 & 0 \\ 0 & 0 \end{pmatrix} \frac{1}{\sqrt{2}} \begin{pmatrix} 1 & -1 \\ 1 & 1 \end{pmatrix}, \tag{6.31}$$

---

[22] Both subspaces of the transformation are vector spaces since both contain $\mathbf{0}$.

we again find S is rank 1 with one dynamic mode,[23] $\mu_1 = \begin{pmatrix} \frac{1}{\sqrt{2}} \\ -\frac{1}{\sqrt{2}} \end{pmatrix}$, as in the last section. Applying μ_1 to S and multiplying both sides by $\sqrt{2}$, we find

$$S = V\Sigma^{1/2}U^\dagger = \sum_{k \in R} \sqrt{\varsigma_k}\, v_k \mu_k^\dagger, \quad \text{where } R = 1;$$

$$\sqrt{2}S\mu_1 = \sqrt{2\varsigma_1}\, v_1;$$

$$S\begin{pmatrix} 1 \\ -1 \end{pmatrix} = 2\begin{pmatrix} -1 \\ 1 \end{pmatrix}.$$

This result tells us how the reaction proceeds. The active flux mode of S, which includes f_1 and f_2, acts to change concentrations $x_1 = [AB]$ and $x_2 = [BA]$ in opposite directions at twice the flux magnitude. For example, if $f_1 = 0.9$ and $f_2 = 1.1$, then $\dot{x}_1 = 0.2$ and $\dot{x}_2 = -0.2$, showing the dynamic changes. In words, a 10% change in flux generates a 20% change in concentration rate. The null spaces, which are given by the second columns of V and U whose singular value is zero, define conservation relationships [78]. When $f_1 = f_2$, then $\dot{x} = 0$ and the system is in equilibrium.

We now explore properties of the linear transformation in (6.29) for a somewhat larger system of equations.

Example 6.9.1. *Chemical system: Consider the general chemical system described by the graph in Figure 6.8a, which includes three compounds $x = (x_1, x_2, x_3)^\top$ and six flux values $f = (f_1, f_2, f_3, f_4, f_5, f_6)^\top$. Adopt the linear model $\dot{x} = Sf$ and assume the system is in equilibrium. (a) Determine the stoichiometric matrix. (b) Solve for f. (c) Illustrate the functional pathways in Figure 6.8b from the solution to part (b). (d) Assume the experimentally observed pathways in Figure 6.8c and find a second transformation matrix that transforms your solution in part (c) to that illustrated in Figure 6.8c.*

Solution

(a) S *is a* 3×6 *matrix where the number of rows is determined by the number of compounds and the number of columns is the number of fluxes. Because we see* f_1 *entering* x_1, S_{11} *is set to 1. Since* f_1 *does not interact directly with* x_2 *or* x_3, *then* $S_{21} = S_{31} = 0$. *As a result, the first column of* S *is* $\begin{pmatrix} 1 \\ 0 \\ 0 \end{pmatrix}$. *Those fluxes exiting a compound are set to* -1. *Continuing with all six fluxes, we find*

---

[23] The columns of U are the orthonormal eigenvectors in the vector space of f. The columns of V are eigenvectors in the vector space of $\dot{x}$. $\mu_1^\dagger$ is the first row of $U^\dagger$ in (6.31).

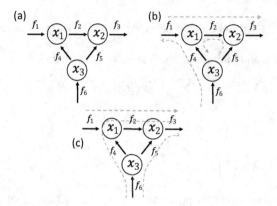

Figure 6.8 Illustrations associated with Example 6.9.1. (b) is the solution to part (c) of the example, and (c) is the solution to part (d).

$$S = \begin{pmatrix} 1 & -1 & 0 & 1 & 0 & 0 \\ 0 & 1 & -1 & 0 & 1 & 0 \\ 0 & 0 & 0 & -1 & -1 & 1 \end{pmatrix} \xrightarrow[R1=R1+R3,\, R3=-R3]{R1=R1+R2} \begin{pmatrix} 1 & 0 & -1 & 0 & 0 & 1 \\ 0 & 1 & -1 & 0 & 1 & 0 \\ 0 & 0 & 0 & 1 & 1 & -1 \end{pmatrix}.$$

$$\qquad\qquad\qquad\qquad\qquad\qquad\qquad\qquad\qquad\qquad\qquad\uparrow\qquad\quad\uparrow\ \uparrow\qquad.$$

$$(6.32)$$

The second form of (6.32) places S in reduced echelon form, which we find in MATLAB via rref(S). With three pivot points, the rank of S is 3 (a 3 × 6 matrix cannot have a rank greater than 3). We have an under-determined system because there are three equations and six unknowns. The three free variables are indicated in (6.32) by the arrows at columns 3, 5, and 6.

(b) Writing out the system of equations from $\dot{\mathbf{x}} = \mathbf{Sf} = \mathbf{0}$ for the equilibrium state, we find

$$\dot{x}_1 = f_1 - f_2 + f_4 = 0$$
$$\dot{x}_2 = f_2 - f_3 + f_5 = 0$$
$$\dot{x}_3 = -f_4 - f_5 + f_6 = 0.$$

Now writing expressions for each flux in terms of other fluxes, we set the free variables to arbitrary scalar constants r, s, and t, and express all fluxes in terms of those constants:

$$f_1 = f_2 - f_4 = r - s - (-s + t) = r - t$$
$$f_2 = f_3 - f_5 = r - s$$
$$f_3 = r \ \ (free)$$
$$f_4 = -f_5 + f_6 = -s + t$$

$$f_5 = s \quad (\text{free})$$
$$f_6 = t \quad (\text{free})$$

$$\mathbf{f} = r \begin{pmatrix} 1 \\ 1 \\ 1 \\ 0 \\ 0 \\ 0 \end{pmatrix} + s \begin{pmatrix} 0 \\ -1 \\ 0 \\ -1 \\ 1 \\ 0 \end{pmatrix} + t \begin{pmatrix} -1 \\ 0 \\ 0 \\ 1 \\ 0 \\ 1 \end{pmatrix}.$$

(c) Expressing **f** *as the sum of the free variables, we find functional pathways through the network. Following each of the three terms in the solution for* **f**, *we can plot the pathways as shown in Figure 6.8b. We can do all of this in* MATLAB *via SVD, but the results are less transparent than those computed by hand.*

(d) The functional modes illustrated in Figure 6.8b are just one of many possible sets of three. Each is described by an eigenvector, so the number of modes is unique even though the exact modal pathways are not unique. We may need to provide more details to obtain an acceptable solution. For example, only positive fluxes are possible experimentally, such as those shown in Figure 6.8c. So which transformation matrix **A** *can be found that transforms the solution diagrammed in Figure 6.8b to that in Figure 6.8c? That is, find* **A** *that satisfies*

$$\mathbf{F}_1 \mathbf{A} = \mathbf{F}_2$$

$$\begin{pmatrix} 1 & 0 & -1 \\ 1 & -1 & 0 \\ 1 & 0 & 0 \\ 0 & -1 & 1 \\ 0 & 1 & 0 \\ 0 & 0 & 1 \end{pmatrix} \mathbf{A} = \begin{pmatrix} 1 & 0 & 0 \\ 1 & 0 & 1 \\ 1 & 1 & 1 \\ 0 & 0 & 1 \\ 0 & 1 & 0 \\ 0 & 1 & 1 \end{pmatrix},$$

where $\mathbf{A} \in \mathbb{R}^{3 \times 3}$. *Because the matrix of pathways,* $\mathbf{F}_1$, *is rectangular, we cannot find* **A** *using* $\mathbf{A} = \mathbf{F}_1^{-1}\mathbf{F}_2$. *We can, however, apply a pseudoinverse to find* $\mathbf{A} = \mathbf{F}_1^+ \mathbf{F}_2$ *via (6.21).*

Using MATLAB, $\mathbf{F}_1^+ = \frac{1}{4} \begin{pmatrix} 1 & 1 & 2 & 0 & 1 & 1 \\ 0 & -1 & 1 & -1 & 2 & 1 \\ -1 & 0 & 1 & 1 & 1 & 2 \end{pmatrix}$

and $\mathbf{A} = \begin{pmatrix} 1 & 1 & 1 \\ 0 & 1 & 0 \\ 0 & 1 & 1 \end{pmatrix}$, *which gives the pathways graphed in Figure 6.8c. The first solution applied a* general *linear basis while the second solution is an example of a* convex *basis in which the solution space is constrained to match physical constraints.*

6.10   Problems

6.1   Let $\mathbf{A}$ and $\mathbf{B}$ be unitary matrix operators and let c be a complex constant. Find the following:

(a) $\mathbf{A}^\dagger \mathbf{A}$

(b) $\mathbf{A}^\dagger \mathbf{A}^{-1}$

(c) $(c\mathbf{A}\mathbf{A}^\dagger)^\dagger$

(d) $\begin{pmatrix} \mathbf{A} & \mathbf{0} \\ \mathbf{0} & \mathbf{B} \end{pmatrix}^{-1}$

6.2   Consider $\mathbf{g}_1 = \mathbf{Hf}_1$ and $\mathbf{g}_2 = \mathbf{Hf}_2$ as describing the measurement data in $\mathbb{V}$ resulting from operator $\mathbf{H}$ being applied to two object vectors $\mathbf{f}_1$ and $\mathbf{f}_2$ in $\mathbb{U}$. Each acquisition may be written as a summation using $g_i[m] = \sum_{n=1}^{N} H[m,n] f_i[n]$ for $i = 1, 2$. Also note the various forms used to express the inner product of vectors $\mathbf{a}$ and $\mathbf{b}$, $\mathbf{a}^\dagger \mathbf{b} = \langle \mathbf{a}, \mathbf{b} \rangle \triangleq \sum_{n=1}^{N} a^*[n] b[n]$, where the last form explicitly shows that a complex conjugate of the first vector is implied by the other two.

Barrett and Myers [2] defined properties of adjoint operator $\mathbf{H}^\dagger$ by equating the inner products $\langle \mathbf{g}_2, \mathbf{Hf}_1 \rangle = \langle \mathbf{H}^\dagger \mathbf{g}_2, \mathbf{f}_1 \rangle$, where both sides result in scalars. The left side involves data vectors in $\mathbb{V}$, while the right side involves the corresponding object vectors in $\mathbb{U}$. When you write this out, you will see that $[\mathbf{H}^\dagger][n, m] = H^*[m, n]$, which is the expression used to find (6.6). Show this definition is true.

6.3   Show that $\langle \mathbf{x}, \mathbf{y} \rangle \mathbf{z} = \mathbf{z}\mathbf{x}^\dagger \mathbf{y}$, where $\mathbf{x}$ and $\mathbf{y}$ are $N \times 1$ and $\mathbf{z}$ is $M \times 1$.

6.4   A discrete-to-discrete transformation is given by $\mathbf{y} = \mathbf{Ax}$. A continuous-to-discrete transformation is given by $\mathbf{y} = \mathcal{A}\{x(t)\}$. The corresponding continuous-to-continuous transformation is $y(t) = \mathcal{A}\{x(t)\}$. Allow $\mathbf{A}$ and $\mathcal{A}$ to be complex. Show the following quantities as sums or integrals without assuming time invariance. Note that $\mathbf{x}$ lives in a space different from $\mathbf{y}$.

(a) $\mathbf{y} = \mathbf{Ax}$

(b) $\mathbf{x} = [\mathbf{A}^\dagger \mathbf{y}]$

(c) $\mathbf{y} = \mathcal{A}\{x(t')\}$

(d) $x(t') = [\mathbf{A}^\dagger \mathbf{y}](t')$

(e) $y(t) = \mathcal{A}\{x(t')\}$

(f) $x(t') = \mathcal{A}^\dagger \{y(t)\}$

6.5   Beginning with temporal acquisition equation $\mathbf{g} = \mathbf{Hf} + \mathbf{e}$ for a noisy LTI instrument, apply (6.7) and (6.8) to prove the Fourier Convolution Theorem.

6.6   Show that the Fourier-operator matrix $\mathbf{Q} \in \mathbb{C}^{4 \times 4}$ is unitary.

6.7   Compute analytically the 2-D CT-FT $F(u, v)$ for $f(x, y) = A \operatorname{rect}((x - x_0)/X_0) \, B \operatorname{rect}((y - y_0)/Y_0)$.

6.8 (a) Given $\mathbf{g} = \mathbf{Hf}$, find the eigenvalues and eigenvectors (by hand) for the following measurement matrix:

$$\mathbf{H} = \begin{pmatrix} 7 & 0 & -3 \\ -9 & -2 & 3 \\ 18 & 0 & -8 \end{pmatrix}.$$

(b) $\mathbf{H} \in \mathbb{R}^{N \times N}$ is diagonalizable if it is full rank. In (a), you will have shown it is diagonalizable. Now determine whether the eigenvector matrix $\mathbf{U}$ is unitary. Reconstruct $\mathbf{H}$ from $\mathbf{\Lambda}$ and $\mathbf{U}$ to test your answer.

6.9 To solve Problem 6.8 with a different approach, use MATLAB to show that $\mathbf{H}_1$ here has a unitary eigenvector matrix, such that $\mathbf{H}_1 = \mathbf{U} \mathbf{\Lambda} \mathbf{U}^\dagger$:

$$\mathbf{H}_1 = \begin{pmatrix} 3 & 0 & -1 \\ 0 & 1 & 0 \\ -1 & 0 & 2 \end{pmatrix}.$$

6.10 Analytically decompose $\mathbf{H} = \begin{pmatrix} 3 & 1 & 1 \\ -1 & 3 & 1 \end{pmatrix}$ using SVD and find singular components $\mathbf{U}$, $\mathbf{\Sigma}^{1/2}$, and $\mathbf{V}$.

6.11 From the following equation, find stoichiometric matrix $\mathbf{S}$ and apply SVD analysis to discover the functional modes. Hint: The answer is obvious. The goal is to make sure you understand the process.

$$x_1 + x_2 \underset{f_2}{\overset{f_1}{\rightleftharpoons}} x_3 \tag{6.33}$$

6.12 Consider the graph in Figure 6.9. (a) Find the stoichiometric matrix $\mathbf{S}$. (b) Since $\mathbf{S}$ is square, we can apply eigenanalysis via MATLAB. Any eigenvectors corresponding to zero eigenvalues are labeled as homogeneous pathways through the system

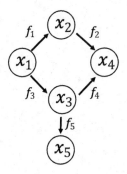

Figure 6.9 Figure for Problem 6.12.

because they do not pass through the system. Identify the homogeneous pathways and plot them on the figure.

6.13 Refer to the discussion and the examples of the *matrix condition number* in Appendix A for this problem. The examples in that section describe the importance of having a stable matrix $\mathbf{H}$ when $\mathbf{g} = \mathbf{Hf} + \mathbf{e}$ is inverted to estimate $\hat{\mathbf{f}} = \mathbf{H}^{-1}\mathbf{g}$.

First, form a random matrix $\mathbf{H} \in \mathbb{R}^{N \times N}$ assuming the eigenvalues are `[N*rand(1,(N-1)),0.01]`. Also assume the eigenvector matrix is given by `randn(N)`. Form $\mathbf{H}$ from these components and show it is full rank but poorly conditioned.

Next, create random data vectors `g1 = rand(N,1)` and `g2 = g1 + 0.1* rand(N,1)`, where g2 has more independent noise than g1. Compare the relative variations in $\hat{\mathbf{f}}$, i.e., $\|\hat{\mathbf{f}}_1 - \hat{\mathbf{f}}_2\|_2 / \|\hat{\mathbf{f}}_1\|_2$, with those in $\mathbf{g}$. Since $\mathbf{H}$ and $\mathbf{g}$ are both stochastic, you will have to repeat this estimate many times and average the results. What do the results tell us about how the condition of the matrix influences errors in inverse-problem estimates, i.e., $\hat{\mathbf{f}}$?

6.14 Demonstrate that the restricted isometry property holds for the data in Example 6.8.1. Specifically, show that the conditions described by (6.24) are true for $\mathbf{\Phi} \in \mathbb{R}^{256 \times 1,024}$ and for $\mathbf{\Phi} \in \mathbb{R}^{256 \times S}$ where $S < 1,024$.

6.15 In Example 6.8.1, we reduced sampling from $N = 1,024$ to $M = 256$, a decrease of only a factor of 4. Use the code from that example to plot the relative error variable `err_iter_ISC` for increasingly sparse sampling. Use test values of $M = 256, 128, 64, 32, 16$. Note that you may need to adjust β and the number of iterations as M changes. Also, you may need to run the program a few times for each value of M since the results will vary. What conclusions can you draw from the error plots at each value of M?

7 Projection Radiography

This chapter discusses an important application of the analytical tools developed in previous chapters – medical x-ray imaging. Although we briefly touch on the imaging technology and the physics of photon interactions with matter in this chapter, I assume readers have some familiarity with these processes. Background information is provided in several texts, e.g., [23, 64]. Our focus here is on modeling the measurement process so we may demonstrate quantitative assessments of results. These examples may serve as templates for analyzing other biomedical measurement technologies.

7.1   Acquisition-Stage Processing: Modeling Photon Detection

The acquisition geometry for a typical projection x-ray imaging system is shown in Figure 7.1c. A patient is positioned between an x-ray source and a detector that records the transmitted photon intensity. Photons not absorbed by the patient are mostly absorbed by the detector, which forms a *projection* of a 3-D patient onto a 2-D detector. The object/patient property that couples to the photon field is the *linear attenuation coefficient* $\mu(\mathbf{x}) \triangleq \mu(Z(\mathbf{x}), \rho(\mathbf{x}), \epsilon)$. Spatial variations in tissue attenuation result from spatial variations in atomic number Z, mass density ρ, and photon energy for the pth photon, $\epsilon_p = h\nu$. The variation patterns described by $\mu(\mathbf{x})$ generate *object contrast*.

The ideal measurement device includes an intense *point source* of x-rays delivered during a short time interval and small, efficiently absorbing, uniformly responding detector elements. Ideal features create the *quantum-limited* acquisition seen in Figure 7.1b for a uniformly absorbing object.[1] The quantum-limited state is highly desirable, because it implies a near-perfect imaging system where performance is limited only by photon statistics. Here we have the highest spatial resolution possible for linear measurements of this Poisson process (although there is nothing in this object to resolve). Distributing x-rays uniformly over patients makes it more likely that fluctuations in the recorded photon density (image contrast) are caused by differential tissue absorption (object contrast).

---

[1] In this chapter, the simulations involve acquisition data, rather than image data that have gone through display-stage processing. In acquisition data, bright regions in the grayscale data describe the photons detected. A typical radiograph records high photon counts as dark regions because of prior history using screen-film detectors.

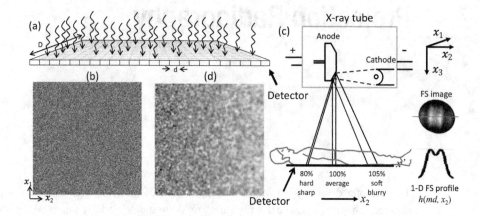

Figure 7.1 (a) A planar detector array in air irradiated by a uniform field of x-ray photons is simulated. (b) Acquired data from a 200×200 element detector are simulated using a remote point source, an LSI impulse response of negligible extent, and no display-stage processing. (c) In practice, a shift-varying impulse response is found using a rotating-anode source (drawing not to scale). The focal spot image and 1-D profile are shown. The FS projection onto the anode side of the detector is sharper with lower photon intensity and higher average photon energy (beam hardening). The FS projection onto the cathode side of the detector is wider with higher photon intensity and lower average photon energy (beam softening). (d) We simulated a flat-field irradiation of the detector using an exaggerated shift-varying impulse response h in (c). The magnitude of these effects in (d) is exaggerated.

A poorly designed system improperly records the same flat incident x-ray field with the variations seen in Figure 7.1d. Our goal in this chapter is to analyze images, learn to assess *measurement quality*, and thereby understand what might potentially compromise task performance. Realistic radiographic systems include elements of geometric distortions (finite focal spot size, patient motion, magnification), detector distortions (non-uniform element response, finite detector-element size), influences from patient scattering, and acquisition noise (both quantum and detector). In this chapter, we model and analyze data to reveal basic engineering properties of a radiographic imaging system such as *contrast*, *noise*, and *spatial resolution* from which the quality of the recording is assessed.

7.1.1   Stochastic Impulse Response Function

To simulate data from an x-ray imaging device, we interpret the acquisition equation $\mathbf{g} = \mathcal{H}\mathbf{f}$ in the appropriate physical context. We begin by modeling a point source of x-ray photons located far from an ideal plane-detector array. The detector is exposed to a *wide-sense stationary* (WSS) random field of photons across its surface – the "Poisson rainfall" illustrated in Figure 7.1a that results in the flat-field, quantum-limited image shown in Figure 7.1b. The photon distribution (q for quanta) is a function of position $\mathbf{x}$ and spectral energy ϵ,

$$q(\mathbf{x}, \epsilon) = \frac{1}{A_d} \sum_{p=1}^{N_p} \epsilon_p \, \delta(\mathbf{x} - \mathbf{x}_p). \tag{7.1}$$

The total number of photons incident on and absorbed by total detector area A_d for a single exposure is N_p, a Poisson variable. The locations are determined by 2-D *uniform random variable*, $\mathbf{x}_p = (x_1, x_2)_p^\top$, with pdf $p(\mathbf{x}_p) = 1/A_d$, $\forall (x_1, x_2) \in A_d$. As shown in (7.1), the pth photon is characterized by both position and energy. An example photon field is simulated in the MATLAB code shown later in this section, where we have simplified the model for this discussion by ignoring energy influences; i.e., $q(\mathbf{x}, \epsilon) = q(\mathbf{x})$.

When detector noise, contrast transfer,[2] and blur are comparatively small influences, target visibility is limited primarily by the randomness of the illuminating field. Let square detector area $A_d = D^2$ be an array of $N_d \times N_d$ square elements, each with area d^2 that records the incident photon intensity.[3] For convenience, let N_d be an odd integer and $N_d^2 = (D/d)^2$.

To model this measurement, we couple stochastic photon counts $q(\mathbf{x})$ with the detector-array geometry to form the *stochastic impulse response*,

$$h(n, m, \mathbf{x}) \triangleq h(nd, md, x_1, x_2) = h' \, q(x_1, x_2) \, \text{rect}\left(\frac{x_1 - nd}{d}\right) \text{rect}\left(\frac{x_2 - md}{d}\right), \tag{7.2}$$

where the range of n and m for detector area A_d is $0, \pm 1, \pm 2, \ldots, (N_d - 1)/2$. h' is a constant with units, for example, of [volts/quanta]. The four independent variables of h span the detector surface in continuous object and discrete data spaces: x_1 and x_2 are continuously varying positions within A_d, while nd and md are center positions of the detector elements where intensities are recorded.

Realizations of h are used to model single exposures for simulations, while overall system properties are characterized by its ensemble mean,

$$\mathcal{E}\{h(n, m, \mathbf{x})\} = h' \, \bar{q} \, \text{rect}\left(\frac{x_1 - nd}{d}\right) \text{rect}\left(\frac{x_2 - md}{d}\right), \tag{7.3}$$

where $\bar{q} = \mathcal{E}\{q(\mathbf{x})\}$ is the ensemble mean of x-ray fluence for the exposure time. *Flux* describes the number of photons per area per time, with units of [mm^{-2} s^{-1}]. The integral of flux over exposure time is the photon *fluence*, the number of photons per area with units of [mm^{-2}], and the integral of fluence over detector area is N_p, a unitless Poisson random variable. In this case, h has units of [volts/mm^2].

---

[2] Contrast transfer, defined in Section 7.2, refers to the transfer of object contrast into image contrast.

[3] The x-ray intensity incident on a detector element is the number of photons per element area per unit time, weighted by the energy of each photon. In this section, we assume all photons have the same energy, so intensity is proportional to photon flux, which is the incident number per element area per time. The concept of flux here is analogous to that described in Section 6.9 for chemical reactions.

7.1.2   Latent Image Formation

The acquisition stage of an x-ray imaging system yields latent "image" **g**, which is the data acquired before it is displayed. We know from Section 4.3 that a deterministic linear shift-invariant (LSI) system can be modeled as a convolution between h and f, which we obtain here by ensemble averaging. However, we cannot say that a stochastic h for a single exposure has a shift-invariant response. The acquisition equation adapted for projection radiography, assuming a linear continuous-to-discrete transformation, is

$$g[n,m] = \int_{-\infty}^{\infty} d\mathbf{x}\, h(n,m,\mathbf{x})\, f(\mathbf{x}) \quad n,m = 0, \pm 1, \pm 2, \ldots, (N_d - 1)/2. \quad (7.4)$$

Here, function f is the object/patient property that couples to the energy field of the measurement. In this example, f describes the net photon attenuation over the source-to-detector distance X_3 (see Figure 7.1c). The *linear attenuation coefficient*, μ [cm^{-1}], is a 3-D function of space, $\mu(x_1, x_2, x_3)$. Using Beer's law for photon attenuation [10], we find

$$f(x_1, x_2) = \exp\left(-\int_0^{X_3} dx_3\, \mu(x_1, x_2, x_3)\right), \quad (7.5)$$

where the source-to-detector distance, X_3, is large enough to assume incidence along the x_3 axis.

Example 7.1.1. *Flat-field image: To see how the model works, we first simulate the flat irradiation field of Figure 7.1b for a thin constant attenuation layer, $\mu(\mathbf{x}) = \mu_0$. From (7.5), $f(\mathbf{x}) = e^{-\mu_0 X_3} \lesssim 1$. Combining that result with (7.2)–(7.4), the quantum-limited acquisition yield the random variable,*

$$g[n,m] = C_0 \int_{(n-1/2)d}^{(n+1/2)d} dx_1 \int_{(m-1/2)d}^{(m+1/2)d} dx_2\, q(x_1, x_2), \text{ for}$$

$$n, m = 0, \pm 1, \pm 2, \ldots, (N_d - 1)/2, \quad (7.6)$$

where $C_0 = h'\, e^{-\mu_0 X_3}$ is a constant. It is reasonable to assume the Poisson spatial point process $g[n,m]$ for this simple measurement is ergodic, so the spatial average and ensemble averages are equal:

$$\langle g \rangle = \frac{1}{N_d^2} \sum_{n,m \in A_d} g[n,m] = \mathcal{E}g[n,m] = \mathcal{E}g = C_0\, A_d\, \bar{q}. \quad (7.7)$$

In this example, we have neglected detector noise, photon scattering effects, sources of blur, and dead spaces between detector elements, which are all present in practice. The acquisition displayed in Figure 7.1b is for a 20 × 20 mm^2 detector (very small) with array elements having an area of 0.1 × 0.1 mm^2. The photon flux is 1,000 mm^{-2} ms^{-1}. The first section of the MATLAB code provided here generated this constant-attenuation image for a 1-ms exposure time.

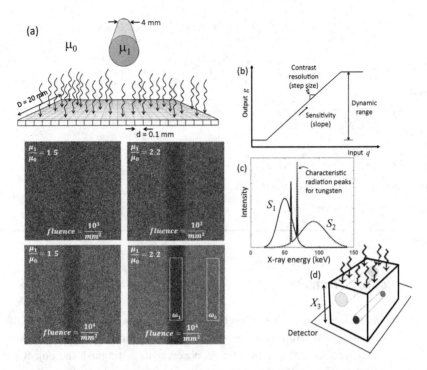

Figure 7.2 (a) Four x-ray image simulations with 4-mm-diameter cylindrical inclusions for the simple quantum-limited system depicted in Figure 7.1a and 7.1b. The attenuating cylinders have higher density such that the linear attenuation coefficient is either $\mu_1 = 1.5\mu_0$ or $\mu_1 = 2.2\mu_0$, where μ_0 is the attenuation coefficient for the background. Photon fluence is either 10^3 or 10^4 for a fixed exposure time. Cylinder visibility increases with either fluence or object contrast, μ_1/μ_0. Boxes in the lower-right image indicate regions selected for contrast transfer factor measurements γ in Figure 7.3. (b) Illustration of input–output quantities that define contrast measures. (c) An example of two x-ray spectra S_1 acquired at 80 kVp and S_2 at 140 kVp that are applied in energy-selective imaging in Section 7.2. (d) The phantom geometry used in Example 7.2.1.

Example 7.1.2. *Imaging object contrast: Adding spatial attenuation variability (object contrast), we can generate image contrast. The second part of the MATLAB code forms projection images of cylindrical inclusions with different attenuation values (μ_1); see Figure 7.2a. Comparing the two images in the left column of Figure 7.2a, we see that adding photons (increasing radiation exposure) improves the visibility of the low-contrast cylinder. These results hint at a relationship between x-ray exposure and task performance.*

```
%%%%%%%%%%%%%%%% Generate random photon fields for Figs. 7.1b and 7.2 %%%%%%%%%%%%%%%%%%%%%%%%%%
dx=0.1;X0=20;qpa=10000;        % 20x20 mm detector, 0.1-mm elements, qpa = photons/mm^2
x=0:dx:X0;xl=length(x);lam=qpa*X0^2;Np=poissrnd(lam); %Np=total photon number
qx=X0*rand(Np,1);qy=X0*rand(Np,1);q=zeros(xl); % spatial coordinate of each photon
for j=1:Np                    % uniformly distribute all Np photons
    a = ceil(qx(j)/dx);b = ceil(qy(j)/dx);q(a,b) = q(a,b)+1;
end
hp=q(1:xl-1,1:xl-1);          % stochastic impulse response with no blur
imagesc(hp);colormap gray,axis square % result in Fig. 7.1b
```

```
figure;histogram(reshape(hp,(xl-1)^2,1),20) % following line checks whether pmf is Poisson
% sum(sum(hp))/X0^2 should be approximately qpa
%
% form cylindrical inclusion; generate images g for Fig. 7.2
C=1.5;                      % contrast ratio set by you
a=ones(xl)/xl;N2=floor((xl-1)/2);R=floor(xl/10); % R=2 mm radius in (x2,x3) plane
for j=1:xl;
    for i=1:xl;             % make a circle of diameter 4 mm for the cylinder
        if j<N2+sqrt(R^2-(i-N2)^2) && j>N2-sqrt(R^2-(i-N2)^2);
            a(j,i)=a(j,i)*C;
        end;
    end;
end;                       % plot the center line of a to see linear atten coeff profile
ff=exp(-sum(a));f=ff(1:xl-1)/ff(1); %figure;plot(A)
for j=1:xl-1
    g(j,:)=hp(j,:).*f;      % equation (7.4) with no blur
end
figure;imagesc(g);colormap gray;axis square;axis off
```

7.2   Contrast Transfer

Object contrast refers to differences in object properties that couple to the measurement energy. *Measurement contrast* describes how object-property variations appear in displayed data as required for an observer to distinguish the object from its surroundings. The goal of detection through measurement is to efficiently transfer object contrast into measurement contrast (see Figure 7.2a). The first column of images in Figure 7.2a is from an object with lower object contrast than the second column of images, yet all four images display different measurement contrasts.

Assume the plot of Figure 7.2b describes the contrast sensitivity of the ideal detector in our example. Here the input describes photon density at the detector q and the output is the voltage response of the detector from that input, g. The *dynamic range (DR)* is the maximum range of detector responses for the full range of input, three or four orders of magnitude for typical x-ray exposures. *Contrast resolution (CR)* is the smallest measurable increment of output change. Contrast resolution is ultimately limited by the bit depth of the digital recording device. For example, if the detector can record $2^{12} = 4,096$ distinct output levels over the DR, then $CR = 2^{-12}$ DR, though we often just say $CR = 12$ bits. Noise levels and limited detector sensitivity further limit DR and CR. The instantaneous slope of the input–output curve (the response in Figure 7.2b is linear, but that is not always the case) is a measure of device responsiveness or *sensitivity* of the detector in units of [V/quanta]. Some acquisition sensitivity limitations can be compensated through display-stage processing now that digital detectors have become standard equipment.

The *efficiency* by which an imaging system transfers contrast from object f into data g is quantified by the *large-area contrast transfer factor*, γ. Small-area contrast transfer is classified as spatial resolution; see Section 7.4. Referring to the user-defined ω boxes in Figure 7.2a (lower right image), the object contrast for the target input to the device, C_{in}, is defined based on the mean object properties in ω_1 and the adjacent region ω_0. Because the object property that couples to x-rays is the linear attenuation coefficient μ, input contrast is found from sample ensemble means via (3.19) that are measured or assumed; i.e.,

$$C_{in} = \frac{\bar{\mu}_1 - \bar{\mu}_0}{\bar{\mu}_1 + \bar{\mu}_0}. \tag{7.8}$$

Contrast output from the device, C_{out}, is defined based on the sample spatial means of measurements g_i made in the corresponding regions ω_1 and ω_0. The sample spatial mean obtained via (3.24) is $\langle g \rangle_i = \sum_{n,m \in \omega_i} g[n,m]/N_{\omega_i}$, where N_{ω_i} is the number of independent samples in region ω_i, for $i = 0, 1$. Measurement contrast output is

$$C_{out} = \frac{\langle g \rangle_1 - \langle g \rangle_0}{\langle g \rangle_1 + \langle g \rangle_0}. \tag{7.9}$$

Spatial means apply even if object contrast is not constant in each region. The contrast transfer factor is the absolute value of the ratio

$$\gamma = \left| \frac{C_{out}}{C_{in}} \right|. \tag{7.10}$$

This equation assumes the input–output curve in Figure 7.2b is linear over the dynamic range. Otherwise, γ becomes a function of input contrast $\gamma = |dC_{out}/dC_{in}|_{C_{in}}$, where the instantaneous slope is applied at values of C_{in} specific to the task at hand.

We adapted the simulation code from Section 7.1 to compute γ as a function of C_{in}. The results are shown in Figure 7.3a. The contrast-transfer efficiency in this simulation increases linearly for $0.09 \leq C_{in} \leq 0.33$. The γ measurement deviates from linear at lower C_{in} values because of the singularity at the origin. A singularity at $C_{in} = 0$ distorts values near zero. This curve displays the low-contrast dynamic range of the system.

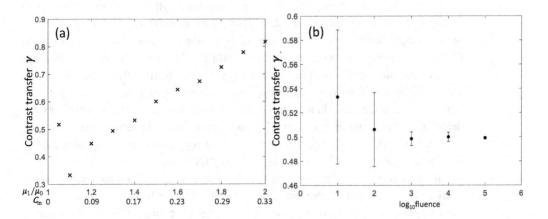

Figure 7.3 (a) Measurements of contrast transfer γ using the simulation model for a fixed photon fluence of 1,000 mm^{-2}. The abscissa is plotted in terms of the object contrast μ_1/μ_0 and the input contrast C_{in}. (b) γ measurements versus photon fluence for $C_{in} = 0.130$. The error bars denote sample standard deviations of γ measurements for 20 trials.

In Figure 7.3b, we fixed $C_{in} = 0.13$ (where $\mu_1/\mu_0 = 1.3$) and computed γ as a function of photon fluence. We find the mean γ estimate does not change, as expected, but the uncertainty in the estimate varies significantly. That result makes sense if you consider photons as *degrees of freedom* available to the observer striving to visualize objects of diagnostic value. The images in each column of Figure 7.2a have the same input contrast; the difference in visibility with exposure is due to the number of degrees of freedom available to observers to decide how visible the cylinder appears. Differences in cylinder visibility from different exposures reflect the different SNR values, which for Poisson noise under quantum-limiting conditions is SNR = signal energy/noise energy $= N_d$.

Before addressing noise properties, we explore a method for contrast quantification that brings photon energy into the analysis.

7.2.1   Dual-Energy Radiography

Medical imaging is approximately 120 years old, and during that time the primary goal has been to generate image contrast that renders visible biological anatomy and function as noninvasively as possible. Physicians reading displayed data interpret the results to direct patient care. Recently, significant advances in *machine learning* have employed artificial neural networks that learn to interpret the available measurement contrast to improve the medical decision-making process [84]. Other approaches enhance object contrast in the body through the use of injectable media or strive to more clearly separate components of intrinsic contrast via basis decomposition methods.

An example of the latter approach was proposed in 1976 by Alvarez and Macovski [3]. They developed a way of increasing the "specificity of image contrast" in radiography by identifying physical sources of object contrast appearing in images. These researchers noted that the x-ray linear attenuation coefficient $\mu(\mathbf{x})$ responsible for object contrast in patients [see (7.5)] is well approximated by the sum of two energy-dependent interaction mechanisms, *photoelectric absorption* $\mu_P(\mathbf{x})$ and *Compton scattering* $\mu_C(\mathbf{x})$. Data acquired at different energies can be *quantitatively* parsed according to the photon–tissue interaction mechanism that gave rise to the data, thereby revealing new diagnostic details about tissue properties. At the time when Alvarez and Macovski introduced their method, however, energy-sensitive x-ray detectors were not routinely employed in radiography; only the total incident intensity was measured. Consequently, *detected* photon energy information was discarded in both projection radiography and computed x-ray tomography.

Emitted energy considerations have always influenced radiographic techniques, as x-ray tube voltage and current are set by a technician to influence the relative contribution of the two attenuation mechanisms. Photoelectric interactions occur primarily

---

[4] Because Compton interactions occur between x-rays and orbital electrons, it is more accurate to say that higher *electron* density increases Compton scattering. Mass density follows electron density in body tissues.

at lower diagnostic photon energies and are highly sensitive to atomic number (Z) differences in tissues; e.g., bone versus soft tissue contrast is strongest at diagnostic x-ray energies less than 30 keV. Conversely, Compton scattering interactions occur primarily at higher photon energies, more than 30 keV, and are most sensitive to mass-density (ρ) differences.[4] For this reason, if we wish to see a fracture in cortical bone, where it is important to discriminate between calcified tissues and fluid, we would typically set the peak kilovoltage (kVp) on the x-ray tube to about 60 kVp, which gives an average bremsstrahlung[5] spectrum value of 20–30 keV where photoelectric interactions dominate. However, to see a soft-tissue mass in the lungs that is surrounded by air and bony structures, higher photon energies from a tube voltage of 100–110 kVp give the clearest lesion contrast, as Compton interactions emphasize tissue-density contrast.

Example 7.2.1. *Dual-energy model: Alvarez, Macovski, and their team [3, 65] found they could separate $\mu(\mathbf{x})$ into photoelectric $\mu_P(\mathbf{x})$ and Compton $\mu_C(\mathbf{x})$ components by decomposing two acquisitions obtained at different photon energies. The principal components obtained from these two measurements could be used to emphasize regions of varying Z and ρ that are interpretable medically. This notion was based on the two energy-dependent terms in the equation for the linear attenuation coefficient,*

$$\mu(\mathbf{x}) \simeq [\mu_C + \mu_P](\mathbf{x}) = \rho N_A \frac{Z}{A} \left\{ f_C(\epsilon) + f_P \frac{Z^{3.8}}{\epsilon^{3.2}} \right\} (\mathbf{x}). \qquad (7.11)$$

In this equation, $N_A = 6.022 \times 10^{23}$ [molecules/mole] is Avadogro's number, ρ [g/cm³] is mass density, $Z/A \simeq 1/2$ is the ratio of atomic number to atomic mass of atoms in the body (except for hydrogen, where $Z/A = 1$), and ϵ represents photon energy hv [keV]. In the 20 keV $\le \epsilon \le$ 140 keV range of most body imaging, Macovski [68] approximated the Klein–Nishina equation[6] as $f_C(\epsilon) = 0.6 \times 10^{-24} \exp[-0.0028(\epsilon - 30)]$ and the photoelectric constant as $f_P = 9.8 \times 10^{24}$. Overall attenuation μ has units of cm⁻¹ and is a function of position $\mathbf{x}$ because ρ, Z, and ϵ all vary with position in the body, as demonstrated later in this section.

We simulated the phantom in Figure 7.2d to contain two cylindrical inclusions in a muscle-tissue background where $\rho = 1.06$ g/cm³ and $\bar{Z} = 6$ represent properties of muscle with the value for carbon. The 4-mm-diameter cylinder of muscle tissue has $\rho = 1.59$ g/cm³ and $Z = 6$, where the density is 50% higher than the background, simulating a lesion. The 2-mm-diameter cylinder has $\rho = 1.85$ g/cm³ and $Z = 20$ which are the density of bone and the atomic number of calcium, respectively. Data were acquired using the two x-ray tube spectra[7] shown in Figure 7.2c.

---

[5] Literally, "braking radiation." These x-rays are produced by decelerating electrons in a vacuum tube as they pass near the nuclei of anode metals such as tungsten or molybdenum.

[6] The Klein–Nishina equation describes the ratio of scattered to incident photons at energies where photons are scattered from single "free" orbital electrons.

[7] These spectra are approximations of the Siemens Somatom Drive dual-source CT at 80 kV and 140 kV plus 0.4 mm of additional Sn filtration, as reproduced from [36].

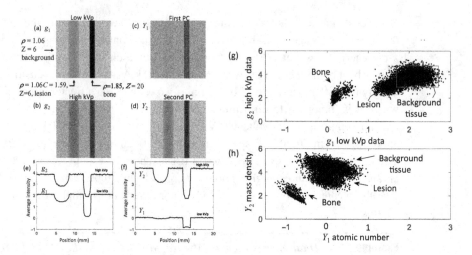

Figure 7.4 Results from Example 7.2.1. (a) and (b) are, respectively, the low- and high-energy x-ray acquisitions g_1, g_2 via (7.12), which are found by applying photon spectra S_1, S_2 from Figure 7.2c. Averaging the rows of data in (a) and (b), we obtain the profiles shown in (e). (c) and (d) are the first and second principal component images Y_1 and Y_2, respectively. Their profiles are shown in (f). (g) and (h) are the scatter plots for the data in (a)–(d). Notice how the high Z (bone) and high ρ (lesion) data cluster relative to the background data.

Let S_1 be the low-energy incident x-ray spectrum that results in acquisition g_1 and let S_2 be the high-energy spectrum that yields g_2. Then [68]

$$g_1(x_1, x_2) = \int_{\epsilon \in S_1} d\epsilon \, S_1(\epsilon) \exp\left(-\int_{x_3=0}^{X_3} dx_3 \, \mu(\mathbf{x}, \epsilon)\right)$$

$$g_2(x_1, x_2) = \int_{\epsilon \in S_2} d\epsilon \, S_2(\epsilon) \exp\left(-\int_{x_3=0}^{X_3} dx_3 \, \mu(\mathbf{x}, \epsilon)\right). \tag{7.12}$$

Returning now to Example 5.1.1, we find the basis set from an eigen-decomposition of the covariance matrix of the acquisition, $\mathbf{K}_g$. To do this, we first reshape the acquired data g_1 and g_2 into vectors that form columns of matrix $\mathbf{G}$; rows are samples from the two acquisitions. That is, $g_j \in \mathbb{R}^{N \times N} \longrightarrow \mathbf{g}_j \in \mathbb{R}^{N^2 \times 1}$ yields $\mathbf{G} = (\mathbf{g}_1, \mathbf{g}_2) \in \mathbb{R}^{N^2 \times 2}$. Computing the 2×2 covariance matrix via (5.1), $\mathbf{K}_g \propto \mathbf{G}^\top \mathbf{G}$, we decompose it to find $\mathbf{K}_g = \mathbf{U}^\dagger \mathbf{D}_g \mathbf{U}$ as explained in (5.2). In MATLAB, this operation is achieved via [U,Dg]=eig(Kg); *. Finally, the orthogonal basis vectors are the columns of $\mathbf{Y} = \mathbf{G}\mathbf{U}$. We reshape $\mathbf{Y} \in \mathbb{R}^{N^2 \times 2} \longrightarrow Y_1(x_1, x_2), Y_2(x_1, x_2) \in \mathbb{R}^{N \times N}$. Equations (7.11) and (7.12) and the analysis in this paragraph are coded next. The results are displayed in Figure 7.4.*

```
%%%%%%%%%%%%%%%%%%%%%%%%%%%%%%%%%%%%%%   Code for Example 7.2.1   %%%%%%%%%%%%%%%%%%%%%%%%%%%%%%%%%%%%%%
dx=0.1;X0=20;x=0:dx:X0;xl=length(x);      % detector area=20x20mm; phantom vol=20x20x20mm
qpa=10000;Np=qpa*X0^2;                    % qpa = photons/mm^2; Np=total photons/exposure
%
% Make a spectrum
hv=20:140;m1=50;s1=10;m2=92;s2=18;z1=6;z2=20;   % hv=energy [keV]; spectral means, stds, atomic #s
% S1=exp(-(hv-m1).^2/(2*s1^2))/(s1*sqrt(2*pi));  % use only to approximate spectrum for display
% S2=exp(-(hv-m2).^2/(2*s2^2))/(s2*sqrt(2*pi));  % bremsstrahlung
```

```
% S1(59-19)=2*S1(59-19);S1(68-19)=3*S1(68-19);S2(59-19)=8*S2(59-19);S2(68-19)=8*S2(68-19); %char rad
% S1(80-19:end)=0;        % for 80 kVp, no hv>80 keV -- limit normal pdf to have hard cutoffs
% S1=S1/sum(S1);S2=S2/sum(S2);
% plot(hv,S1,hv,S2)
SS=randn(Np,2);SS(:,1)=s1*SS(:,1)+m1;SS(:,2)=s2*SS(:,2)+m2; % assign energy value to each photon
for j=1:Np                                   % SS=photon energies from high,low energy exposures
    if SS(j,1) <20 || SS(j,1) > 140;         % limit to 20-140 keV bremmstrahlung
        SS(j,1) = 59;                        % assign photons outside range to char rad peaks
    end
    if SS(j,2) <20 || SS(j,2) > 140;
        SS(j,2) = 68;
    end
end
%
% Form two cylindrical inclusions
C=1.5;                                       % soft-tissue cylinder contrast ratio (density ratio wrt bkg)
Ag2=6.022e23/2;                              % Avagodro's number times Z/A(=1/2)
fc=Ag2*0.6e-24;                              % Compton constant
fp=Ag2*9.8e-24;                              % PE constant (Macovski's book, pp. 33-34)
a=ones(xl);                                  % arbitrary background tissue attenuation
N2=floor((xl-1)/3);R2=floor(xl/10);          % N2 position, R2=2.0 mm lesion radius
N3=floor(2*(xl-1)/3);R3=floor(xl/20);        % N3 position, R3=1.0 mm bone radius
%
af=zeros(xl,3);
for j=1:xl;                                  % make two circles of diameter 4 mm and 2 mm for cylinders
    for i=1:xl;                              % af is used & a is not used except to show cross-section geometry
        if j<N2+sqrt(R2^2-(i-N2)^2) && j>N2-sqrt(R2^2-(i-N2)^2);
            a(j,i)=a(j,i)*0.7; af(j,2)=af(j,2)+1;  % af2=vertical pixels that are lesion
        end
        if j<N3+sqrt(R3^2-(i-N3)^2) && j>N3-sqrt(R3^2-(i-N3)^2);
            a(j,i)=0.1; af(j,3)=af(j,3)+1;         % af3=vertical pixels that are bone
        end
    end;
end;
for j=1:xl
    af(j,1)=1.06*(xl-af(j,2)-af(j,3))/xl;    % mass density x fraction of vert bkg pixels
    af(j,2)=1.06*af(j,2)/xl;                 % mass density x fraction of vertical pixels that are lesion
    af(j,3)=1.85*af(j,3)/xl;                 % mass density x fraction of vertical pixels that are bone
end
%
qx=X0*rand(Np,1);qy=X0*rand(Np,1);g1=zeros(xl);
% uniform random spatial coordinate for each photon (aa,bb)
for j=1:Np                                   % place all Np photons...
    aa = ceil(qx(j)/dx);                     % ...in the imaging plane
    bb = ceil(qy(j)/dx);
    g1(aa,bb)=g1(aa,bb)+exp(-fc*exp(-0.0028*(SS(j,1)-30))*(af(bb,1)+C*af(bb,2)+af(bb,3))*X0) ...
        *exp(-fp*((z1^3.8/(SS(j,1)^3.2)*(af(bb,1)+C*af(bb,2)))) ...
        +(z2^3.8/(SS(j,1)^3.2)*af(bb,3)))*X0); % calculate # photons transmitted at low energy
end
qx=X0*rand(Np,1);qy=X0*rand(Np,1);g2=zeros(xl);
for j=1:Np                                   % place all Np photons...
    aa = ceil(qx(j)/dx);                     % ...in the imaging plane
    bb = ceil(qy(j)/dx);
    g2(aa,bb)=g2(aa,bb)+exp(-fc*exp(-0.0028*(SS(j,2)-30))*(af(bb,1)+C*af(bb,2)+af(bb,3))*X0) ...
        *exp(-fp*((z1^3.8/(SS(j,2)^3.2)*(af(bb,1)+C*af(bb,2)))) ...
        +(z2^3.8/(SS(j,2)^3.2)*af(bb,3)))*X0); % calculate # photons transmitted at high energy
end
figure;g11=g1(1:200,1:200);g22=g2(1:200,1:200);
ming1=min(min(g11));maxg1=max(max(g11));
subplot(2,2,1);imagesc(g11);colormap gray;axis square;axis off
subplot(2,2,3);imagesc(g22);colormap gray;axis square;axis off
%%
% Here I compute principal components and Y1=rho and Y2=Z images
G1=reshape(g11,200*200,1);G2=reshape(g22,200*200,1);
G=[G1 G2];K=G'*G;[U,D]=eig(K);Y=G*U;                     % PCA
Y1=reshape(Y(:,1),200,200);Y2=reshape(Y(:,2),200,200);
subplot(2,2,2);imagesc(-Y1);colormap gray;axis square;axis off    % change grayscale polarity
subplot(2,2,4);imagesc(Y2);colormap gray;axis square;axis off
%scatterplot(G,100);scatterplot(Y,100)
figure;subplot(2,1,1);plot(G(1:5:end,1),G(1:5:end,2),'k.')
subplot(2,1,2);plot(-Y1(1:5:end),Y2(1:5:end),'r.')
```

As you read through the code, note two things. First, photon energies are drawn from normal pdfs; i.e., $S_i \sim \mathcal{N}(\mu_i, \sigma_i^2)$, where realizations that are out of range are given a characteristic-photon energy value. Second, acquisitions g_1 and g_2 are formed one photon at a time using a Monte Carlo *approach. I selected a large number of photons for the acquisitions, N_p. Each photon has an energy randomly selected from its spectrum and then that photon is randomly positioned in the detector plane (x_1, x_2). As each photon emerges from the phantom of thickness X_3, the fractional probability of transmission is recorded by a detector element. Accumulating those fractions after considering all N_p photons yields the number of counts per detector element. This process is run for both spectra. A great number of simplifying assumptions were made for the purpose of providing readers with this means of numerical experimentation.*

Figure 7.4a, 7.4b, and 7.4e show that low-energy acquisition g_1 is more attenuated overall than is high-energy acquisition g_2. Also, the relative attenuation of "bone" versus "lesion" is greater at low energy. These findings are consistent with the well-known effects on image contrast from energy-sensitive photoelectric and Compton attenuation mechanisms. The outputs from each of the $200 \times 200 = 40,000$ detector elements at both acquisition energies are plotted against each other in the scatter plot (Figure 7.4g). We can see how the pixel values cluster for the three types of absorbers (background, bone, lesion).

Computing the principal components effectively rotates and shifts these clusters in Figure 7.4h, where now the lesion and background have no contrast in Y_1. Indeed, if you examine Y_1 in Figure 7.4c, the larger inclusion almost disappears. Y_1 is the "atomic number image" in the sense that it displays contrast for regions of varying Z and suppresses ρ variations. Conversely, Y_2 is the "density image," where both inclusions appear because they both vary in ρ with respect to the background.

Example 7.2.1 is intended to demonstrate how mechanisms of a measurement may provide a physical basis for obtaining specific information. It is not an example of a clinical method. Clinical methods completely separate bone from soft-tissue variations, as demonstrated by Lehmann et al. [65]. An additional calibration step is required because the photoelectric–Compton "basis" is not really a basis, but rather a *frame*; i.e., the two vectors span the space but are not linearly independent. Frames are discussed in Appendix A. We now leave the energy-selective discussion and return to intensity acquisitions and a discussion of radiographic noise.

7.3   Noise Power Spectrum

Even without acquisition noise from the detector, the simulated x-ray images shown in Figures 7.2 and 7.4 illustrate that quantum noise can partially mask the visibility of low-contrast objects. The traditional measure of WSS noise from normal and Poisson processes is the *noise power spectral density (NPS)*, written in symbols as $S_e(u)$ (see Section 5.12). Often we don't know the detailed nature of noise sources contributing to the acquired signal g. A practical approach is to identify a region in the image, such as ω_i in Figure 7.2, where object function f is relatively constant, and then to measure noise simply as

$$e[n,m] = g[n,m] - \langle g \rangle, \tag{7.13}$$

which includes any deviations from the sample mean. Although overall bias is not usually considered noise, it contributes to measurement error depending on whether error is defined using spatial averages via (7.13) or ensemble means, $e[n,m] = g[n,m] - \mathcal{E}g[n,m]$.

NPS may be estimated along the x_1 axis of the $N_d \times N_d$ detector from the magnitude of short-time Fourier transforms computed along x_1 and then averaged over the x_2 axis. The STFT in (5.34) is needed here because we are analyzing a subregion. Because the 2-D STFT is separable, applying (5.35) and (7.13) gives the noise power spectrum

$$\hat{S}_e[k; N_d] = \frac{1}{N_d} \sum_{m=1}^{N_d} \left| \hat{E}[k,m; N_d] \right|^2, \quad \text{NPS along } x_1 \text{ axis} \tag{7.14}$$

where the STFT of the noise trace is

$$\hat{E}[k,m; N_d] = \mathcal{F}_{x_1}\{e[n,m]\} = \frac{1}{N_d} \sum_{n=1}^{N_d} e[n,m] \, e^{-i2\pi kn/N_d}.$$

The NPS along the x_2 axis is

$$\hat{S}_e[\ell; N_d] = \frac{1}{N_d} \sum_{n=1}^{N_d} \left| \hat{E}[n,\ell; N_d] \right|^2, \quad \text{NPS along } x_2 \text{ axis} \tag{7.15}$$

where

$$\hat{E}[n,\ell; N_d] = \mathcal{F}_{x_2}\{e[n,m]\} = \frac{1}{N_d} \sum_{m=1}^{N_d} e[n,m] \, e^{-i2\pi km/N_d}.$$

In this context, separability means the 2-D spectrum is formed from the product of the 1-D spectra, $\hat{S}_e[k,\ell; N_d] = \frac{1}{N_d^2} \sum_{m=1}^{N_d} \sum_{n=1}^{N_d} \left(\left| \hat{E}[k,m; N_d] \right| \left| \hat{E}[n,\ell; N_d] \right| \right)^2$. The 1-D NPS for the image in Figure 7.1b along the x_1 and x_2 axes are shown separately in Figure 7.5a. Both spectra are flat across the available bandwidth and equal, which is an example of spectrally *white noise* – in this case, generated by a Poisson process. We explore power spectral estimation in Problem 8.6 in Chapter 8.

If the noise measured between adjacent detector elements is correlated, e.g., when the focal spot is not a point, we find *colored-noise* spectra like those shown in Figure 7.5b. Color is a reference to the optical spectrum, such that low-frequency noise is labeled "red" and high-frequency noise is labeled "blue." When image noise is spectrally white, human observers find it less limiting for visual detection tasks

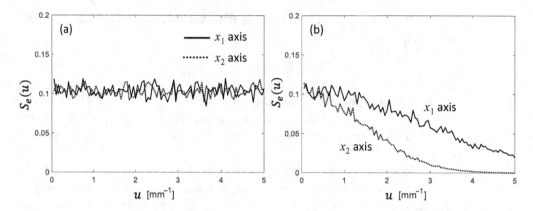

Figure 7.5 (a) Noise power spectral density (up to Nyquist frequency) for the flat-field image in Figure 7.1b as a function of the two axes. This is a *white-noise spectrum* for the uncorrelated samples of quantum-limited imaging that is the same for both axes. Note that noise variance was normalized so that $\text{var}(e) = \int_{-5}^{5} du \, S_e(u) = 10 \,\text{mm}^{-1} S_0 \simeq 1 \,\text{V}^2$, since $S_0 \simeq 0.1 \,\text{V}^2\text{mm}$. (b) NPS for colored-noise samples. When measurement samples are uniformly correlated so the noise remains a WSS random process, high-frequency components are lost. In this example, noise along the x_1 axis has twice the "effective" bandwidth of that along x_2, where $BW_{eff} = \int_{-\infty}^{\infty} du \, S_e(u)/\text{max} S_e$. Spatial axes x_1, x_2, x_3 are defined in Figure 7.1.

than colored noise [116]. This is because humans have difficulty decorrelating noise visually while also trying to differentiate noise from embedded signals. We will see in Section 7.5 that predictions of visual task efficiency consider the amount of noise power in the frequency channels of the signal.

7.4   Spatial Resolution

Spatial/temporal resolution versus contrast resolution.
Spatial resolution describes the smallest *spatial separation* between distinct objects that can be reliably detected by observers of the measurement data. Similarly, temporal resolution describes the smallest discernable separation between distinct objects in time.

In *contrast resolution analysis*, we quantify the smallest discernable difference in *ensemble-averaged amplitude* by studying large homogeneous spatial regions many times larger than the impulse response. In this way, we intentionally eliminate the effects of spatial resolution and noise limitations in assessing amplitude changes. In *spatial resolution analysis*, we quantify discernable differences between *high-contrast spatially proximal targets* using ensemble-averaged measurement data to avoid the influences of contrast resolution and noise. In both types of analyses, we work to isolate the property to be measured by averaging data in ways that minimize the effects of other properties. Every component of a measurement system can add blur to the data, but the following analysis of x-ray imaging systems simplifies the measurement model by considering only those limitations imposed by a finite x-ray tube focal spot.[8]

Experimental considerations.
Up to now, we have considered an idealized point source of x-rays located far from the object and detector. In reality, x-rays emerge from an anode patch about 0.1–4 mm^2 in area within the x-ray tube positioned approximately 0.5–1 m from the detector. A photon spectrum with energy up to X keV is generated when fast electrons emerging from the cathode (–) are accelerated in a vacuum at X kV peak (kVp) potential to the anode (+). Electrons are focused along their path to strike a rotating-disk anode at the *focal spot (FS)*. The sudden deceleration converts a small percentage of their kinetic energy into a broad spectrum of photon energies (bremsstrahlung). Unfortunately, most of the kinetic energy absorbed goes into heating the anode metal.

We want the smallest possible FS so that $h(\mathbf{x}, \mathbf{x}') \rightarrow \delta(\mathbf{x} - \mathbf{x}')$, which is both compact and spatially invariant. However, reducing the FS area increases the risk of melting the anode metal from intense heating. The beveled surface on the rotating anode mitigates the heating problem somewhat with an acceptable FS size, but it varies the FS size, photon flux, and beam energy across the tube axis dimension of the detector area. The ensemble impulse response becomes shift varying so that spatial resolution, image contrast, and noise properties vary across the field of view. This is is necessary for long tube life, but it is bad news for image quality. For example, note the exaggerated exposure simulation in Figure 7.1d generated by applying an FS with strongly linear shift-varying properties, and compare it to the simulated flat-field exposure from a point source in Figure 7.1b using the same detector. In practice, a spatially varying impulse response can be so great that the visibility of pathology in the x-ray image depends on how a patient is positioned in the field! We will describe how to quantify spatial resolution for these non-quantum-limited, linear shift-varying (LSV) conditions. First, however, we describe the simpler measurement of spatial resolution for a 1-D LSI source-detector system.

7.4.1   Modulation Transfer Measurements for LSI Systems

Concepts.
This section describes the *modulation transfer* concept developed to quantify spatial resolution as a function of spatial frequency. To simplify the analysis, consider a 1-D detector array of length D and the 1-D LSI focal-spot impulse-response function shown in the first row of plots in Figure 7.6. This 1-D model has a double peak common with x-ray tube focal spots. We can adapt (7.2) to include the 1-D FS by allowing h' to be a function of position as follows:

$$h(n, x) = h'(nd - x)\, q(x)\, \mathrm{rect}\left(\frac{x - nd}{d}\right), \quad \text{for } n = 0, \pm 1, \pm 2, \ldots, (N_d - 1)/2.$$

$$(7.16)$$

---

[8] For medical x-ray imaging, where objects range in size between 50 μm and 50 cm, we can ignore diffraction and turn to geometric optics for discussions of spatial resolution. X-ray diffraction is an informative phenomenon below the micrometer scale that is not considered here.

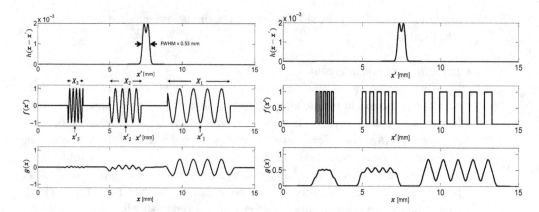

Figure 7.6 The top plots display the 1-D LSI impulse response $h(x - x')$ of size given by FWHM = 0.53 mm. Two object functions $f(x')$ are examined in the middle plots: On the left are three sine-wave bursts with wavelengths 0.24 mm, 0.48 mm, and 0.96 mm at three positions along the detector. These object patterns appear at spatial frequencies 4.2 mm^{-1}, 2.1 mm^{-1}, and 1.04 mm^{-1}, respectively. On the right are three square-wave patterns at 0.24 mm/cycle, 0.48 mm/cycle, and 0.96 mm/cycle. The corresponding detector recordings $g(x) = [h * f](x)$ are shown below each of the object patterns. Modulation transfer efficiencies as a function of frequency for sine-wave and square-wave inputs quantify spatial resolution.

In (7.16), we note the psf h is shift varying because it is stochastic via $q(x)$, even though the underlying impulse response h' is LSI. Because the ensemble average of h is LSI, below we apply a Fourier basis to implement an eigen-decomposition.

If we input an eigenfunction of the system into this linear measurement device, we expect to see the same eigenfunction at the output, except that it will be scaled by a complex constant, which is the corresponding eigenvalue. Imagine it is possible to create a sinusoidally varying test object of the form

$$f(x) = \frac{B}{D} \cos(2\pi u_1 x), \qquad (7.17)$$

where B is a constant to be determined and we choose spatial frequency $u_1 = k_1/D$ for integer $k_1 > 0$ as the modulation frequency. Recall that a detector of length D is "sampled" on the interval d. u_1 is selected to be a harmonic of the detector's fundamental frequency at $u_0 = 1/D$. The measurement of this object, in additive WSS zero-mean acquisition noise $e[n]$, is

$$g[n] = \int_{-\infty}^{\infty} dx\, h(nd, x)\, f(x) + e[n]$$

$$= \frac{B}{D} \int_{(n-1/2)d}^{(n+1/2)d} dx\, h'(nd - x) q(x) \cos(2\pi k_1 x/D) + e[n].$$

Noise is not of interest for this system's characterization, so we take the spatial mean of the output. We also apply a variable substitution $y = nd - x$,

$$
\mathcal{E}\{g[n]\} = \frac{B\bar{q}}{D} \int_{(n-1/2)d}^{(n+1/2)d} dx \, h'(nd - x) \Re \left\{ e^{-i2\pi k_1 x/D} \right\}
$$

$$
= -\frac{B\bar{q}}{D} \Re \left\{ \left[\int_{d/2}^{-d/2} dy \, h'(y) e^{-i2\pi k_1(nd-y)/D} \right] \right\}
$$

$$
= B\bar{q} \cos \left(\frac{2\pi k_1 nd}{D} \right) \Re \left\{ \frac{1}{D} \int_{-d/2}^{d/2} dy \, h'(y) e^{i2\pi k_1 y/D} \right\}
$$

$$
= B\bar{q} \cos \left(\frac{2\pi k_1 nd}{D} \right) \Re \left\{ \hat{H}'^*[k_1] \right\} = \hat{G}[k_1] \cos \left(\frac{2\pi k_1 nd}{D} \right). \quad (7.18)
$$

The integral in the third line is the short-time conjugate Fourier transform over detector element area d^2 that we label as $\hat{H}'^*[k_1]$.

Also from (7.18), we find for real symmetric h,

$$
\hat{G}[k_1] = B\bar{q} |\hat{H}'[k_1]| = \bar{q} \frac{|\hat{H}'[k_1]|}{\hat{H}[0]}, \quad (7.19)
$$

if we set arbitrary constant $B = 1/\bar{q}\hat{H}'[0] = 1/\hat{H}[0]$. Equation 7.19 is an eigenvalue of the acquisition system because the cosine input in (7.17) also appears at output (7.18) except for constants.

Substituting $|\hat{H}[k_1]| = \bar{q}|\hat{H}'[k_1]|$, where $\hat{H}'$ is the *modulation transfer* amplitude and $\bar{q}$ is the mean fluence at spatial frequency k_1/D, we have

$$
MT[k_1] = \frac{|\hat{H}[k_1]|}{\hat{H}[0]}. \quad (7.20)
$$

Stepping through other harmonic frequencies, we can estimate $MT[k]$ over the detector bandwidth without knowledge of the impulse response of the system. In essence, we are measuring the fractional transfer of sinusoidal modulation by the measurement device at spatial frequencies in the useful bandwidth.

Interpretation.
MT is a laboratory *measurement*; it is a real-valued scalar between 0 and 1. If $MT[k_1] \simeq 1$, the recording transfers all modulation at that frequency without loss. However, if $MT[k_n] \simeq 0$, no modulation is transferred by the measurement because the eigenvalue is zero for the eigenfunction at frequency k_n/D. The second and third rows in the left column of Figure 7.6 are simulation examples of 1-D input and output modulations, respectively. MT is the relative change in measured output modulation amplitude.

7.4.2 Modulation Transfer Function

Spatial resolution for LTI devices can be easily *predicted* at all frequencies without measurement if one knows $h'[n]$ a priori.[9] The impulse response in the preceding

---

[9] MT denotes modulation transfer *measurements* and MTF indicates MT *predictions* from h.

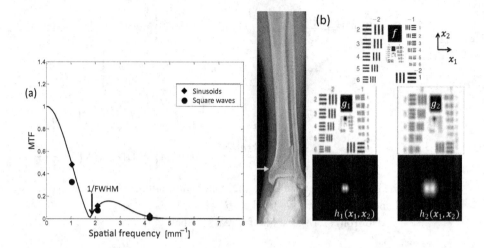

Figure 7.7   (a) Measured (MT points) and predicted (MTF line) values for the 1-D impulse response in Figure 7.6 are shown. Measured MT via $(g_{max} - g_{min})/(f_{max} - f_{min})$ for sine-wave (◆) and square-wave (●) objects are shown. The inverse of the full-width-at-half-maximum (FWHM) FS size is shown on the frequency axis to be near the zero point of the MTF curve. Close inspection of Figure 7.6 reveals a 180° phase shift that inverts the modulation amplitude at the two highest-frequency sinusoids, 2.1 mm^{-1} and 4.2 mm^{-1}. Both frequencies are in the MTF side lobe. (b) A 2-D test pattern (top) is imaged, with the 2-D impulse responses shown in the bottom row. Impulse responses are shown at four times their actual size. The radiograph of a bone fracture demonstrates the need for high spatial resolution in some clinical tasks.

example was for the focal spot. The expression for MTF is another form of (7.20) expanded to include all frequencies in the bandwidth. The *modulation-transfer function* at frequencies k/D and $k = 0, 1, 2, \ldots, N_d$ is

$$\text{MTF}[k] = \frac{|\mathcal{F}\{h'\}[k]|}{\mathcal{F}\{h'\}[0]} = \lim_{N_d \to \infty} \frac{\frac{\bar{q}}{\sqrt{N_d}} \left| \sum_{n=0}^{N_d-1} h'[n] e^{-2\pi k n/N_d} \right|}{\frac{\bar{q}}{\sqrt{N_d}} \sum_{n=0}^{N_d-1} h'[n]} = \frac{|H[k]|}{H[0]}. \quad (7.21)$$

The MTF curve predicted by (7.21) is plotted in Figure 7.7a as a solid line. We may summarize the curve by stating the value of D/k at which MTF[k] first falls to 0.1. For example, a 10% MTF criterion predicts a 1-D spatial resolution of approximately $1/(1.65 \text{ mm}^{-1}) = 0.61$ mm, which is fairly close to the FWHM estimate of the impulse response of 0.53 mm. Summary measures are convenient, but spatial resolution is fully characterized by the entire MTF curve. In this example, we selected transform frequencies based on geometric properties of the discrete detector, which is convenient analytically but is not an experimental requirement.

7.4.3   Practical MT Measurements

For 50 years [83, 118], spatial resolution has been measured for medical imaging devices using test objects that offer a range of short-length eigenvector inputs or their approximations. A spatial-domain view of MT measurements is to measure the amplitude of the input and output waveforms over a range of frequencies to compute

$$\text{MT}[k] = \frac{|g_{max} - g_{min}|}{|f_{max} - f_{min}|}, \quad \text{for } f(x) = \frac{B}{D} \cos(2\pi u_k x) \, \text{rect}\left(\frac{x - x_k}{X_k}\right), \quad (7.22)$$

where x_k is the central location and X_k is the length of the sinusoid at the kth frequency, as noted in Figure 7.6. Examining the bottom left plot of Figure 7.6, we see the influence of end effects for the sinusoidal segments approximating eigenvectors. Such end effects should be avoided when determining output modulation amplitudes. Figure 7.7a depicts MT measurements at the three spatial frequencies of the sine-wave bursts shown in the left column of Figure 7.6. The measured modulation transfer points (diamonds) agree well with the predictions (solid line) at the burst frequencies. Because it can be difficult to build a sinusoidally attenuating test object for measurements in the lab, square-wave patterns are often used for this purpose.

In theory, the best measurement would be to image a point absorber. A point in space provides equal stimulus to all spatial frequencies, so in one acquisition we would obtain all possible MT frequency measurements, provided our device responds linearly. Unfortunately, realistic "point absorbers" provide low input signal energy, so the SNR of the acquisition is often too low to ignore the noise effects. If we were able to reliably detect the input energy removed from a spatial point absorber, our device would be unlikely to respond linearly. Alternatively, a sharp-edged attenuation line can provide an edge-response acquisition [83]. A derivative of that *edge response* yields a *line response* that serves as an impulse response along one axial dimension. Taking spatial derivatives of noisy data amplifies the noise, yet the results can be quite good for high exposures yielding high-SNR conditions. Applying test objects with input modulation at individual frequencies is the highest-SNR estimate of MT values.

7.4.4 Line-Pair Test Objects

Square-wave line pairs are relatively easy to manufacture, so line-pair test objects are common (Figure 7.6). However, if you expand a square-wave object vector in a Fourier series,[10] you will find nonzero coefficients at the fundamental inverse period of the square wave u_1 and at odd harmonics $3u_1, 5u_1, \ldots$, where a linear measurement device with a realistic bandwidth is unlikely to respond. For example, a square wave at 2.1 line pairs per mm (lp/mm) responds at fundamental sinusoidal frequency 2.1 mm^{-1} and at odd harmonics – 6.3 mm^{-1}, 10.5 mm^{-1}, etc. We can reconstruct the square-wave object exactly by summing those sinusoids after weighting each by its coefficient. It should come as no surprise, then, that MT estimates for a square-wave modulation (shown by the dots in Fig. 7.7) are smaller than the predicted MTF values, since they are only responding to the fundamental frequency.

Shown in Figure 7.7b are two 2-D impulse responses similar to that from an x-ray tube and their effect on a standard 2-D test pattern (labeled **f** in top right). The effects,

---

10 square wave $= \frac{4}{\pi}\left[\sin(2\pi t/D) + \frac{\sin(6\pi t/D)}{3} + \frac{\sin(10\pi t/D)}{5} + \cdots + \frac{\sin(2\pi(2k-1)t/D)}{(2k-1)} + \cdots\right],$
$k = 1, 2, 3 \ldots$

labelled as g_1 and g_2 in the figure, appear when the impulse responses differ by a factor of 2. The MATLAB code used to generate 2-D impulse-response functions for an LSI system is given here.

```
%%%%%%%%%%%%%%%%%%%% Script to generate focal spots shown in Fig. 7.7b %%%%%%%%%%%%%%%%%%%%%%%%%%%%
%%%%%%% To scale focal spot, change s1,s2, s3 and adjust mean values along the 3-Gaussian axis %%%
y=1:2001;hh=zeros(2001);s1=33;s2=50;s3=60;          % y in um; synthesize h as weighted Gaussians
h=exp(-(y-951).^2/(2*s1^2))+0.3*exp(-(y-1001).^2/(2*s2^2))+exp(-(y-1051).^2/(2*s1^2));
for j=1:2001
    hh(:,j) = exp(-(y-1001).^2/(2*s3^2))*h(j)        % 2-D impulse response h
end
ha=sum(sum(hh));hh=hh/ha;                             % normalize the area of 2-D h to 1
h1=downsample(hh,10);h2=downsample(h1',10);h3= h2';h4=h3(96:106,96:106);
subplot(1,2,1);imagesc(hh(901:1101,901:1101));colormap(gray);axis square;
subplot(1,2,2);imagesc(h3);axis square               % downsample as needed to match f
```

To run this code, locate a test pattern like the one in Figure 7.7b by conducting a web search and save it as the file "fun1.gif." Alternatively, examine the code in Problem 7.1. The following code uses the $h(x_1, x_2)$ file given here and a 2-D convolution to "image" the pattern with low blur. You can always rescale h to vary the FS resolution.

```
%%%%%%%%%% Code to blur test patterns in Fig. 7.7b with the FS generated above %%%%%%%%%%%%%
Q=imread ('fun1.gif');ff=double(Q);            % imagesc(ff);colormap(gray);axis square
for j=1:720
    for k=1:960
        if ff(j,k)<100 ff(j,k)=0;
        elseif ff(j,k)>200 ff(j,k)=256;         % rescale amplitude
        end
    end
end
g=conv2(ff,h3,'same');                          % create image
figure;subplot(1,2,1);imagesc(ff);colormap(gray);axis square;
subplot(1,2,2);imagesc(g);axis square
```

Measurement systems generally shift the magnitude and the phase of the input modulation, although the latter is ignored by modulation transfer measurements. If we are interested in phase changes, we can measure the complex-valued *optical-transfer function* $OTF(u) = H(u)/H(0)$ that predicts changes to both amplitude and phase. The modulation-transfer function $MTF = |OTF|$ and the *phase-transfer function* $PTF = \tan^{-1}(\Im\{OTF\}/\Re\{OTF\})$ combine such that $OTF(u) = MTF(u) \times \exp(i\,PTF(u))$.

The analysis of spatial resolution presented here also applies to data for which the independent variable is time. The challenge experimentally is to develop a time-varying sinusoidal-like modulation function to measure *temporal resolution*. That task is easy to do in simulation.

7.4.5   Modulation Transfer for LSV Systems

Realistic systems have impulse responses that change to some degree in space and/or time. These linear systems have shift-varying properties (LSV). Specifically, $h(x, x')$ depends on x and x' individually, and not just on their difference $x - x'$. Convolution and Fourier analyses no longer strictly apply to these systems, so we turn to SVD for analysis tools.

Figure 7.8 displays an impulse-response function with a large degree of shift variability, exaggerated to make the effects obvious. The LSV nature of the impulse

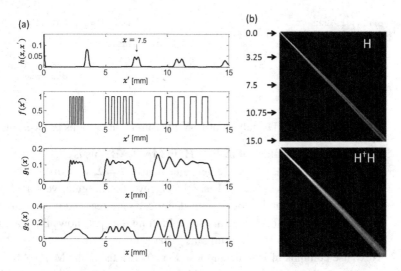

Figure 7.8 (a) The LSV 1-D impulse-response function shown at the top is narrower on the anode side of the detector (0 mm) than on the cathode side (15 mm). Only half of the h function is shown at either extreme since **H** is not circulant. Imaging the same square-wave test pattern shown in Figure 7.6, we find different output results depending on whether we position the test object $f(x')$ oriented from anode to cathode $g_1(x)$ or from cathode to anode $g_2(x)$ because of the shift-varying system matrix **H**. (b) Images of **H** showing how the impulse response changes with position. The numbers correspond to the h functions plotted in (a). Also shown is compound matrix $\mathbf{H}^\dagger\mathbf{H}$ used in the SVD analysis.

response could be caused by FS properties with the effects that are illustrated in Figure 7.1d. We select system matrix **H**, whose rows are the impulse responses at each value of x along the detector, to be a square matrix. Square **H** is not necessary for SVD analysis, as we saw in Section 6.5. Imaging the 1-D square-wave test pattern with this 1-D acquisition operator, we see in Fig. 7.8 that g_1 and g_2 are different only because of how the test object is placed relative to the impulse response. For the data of g_1, the high resolution side is on the left side of the plot; in g_2, the high resolution side is on the right. This degree of variability in h would be unacceptable in practice. We can apply the results of Section 6.6 to analyze the "spatial resolution" of this system. I use quotes because there are no simple ways to specify a measurement that varies with position and orientation in the field. The MTF concept described earlier is difficult to apply when the measurement system is clearly shift varying.

Some of the code used to generate the plots of Figure 7.8 follows.

```
%%%%%%%%%%%%%%%%%%%%%%% Script used to generate Fig. 7.8 results %%%%%%%%%%%%%%%%%%%%%%%%%%%%%%%
y=0:10:15000;yp=y/1000;H=zeros(1501);              % y-axis is in micrometers, yp in mm
aa=0.6+0.08*yp;a=aa/aa(751);q=(2*a)-1;             % position, IR scaling, q=mean,a^2=var
for j=1:1501                        % noncirculant, Toeplitz, shift-varying system matrix
  H(j,:)=exp(-(y-q(j)*7350).^2/(2*(a(j)*100)^2))/(sqrt(2*pi)*a(j)*10)+...
  0.3*exp(-(y-q(j)*7500).^2/(2*(a(j)*150)^2))/(sqrt(2*pi)*a(j)*15)+...
  exp(-(y-q(j)*7650).^2/(2*(a(j)*100)^2))/(sqrt(2*pi)*a(j)*10);
end
h=H(1,:)+H(326,:)+H(751,:)+H(1076,:)+H(1501,:);    % plot 5 of the 1-D IR functions
subplot(4,1,1);plot(yp,h)
```

```
f=zeros(size(yp));f(211:222)=1;f(235:246)=1;f(259:270)=1;f(283:294)=1;f(307:318)=1;
f(501:524)=1;f(549:572)=1;f(597:620)=1;f(645:668)=1;f(693:716)=1;
f(901:948)=1;f(997:1044)=1;f(1093:1140)=1;f(1189:1236)=1;f(1285:1332)=1;
subplot(4,1,2);plot(yp,f);
g1= 0.1*H*f';subplot(4,1,3);plot(yp,g1);              % image with f positioned as in plot
g2= 0.1*H*f(end:-1:1)';subplot(4,1,4);plot(yp,g2(end:-1:1)); % invert f and image
figure;subplot(1,2,1);imagesc(H);axis square;
subplot(1,2,2);imagesc (H'*H);axis square
```

One approach to SVD analysis of LSV measurement systems proceeds as follows:

1. Estimate the eigenvalues of real $\mathbf{H}^\top \mathbf{H}$. In MATLAB, this is
   [U, S]=eig(H'*H), where eigenvalue matrix S $\triangleq \Sigma$ with diagonal elements
   ς_k, and the columns of matrix $\mathbf{U}$ are the eigenvectors/singular vectors. Because
   $\mathbf{H}^\top \mathbf{H}$ is Hermitian and positive semi-definite, ς_k are real and nonnegative. The
   diagonal elements of $\Sigma^{1/2}$ are also real. Plotting diag(Σ) shows there are many
   zero eigenvalues, so $\mathbf{H}$ is not full rank. That is no surprise.
2. Next, use columns of $\mathbf{U}$ to decompose $\mathbf{f}$ via (6.14), using the MATLAB code
   A= U'*f. Be sure the columns of $\mathbf{U}$ are orthonormal (must they be for this
   approach to work) by taking the inner products to find $\mu_j^\dagger \mu_k = \delta_{jk}$. The diagonal
   elements of $\mathbf{A}$ are α_k coefficients.
3. Now $\mathbf{B} = \mathbf{A}\Sigma^{1/2}$, or from the diagonal elements of $\mathbf{B}$, $\beta_k = \alpha_k\sqrt{\varsigma_k}$.
4. You may wish to verify that $\mathbf{g} = \mathbf{Hf} = \mathbf{AHU}$. I did this in the bottom plot of
   Figure 7.9. Be careful when using (6.18) to compute ν_k, since some values of
   $\sqrt{\varsigma_k} \simeq 0$.

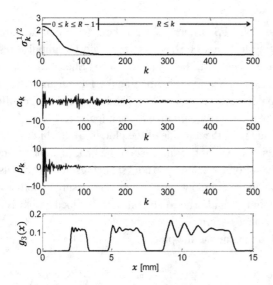

Figure 7.9 Plots of the largest 500 out of 1,501 singular values from SVD analysis. $\varsigma_k^{1/2}$
describe the measurement system (this is the equivalent of unscaled MTF applied to LSV
instruments), α_k describe the square-wave object function in Figure 7.8, and β_k describe the
measurement data. At the bottom is a plot of the entire measurement vector $\mathbf{g}_3$ reconstructed
from singular values. $\mathbf{g}_3$ is identical to $\mathbf{g}_1$ in Figure 7.8, which it must be since $\mathbf{f}$ is the same,
thus verifying the calculations. Note that $\sigma^{1/2} = \varsigma^{1/2}$ (font limitations!).

```
%%%%%%%%%%%%%%%%%%%%%%%%%%%% Script to generate Fig. 7.9 %%%%%%%%%%%%%%%%%%%%%%%%%%%%%%%%%%%
[U,D]=eig(H'*H);              % note that U'*U = I and diag(D) in increasing order
sd=sqrt(diag(D));subplot(4,1,1);
plot(sd(end:-1:end-500));hold on;    % view sqrt of largest eigenvalues of H'H.
A= U'*f';B=A.*sd;Vg=H*U;g3=Vg*A*0.1;subplot(4,1,2);plot(A(end:-1:end-500));
subplot(4,1,3);plot(B(end:-1:end-500));subplot(4,1,4);plot(yp,g3)
```

Let's examine the coefficients $\varsigma_k^{1/2}$, α_k, and β_k shown in Figure 7.9. Only the largest 500 coefficients are plotted because the last 1,000 points are negligible. While they are not Fourier coefficients, singular spectra do describe essential properties of the measurement process at different spatial scales. The α-coefficient spectrum defines properties of the object that can be measured. The $\varsigma^{1/2}$-coefficient spectrum defines the sensitivity of the measurement system to α coefficients. The *Hadamard product* of the two spectra gives the β-coefficient spectrum that quantifies the extent to which the object properties are captured by the measurement data. All this is given by the first R values of the spectra [$R = \mathrm{rank}(\mathbf{H})$], which define the sensitivity of the measurement device to the object – the bandwidth of $\mathbf{H}$. Object coefficients are all real, but may be either positive or negative. Nonzero object coefficients extend over as many as 300 coefficients in this example, more than the bandwidth of system coefficients (about 120). Thus, we know there exists a null space (Section 6.4) for which object properties cannot be captured. In the null space, we can find $\mathbf{f}_1$ and $\mathbf{f}_2$ that both give the same $\mathbf{g}$ (measurements are not one-to-one). The β spectrum shows that this system amplifies some object coefficients and attenuates others; these are the distortions of the object by the measurement that appear in the data.

7.5   Relating Quality Measures to Task Performance

This chapter began with a description of a mathematical model for simulating the acquisition stage of radiographic images. We estimated statistical properties of the acquired data to summarize the contrast, noise, and resolution features characteristic of the measurement process. This section assembles those elements through the Wagner–Brown [114] approach to decision theory, which defines task performance for LTI/LSI imaging instruments. It also establishes connections with information-theoretic descriptors such as noise-equivalent quanta (NEQ) that encompasses the frequency-dependent descriptors of MTF and NPS discussed earlier in this chapter. The Wagner–Brown approach yields a single expression that answers the question, *How well does this measurement achieve its intended tasks?* Their methods, which have been adopted by scientific organizations [54] and formally integrated into the foundations of image science [10], may be extended to other measurement modalities.

The Wagner–Brown approach combines a spatial-frequency-domain representation of specific imaging tasks with contrast transfer, noise power, and spatial resolution properties of the measurement device. This combination yields a scalar quantity that quantifies the best-possible classification performance for that task. *Classification* involves hypothesis testing (Chapter 8) when a countable number of potential decision outcomes exist [10]. For example, is blood flow enhanced in regions surrounding a tumor? Potential outcomes comprise the null hypothesis (flow in the normal range) or two alternative hypotheses (flow is suppressed or enhanced). By comparison, *estimation* entails hypothesis testing with an infinite number of potential outcomes. For example, what is the blood flow measured in [mL/min/g] in a region near a tumor? The answer to this question must be a nonnegative real number.

The unifying element in the Wagner–Brown approach is the statistical concept of the *ideal observer*,[11] which provides engineers with rigorous strategies for designing instruments and evaluating task performance. An ideal observer is generally an algorithm capable of applying all available statistical information from a data set regarding a task. The ideal observer achieves the upper bound of decision performance because it fully uses that information and is limited only by statistical uncertainty. Ideal observers minimize *Bayes' risk* [110] by considering decision errors and costs associated with each error, while averaging over not only the data samples but all of the data possible for that measurement. [For example, see (9.12) in Section 9.3.]

Ideal-observer responses achieve the greatest true-positive decision rate at all possible false-positive decision rates [110]. We can then compare the ideal-observer performance with that of an expert human or machine classifier via ROC analysis (Section 8.3) to estimate the *task efficiency*. Low efficiency is a rigorous criterion and a prime motivation for resource allocation aimed at improving measurement methods. The difficulty with the ideal observer concept is that the objects behind the alternative decisions must be known exactly or statistically [10], which often limits tasks to simple abstractions of those encountered clinically.

The next section describes SNR_I, which is the signal-to-noise ratio for the ideal observer's test statistic; e.g., see (13.132) in [10]. SNR_I relates ideal performance with key engineering parameters. That relationship is important because it points to statistical measurements that provide rigorous design and evaluation criteria, some of which are also described in the next section. SNR_I is an information-theoretic measure of performance because this test statistic is based on a log-likelihood ratio. It has been shown [10, 73] that SNR_I is related to *relative entropy* that is described in (3.67).

7.5.1   SNR$_I$, NEQ, and DQE

The task Wagner and Brown [114] selected was the discrimination of known, low-contrast patterned variations in the properties for object f_1 compared with the properties for object f_0 by viewing many image pairs formed from those two objects, $\{g_1, g_0\}$. The subscripts indicate the two classes of objects between which observers are asked to decide. This basic classification task is known as *binary detection*. Because the visual data are stochastic, so are the test statistics measuring observer responses. In consequence, many data sets must be examined and responses averaged to characterize performance. An example of one alternative image pair is given in Figure 7.10.

[11] The concept of "optimal detectors," which was originally developed for radar systems [74], has been central to the development of many detection systems and in studies of biosensory mechanisms [10, 47, 70, 104, 110].

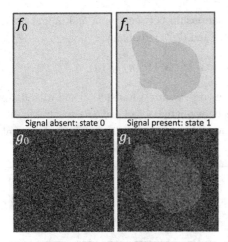

Figure 7.10 (a) Alternative objects (top row) from which images are formed (one of the many image sets is shown in the bottom row). Image pairs are presented to observers who are shown the exact alternative objects and must decide which of the two randomized images belongs to state 1. This is the two-alternative forced-choice (2AFC) observer paradigm [47]. Discrimination in this example is simple to achieve. The observer challenge is increased by reducing object contrast and/or size so that detection becomes limited by noise or spatial resolution.

The *input signal power* is defined in the frequency domain using the squared magnitude of the transformed difference between the two known alternative object functions, $|\Delta F(\mathbf{u})|^2 = |\mathcal{F}\{f_1(\mathbf{x}) - f_0(\mathbf{x})\}|^2$. The input template for the observer is established in the frequency domain where data samples are uncorrelated.

The *input noise power* is defined by the noise power spectral density $S_{in}(u)$ (see Section 5.12 and Section 7.3). Therefore the scalar-valued signal-to-noise ratio that is based on the ideal-observer test statistic [not the point SNR of (4.4)] is related to the input signal power and input noise power using [54]:

$$\text{SNR}_I = \int_{-\infty}^{\infty} d\mathbf{u} \, \frac{|\Delta F(\mathbf{u})|^2}{S_{in}(\mathbf{u})} = \int_{-\infty}^{\infty} du \int_{-\infty}^{\infty} dv \, \frac{|\Delta F(u,v)|^2}{S_{in}(u,v)}. \tag{7.23}$$

We use SNR_I instead of SNR_I^2 in Wagner's expression because we define the signal-to-noise ratio as a ratio of signal and noise energy instead of amplitude. $|\Delta F|^2$ has units of the [area]^2, S_{in} has units of [area], and $d\mathbf{u}$ has units of [area^{-1}], so SNR_I is unitless as it must be. Wagner called SNR_I the *area-wise or lesion SNR* because it is averaged over a region in space-time.

$|F(u)|^2$ can be intuited or measured, but often $S_{in}(u)$ is difficult to measure. However, *output noise power*, $S_{out}(u)$, is straightforward to measure. Using the methods described in Section 5.13, the relationship between input and output noise power is found by combining (5.35), (7.10), and (7.21) to obtain

$$S_{out}(u) = \bar{q}^2 \, \gamma^2 \, \mathrm{MTF}^2(u) \, S_{in}(u).$$

The squares of mean photon fluence $\bar{q}$, contrast transfer γ, and modulation transfer MTF define scaling relationships between the object and data spectra, in the sense that larger values amplify the input power that appears in the output power. Combining this result with (7.23), we find

$$\mathrm{SNR}_I = \int_{-\infty}^{\infty} d\mathbf{u} \, |\Delta F(\mathbf{u})|^2 \frac{\bar{q}^2 \, \gamma^2 \, \mathrm{MTF}^2(\mathbf{u})}{S_{out}(\mathbf{u})} = \int_{-\infty}^{\infty} d\mathbf{u} \, |\Delta F(\mathbf{u})|^2 \, \mathrm{NEQ}(\mathbf{u}),$$

$$\text{(Area-wise SNR)} \qquad (7.24)$$

where

$$\mathrm{NEQ}(\mathbf{u}) = \frac{\bar{q}^2 \, \gamma^2 \, \mathrm{MTF}^2(\mathbf{u})}{S_{out}(\mathbf{u})} \qquad (7.25)$$

is the *noise-equivalent quanta* with units of $\bar{q}^2$, which are [(number of quanta)^2/ detector area] or simply [area]$^{-1}$ for 2-D data. If the data are 1-D or 3-D, the units become [length]$^{-1}$ or [volume]$^{-1}$, respectively. NEQ at spatial frequency $\mathbf{u}$ is a count of the number of photons per area at the input of a perfect acquisition system that yields the same output noise at $\mathbf{u}$ as the system being analyzed. More generally, NEQ is the density of information displayed that is available to a decision maker over the measurement bandwidth. That is, each quantum of information is a *degree of freedom* available to the observer. When that measurement device is applied to a task, as in the last form of (7.24), we can predict the best-possible performance via SNR_I.

NEQ was originally proposed by Rodney Shaw to describe photographic quality [98], but was quickly adopted for medical imaging [115]. It combines the essential engineering properties of a measurement device: patient exposure, contrast transfer, spatial resolution, and noise power into a single task-independent "quality" metric. NEQ leverages the properties of a Fourier basis to separate samples of system properties into orthogonal frequency channels matched to those of the task, $|\Delta F(\mathbf{u})|^2$.

Equation (7.24) may be interpreted as saying the best-possible measurement performance is achieved when the device is highly responsive to spatial frequencies of a task with low noise power. It is not the total noise energy compared with the total signal energy that determines task performance, but rather the noise *in the frequency channels of task signals* that determines ideal performance. Because performance is task dependent, it makes little sense to say that one device is better than another until we have challenged both with measuring the same object.

Imaging with ionizing radiation requires the medical community to balance the potential benefits of an exam with the somatic and genetic risks to a patient and the population. The *detective quantum efficiency* (DQE) is one measure that describes how well an x-ray imaging system uses photons to acquire information:

$$\text{DQE}(\mathbf{u}) = \frac{\text{SNR}_{out}}{\text{SNR}_{in}} = \frac{\text{NEQ}(\mathbf{u})}{\bar{q}}.$$

The concepts developed by Wagner and Brown have been extended to other medical imaging systems [55, 73, 108, 119] and adopted within standards of practice [54]. They form a coherent basis for evaluating the acquisition stage of medical imaging devices.

7.6 Display-Stage Processing

7.6.1 Restoration with Wiener Deconvolution Filter

Up to this point, the emphasis has been on acquisition-stage processing. For display-stage processing, two prominent areas of significant intellectual development are *image reconstruction* [10, 58] and *data restoration* [5, 42]. Between classic texts and recent review papers [87], interested readers will find much information available on image reconstruction. This section describes one standard technique of image restoration that applies broadly to biomedical measurements. Examples are provided to illustrate concepts.

The objective of display-stage processing is to represent the object function to the observer based on the acquired data, i.e., $\hat{\mathbf{f}} = \mathcal{O}\mathbf{g} = \mathcal{O}\{\mathbf{Hf} + \mathbf{e}\}$. Assume (a) acquired data have formed an image as in Figure 7.7, (b) $\mathbf{H} \in \mathbb{R}^{N \times N}$, and (c) acquisition noise and the impulse response of the linear measurement device have corrupted the end result. As discussed in Section 6.4, noise and blur create null spaces that complicate the straightforward inverse problem $\hat{\mathbf{f}} = \mathbf{H}^{-1}\mathbf{g}$ so that this effort fails completely or produces unsatisfactory or unstable results. One popular approach to solving ill-posed *inverse problems* is to find a reduced-rank approximation of $\mathbf{H}$ that improves conditioning (Appendix A) while delivering acceptable clinical performance [7]. In this section, we will examine one simple approach that employs a pseudoinverse.

One approach is *Tikhonov regularization*, which involves *constrained least-square optimization* via the *method of Lagrange multipliers*. Constrained optimization is introduced in Appendix E and discussed briefly in Section 6.8. The goal is to select $\hat{\mathbf{f}}$, an estimate of object function $\mathbf{f}$, by minimizing the squared ℓ_2 norm $||\mathbf{A}\hat{\mathbf{f}}||_2^2$ for some suitably chosen matrix $\mathbf{A}$. The solution is also subject to (or constrained by) $||\mathbf{g} - \mathbf{Hf}||_2^2 = ||\mathbf{e}||_2^2$, which is a consequence of the acquisition model, $\mathbf{g} = \mathbf{Hf} + \mathbf{e}$. The strength of the Lagrange multiplier approach is its ability to constrain solutions to the minimization of $||\mathbf{A}\hat{\mathbf{f}}||_2^2$ without specifying exactly what $\mathbf{A}$ needs to be, such that this method offers significant flexibility.

This goal is expressed mathematically through regression analysis (Section 3.12) resulting in objective function Θ. Minimizing Θ with respect to $\hat{\mathbf{f}}$ and then solving for $\hat{\mathbf{f}}$, we obtain

$$\Theta = \underset{\hat{\mathbf{f}}}{\operatorname{argmin}} \left[||\mathbf{A}\hat{\mathbf{f}}||_2^2 + \gamma' \left(||\mathbf{g} - \mathbf{H}\mathbf{f}||_2^2 - ||\mathbf{e}||_2^2 \right) \right],$$

$$\frac{\partial \Theta}{\partial \hat{\mathbf{f}}} = 0 = 2\mathbf{A}^\top \mathbf{A}\hat{\mathbf{f}} - 2\gamma' \mathbf{H}^\top (\mathbf{g} - \mathbf{H}\hat{\mathbf{f}}) = \left(\frac{1}{\gamma'}\mathbf{A}^\top \mathbf{A} + \mathbf{H}^\top \mathbf{H} \right)\hat{\mathbf{f}} - \mathbf{H}^\top \mathbf{g},$$

$$\hat{\mathbf{f}} = \left(\mathbf{H}^\top \mathbf{H} + \gamma \mathbf{A}^\top \mathbf{A} \right)^{-1} \mathbf{H}^\top \mathbf{g}. \tag{7.26}$$

Here we have assumed signal-independent noise, redefined the Lagrange multiplier as $\gamma \triangleq 1/\gamma'$, and applied methods from Appendix A to compute the derivatives.

One interpretation of $\mathbf{A}$ is as a Bayesian prior [10] with a zero-mean iid multivariate normal distribution and covariance matrix $\mathbf{K}_g = \operatorname{var}(g_n)\mathbf{I}$. The idea is that additional assumptions provided by a prior improve conditioning and thereby narrow the possible solution space. This Tikhonov-regularized solution is the most probable solution according to Bayes' theorem.

Others [5, 42] have proposed a different statistical interpretation of $\mathbf{A}$. First, assume LSI $\mathbf{H}$, and let $\mathbf{H}\mathbf{f}$ and $\mathbf{e}$ be signal and noise samples drawn from separate WSS random processes. From Section 3.7, we find the corresponding autocorrelation matrices,

$$\phi_{Hf} = \mathcal{E}\{\mathbf{H}(\mathbf{f}\mathbf{f}^\top)\mathbf{H}^\top\} = \mathbf{K}_{Hf} + \overline{\mathbf{H}\mathbf{f}} \qquad \text{and} \qquad \phi_e = \mathcal{E}\{\mathbf{e}\mathbf{e}^\top\} = \mathbf{K}_e + \bar{\mathbf{e}}.$$

Next, associate ϕ_{Hf}^{-1} with $\mathbf{A}^\top$ and ϕ_e with $\mathbf{A}$,

$$\mathbf{A}^\top \mathbf{A} = \phi_{Hf}^{-1}\phi_e, \tag{7.27}$$

and substitute (7.27) into (7.26) to find the spatial-domain solution,

$$\hat{\mathbf{f}} = \left(\mathbf{H}^\top \mathbf{H} + \gamma \phi_{Hf}^{-1}\phi_e \right)^{-1} \mathbf{H}^\top \mathbf{g}. \tag{7.28}$$

Interpretation of (7.28) is clearest in the frequency domain. The equivalent expressions for autocorrelation matrices ϕ_{Hf} and ϕ_e describing signal and noise in the Fourier domain are, respectively, the power spectral density matrices, $\mathbf{S}_{Hf}$ and $\mathbf{S}_e$. We find these matrices by applying (6.8) and the *Wiener–Khintchine theorem* from Section 5.12.2:

$$\mathbf{S}_{Hf} = \mathbf{Q}^\dagger \phi_{Hf}\mathbf{Q}, \qquad \mathbf{S}_{Hf}^{-1} = \mathbf{Q}^\dagger \phi_{Hf}^{-1}\mathbf{Q}, \qquad \text{and} \qquad \mathbf{S}_e = \mathbf{Q}^\dagger \phi_e\mathbf{Q}. \tag{7.29}$$

The spectral matrices are diagonal so that $\mathbf{S}_{Hf}[k,\ell] \rightarrow \operatorname{diag}(\mathbf{S}_{Hf})$ and $\mathbf{S}_e[k,\ell] \rightarrow \operatorname{diag}(\mathbf{S}_e)$ after reshaping. Also, the spatially averaged SNR is the ratio of matrix traces. From (4.4),

$$\text{SNR} = \frac{\operatorname{tr}\mathbf{S}_{Hf}}{\operatorname{tr}\mathbf{S}_e}. \tag{7.30}$$

Expressing (7.28) in the frequency domain, with help from (6.8) and (7.29), we find

$$\hat{\mathbf{f}} = \left(\mathbf{H}^\top \mathbf{H} + \gamma \boldsymbol{\phi}_{Hf}^{-1} \boldsymbol{\phi}_e \right)^{-1} \mathbf{H}^\top \mathbf{g}$$

$$= \left(\mathbf{Q}\boldsymbol{\Lambda}^*\mathbf{Q}^\dagger \mathbf{Q}\boldsymbol{\Lambda}\mathbf{Q}^\dagger + \gamma \mathbf{Q}\mathbf{S}_{Hf}^{-1}\mathbf{Q}^\dagger \mathbf{Q}\mathbf{S}_e\mathbf{Q}^\dagger \right)^{-1} \mathbf{Q}\boldsymbol{\Lambda}^*\mathbf{Q}^\dagger \mathbf{g}$$

$$= (\mathbf{Q}(\boldsymbol{\Lambda}^*\boldsymbol{\Lambda} + \gamma \mathbf{S}_{Hf}^{-1}\mathbf{S}_e)\mathbf{Q}^\dagger)^{-1}\mathbf{Q}\boldsymbol{\Lambda}^*\mathbf{Q}^\dagger \mathbf{g}$$

$$= \mathbf{Q}(\boldsymbol{\Lambda}^*\boldsymbol{\Lambda} + \gamma \mathbf{S}_{Hf}^{-1}\mathbf{S}_e)^{-1}\boldsymbol{\Lambda}^*\mathbf{Q}^\dagger \mathbf{g}$$

and $\quad \mathbf{Q}^\dagger\hat{\mathbf{f}} = (\boldsymbol{\Lambda}^*\boldsymbol{\Lambda} + \gamma \mathbf{S}_{Hf}^{-1}\mathbf{S}_e)^{-1}\boldsymbol{\Lambda}^*\mathbf{Q}^\dagger \mathbf{g}.$ $\hfill (7.31)$

Applying (6.8) to (7.31) and reshaping vectors, the matrix result is expressed as a 2-D DFT [42],

$$\hat{F}[k,\ell] = \left[\frac{H^*[k,\ell]}{|H[k,\ell]|^2 + \gamma \left(\frac{S_e[k,\ell]}{S_{Hf}[k,\ell]} \right)} \right] G[k,\ell] = W[k,\ell]\, G[k,\ell], \hfill (7.32)$$

where $W[k,\ell]$ is the *Wiener deconvolution filter*.

If the signal power at spatial frequency channel $[k,\ell]$ is much larger than that of the noise power, then $S_e[k,\ell]/S_{Hf}[k,\ell] \simeq 0$ and we find from (7.32) that $W[k,\ell] \simeq 1/H[k,\ell]$. Hence the filter is responding like an *inverse filter*. Great! However, if the signal power at $[k,\ell]$ is much smaller than the noise power at $[k,\ell]$, then $S_e[k,\ell]/S_{Hf}[k,\ell] \gg |H[k,\ell]|^2$ and (7.32) yields $W[k,\ell] \propto H^*[k,\ell]$. Now the filter is behaving like a *matched filter* to suppress noise. We see that $W[k,\ell]$ is able to adapt to population statistics at each frequency through measurements of signal and noise power.

$W[k,\ell]$ responds optimally if we know the spectral densities of signal and noise parent populations and if we constrain solutions by selecting the value of γ that minimizes $||\mathbf{g} - \mathbf{Hf}||^2 - ||\mathbf{e}||^2$. That's a pretty tall order because we rarely have this much prior knowledge about signal and noise parent distributions.

Approximations to the Wiener deconvolution filter can be very helpful even if they are suboptimal. For example, assuming that $\gamma = 1$ and the signal and noise processes are spectrally white, then from (7.30), $S_e[k,\ell]/S_{Hf}[k,\ell] = (\text{SNR})^{-1}\mathbf{I}$. When I made the further unjustified assumption that SNR = 1 in Example 4.2.1, we saw that the result was imperfect but task performance was improved. If you recall from Figure 4.4, the object presented three rectangular pulses of equal amplitude but varying lengths. The simplified deconvolution could not restore pulse durations, but the amplitudes varied with pulse area, so interpretation was improved. Remember, task performance is the evaluation objective. The point being made in this section is that matrix $\mathbf{A}$ is flexible and nonparametric, so Tikhonov regularization can have many forms depending on how much prior information you have.

7.6.2   Examples

Beginning with the acquisition equation, $\mathbf{g} = \mathbf{Hf} + \mathbf{e}$, we can compare alternatives for estimating $\hat{\mathbf{f}}$ using the inverse filter $\mathbf{H}^{-1}$, the pseudoinverse $\mathbf{H}^+$ (reduced-rank approximation), and a time-domain approximation to the Wiener deconvolution filter **W** (Tikhonov regularization). To do so, let's examine a toy problem.

Example 7.6.1. *Comparing inversion methods: Let*

$$\mathbf{H} = \begin{pmatrix} 1 & 2 \\ 2 & 1 \end{pmatrix}, \quad \mathbf{f} = \begin{pmatrix} 30 \\ 20 \end{pmatrix}, \quad \mathbf{e} = \begin{pmatrix} -3 \\ 2 \end{pmatrix} \text{ so that } \mathbf{g} = \mathbf{Hf} + \mathbf{e} = \begin{pmatrix} 67 \\ 82 \end{pmatrix}.$$

Find $\hat{\mathbf{f}}_1 = \mathbf{H}^{-1}\mathbf{g}$, $\hat{\mathbf{f}}_2 = \mathbf{H}^{+}\mathbf{g}$, *and* $\hat{\mathbf{f}}_3 = \mathbf{Wg}$ *and estimate the errors for each estimate.*

Solution
The point SNR is $((30^2 + 20^2)/((-3)^2 + 2^2)) = 100$ *or 20 dB. We can apply a time-domain version of the Wiener filter, where* $\hat{\mathbf{f}} \simeq \left(\mathbf{HH}^{\dagger} + \frac{1}{\sqrt{(SNR)}}\mathbf{I}\right)^{-1}\mathbf{H}^{\dagger}\mathbf{g}$. *Since SNR is defined for energy terms, we take the square root to scale amplitudes. Let's use* MATLAB *despite the simplicity of calculations.*

```
fh=zeros(2,4);                              % original f in 4th column; estimates in others
H=[1 2;2 1];f=[30;20];e=[-3;2];g=H*f+e;    % generate noisy data
fh(:,4)=f;fh(:,1)=H\g;                      % inverse estimate
fh(:,2)=pinv(H)*g;                          % pseudoinverse estimate
invSNR=0.1;                                 % this is 1/sqrt(SNR)
fh(:,3)=(H*H'+invSNR*eye(2))\H'*g;          % Wiener filtered estimate
err=zeros(3,1);                             % initialize error array
for j=1:3                                    % MSE via L-2 norms
    err(j)=norm(fh(:,j)-fh(:,4));
end;fh,err                                   % report results
```

For $\mathbf{f} = \begin{pmatrix} 30 \\ 20 \end{pmatrix}$, *we find*

$$\hat{\mathbf{f}}_1 = \begin{pmatrix} 32.33 \\ 17.33 \end{pmatrix}, \quad \hat{\mathbf{f}}_2 = \begin{pmatrix} 32.33 \\ 17.33 \end{pmatrix}, \quad \text{and } \hat{\mathbf{f}}_3 = \begin{pmatrix} 31.38 \\ 17.74 \end{pmatrix}.$$

The errors $\|\hat{\mathbf{f}}_j - \mathbf{f}\|$ *are, respectively, 3.5, 3.5, and 2.7. The inverse and pseudoinverse give the same results, as expected for well-conditioned full-rank matrix* **H**, *and the Wiener filter result is more accurate because we applied our knowledge of the noise power.*

Now, eliminate noise and try again. For $\mathbf{f} = \begin{pmatrix} 30 \\ 20 \end{pmatrix}$, *we find*

$$\hat{\mathbf{f}}_1 = \begin{pmatrix} 30.00 \\ 20.00 \end{pmatrix}, \quad \hat{\mathbf{f}}_2 = \begin{pmatrix} 30.00 \\ 20.00 \end{pmatrix}, \quad \text{and } \hat{\mathbf{f}}_3 = \begin{pmatrix} 29.27 \\ 20.18 \end{pmatrix}.$$

The matrix inverse and pseudoinverse give the same zero-error accurate result, while the Wiener filter result has error 0.75. If we change $1/\sqrt{SNR}$ *from 0.1 to 0.0 (as we should), then all three methods give the same accurate result.*

7.7   Problems

7.1   Use the following code to generate the 6 mm × 6 mm test pattern object. This 2-D software object has a 1 μm pixel size, so the output is 6,001 × 6,001 pixels. The

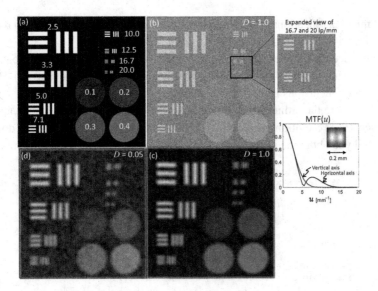

Figure 7.11 Problem 7.1. (a) Test pattern object with labels. Line-pairs values in lp/mm and disk numbers are object contrast fractions. (b) Quantum-limited image with 10^6 photons (100% exposure). (c) Same as (b) but applying the 0.2-mm FS. (d) Same as (c) but using 50,000 photons (5% exposure). The horizontal and vertical axis MTF curves are shown. D is photon exposure fraction.

test object includes high-contrast line pairs from 2.5 to 20 lp/mm. This approach is one way to visualize the effects of limited spatial resolution. Also included are four 1.4-mm-diameter disks to observe contrast resolution limitations. The amplitude of the test-pattern object ranges from 0 to 1, simulating the linear attenuation coefficient patterns generating image contrast from spatially varying photon absorption. Object regions with zero amplitude absorb no photons, while those with amplitude 1 absorb the most photons. I illustrate my solution to this problem in Figure 7.11, and ask you to do the same as well as answer a few questions.

(a) Use the equations and code given in Section 7.1 to simulate a quantum-limited radiograph of the test object. Do this by uniformly distributing 10^6 photons over the 6×6 mm object area. In this part of the problem, the focal-spot size is assumed to be a point, so there is no blur – only photon noise. Discuss how the effects of quantum noise influence target visibility throughout the test pattern. In particular, discuss the effects on your ability to resolve sharp edges.

(b) Use the three-Gaussian 2-D focal spot described in Figure 7.7b. Scale it spatially to have a horizontal width about 0.2 mm and an area of 1.0. Typical mammographic systems have 0.3–0.6 mm focal spots, so this x-ray tube is appropriate for high-resolution imaging. Assuming the impulse response is LSI, apply it to the result in part (a) of this problem and describe how the spatial resolution of the system changes based on visual inspection of the line-pair image combined with the MTF curves for that focal spot. What size breast microcalcifications (small high-contrast targets) would you expect to routinely see with this system?

(c) Reduce the number of photons to 50,000/(36 mm^2). That is, reduce patient exposure to 5% in part (b) and apply the impulse response in part (b). How do your conclusions from part (b) change?

(d) What are the influences of the focal spot and patient exposure on the apparent contrast resolution, as observed from the low-contrast circles? What does that tell you about the utility of low-exposure imaging for breast cancer imaging?

```
%%%%% Test pattern object function: includes line pairs and low-contrast disks %%%%%%%
dx=0.001;X0=6;x=0:dx:X0;N=length(x);f=zeros(N); % 6-mm region cut into 0.001-mm elements
                                          % The following generated 0.4 mm/lp or 2.5 lp/mm
f(701:901,501:1401)=1;f(1101:1301,501:1401)=1;f(1501:1701,501:1401)=1;
f(701:1701,1801:2001)=1;f(701:1701,2201:2401)=1;f(701:1701,2601:2801)=1;
                                          % 0.3 mm/lp or 3.3 lp/mm
f(2501:2651,501:1251)=1;f(2801:2951,501:1251)=1;f(3101:3251,501:1251)=1;
f(2501:3251,1551:1701)=1;f(2501:3251,1851:2001)=1;f(2501:3251,2151:2301)=1;
                                          % 0.2 mm/lp or 5.0 lp/mm
f(4001:4101,501:1001)=1;f(4201:4301,501:1001)=1;f(4401:4501,501:1001)=1;
f(4001:4501,1201:1301)=1;f(4001:4501,1401:1501)=1;f(4001:4501,1601:1701)=1;
                                          % 0.14 mm/lp or 7.1 lp/mm
f(5001:5071,501:851)=1;f(5141:5211,501:851)=1;f(5281:5351,501:851)=1;
f(5001:5351,991:1061)=1;f(5001:5351,1131:1201)=1;f(5001:5351,1271:1341)=1;
                                          % 0.1 mm/lp or 10.0 lp/mm
f(701:751,4101:4351)=1;f(801:851,4101:4351)=1;f(901:951,4101:4351)=1;
f(701:951,4501:4551)=1;f(701:951,4601:4651)=1;f(701:951,4701:4751)=1;
                                          % 0.08 mm/lp or 12.5 lp/mm
f(1501:1541,4101:4301)=1;f(1581:1621,4101:4301)=1;f(1661:1701,4101:4301)=1;
f(1501:1701,4451:4491)=1;f(1501:1701,4531:4571)=1;f(1501:1701,4611:4651)=1;
                                          % 0.06 mm/lp or 16.7 lp/mm
f(2001:2031,4101:4251)=1;f(2061:2091,4101:4251)=1;f(2121:2151,4101:4251)=1;
f(2001:2151,4401:4431)=1;f(2001:2151,4461:4491)=1;f(2001:2151,4521:4551)=1;
                                          % 0.05 mm/lp or 20.0 lp/mm
f(2501:2526,4101:4226)=1;f(2551:2576,4101:4226)=1;f(2601:2626,4101:4226)=1;
f(2501:2626,4351:4376)=1;f(2501:2626,4401:4426)=1;f(2501:2626,4451:4476)=1;
for j=1:N                                 % generate the low-contrast disks
    for i=1:N
        if j<3500 + sqrt(700^2-(i-3400)^2) && ...
                j>3500 - sqrt(700^2-(i-3400)^2);
            f(j,i) = f(j,i)+0.1;    %10% contrast
        end
        if j<3500 + sqrt(700^2-(i-5000)^2) && ...
                j>3500 - sqrt(700^2-(i-5000)^2);
            f(j,i) = f(j,i)+0.2;   %20% contrast
        end
        if j<5100 + sqrt(700^2-(i-3400)^2) && ...
                j>5100 - sqrt(700^2-(i-3400)^2);
            f(j,i) = f(j,i)+0.3;   %30% contrast
        end
        if j<5100 + sqrt(700^2-(i-5000)^2) && ...
                j>5100 - sqrt(700^2-(i-5000)^2);
            f(j,i) = f(j,i)+0.4;   %40% contrast
        end
    end
end
end                                       % takes a couple minutes to run
imagesc(f);colormap gray, axis square
save('testpattern','f');                  % save for use in problem
```

7.2 Notice that the nonzero singular values, $\varsigma_k^{1/2}$, for the measurements described in Figure 7.9 extend to only about $k = 150$. That means it might be possible to downsample the data vector by a factor of 10 (from 1,500 to 150) and still obtain nearly the same result, $g_3(x)$. Try it and describe what you find.

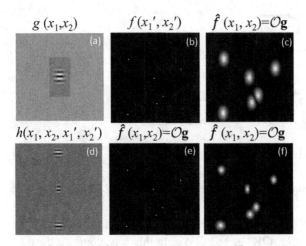

$g(x_1,x_2)$ $f(x_1',x_2')$ $\hat{f}(x_1,x_2)=\mathcal{O}\mathbf{g}$

(a)   (b)   (c)

$h(x_1,x_2,x_1',x_2')$ $\hat{f}(x_1,x_2)=\mathcal{O}\mathbf{g}$ $\hat{f}(x_1,x_2)=\mathcal{O}\mathbf{g}$

(d)   (e)   (f)

Figure 7.12 Problem 7.3: The noiseless LSI ultrasound system with the impulse response shown in (a) was applied to point reflectors in (b). The B-mode image is displayed in (c) via $\hat{f}(x_1,x_2)=\mathcal{O}\mathbf{g}$. Problem 7.4: (d)–(f) are the equivalent of (a)–(c) for the LSV system.

7.3 Recall the ultrasonic image simulation code from Chapter 4, Example 4.7.1. The top row of Figure 7.12 shows an LSI impulse response h for B-mode imaging in (a) and a resolvable field of point reflectors f in (b). The resulting B-bode image of (b) is displayed in (c) after applying the demodulation methods described in Appendix C.

(a)   Beginning with the code in Example 4.7.1, generate a circulant system matrix to implement $\hat{\mathbf{f}}=\mathcal{O}\{\mathcal{H}\{f(x_1,x_2)\}\}=\mathcal{O}\{\mathbf{g}\}$ and reproduce the acquisition-noise-free B-mode image in Figure 7.12c. Display-stage operator $\mathcal{O}$ implements echo demodulation and scan conversion only.

(b)   Apply the pseudoinverse operator before $\mathcal{O}$ and recover $\mathbf{f}$ using $\hat{\mathbf{f}}=\mathcal{O}\{\mathbf{H}^+\mathbf{g}\}$.

(c)   Generate a random scattering field with a circular void as in Example 4.7.1. Compute both $\mathcal{O}\{\mathbf{g}\}$ and $\mathcal{O}\{\mathbf{H}^+\mathbf{g}\}$.

7.4 (a)   Repeat the noise-free simulations in Problem 7.3 with the LSV impulse response in Figure 7.12d. Pulse $h(x_1,x_2)$ has constant properties along vertical axis x_1 but varies horizontally, being wider at the top and bottom than at the center, as would be expected for a fixed focus array. Generate the LSV system matrix and use $\mathbf{g}=\mathbf{H}\mathbf{f}$ to reproduce the result of Figure 7.12f.

(b)   Apply the pseudoinverse operator to $\mathbf{g}$ before displaying and comparing with the result of Problem 7.3, part (b).

(c)   As in Problem 7.3c, generate a random scattering field with a circular void and compute both $\mathcal{O}\{\mathbf{g}\}$ and $\mathcal{O}\{\mathbf{H}^+\mathbf{g}\}$.

7.5 Apply SVD analysis to the LSV $\mathbf{H}$ matrix in Problem 7.4. Do this at 10 MHz and 20 MHz, each with a fractional bandwidth of 73%. Plot the singular values on a linear and semilog scale to see the relative advantages of each frequency for the bandwidth (and therefore for spatial resolution).

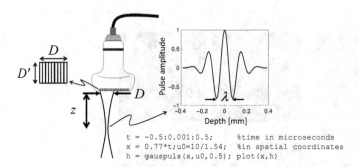

Figure 7.13 Illustration of experiment described in Problem 7.7. The pulse axis is x near z.

7.6   Returning to the random scattering field, now add noise at 20 and 100 dB to **g** before computing $\mathcal{O}\mathbf{g}$, $\mathcal{O}\mathbf{H}^+\mathbf{g}$, and $\mathcal{O}\mathbf{W}\mathbf{g}$. In this way, compare the different image restoration approaches, including the pseudoinverse and Wiener deconvolution filter methods, on data with different amounts of noise.

7.7   Optical and acoustic imaging systems are designed similarly in the sense that both generate beams of diffracting waves that limit spatial resolution, and both have limited penetration into tissue because of attenuation. The lateral spatial resolution (LR) for both is governed by the *Rayleigh criterion*, which is roughly $\text{LR} = \lambda_0 z/D$, where λ_0 is the wavelength at the center frequency of the pulse, z is the focal length of the focused beam, and D is the in-plane aperture length as in Figure 7.13.

Assume we have an acoustic imaging system capable of generating pulse frequencies between 5 and 25 MHz. An $f_0 = 5$ MHz pulse has a wavelength of $\lambda_0 = c/f_0 = (1.54\,\text{mm}/\mu\text{s})/5\,\text{MHz} = 0.308$ mm. We have active-aperture dimensions of $D = 15$ mm in plane, $D' = 10$ mm out of plane, and a focal length $z = 30$ mm. In addition, the attenuation coefficient constant is $\alpha_0 = 0.5$ dB/cm/MHz. At 5 MHz, the reduction in amplitude found after penetrating 30 mm into and out of tissue is $A/A_0 = \exp(-0.5\,\text{dB/cm/MHz} \times (1\,\text{Np}/8.686\,\text{dB}) \times 6\,\text{cm} \times 5\,\text{MHz}) = 0.178$, which is a 15 dB loss in pulse energy from attenuation in tissue. The 8.686 factor converts dB into Np, which is a natural logarithm equivalent with no SI unit equivalent.

To perform adequate breast imaging, you need to see 30 mm into the patient with $f/2$ (in-plane) focusing without losing more than 40 dB from attenuation. High-frequency probes give the highest spatial resolution but may not have the echo SNR required to penetrate 30 mm into tissue because of attenuation losses. (a) What probe frequency gives a reasonable compromise? (b) What lateral in-plane and out-of-plane spatial resolutions can we expect at that frequency?

7.8   The object function for medical sonography, $f(\mathbf{x})$, describes how pressure pulse energy is scattered by spatial fluctuations in tissue acoustic impedance along the pulse propagation path. In discrete form, $\mathbf{f} \in \mathbb{R}^{N \times 1}$, where vector components are stationary, zero-mean, iid multivariate-normal (MVN) random variables. The development in Section 3.9 showed us that

$$\mathbf{f} \sim \text{MVN}(\mathbf{0}, \mathbf{K}_f).$$

(a) Write out the pdf expression $p(\mathbf{f})$ in simple form.

(b) The acquisition stage includes additive signal-independent noise that is also zero-mean, iid MVN with covariance matrix $\mathbf{K}_e = \sigma_e^2 \mathbf{I}$. Noise variance σ_e^2 and object variance σ_f^2 are generally different. Equation 4.1 is the measurement equation for full-rank acquisition matrix, $\mathbf{H} \in \mathbb{R}^{N \times N}$. Derive $p(\mathbf{g})$.

7.9 In Problem 7.8, you were asked to find statistical properties of measurement data $g(t)$ from knowledge of object $f(t)$ and noise $e(t)$ processes. This problem continues the image formation process to follow statistical properties from detection through to the formation of a B-mode image. The discussion of this process includes a short description of *speckle formation* (Figure 7.14) and the physics of detectors capable of coherently sensing mechanical wave energy. This treatment applies to any coherent imaging system where the fractional bandwidth is less than 100%. In medical imaging, that includes ultrasonic, optical, and magnetic resonance imaging methods.

Scattered waves generated by sound pulse–tissue interactions travel back to the transducer, where they are *coherently summed* over the detector surface during the process of forming echo signals $\mathbf{g}$. See Figure 7.14b for an illustration. Coherent summation means that the amplitude and phase of waves occurring at the detector surface influence the response, so that phase cancellation is possible. Reflected waves from the jth scatterer are represented as a complex *phasor*, $a_j(t) \, e^{i\phi_j(t)}$, characterized by amplitude $a_j(t)$ and phase $\phi_j(t)$. The net instantaneous echo value results from a vector sum of reflected-wave phasors. Correlations in echo signal $\mathbf{g}$ arising from spatiotemporal correlations that are formed during pulse–reflector interactions appear in the image as *speckle* [44, 117].

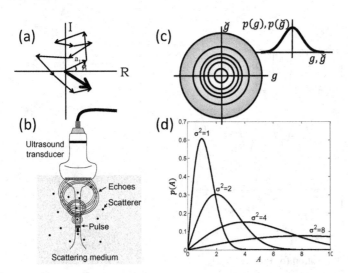

Figure 7.14 The *random walk* of the phasors depicted in (a) arises from echo waves, as shown in (b). The net phasor in (a) (bold arrow) is the sum of component phasors representing scattered waves at an instant of time. Repeating the process many times at different spatial positions and creating a histogram of the results gives the circular-Gaussian probability density of the analytic signal in (c). Echo envelope pdfs $p(A)$ are shown in (d) for four values of σ^2.

Only real-valued voltage signals, $g(t)$, are recorded. Yet the complex-valued phasor-sum representation reminds us that echo signals encode both amplitude and phase information, which reveal scatterer reflectivity and location, respectively. The complex form of the recorded echo is the *analytic signal*, $g_a(t)$, described in Appendix C. It can be written three ways, each offering a different perspective:

$$g_a(t) = \sum_{j=1}^{J} a_j(t)\, e^{i\phi_j(t)} = g(t) + i\breve{g}(t) = A(t)e^{i\theta(t)}. \qquad (7.33)$$

Of course, $g(t) = \Re g_a(t)$. The analytic signal for an echo-data time series may be expressed as the sum of complex phasors (Figure 7.14a) or as the sum of the real recorded signal $g(t)$ and imaginary $i\breve{g}(t)$ (Figure 7.14c), where $\breve{g}(t)$ is the *Hilbert transform* of $g(t)$ (see Appendix C). It is convenient to write the latter form as amplitude $A(t)$ and phase $\theta(t)$ of the analytic signal. $A(t)$ is important because it becomes the B-mode image signal $b(\mathbf{x})$ output from the display stage. Derive $p(A, \theta)$ and then $p(A)$ via $p(A) = \int d\theta\, p(A, \theta)$ for $p(\theta) \sim \mathcal{U}(0, 2\pi)$ (uniformly distributed over 2π).

7.10   Simulate the *random walk*[12] process via (7.33). Assume $a_j(t) = 1$ is constant. Show that $g(t)$ and $\breve{g}(t)$ are identically distributed normal random variables and that the envelope $A(t)$ is a Rayleigh random variable with parameter σ^2 equal to the variances of g and $\breve{g}$.

7.11   Object function $f(t)$ in Figure 7.15 is measured by two LTI devices that yield $g_1(t)$ and $g_2(t)$. Find the impulse-response functions that might have generated such different measurement data for the same object. Plot $|H(u)|/|H_{max}|$ for both impulse responses, which are essentially the MTF curves where $|H_{max}|$ replaces $H(0)$ in (7.21).

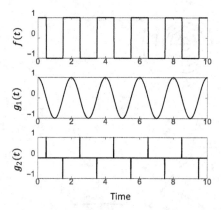

Figure 7.15   Illustration for Problem 7.11.

---

[12] The term "random walk" and the distribution of distances covered in a fixed number of equal steps (now called the Rayleigh distribution) were described in a one-paragraph note in *Nature* [88].

8   Statistical Decision-Making

In previous chapters, the focus was on representing and analyzing the elements of measurement that lead to acceptable task performance. In this chapter, we discuss metrics to guide decision-making related to experimental design and performance evaluation. Because object properties, noise, and measurement errors are naturally stochastic, these types of decisions are inherently statistical. This chapter assumes readers are familiar with the concepts discussed in Chapters 2 and 3.

8.1   Experimental Design

We can predict (before measuring) the number of samples needed to achieve confidence in a measurement result within stated error bounds if we have some statistical information provided from preliminary experiments. Research proposals require investigators to conduct power calculations that determine the number of samples necessary to reach meaningful conclusions. They want this number identified before resources are allocated, and as a taxpayer I'll bet you are glad they ask! Before summarizing concepts in hypothesis testing in this section, we investigate some ensemble properties of first-order sample statistics. Since the sample mean and variance are random variables, we must ask how close to the population statistics a sample statistic might be for sample size N.

8.1.1   Ensemble Mean of the Sample Mean

Assume a wide-sense stationary univariate normal (WSS-UVN) process $X \sim p(x; \mu, \sigma^2)$. We know from Section 3.4 that the *ensemble mean* and the *sample mean* (in time or space) are, respectively,

$$\mathcal{E}X = \int_{-\infty}^{\infty} dx \, x \, p(x) = \mu \quad \text{and} \quad \bar{X} = \frac{1}{N} \sum_{n=1}^{N} x_n,$$

where N is the *sample size*. What is the ensemble mean of the sample mean? Remember, ensemble means are considered deterministic because this average is taken over the complete event space. Acknowledging the linearity of the ensemble operator, the stochastic first moment is

$$\mathcal{E}\bar{X} = \mathcal{E}\left\{\frac{1}{N}\sum_{n=1}^{N}x_n\right\} = \frac{1}{N}\left(\sum_{n=1}^{N}\mathcal{E}\{x_n\}\right)$$

$$\mathcal{E}\bar{X} = \frac{1}{N}\left(\sum_{n=1}^{N}\mu\right) = \frac{1}{N}N\mu = \mu. \quad \text{(Ensemble mean of sample mean)} \quad (8.1)$$

The sample mean is an unbiased estimate of the ensemble mean. That's an important point! Let's see how the precision of the sample mean changes as we add more data.

8.1.2   Ensemble Variance of the Sample Mean

We know from Section 3.4 that variance is the second central moment,

$$\text{var}X = \int_{-\infty}^{\infty} dx\,(x-\mu)^2\,p(x).$$

Consequently, the ensemble variance of the sample mean is

$$\text{var}\bar{X} = \text{var}\left(\frac{1}{N}\sum_{n=1}^{N}x_n\right) = \text{var}\left(\frac{x_1}{N} + \cdots + \frac{x_n}{N} + \cdots + \frac{x_N}{N}\right)$$

$$= \frac{1}{N^2}\left(\text{var}x_1 + \cdots + \text{var}x_n + \cdots + \text{var}x_N\right)$$

$$\text{var}\bar{X} = \frac{1}{N^2}\sum_{n=1}^{N}\sigma^2 = \frac{\sigma^2}{N} = \sigma_\mu^2. \quad \text{Ensemble variance of sample mean} \quad (8.2)$$

We refer to $\sqrt{\text{var}\bar{X}} = \sigma_\mu = \sigma/\sqrt{N}$ as the *standard deviation of the mean*, σ_μ, or the *standard error*. Selecting σ or σ_μ to use as error bars in plots will depend on the message you wish to convey to readers. σ is a measure of the uncertainty expected in the next measurement made. In contrast, σ_μ is a measure of the uncertainty in the sample mean from the N samples included in the estimate. Note that $\lim_{N\to\infty}\sigma^2 = \sigma^2$, but $\lim_{N\to\infty}\sigma_\mu^2 = 0$.

You might be wondering how N^2 was factored out of the equations leading to (8.2). Recall that variance is quadratic in X, so removing one power of $1/N$ from the argument of $\text{var}(x_n/N)$ gives $\text{var}(x_n)/N^2$, in the same way that $(x/N)^2 = x^2/N^2$.

8.1.3   Ensemble Mean of the Sample Variance

Recall from (3.27) the expression for the mean-squared error (MSE) of our WSS-UVN process. Adding to that expression the results of (8.1) and (8.2) we have

$$\mathcal{E}\{X^2\} = \text{var}X + (\mathcal{E}X)^2 = \sigma^2 + \mu^2$$

$$\mathcal{E}\{\bar{X}^2\} = \text{var}\bar{X} + \left(\mathcal{E}\bar{X}\right)^2 = \frac{\sigma^2}{N} + \mu^2. \tag{8.3}$$

Now, beginning with the expression for sample variance,

$$s^2 = \frac{1}{N-1} \sum_{n=1}^{N}(x_n - \bar{x})^2,$$

$$\mathcal{E}s^2 = \frac{1}{N-1}\mathcal{E}\left\{\sum_{n=1}^{N} x_n^2 - 2\bar{x}\sum_{n=1}^{N} x_n + \sum_{n=1}^{N}(\bar{x})^2\right\}.$$

Multiplying the middle term in brackets by N/N, and from the linearity of $\mathcal{E}$,

$$\mathcal{E}s^2 = \frac{1}{N-1}\left(\mathcal{E}\left\{\sum_{n=1}^{N} x_n^2\right\} - N\mathcal{E}\{\bar{X}^2\}\right)$$

$$= \frac{1}{N-1}\left(\sum_{n=1}^{N}\mathcal{E}\left\{X^2\right\} - N\mathcal{E}\{\bar{X}^2\}\right)$$

$$= \frac{N}{N-1}\left(\mathcal{E}\left\{X^2\right\} - \mathcal{E}\{\bar{X}^2\}\right).$$

Combining this result with (8.3), we find

$$\mathcal{E}s^2 = \frac{N}{N-1}\left(\sigma^2 + \mu^2 - \frac{\sigma^2}{N} - \mu^2\right) = \frac{N}{N-1}\left(\frac{N-1}{N}\sigma^2\right) = \sigma^2.$$

Consequently, the ensemble mean of sample variance s^2 is

$$\mathcal{E}s^2 = \lim_{N\to\infty} \frac{1}{N-1} \sum_{n=1}^{N}(x_n - \bar{x})^2. \quad \text{(Ensemble mean of sample variance)} \quad (8.4)$$

$\mathcal{E}s^2$ is an unbiased estimate of var $X = \sigma^2$. With these tools, we're ready to explore hypothesis testing.

8.1.4   Effects of Sample Size

The preceding discussion suggests that using sample means instead of individual measurements to make classification decisions can narrow distributions of a test statistic. Consequently, as long as the means of normally distributed data are distinct,

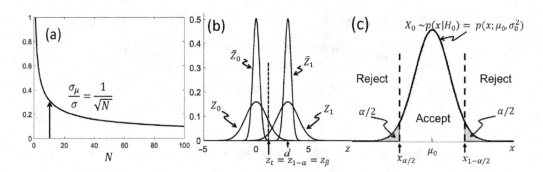

Figure 8.1 (a) The ratio of standard error σ_μ to standard deviation as a function of sample size N. (b) Measurements X_0 and X_1 are converted to standard normal variables Z_0 and Z_1, which overlap. Consequently, significant classification errors are produced using individual deviates z as test statistics to decide at threshold z_t whether members belong to group 0 or 1. If instead we use sample means $\bar{X}_0, \bar{X}_1$ with sample size $N = 10$ as a test statistic, then $\bar{Z}_0$ and $\bar{Z}_1$ yield significantly fewer classification errors at z_t. (c) A univariate normal (UVN) pdf and associated thresholds for testing the null hypothesis H_0 are shown. The likelihood of falsely rejecting the null hypothesis at thresholds $x_{\alpha/2}$ and $x_{1-\alpha/2}$ is given by two-sided Type I error probability α, which is the sum of shaded areas. Note distances $\mu_0 - x_{\alpha/2} = x_{1-\alpha/2} - \mu_0$ for the symmetric thresholds. Since we know nothing about the alternative hypothesis, we are unable to predict other errors.

these groups can be correctly classified with minimal error given the resources and opportunity to gather enough statistically independent data.

For example, Figure 8.1a shows that the standard error for a sample size of $N = 10$ is only $100/\sqrt{10} = 32\%$ of the standard deviation; i.e., $\sigma_\mu \simeq 0.32\sigma$. Because of (8.1), we know the sample mean is an unbiased estimate of the population mean, so increasing the sample size from 1 to 10 significantly decreases the overlap between adjacent distributions as shown in Figure 8.1b, but there is no shift in the means. The price of reducing the decision-error probability to this degree is a ten-fold increase in acquisition data costs.

Figure 8.1b displays the standard normal distribution $Z_0 \sim \mathcal{N}(0,1)$ that overlaps $Z_1 \sim \mathcal{N}(d,1)$ for $d = 3$. These are the distributions for sample size $N = 1$ (no measurement averaging). The horizontal axis z is the *normal deviate*, which is related to individual measurements x by $z = (x - \mu)/\sigma_\mu$. Of course, $\sigma_\mu = \sigma$ for $N = 1$. Converting $X \to Z$ focuses attention on the *intrinsic discriminability* between two groups irrespective of moment specifics. The ratio d/σ_μ is the *detectability index* d' for binormal alternative pdfs. In this example, $d' = 3$ states that the means are separated by three common standard errors. The *common standard error* is $(\sigma_{\mu 1} + \sigma_{\mu 2})/2$ when variances for the alternative pdfs are unequal.

Figure 8.1b also shows the distributions $\bar{Z}_0 = \mathcal{N}(0, 1/\sqrt{N})$ and $\bar{Z}_1 \sim= \mathcal{N}(d, 1/\sqrt{N})$, where $\bar{x}_i = \sum_{j=1}^{N} x_{ij}/N$ for $i = 0, 1$, $\bar{z}_i = (\bar{x}_i - \mu_0)/(\sigma/\sqrt{N})$; in the figure we set $N = 10$. If we can afford the cost and risk associated with generating enough data to acquire and average 10 independent samples, we gain significantly in terms of separability of the distributions as $d\sqrt{1}/\sigma = 3 \longrightarrow d\sqrt{10}/\sigma = 9.5$. Statistical independence is necessary to count each measurement as a full degree of freedom.

8.2   Hypothesis Testing

Two-sided Type I error probability. Let's say that physiological measurement X made on a healthy population is given as a UVN random variable $X_0 \sim p(x; \mu_0, \sigma_0^2)$ (see Figure 8.1c). For example, let X_0 be the core body temperature of adults who are considered healthy if their temperature is within the acceptable range. Measurements falling above the hyperthermic threshold $x_{1-\alpha/2}$ or below the hypothermic threshold at $x_{\alpha/2}$ indicate an unhealthy patient. The *null hypothesis* H_0 for each patient measurement, x, is that the individual is a member of the healthy group, labeled with subscript 0. If x falls outside the acceptable range, we reject the null hypothesis and identify that sample as belonging to the alternative group, labeled with subscript 1.

We chose to set the decision thresholds symmetrically about the population mean μ_0 in the absence of reasons to do otherwise. Whatever thresholds we select, we know that classification errors are possible. Because we know the distribution of healthy temperatures, we can predict the *probability of Type I error*, α. Type I errors are the *false-positive decisions* occurring when a measurement indicates that a healthy person (H_0) is unhealthy (H_1). The null hypothesis is not rejected when $x_\alpha \leq x \leq x_{1-\alpha}$ in Figure 8.1c. The probability that rejecting the null hypothesis is an error (false positive) is given by the shaded areas in the plot:

$$\alpha = \int_{-\infty}^{x_{\alpha/2}} dx\, p(x) + \int_{1-x_{\alpha/2}}^{\infty} dx\, p(x)$$

$$= \underbrace{P(x_{\alpha/2})}_{\alpha/2} + \underbrace{1 - P(x_{1-\alpha/2})}_{\alpha/2}. \quad \text{(Two-sided Type I error probability)}$$

For symmetric thresholds in two-sided tests, $(\mu_0 - x_{\alpha/2}) = (x_{1-\alpha/2} - \mu_0)$. $P(x)$ is the cumulative distribution function (CDF) from Section 3.1. Although we are free to select the thresholds, we must realize that our choice also determines error probability α.

One-sided Type I error probability. Assigning errors to the right and left tails of the pdf, as in Figure 8.1c, assumes measurements on both sides of the distribution are possible. In our body temperature example, if the test population is unlikely to have hypothermic members but some individuals may have a fever from a viral infection, then a one-sided Type I error probability is more relevant:

$$\alpha = \int_{1-x_\alpha}^{\infty} dx\, p(x) = 1 - P(x_{1-\alpha}). \quad \text{(One-sided Type I error probability)}.$$

In either case, the total Type I error probability is α, a unitless scalar where $0 \leq \alpha \leq 1$.

The horizontal axis in Figure 8.1c extends over the range of temperature measurements that are feasible in a specific experiment. To use statistical tables or statistical packages, we convert these null hypothesis data x to the standard normal axis z, where thresholds are

$$z_\alpha = (x_\alpha - \mu_0)/\sigma \quad \text{and} \quad z_{1-\alpha} = (x_{1-\alpha} - \mu_0)/\sigma, \tag{8.5}$$

and then $z_\alpha = -z_{1-\alpha}$. If x is a UVN measurement with units, z is a unitless, standard-normal version of that same measurement.

If we conduct a hypothesis test on groups of individuals, where we know the UVN parameters and the test statistic is the sample average of group members, $\bar{x} = (1/N)\sum_{j=1}^{N} x_j$, then the threshold becomes $z_\alpha = (\bar{x}_\alpha - \mu_0)/(\sigma/\sqrt{N})$.

Example 8.2.1. *Body temperature: The mean body temperature in the healthy adult human population is $98.2 \pm 1.5°F$. We assume we know the mean and variance of the population because the measurement ensembles used to determine $\mu \pm \sigma$ are very large. (a) What is the probability of finding one healthy patient with a body temperature between $97°F$ and $99°F$?*

Solution

The sample size is $N = 1$. Use the probability statement in z to compute the CDFs.

$$\Pr(97 \le \bar{x} \le 99)_{N=1} = \Pr\left(\frac{97 - \mu}{\sigma/\sqrt{N}} \le z \le \frac{99 - \mu}{\sigma/\sqrt{N}}\right)$$

$$= P\left(\frac{99 - 98.2}{1.5}\right) - P\left(\frac{97 - 98.2}{1.5}\right)$$

$$= P(0.53) - P(-0.80) = 0.702 - 0.212 = \boxed{0.49}$$

There is a 49% probability that the next healthy patient entering the clinic has a temperature between $97°F$ and $99°F$.

(b) What is the probability of finding the average temperature in that range for a group of 10 patients?

Solution

Now the sample size is $N = 10$.

$$\Pr(97 \le \bar{x} \le 99)_{N=10} = P\left(\frac{99 - 98.2}{1.5/\sqrt{10}}\right) - P\left(\frac{97 - 98.2}{1.5/\sqrt{10}}\right)$$

$$= P(1.69) - P(-2.53) = 0.954 - 0.0057 = \boxed{0.95}$$

The chance of finding a group of 10 patients whose average temperature is in that range increases to 95%.

One form of the MATLAB code providing the numerical answers to these problems is

```
Pr1=cdf('norm',0.53,0,1)-cdf('norm',-0.80,0,1)
Pr10=cdf('norm',1.69,0,1)-cdf('norm',-2.53,0,1)
```

The threshold values along the z axis and areas are illustrated in Figure 8.2a. These thresholds are not symmetric about the population mean. The nonlinear change of probability resulting from a change in threshold can make it difficult to intuit the relationship. For this reason, software packages and statistical tables are needed.

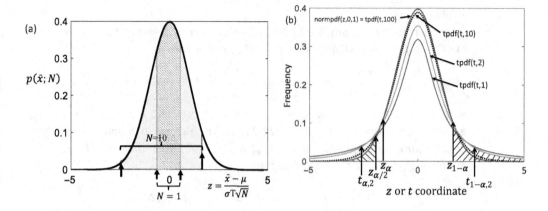

Figure 8.2 (a) Illustration associated with Example 8.2.1 for asymmetric thresholds and sample sizes $N = 1$ (cross-hatched area) and $N = 10$ (gray area). (b) Comparisons of t-statistics in Example 8.2.2 for various degrees of freedom ϱ compared with those for standard normal variable z. The curves indicate the MATLAB functions used to generate them. Note that $z_\alpha \simeq t_{\alpha, \varrho} \; \forall \; \alpha$ provided $\varrho > 100$.

8.2.1   The t-Statistic

In the preceding discussion, we knew, a priori, the population variance of our univariate normal process. The test statistic for hypothesis testing is $z = (\bar{X} - \mu)/(\sigma/\sqrt{N})$, which has a standard normal pdf, $\mathcal{N}(0, 1)$.

However, if σ^2 is unknown, we must estimate it from the data using sample variance, $s^2 = \sum_{j=1}^{N}(x - \bar{x})^2/\varrho$ with $\varrho = N - 1$ *degrees of freedom*. The relevant test statistic is now $t = (\bar{X} - \mu)/(s/\sqrt{N})$, which is not well represented by a standard normal pdf because it is a ratio of two random variables, $\bar{X}$ and s. The exception is when N is large enough that $s \to \sigma$, as we found in (8.4). When σ is unknown and the sample size is small, the test statistic follows a t-variable distribution [13, 15], a long-tailed symmetric pdf about zero that slowly morphs into a standard normal pdf as $N \to \infty$ (Figure 8.2b).

The testing threshold is $t_{\alpha, \varrho}$ with parameters α (Type I error probability) and $\varrho = N - 1$ (degrees of freedom). We may gain intuition about the t pdf through the following example.

Example 8.2.2. *t-Statistics: (a) Find one-sided thresholds for the standard normal coordinate at probabilities $\alpha = 0.05$, $\alpha/2 = 0.025$, and $1 - \alpha = 0.95$. (b) Find the one-sided thresholds for the t-pdf coordinate at probabilities $\alpha = 0.05$ and $1 - \alpha = 0.95$, both at sample size $N = 3$. Since z and t are both found from measurement x, we may compare the t and z thresholds on the same axis.*

Solution

To answer these questions, we first translate the various forms of the equations into MATLAB *functions. Denoting* α *as* a *and* z_α *as* za, *then for*

$$\alpha = \int_{-\infty}^{z_\alpha} dz\, p(z) = P(z_\alpha) = \int_{z_{1-\alpha}}^{\infty} dz\, p(z) = 1 - P(z_{1-\alpha}),$$

we can find α *from* z_α *using* a = cdf('norm',za,0,1) *and* z_α *from* α *using* za = icdf('norm',a,0,1). *The* MATLAB *statements follow the shorthand notation for inverse operators* $\alpha = P(z_\alpha)$ *and* $z_\alpha = P^{-1}(\alpha)$, *where P is CDF.*

Similarly, denoting t_α *as* ta, *then for*

$$\alpha = \int_{-\infty}^{t_{\alpha,\varrho}} dt\, p(t;N) = \int_{t_{1-\alpha,\varrho}}^{\infty} dt\, p(t;N),$$

the MATLAB *statements are* a = tcdf(t,N-1) *and* ta = tinv(a,N-1).

(a) Finding z_α, $z_{\alpha/2}$, *and* $z_{1-\alpha}$,

```
za=icdf('norm',0.05,0,1)    = -1.645
zao2=icdf(norm,0.025,0,1)   = -1.960
z1ma=icdf('norm',0.95,0,1)  =  1.645
```

These are indicated in Figure 8.2b on the standard normal curve.

(b) The t coordinates $t_{\alpha,2}$ *and* $t_{1-\alpha,2}$ *for* $\varrho = N - 1 = 2$ *degrees of freedom are*

```
ta2=tinv(0.05,2)  = -2.920
t1ma2=tinv(0.95,2) =  2.920
```

These threshold values are indicated in Figure 8.2b, where we plot t-distributions for sample sizes of $N = 2, 3, 11, 101$ giving $\varrho = 1, 2, 10, 100$, respectively. Note that the t pdf for a sample size of 101 is nearly identical to the standard normal pdf. However, we will show in the end-of-chapter problems that $z_\alpha = t_{\alpha,\varrho}$ for $\varrho \gtrsim 20$. Because the sample mean and variance are unbiased estimates of the population mean and variance, the convergence of the t and standard normal pdfs at large N is not surprising. However, when $\varrho < 10$ (which is where many real-life experiments take place), the t pdf has significantly longer tails compared to that for the standard normal variable z. Consequently, decision thresholds for a given Type I error probability are significantly larger for t pdfs compared with standard normal pdfs when $\varrho < 10$. For example, at $N = 3$, the threshold for a 5% error Type I probability is 78% larger if we must estimate the variance from the data compared to situations where when we know the population variance a priori. This large difference influences the performance of hypothesis testing and associated costs of validating diagnostic instruments.

8.2.2   Power Analysis

We recorded two groups of data whose histograms are plotted in Figure 8.3a as group 0 and group 1. Each data point in the distribution is the sample average of N independent measurements. If these data were collected without knowledge of the presence of two

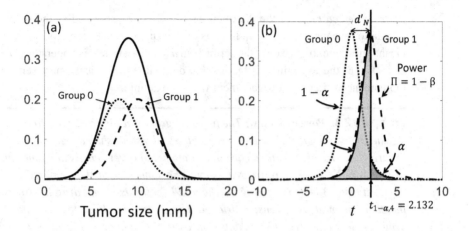

Figure 8.3 Illustration of power calculation problem from Example 8.2.3. (a) Histograms of the tumor size data, X. (b) Plot of the t-distributions after converting $x \rightarrow t$ for $N = 5$. The shaded areas are the Type I and Type II error probabilities, α and β, respectively.

distinct groups, the histogram would be given by the solid line plotted in Figure 8.3a. That histogram is not a normal distribution, but it is close enough to assume these data may be represented by a single normal distribution. Let's test the null hypothesis H_0 that all the data come from one group. The alternative hypothesis H_1 is that there are two distinct groups with different distributions, $X_0 = \mathcal{N}(\mu_0, \sigma_0)$ and $X_1 = \mathcal{N}(\mu_1, \sigma_1)$. We don't know the population statistics, so we apply sample statistics and convert the measurement data axis x into a t-distribution axis, as in Figure 8.3b, given degrees of freedom $\varrho = N-1$. As before, α is the probability of a Type I error, which occurs if we mistakenly reject the null hypothesis and claim there are two populations when there is only one. Because the alternative hypothesis has defined distributions, we can also define β as the probability of a *Type II error.* Type II errors arise when we mistakenly do not reject the null hypothesis when H_1 is true. $1 - \alpha$ is the probability of correctly detecting a single distribution when that condition is true, and $1 - \beta$ is the probability of correctly detecting two distributions when that condition is true. Whereas α and β are the two error probabilities, $1 - \alpha$ and $1 - \beta$ are the complementary probabilities of correct classifications.

The *statistical power* of the test, denoted as Π (capital Greek letter pi, not Roman numeral II), is the probability that statistical significance is achieved when H_0 is correctly rejected – the probability of correctly detecting two distributions. Quantitatively, $\Pi = 1 - \beta$ is the *sensitivity* of the test. When statistical power is maximized within the available resources, we have a well-designed experiment that is as sensitive as it can be to the presence of two distributions.

Four factors influence Π: (1) *sample size N*; (2) the *distribution* of the test statistic; (3) the *significance level* of the test, α, which we are free to choose; and (4) the intrinsic *detectability* of the distributions, $d'_N = d/\sigma_\mu$, defined as the separation of population means $d = \mu_1 - \mu_0$ in units of the square root of common variance of the mean. Subsequently, we omit the script N in d'_N and assume it is understood. Although $\Pi = 1 - \beta$, the experimental factors that determine the measurement sensitivity are more nuanced. Let's investigate them with an example.

Example 8.2.3. *Power of a test: The two data groups in Figure 8.3 are from tumor-size measurements in mice. X_0 mice were treated with established drug A, which is known to shrink the size of implanted cancerous tumors but is very costly and time consuming to produce. X_1 mice were treated exactly the same except that they received a new low-cost drug B instead of drug A. The null hypothesis is that drug B shrinks tumors the same amount as expensive drug A. The alternative hypothesis is that low-cost drug B is less effective at shrinking mouse tumors compared to drug A. Careful! The "positive" test result (i.e., the drug is working as advertised) is the null hypothesis in this example.*

We measured tumor sizes of $N = 5$ animals in each group, pairwise subtracted those measurements $d_j = x_{1j} - x_{0j}$ for $1 \le j \le N$, and averaged the results $\bar{d} = \sum_{j=1}^{N} d_j / N$. The sample mean $\bar{d}$, sample standard deviation s, and sample size N combine to form the test statistic, $\bar{d}/(s/\sqrt{N})$.[1] Since the population statistics are unknown, a one-sided t-test is applied with $\varrho = 4$ degrees of freedom to measure Π. The null hypothesis H_0 states that $\mathcal{E}\{d\} = 0$, while the alternative hypothesis H_1 is $\mathcal{E}\{d\} > 0$. An acceptable significance level of Type I error probability is $\alpha = 0.05$, so that $t_{\alpha,\varrho} = t_{0.05,4} = 2.132$.

H_0 will be rejected and H_1 adopted if $t > 2.132$. When two distributions are detected, we say drug B has failed to perform as designed – a negative result. In addition, we need statistical power greater than 0.80 to feel we have sufficient sensitivity. Significance level α and power $1 - \beta$ are determined by the decision maker based on the specifications of the classification task (in this case, the drug company working with oncologists). What conclusions can we reach from these data?

In terms of the four factors, (1) $N = 5$, (2) $t_{0.05,4}$ distribution, (3) $\alpha = 0.05$, and (4) because we don't know the detectability,[2] we will study power as a function of the separability variable $\theta = \mathcal{E}\{d\}$. Statistical power, $\Pi(\theta)$, is parameterized by N, α that determine the decision threshold, giving

$$\Pi(\theta; \alpha, N) = \Pr\left(t > t_{\alpha,\varrho} \,\middle|\, \mathcal{E}\{\bar{d}\} = \theta\right) \quad \text{for } \theta \ge 0$$

$$= \Pr\left(\frac{\bar{d}}{s/\sqrt{N}} > 2.132 \,\middle|\, \mathcal{E}\{\bar{d}\} = \theta\right)$$

---

[1] If drug B shrinks tumors more than drug A without additional side effects (great news!), d is negative. If a measurement gives $d_j < 0$, we set $d_j = 0$.

[2] For simulations, we know the detectability a priori. A t-distribution has variance $\sigma^2 = \varrho/(\varrho - 2)$ provided $\varrho > 2$. Here, $\varrho = 4$ and $\sigma = \sqrt{2}$. The simulation sets $\mathcal{E}\{d\} = 2$, so $\mathcal{E}\{d\}/\sigma \sim \sqrt{2}$ (see Figure 8.3b). Experiments do not provide this kind of detailed information.

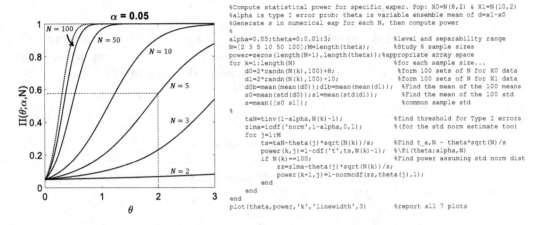

```
%Compute statistical power for specific exper. Pop: X0=N(8,2) & X1=N(10,2)
%alpha is type I error prob; theta is variable ensemble mean of d=x1-x0
%Generate s in numerical exp for each N, then compute power
%
alpha=0.05;theta=0:0.01:3;                    %level and separability range
N=[2 3 5 10 50 100];M=length(theta);         %Study 6 sample sizes
power=zeros(length(N+1),length(theta));       %appropriate array space
for k=1:length(N)                             %for each sample size...
    d0=2*randn(N(k),100)+8;                   %form 100 sets of N for X0 data
    d1=2*randn(N(k),100)+10;                  %form 100 sets of N for X1 data
    d0b=mean(mean(d0));d1b=mean(mean(d1));    %Find the mean of the 100 means
    s0=mean(std(d0));s1=mean(std(d1));        %Find the mean of the 100 std
    s=mean([s0 s1]);                          %common sample std
%
    taN=tinv(1-alpha,N(k)-1);                 %find threshold for Type I errors
    z1ma=icdf('norm',1-alpha,0,1);            %(for the std norm estimate too)
    for j=1:M
        ts=taN-theta(j)*sqrt(N(k))/s;         %Find t_a,N - theta*sqrt(N)/s
        power(k,j)=1-cdf('t',ts,N(k)-1);      %\Pi(theta;alpha,N)
        if N(k)==100;                         %Find power assuming std norm dist
            zz=z1ma-theta(j)*sqrt(N(k))/s;
            power(k+1,j)=1-normcdf(zz,theta(j),1);
        end
    end
end
plot(theta,power,'k','linewidth',3)          %report all 7 plots
```

Figure 8.4 Results from the statistical power calculations in Example 8.2.3 and MATLAB code. The solid lines indicate the power for a t-distribution. The dotted line gives the result for a standard normal distribution for $N = 100$. The dashed lines indicate $\Pi(2; 0.05, 5) = 1 - \beta = 0.57$, the statistical power for the experiment designed in the example. θ is the unknown separability parameter in (8.6).

$$= 1 - \Pr\left(\frac{\bar{d} - \theta}{s/\sqrt{N}} < 2.132 - \frac{\theta}{s/\sqrt{N}} \,\middle|\, \mathcal{E}\{\bar{d}\} = \theta\right)$$

$$= 1 - P\left(2.132 - \frac{\theta}{s/\sqrt{N}}\right). \tag{8.6}$$

The varying separability θ shows us how statistical power depends on detectability variable $\theta\sqrt{N}/s$. Figure 8.4 displays plots of $\Pi(\theta)$ with curves for different N, all at $\alpha = 0.05$. Clearly, sample size is very influential. Increasing N from 2 to 3 to 5 at separability $\theta = 2$ increases power from 7% to 23% to 57%. Studies with greater power show greater reliability in findings because it is less likely that Type II error will occur (at a fixed Type I error). Note that the results at $N = 100$ in Figure 8.4 for standard normal statistics and t-statistics are similar, which is expected as the distributions converge at $N \to \infty$. When $\theta = 0$, X_0 and X_1 have the same mean, and with equal variances, $\Pi(0) = \alpha = 0.05$, the minimum power value for all sample sizes. Clearly, adding a few more mouse studies in this example provides greater confidence in the sensitivity of the study. However, the desire for more statistical power must be balanced with the cost of additional experiments.

To answer our question regarding power for the experiment described, we note from Figure 8.4 that $\Pi = 0.57$. Examining Figure 8.3b, we see that $\beta \simeq 1 - \beta \simeq 0.5$. These two measures won't equal exactly because Π is based on the sample variance s^2. This experiment is underpowered at $\Pi = 0.57$. Increasing the sample size to the $N = 8 - 10$ range generates 80% power. The spectrum of power curves in Figure 8.4 allows experimental designers to explore measurement options for achieving the required error bounds.

8.2.3   Minimum Sample Size

Power analysis is also helpful for calculating the minimum sample size required to detect the alternative hypothesis at defined error rates. For normal populations with unknown means and known and equal variances (Figure 8.1b), the expression for threshold x_t is

$$x_t = x_{1-\alpha} = x_\beta$$

$$= \frac{\sigma}{\sqrt{N}} z_{1-\alpha} + \mu = \frac{\sigma}{\sqrt{N}} z_\beta + \mu + d;$$

$$d = \frac{\sigma}{\sqrt{N}} (z_{1-\alpha} + z_{1-\beta}) \quad \text{because } z_{1-\beta} = -z_\beta; \quad \text{and therefore}$$

$$N = \frac{\sigma^2}{d^2}(z_{1-\alpha} + z_{1-\beta})^2 = \frac{\sigma^2}{d^2}(z_\alpha + z_\beta)^2. \tag{8.7}$$

N is the minimum sample size for these normal populations to achieve acceptable error probabilities α and β. Both of the last two expressions hold because $z_\alpha = -z_{1-\alpha}$ and $z_\beta = -z_{1-\beta}$.

Example 8.2.4. *Sample-size calculation:* *(a) For group detectability of 0.5 and 5% Type I and II errors, find the minimum sample size that achieves this classification performance. (b) If α and β are reduced to 1% but the other parameters remain the same, how does N change? (c) What is the statistical power associated with each study? (d) Plot the associated distributions and indicate the thresholds.*

Solution
(a) Applying the MATLAB functions from Example 8.2.2, we find $z_{1-\beta} = 1.645$. Equation (8.7) gives $N = (\sigma^2/d^2)(2z_{1-\beta})^2 = 44$ (always round up!).
(b) $N = 4(2 \times 2.326)^2 = 87$.
(c) For $\Pi = 1 - \beta$, tests (a) and (b) yield 95% and 99% power, respectively.
(d) Converting back to the original measurement axis, the distributions of test statistic $\bar{X}$ are plotted for $N=1$, 44, and 87 in Figure 8.5. Notice that at a sample size of 1, there is no way that $z_\alpha = z_{1-\beta}$ for a detectability of 0.5.

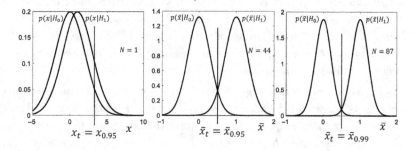

Figure 8.5 Results from Example 8.2.4.

Equation (8.7) shows that the minimum sample size depends on the detectability of the pdf for the alternative hypothesis relative to that for the null hypothesis, d/σ, and the level of sensitivity α and statistical power $1 - \beta$ expected. Finally, if variances are unknown, then (8.7) becomes

$$N = \left(\frac{s}{d}\right)^2 \left(t_\alpha \frac{s_0}{s} + t_\beta \frac{s_1}{s}\right)^2, \quad \text{where } s = \sqrt{s_0^2 + s_1^2}. \tag{8.8}$$

8.3 Receiver Operating Characteristic Analysis

The preceding discussion outlined a method for designing experiments that test binary hypotheses regarding the similarity of two data sets. With this approach, test statistics are formed from measurement data after determining sample sizes based on acceptable error probabilities. In this way, we can define the resources needed to make decisions that will minimize classification errors despite the presence of uncertainty/noise in the measurement data.

For example, consider data from two groups of patients. Group 1 is positive for disease state 1 and group 0 is negative for that disease (healthy). In each patient in both groups, we test for disease 1 using Test A and Test B. Receiver operating characteristic (ROC) analysis is a method for predicting α and β errors for both tests at all threshold values when tasked with a binary decision – in this case, deciding whether a patient is healthy or sick.

An ROC curve is a plot of measurement sensitivity $= 1 - \beta$ as a function of significance $= \alpha$. Significance is often labeled 1-specificity in ROC analysis, where specificity is the true negative probability. Yes, there are many names for the four basic decision probabilities, α, $1 - \alpha$, β, and $1 - \beta$, which can frustrating when you are first introduced to ROC analysis. The *sensitivity* of a diagnostic procedure measures the probability of correctly identifying patients with disease 1 – the true positive fraction (TPF). The *specificity* of a diagnostic procedure measures the probability of correctly identifying patients without disease 1 – the true negative fraction (TNF). More on this in Section 8.4.

8.3.1 Medical Decisions and Associated Errors

Test statistic $X = \{x_0, x_1\}$ for a diagnostic test includes data from the two normally distributed patient groups, $X_0 \sim p(x|H_0) = \mathcal{N}(\mu_0, \sigma_0^2)$ and $X_1 \sim p(x|H_1) = \mathcal{N}(\mu_1, \sigma_1^2)$. The alternative hypothesis H_1 indicates positive patients with data labeled X_1, while the null hypothesis H_0 indicates negative patients with data labeled X_0. The word "negative" is typically associated with the null-hypothesis state and the word "positive" with the alternative hypothesis, although there are exceptions (as seen in Section 8.2). A confirmed diagnosis (ground truth) is established by an independent measurement that the medical community agrees defines the disease state, which often

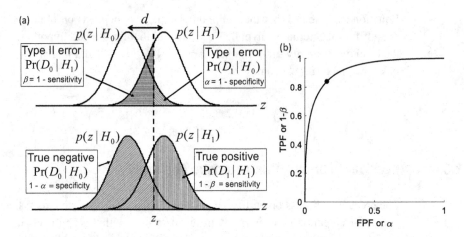

Figure 8.6 (a) Conditional standard normal densities $p(z|H_j)$ for two groups $j = 0, 1$ are plotted. Four probabilities (shaded areas) are defined with respect to threshold z_t. For the z_t value shown, the TPF versus FPF point is plotted by the point in (b). The ROC curve is generated by computing these two probabilities while scanning z_t from $z = \pm\infty$.

comprises pathology from a patient sample. A positive decision D_1 or a negative decision D_0 is made for every unlabeled patient measurement x by comparing its value to threshold x_t according to

$$D_j(x; x_t) \quad \text{for} \quad j = \text{step}(x - x_t). \tag{8.9}$$

In this case, $\mu_1 > \mu_0$. The relationships between the decisions and group labels are illustrated in Figure 8.6. As before, $d = \mu_1 - \mu_0$, $\sigma = \sigma_0 = \sigma_1$, and detectability is d/σ in variate x. However, for deviate $z = \sqrt{N}(x - \mu_0)/\sigma$, separability d equals detectability d' because variance for a standard normal deviate is 1. While d is unknown, σ is known in this case.

Four fundamental decision probabilities are defined in several equivalent forms for normal deviate z, where $z_t = z_{1-\alpha} = z_\beta$:

$$\Pr(D_0|H_0) = \int_{-\infty}^{z_t} dz\, p(z|H_0) = P(z_t|H_0) = P\left(\frac{x_{1-\alpha} - \mu_0}{\sigma/\sqrt{N}}\right)$$

$$= \text{true-negative fraction} = \text{TNF}(z_t) = \text{specificity} = 1 - \alpha$$

$$\Pr(D_1|H_0) = \int_{z_t}^{\infty} dz\, p(z|H_0) = 1 - P(z_t|H_0) = P\left(\frac{\mu_0 - x_{1-\alpha}}{\sigma/\sqrt{N}}\right)$$

$$= \text{false-positive fraction} = \text{FPF}(z_t) = \text{Type I error} = 1\text{-specificity} = \alpha$$

$$\Pr(D_0|H_1) = \int_{-\infty}^{z_t} dz\, p(z|H_1) = P(z_t|H_1) = P\left(\frac{x_\beta - \mu_0 - d}{\sigma/\sqrt{N}}\right)$$

$$= \text{false-negative fraction} = \text{FNF}(z_t) = \text{Type II error} = \beta \tag{8.10}$$

$$\Pr(D_1|H_1) = \int_{z_t}^{\infty} dz\, p(z|H_1) = 1 - P(z_t|H_1) = P\left(\frac{\mu_0 + d - x_\beta}{\sigma/\sqrt{N}}\right)$$

$$= \text{true-positive fraction} = \text{TPF}(z_t) = \text{sensitivity} = 1 - \beta = \text{power}.$$

Bear in mind that α and β are functions of z_t, although that is not explicitly indicated. The error probabilities in (8.10) are calculated when the indices on decision D and hypothesis H are not the same. Between the expressions in (8.10) and their graphical representation in Figure 8.6, you should obtain a clear mental image of decision probabilities. From graphical inspection and the first axiom of probability (2.2), we find

$$\Pr(D_0|H_0) + \Pr(D_1|H_0) = 1 = \Pr(D_0|H_1) + \Pr(D_1|H_1)$$
$$\text{TNF} + \text{FPF} = 1 = \text{FNF} + \text{TPF}.$$

8.3.2   Binormal ROC Curve and AUC for 2AFC Experiments

An ROC curve is a plot of $\text{TPF}(z_t)$ as a function of $\text{FPF}(z_t)$ for $-\infty < z_t < \infty$. These probabilities (when group pdfs are known) or fractions (when group pdfs are approximated by histograms) are the areas to the right of z_t for the underlying binormal distributions. As you may notice from (8.10), both TPF and FPF are obtained by computing 1 minus a CDF, as they are the areas to the right of threshold z_t. When individual distributions are univariate normal, the ROC curves found are referred to as binormal ROC curves. The single value of z_t selected in Figure 8.6 gives just one point on the ROC curve. At $z_t \to \infty$, $\text{FPF} = \text{TPF} = 0$, and at $z_t \to -\infty$, $\text{FPF} = \text{TPF} = 1$. Both TPF and FPF increase as z_t decreases from $\infty \to -\infty$.

A useful scalar summary of an ROC curve is the *area under the curve (AUC)*. The AUC is the probability that a decision maker will correctly select the positive case when considering two randomly selected samples, one from the positive patient group and one from the negative patient group. This decision process describes a *two-alternative forced-choice (2AFC) observer experiment*.

Example 8.3.1. *ROC curves versus separability: Generate* $Z_0 \sim \mathcal{N}(0, 1)$ *and six additional pdfs* $Z_1 \sim \mathcal{N}(d, 1)$ *for* $d = 0, 0.1, 0.5, 1.0, 2.0, 5.0$. *(a) Compare* Z_0 *with each* Z_1 *by computing the ROC curve for each* Z_0, Z_1 *pair. (b) Compute the area under each of the ROC curves. (c) Generate a "continuous" plot of AUC(d) for* $0 \le d \le 5$.

Solution
The code used to solve the problem is given here, and the results are shown in Figure 8.7. When $d = 0$, $p(z|H_0) = p(z|H_1)$ *and the ROC curve is a diagonal line TPF* $=$ *FPF. Consequently, AUC(0) $= 0.5$. This is the signature of two data sets with no separability because the pdfs overlap completely. Near the other extreme, where* $d = 5$, *separability[3] is five times the common standard deviation. We find TPF(FPF)* $\simeq 1 \ \forall \ FPF$ *and AUC(5) $\simeq 1$. A separability of 5 is so large that there are almost no classification errors. These two groups are very well discriminated because the pdfs in Figure 8.7a show very little overlap.*

---

[3] Separability in the normal deviate z is equal to d. Detectability in the normal variate x is d/σ.

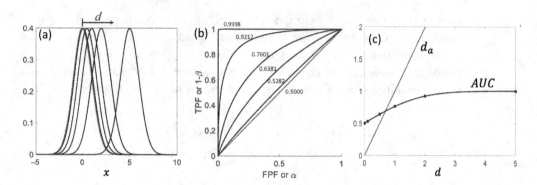

Figure 8.7 (a) $Z_0 = \mathcal{N}(0, 1)$ is plotted using a dotted line. The other pdfs are $Z_1 = \mathcal{N}(d, 1)$ for separabilities $d = 0.0, 0.1, 0.5, 1.0, 2.0, 5.0$. (b) ROC curve generated by comparing Z_0 with each of the six Z_1. The AUC values indicated are in increasing order of d from 0.5 to 1.0. (c) AUC(d) are plotted for $0 \leq d \leq 5$ while specifically indicating points corresponding to the six curves from (b). Also plotted are $d_a = d$ values from (8.11) showing separability can be estimated from the AUC of binormal ROC curves via (8.11). Interested readers may wish to compare the AUC(d) curve with that for relative entropy in (3.14). Both are information-theoretic measurements of the separation between pdfs.

```
%%%%%%%%%%%%%%%%%%%%%%% Script offering the solution in Example 8.3.1 %%%%%%%%%%%%%%%%%%%%%%%%%%%
%%%%%%%%% This code computes ROC curves from CDFs and integrates them to find AUC %%%%%%%%%%%%%
z=-5:0.1:10;Nz=length(z);                        % initialize
d=[0 0.1 0.5 1 2 5];Nd=length(d);                % test six separabilities
%d=0:0.1:5;Nd=length(d);                         % for continuous curve AUC(d) in (c)
Z0=normpdf(z,0,1);Z1=zeros(Nd,length(Z0));       % Z0 is std norm; initialize Z1
TP=zeros(Nd,Nz);FP=zeros(Nd,Nz);AUC=zeros(size(d));   % more initialization
figure(1);plot(z,Z0,'k:','linewidth',2);hold on
for j=1:Nd                                       % for each separability...
    Z1(j,:)=normpdf(z,d(j),1);                   % find Z1 for that d
    if(j ~= 1);plot(z,Z1(j,:),'k','linewidth',2);end   % plots in Fig. 8.7a except d=0
    for k=Nz:-1:1                                 % compute TPF,FPF for this d
        TP(j,Nz-k+1)=cdf('norm',z(k),d(j),1,'upper');  % ROC curve estimation
        FP(j,Nz-k+1)=cdf('norm',z(k),0,1,'upper');
    end
    AUC(j)=trapz(FP(j,:),TP(j,:));               % numerical integral for AUC
end
hold off;xlabel('z')
figure(2);hold on
for j=1:Nd                                       % plot ROC curves in Fig. 8.7b
    if j == 1
        plot(FP(j,:),TP(j,:),'k*','linewidth',2)
    else
        plot(FP(j,:),TP(j,:),'k','linewidth',2)
    end
end
xlabel('FPF or \alpha');ylabel('TPF or 1-\beta');axis square;hold off
```

The ROC curve displays the sensitivity of the test at all possible significance levels; it is the true-positive probability attained at each level of false-positive probability. However you choose to interpret the curve, it offers a comprehensive assessment of the detectability of one group as distinct from the other. The area under a curve has a range of $0.5 \leq \text{AUC} \leq 1$, where $\text{AUC} \simeq 0.5$ indicates a worthless test and $\text{AUC} \simeq 1.0$ indicates a perfectly discriminating test. In practice, many acceptable tests used today have $0.76 \leq \text{AUC} \leq 0.92$, suggesting the typical group detectability of a diagnostic test is $1 \leq d/\sigma \leq 2$. If you look closely at Figure 8.7a, you will see that at $d = 0.1$ the pdfs almost completely overlap. The ROC curve in Figure 8.7b gives AUC = 0.5282, which is a small variation from chance. Above $d = 3$, the ROC curve changes very little.

The literature on *observer performance* [10, 47, 104] defines a detectability index $d' = d/\sigma$, where $\sigma = \sqrt{(\sigma_0^2 + \sigma_1^2)/2}$, which is the same detectability discussed earlier in this chapter. It can be estimated from ROC curves as follows:

$$d_a = 2\text{erf}^{-1}(2\text{AUC} - 1), \tag{8.11}$$

where $\text{erf}^{-1}(\cdot)$ is the inverse *error function*. In MATLAB, use `erfinv`. We can think of detectability d' as an intrinsic property of data and d_a as an estimate of detectability obtained from the ROC curve. d_a applies to any ROC curve regardless of the underlying distribution, while d' is defined for binormal pdfs. Because we have normal pdfs, we see from Figure 8.7c that $d' = d_a$. The d_a index is used to estimate the *task efficiency* of one method relative to another. For example, the efficiency of Test A relative to Test B, η, is [104]

$$\eta = \frac{d_{a,A}^2}{d_{a,B}^2}. \tag{8.12}$$

8.3.3   A Few Details

Pan and Metz [79] describe a straightforward way to estimate the AUC for binormal ROC curves with known variance, although the origin of this method is much older, e.g., [30]. Beginning with (8.10),

$$\text{TPF} = 1 - \beta = 1 - P(z_t|H_1) = P(-z_t|H_1)$$
$$\text{FPF} = \alpha = 1 - P(z_t|H_0) = P(-z_t|H_0), \tag{8.13}$$

where $z_t|H_j = (x_t - \mu_j)/\sigma_j$, $\mu_0 = 0$, $\sigma_0 = 1$, and $\mu_1 = a/b$, $\sigma_1 = 1/b$. Consequently, $a = \mu_1/\sigma_1$ and $b = 1/\sigma_1$. Because threshold x_t is a variable in ROC analysis, let's simplify it to x in the following section:

$$\begin{aligned}
\text{AUC}(d) &= \int_{-\infty}^{\infty} \text{TPF}(x)\, d\text{FPF}(x) = \int_{-\infty}^{\infty} dx\, \text{TPF}(x)\frac{d\text{FPF}(x)}{dx} \\
&= \int_{-\infty}^{\infty} dx\, P(a - bx)\frac{dP(-x)}{dx} = \int_{-\infty}^{\infty} dx\, P(a - bx)p(-x) \\
&= P\left(\frac{a}{\sqrt{1 + b^2}}\right) = P\left(\frac{\mu_1}{\sqrt{\sigma_1^2 + 1}}\right). \tag{8.14}
\end{aligned}$$

We need to try out this method and make sure its results are consistent with our earlier findings. Since (8.14) assumes $\mu_0 = 0$, then $d = \mu_1$. Let $\sigma_1 = 1$; then $\text{AUC}(d) = P(d/\sqrt{2})$. For $d = 1, 2$ and using `cdf('norm',d/sqrt(2),0,1)`, we find AUC $= 0.7602$ and 0.9214, respectively, which match the values found numerically in Figure 8.7b within 2 parts in 10^4.

The sample size for ROC calculations is a more difficult calculation. If we plan to measure AUC values for Tests A and B, we must also be interested in knowing the number of cases that go into each AUC calculation to test the null hypothesis that the tests are from the same distribution. If $\hat{\theta} = AUC_A - AUC_B$, the test statistic is $\hat{\theta}/\sqrt{var_0(\hat{\theta})}$ [75]. The denominator includes the variance of the difference variable $\hat{\theta}$ under conditions where the null hypothesis is true. N_1 is the number of patients with the disease to be included in this study to satisfy our specifications on error probabilities α and β,

$$N_1 = \frac{\left(z_\alpha\sqrt{N_1 var_0(\hat{\theta})} + z_\beta\sqrt{N_1 var_1(\hat{\theta})}\right)^2}{\mathcal{E}\{\hat{\theta}\}_1^2},$$

which bears some similarity to (8.7). $\mathcal{E}\{\hat{\theta}\}_1$ is the ensemble mean and $var_1(\hat{\theta})$ is the variance of $\hat{\theta}$ under H_1. The difficulty with designing realistic observer experiments relates to the need to estimate both variances from the data, which can include covariances,

$$\hat{var}(\hat{\theta}) = \hat{var}(A\hat{U}C_0) + \hat{var}(A\hat{U}C_1) - 2\hat{cov}(A\hat{U}C_0, A\hat{U}C_1).$$

Obuchowski [75] shows how each term is related to parameters a, b in (8.14), each of which is estimated from the data using maximum likelihood methods. If experimental design is of interest, I urge you to read the citations in this section and then connect to the Charles E. Metz website maintained at the University of Chicago via http://metz-roc.uchicago.edu/. This website, which offers programs to which your data can be applied correctly, is a great resource to the biomedical measurement community.

8.4　　Other Performance Metrics

ROC analysis offers rigorous industry standards for rating overall diagnostic performance, but sometimes we wish to evaluate just one aspect of a diagnostic test, such as sensitivity. Here, we list a variety of names and associated probabilities associated with many features found in the literature using set theory.

Earlier sections labeled patients with disease as H_1 with reference to the alternative hypothesis. Now, we label truly positive cases as set W to represent samples "with" disease. H_0 for the null hypothesis identified patients who were truly negative for disease. Here we label negative cases as set W^c to represent the complement of patients with disease (not sick). Previously, patients with a positive test for the disease were labeled as D_1. Here we label them as P for "positive." Those testing negative for the disease and labeled as D_0 in the earlier sections are now labeled by the complement P^c.

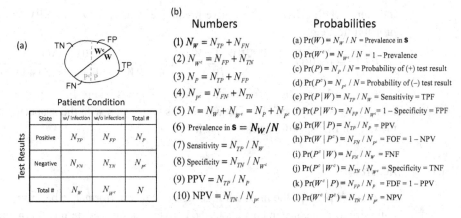

Figure 8.8 (a) Truth table where columns describe states of patient health and rows describe states of test results. N_W is the number of infected patients in S; N_{W^c} is the number in S without infection. N_P is the number of patients testing positive for the infection and N_{P^c} is the number testing negative. $N = N_W + N_{W^c} = N_P + N_{P^c}$. N_{TP} is the number of true-positive test results. N_{FP} is the number of false-positive results. N_{TN} is the number of true-negative results. N_{FN} is the number of false-negative results. PPV is positive predictive value and NPV is negative predictive value. FDF is the false-discovery fraction and FOF is the false-omission fraction.

Let event space S be composed of a population of N patients suspected of being infected with the novel coronavirus that causes COVID-19 illness. All of these patients have symptoms of a respiratory infection, but not all have this particularly nasty virus. The number of COVID-19–infected patients is N_W and the number without this particular virus is N_{W^c}, such that $N_W + N_{W^c} = N$. These numbers are established by applying the Centers for Disease Control and Prevention's (CDC's) testing standards. All of these individuals are also tested with an new, easy-to-use, lost-cost device sold at Walmart for $10. N_P are the number of patients testing positive for the COVID-19 infection, N_{P^c} are the number testing negative with the same device, and $N_P + N_{P^c} = N$. (My apologies for overusing the letter P.)

The number of true-positive patients N_{TP} plus the number of false-negative patients N_{FN} gives the number of infected patients, $N_W = N_{TP} + N_{FN}$. The number of false-positive patients N_{FP} plus the number true-negative patients N_{TN} gives N_{W^c}. The probability that a patient in this study has a positive test result, $\Pr(P)$, may be expressed in terms of conditional probabilities and numbers of patients as follows:

$$\Pr(P) = \Pr(P|W)\Pr(W) + \Pr(P|W^c)\Pr(W^c) = \frac{N_{TP}}{N_W}\frac{N_W}{N} + \frac{N_{FP}}{N_{W^c}}\frac{N_{W^c}}{N} = \frac{N_P}{N}. \tag{8.15}$$

The probability of an error is $(N_{FN} + N_{FP})/N$. Figure 8.8 lists all the probabilities of various states in the manner of (8.15) and most of the labels you will find in the literature. The positive and negative predictive values, PPV and NPV, are posterior probabilities, e.g.,

$$PPV = \Pr(W|P) = \frac{\Pr(P|W)\Pr(W)}{\Pr(P)} = \frac{\Pr(P|W)\Pr(W)}{\Pr(P|W)\Pr(W) + \Pr(P|W^c)\Pr(W^c)}.$$

PPV and NPV are often of greatest interest to diagnosticians. PPV answers this question: What is the probability that a patient is really sick with the suspected disease if the test for that specific disease is positive?

8.4.1   Accuracy

Accuracy is perhaps the most familiar metric of test performance because it is used in daily life to make statements about the correctness of average measurements. Accuracy has a technical definition, which is the total number of correct test responses divided by the total number in the event population, $(N_{TP} + N_{TN})/N$. Accuracy is also the probability of a diseased patient AND a positive test result, $\Pr(WP)$, plus the probability of a nondiseased patient AND a negative test result, $\Pr(W^c P^c)$. By applying Bayes' rule from Section 2.5 and the definitions in Figure 8.8, we see that these statements must be equal:

$$
\begin{aligned}
\text{Accuracy} &= \Pr(WP) + \Pr(W^c P^c) \\
&= \Pr(P|W)\Pr(W) + \Pr(P^c|W^c)\Pr(W^c) \\
&= \text{sensitivity} \times \text{prevalence} + \text{specificity} \times (1 - \text{prevalence}) \\
&= \frac{N_{TP}}{N_W}\frac{N_W}{N} + \frac{N_{TN}}{N_{W^c}}\frac{N_{W^c}}{N} = \frac{N_{TP} + N_{TN}}{N}.
\end{aligned}
\tag{8.16}
$$

We need to be suspicious of accuracy as a reliable quality metric because of how it weights prevalence. Although prevalence shows up in the second line of (8.16), the accuracy value does not always track with our intuition.

An example of the dubious value of accuracy as an evaluation metric can be seen by imagining a completely worthless diagnostic test that simply decides everyone tested is negative regardless of the data obtained. We hope the accuracy of a worthless test is revealed to be low once measurements are taken. First, assume disease prevalence in the event population is $\Pr(W) = 0.50$. Equation (8.16) gives an accuracy of $(0 + 0.50)N/N = 0.50$, which is quite disturbing: A worthless test is still 50% accurate! Second, let the prevalence fall to $\Pr(W) = 0.01$. The accuracy becomes $(0 + 0.99)N/N = 0.99$, which suggests high performance unless you look more closely. The weaknesses of accuracy-based evaluations can be avoided by applying ROC analysis.

8.5   Problems

8.1 The numbers for minimum sample size found in Example 8.2.4 were fairly large, so even if population variances were unknown we might expect to find N values that are about the same (see Figure 8.9). That is, switching from z to t, we

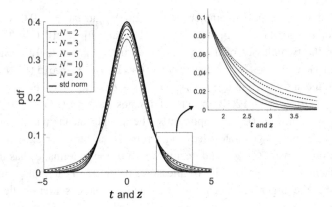

Figure 8.9 Problem 8.1: Comparisons among the standard normal and t pdfs versus N.

may not find much change in N. Let's see. Repeat the analysis using t-distributions instead of standard normal pdfs and compare the minimum sample size calculations for $\alpha = 0.01, 0.05$. Remember, t-distributions change shape with $\rho = N - 1$.

8.2 As the novel SARS-CoV-2 virus that causes COVID-19 illness spreads quickly throughout the world in early 2020, persons with flu-like symptoms and contact with at-risk individuals were evaluated comprehensively by experts, such as CDC personnel in Atlanta. By broad consensus in the United States, CDC findings are considered ground truth regarding this infection.

A fictitious company in the business of developing simple at-home immunoassay kits, called *XbioeX*, proposes an inexpensive, microfluidic-based device that a patient blows into. The device measures trace concentrations of proteins produced by the virus or the body's immune system that are specific to COVID-19. To obtain market approval before public use, company scientists acquire access to data from infected patients and healthy volunteers. Results from the *XbioeX* kits costing $10 are labeled B. These are compared to a standard reverse transcription polymerase chain reaction (RT-PCR) test costing $100 that are labeled A. Both kits have been cross-validated by CDC diagnostic standards.

(a) Compute PPV. The prevalence of infection in one test population is 50%. The sensitivity of Test A is 90% and its specificity is 99%. Test B has a greater sensitivity of 99% but is only 90% specific. Which test offers the best positive predictive value?

(b) How does the PPV estimate change if population prevalence is reduced to 1%?

8.3 Machine observers. Conduct a *two-alternative forced-choice* (2AFC) observer study with a *machine observer*. First create 1,000 image pairs, g_0 and g_1, where pixel values in each image are drawn from a uniform random variable with pdf

$$p(g_j; a_j, b_j) = \begin{cases} \frac{1}{b_j - a_j} & a_j \leq x \leq b_j \\ 0 & \text{otherwise} \end{cases} \quad \text{and} \quad j = 0, 1.$$

Images labeled g_0 have pdf parameters $a_0 = 2.0$ V and $b_0 = 10.0$ V. The pdf parameters for images labeled g_1 are $a_1 = 2.1$ V and $b_1 = 10.1$ V. The difference between the two classes of data is subtle. You may generate 1,000 images in MATLAB via $g_0(x_1, x_2) \in \mathbb{R}^{N \times N}$ using g0=a0+(b0-a0)*rand(N,N,1000). The ensemble mean for each image is $\bar{g}_j = 0.5(a_j + b_j)$ and the ensemble variances are $\text{var}(g_j) = (b_j - a_j)^2/12$. Both sets of images have $\text{var}(g_j) = \sigma^2 = 5.33$ V^2 but $d = \mu_1 - \mu_0 = 0.1$ V. Importantly, the sample means and variances are random variables, which generates decision errors. In an observer experiment, you will examine image pairs (g_0, g_1) and decide which of the two images has the larger mean. The machine observer for this task measures the mean of each image in the pair and decides that the larger mean identifies g_1. The machine observer applies (8.9) to make decisions.

(a) Generate ROC curves and AUCs for the machine observer of these training data when the image sizes are $N \times N$ for $N = 10, 30, 50$. Explain why performance increases significantly for images with more pixels. (You may wish to use histogram and [N,edges]=histcounts(...). The function fliplr can help you integrate functions from right to left.)

(b) Change the distribution parameters to $a_0 = 0, b_0 = 20, a1 = 0.1, b_1 = 20.1$ for the $N = 50$ images, which increases the variance without modifying other image features. How is detection performance affected?

(c) Show that the values of d' input when generating data are closely estimated from the data by measuring d_a found from AUC estimates.

(d) At what threshold value is the minimum total error found?

(e) AUC is the probability of a correct 2AFC response. Show this is true.

8.4 Adapt the code from the solution in Example 8.3.1 so that the pdfs have unequal variances. Display pdfs, ROC curves, and AUCs for $\sigma_0 = 2\sigma_1$ and $2\sigma_0 = \sigma_1$ situations.

8.5 *Confidence intervals (CIs)* give the range over which we can expect to find sample means that fall within an error specification given by $1 - \alpha$, where α is a Type I error probability. If we are interested in estimating the population mean from sample means, CIs tell us what confidence we may have in making this association. The two-sided $100(1 - \alpha)\%$ CI is given by

$$\Pr\left(t_{\alpha/2, N-1} < \frac{\bar{X} - \mu}{s/\sqrt{N}} < t_{1-\alpha/2, N-1}\right) = 1 - \alpha.$$

For $\alpha = 0.05$, we expect 95% of sample means from these measurements to fall within this confidence interval. We can illustrate with the following problem.

A large population of patients has a mean total serum cholesterol value of 220 ± 20 mg/dL. (a) Compute the 95% CI for the mean cholesterol based on 10-patient sample sizes drawn from that population. Assume measurements are normally distributed with unknown mean and variance. (b) How do the CIs change if the standard deviation of the population is known? (c) Simulate the sample means and standard deviations for 10 experiments and show how they fall within the CI bounds that you estimated.

8.6 In Chapter 7, we saw that the noise power spectral-density estimate $\hat{S}_e(u; T_0)$ is an important element in the measurement of image quality. It is the frequency spectrum of noise that enters the NEQ(u) calculation and integrates to give sample variance. Marple [69] shows that the short-time FT estimate of $\hat{S}_n$ is an *inconsistent estimator* because increasing the duration of the time series $2T_0$ does not improve estimates at any one frequency. He shows that for white noise, the sample variance of estimates $s^2(\hat{S}_e(u; T_0)) \rightarrow \mathcal{E}\hat{S}_e(u; T_0)$ as $T_0 \rightarrow \infty$. However, *ensemble averaging* does decrease spectral estimate uncertainty; i.e., $s^2\left(\sum_M \hat{S}_{e,m}(u; T_0)\right) \rightarrow 0$ as $M \rightarrow \infty$.
(a) Show numerically that $\hat{S}_e$ estimates for a spectrally white normal noise source do not converge to the population mean as the duration of the time series increases.
(b) Propose a frequency-domain method to approximate the ensemble estimate S_e in which the uncertainty in spectral estimates at each frequency becomes significantly less than $\mathcal{E}\hat{S}_e(u, T_0)$.

9 Statistical Pattern Recognition with Flow Cytometry Examples

Cytometry is the measurement of morphological and functional properties of individual cells and particulate components [97]. *Flow cytometry* is a technique whereby electrical or optical properties of cellular particles suspended in liquid are probed individually as they flow through a probing field (e.g., Figure 9.1). These systems can also separate structures based on electromechanical properties including size, but the cell sorting feature is not addressed here. Flow cytometry is used clinically for many applications, including characterization of lymphocyte subtypes and measurement of cytokine signaling in immunology. Flow cytometry is used to assess DNA content in many cancers.

One measurement task is to identify and count cells and other biological particles based on their interactions with laser light. The input consists of biological material with or without fluorescent labeling, and the output comprises a set of numbers representing optical-sensor signals. Intrinsic properties include the size of cells and particles that scatter light in different directions [97]. Extrinsic properties are related to fluorescent markers attached to some cells that are stimulated to fluoresce by the incident laser beam.

Large homogeneous structures scatter light predominantly in the forward direction; forward scattering (FSC) occurs over an angular range less than $10°$ relative to the incident light beam axis. *Small cells and objects with fine-scale granular internal structures scatter light preferentially at angles perpendicular to the incident beam axis*; side scattering (SSC) occurs over the broad angular range of $15°–150°$. Sensors are positioned to detect the amount of light scattered in those two directions because the scattering angle does not vary monotonically with size. Test samples may be mixed with control samples having known particle sizes, with or without fluorescent markers, to provide a reference. The number of measurement dimensions is at least two (FSC, SSC), and increases as fluorescent markers are added to identify specific sample structures.

The optical and flow elements of the system are the core of flow cytometry. They determine both SNR and sensitivity, depending on the contrast between cells and surrounding fluid. It is the hydrodynamic focusing elements that place cells individually before the optics that determine the trade-off between measurement speed and spatial resolution. Increasing flow pressure increases sampling rate, but that action

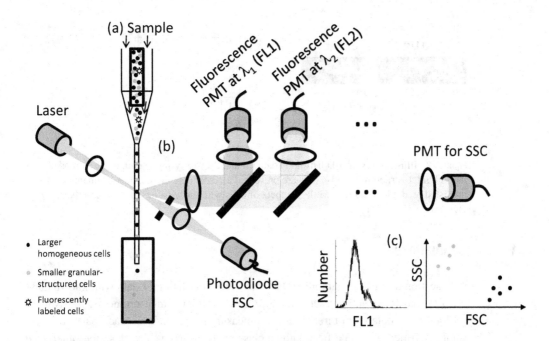

Figure 9.1 Diagram of a basic flow cytometry system shows three components. (a) The biological sample released is captured by a surrounding sheathing fluid flow so that the cell path narrows to single file through the laser light. (b) Forward-scatter (FSC) and side-scatter (SSC) light signals are recorded simultaneously for cells flowing through the beam, along with any fluorescent light (FL1) emission from fluorophore labels at wavelength λ_1. Large structures scatter mostly in the forward direction, while smaller structures scatter mostly to the side. Fluorescent colors can be differentially reflected through dichroic mirrors toward photomultiplier tubes (PMTs) that amplify weak fluorescence and intrinsic SSC light. (c) Electrical pulses from each optical sensor undergo *pulse-height analysis* to be recorded in list mode (Figure 9.2) for display as histograms and scatter plots.

could force cells together, below the spatial resolution of the device, resulting in classification errors.

Compromises are necessary when measuring a wide size spectrum. For example, a mixture of cell sizes may have one type of cell that is much sparser in number than the others. Because the number of cells counted is a Poisson random variable, we count until a set number is reached. That number is determined by the *coefficient of variation*: CV = standard deviation for the counts divided by the mean = $\sqrt{N}/N = N^{-1/2}$. To achieve a 1% CV, 10,000 cells must be counted, which can be challenging for small samples.

Our goal in this chapter is not to model acquisition or display processes, although we do some of that. Instead, our goal is to classify the data being acquired using supervised and unsupervised methods.

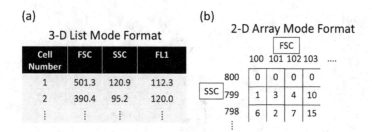

(a)

3-D List Mode Format

Cell Number	FSC	SSC	FL1
1	501.3	120.9	112.3
2	390.4	95.2	120.0
⋮	⋮	⋮	⋮

(b)

2-D Array Mode Format

FSC
100 101 102 103 ....

	100	101	102	103
800	0	0	0	0
SSC 799	1	3	4	10
798	6	2	7	15
⋮				

Figure 9.2 Illustrations of 3-D list-mode data (a) and 2-D array-mode data (b). List mode is a natural graph structure because data are presented as order sets. In (b), measurements values are listed on the left and top, while numbers of cells with those values are listed in the array elements.

9.1   Simulations

This section describes a simple 3-D Monte Carlo simulation of data with known class identity that bears some resemblance to flow cytometry measurements. A trivariate pdf describes data from three simulated measurements made on each particle in the sample. Armed with data from known class populations, we can describe *supervised classification* methods.

First, we describe two data formats with roles in data management and analysis. *List mode* simply lists all measurements made for each cell in the manner of Figure 9.2a. List mode preserves all measurement information for each cell with floating-point precision, and is often displayed in scatter plots where a dot represents one particle measurement in data space. Alternatively, *array mode* arranges data into bins that span the measurement range. As each of M measurements is obtained, the array-mode element representing the bin for that value is incremented by 1. Array mode provides an M-dimensional histogram where summing along dimensional rows yields estimates of marginal densities. Both formats are applied in the following example.

Example 9.1.1. *Flow cytometry data: A sample with a mixture of two cell types is analyzed using flow cytometry. For each of N cells counted, we record M = 3 measurements to classify each as belonging to one of two types, θ_1 or θ_2. While cell numbers are Poisson random variables, the data resulting from optical measurements made on cells are best described as trivariate normal [97]. We used published data (see Figure 6 in [57]) to visually select normal distribution parameters for simulating FSC, SSC, and FL1 flow cytometry data. The simulated experiment mixed a test sample with smaller control cells to which a single-antibody-conjugated fluorophore-reporting probe is introduced. The molecular probe binds primarily to the smaller control cells and sometimes to larger test cells. The control cells can help identify and separate regions in the (FSC, SSC) measurement plane displaying target cells free from the fluorescent probe from those displaying nonspecific target-cell binding. When nonspecific binding of target cells occurs, FSC counts associated with FL1 signals increase. In this way, we learn to recognize nonspecific binding. Simulation*

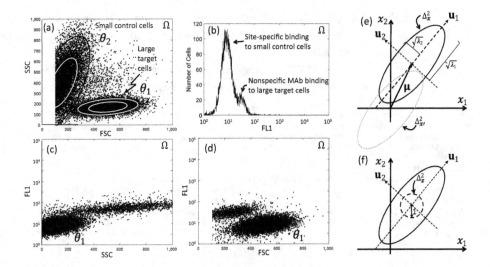

Figure 9.3 Simulated flow-cytometry scatter plots from Example 9.1.1 in planes through 3-D measurement space Ω. In (a), the one- and two-standard-deviation isocontour lines are also indicated. The lowest 10% of the forward scattering data was discarded as unreliable. (b) shows a histogram of FL1 fluorescent signals. The MATLAB code generates (a) only. (e) Eigenvalues and eigenvectors from PCA applied to a representation of the 2σ isocontour line from θ_2 in (a) (see Section 9.2). (f) Isocontour lines Δ from distributions of the original data $\mathbf{X}$ and the WPCT transformed data $\mathbf{Z}$.

results for two trivariate-normal distributions, plotted two data dimensions at a time, are displayed in Figure 9.3 as histograms and scatter plots.

In Figure 9.3a, we used list-mode simulation data of FSC and SSC pulse heights. θ_1 identifies a cluster of 10^4 large test-cell data points, while θ_2 labels a cluster of 10^4 small control-cell data points. Because we know the parent populations, we are able to indicate the one- and two-standard-deviation isocontour lines. These are Mahalanobis distances

$$\Delta_x = \sqrt{(\mathbf{x} - \boldsymbol{\mu})^\top \boldsymbol{\Sigma}_x^{-1/2} (\mathbf{x} - \boldsymbol{\mu})}$$

as defined in (3.57). A third distribution adds 900 noise samples to simulate those seen experimentally. The goal of the simulation is to identify a region in the (FSC, SSC) measurement plane (Figure 9.3a) occupied by larger unbound test cells, the true target of our measurement. As we gate regions to isolate cell components, we can observe Figure 9.3b, 9.3c, and 9.3d to see which elements change, and in this way gain confidence in target cells.

Figure 9.3b shows a small peak with higher fluorescent intensity, indicating the nonspecific binding we hope to eliminate. Figure 9.3c and 9.3d are scatter plots of FL1 versus SSC and FL1 versus FSC, respectively. Viewing these two figures along with Figure 9.3a, it is clear that the low SSC and high FSC cluster labeled θ_1 indicates the *unbound target cells we seek. At this point, a gate is applied to those data. Much of the remainder of this chapter describes methods for classifying and clustering measurement data.*

The following MATLAB *code was used to generate the data shown in Figure 9.3a. The other scatter plots may be generated by changing distribution parameters.*

```
%% Simulating 2-D random data from classes theta1 & theta2 in Ex 9.1.1 (Results for Fig. 9.3a only)
% Parameters: rj is the correlation coefficient, (mxj,myj) is mean vector for thetaj,
% and (sxj,syj) is the std vector for thetaj, where j=1,2
%%%%%%%%%%%%% *** theta 1 data *** %%%%%%%%%%%%%%%
%
close all;NN=zeros(1000);                          % NN is 2-D array mode array
r1=0.3;sx1=120;sy1=40;mx1=500;my1=170;N1=100;      % theta1 data parameters
Z=randn(N1);                                       % N1xN1 measurements
X1=sx1*Z+mx1;                                       % theta1 data along x axis...
Y1=sy1*(r1*Z+sqrt(1-r1^2)*randn(N1))+my1;          % ...along y axis
plot(X1,Y1,'k.');axis([0 1000 0 1000]);hold on     % plot list-mode data; force axis range
for j=1:N1                                          % convert list-mode data to array-mode
    for i=1:N1
        if X1(i,j)>100 && X1(i,j) < 1001 && Y1(i,j) > 1 && Y1(i,j) < 1001;
            NN(floor(Y1(i,j)),floor(X1(i,j))) = NN(floor(Y1(i,j)),floor(X1(i,j))) + 1;
        end
    end
end
% %%%%%%%%%%%%% *** theta 2 data *** %%%%%%%%%%%%%%
%
r2=0.7;sx2=120;sy2=250;mx2=150;my2=400;N2=100;      % theta2 data parameters
Z=randn(N2);                                       % N2xN2 measurements
X2=sx2*Z+mx2;                                       % theta 2 data along x axis...
Y2=sy2*(r2*Z+sqrt(1-r2^2)*randn(N2))+my2;          % ...along y axis
plot(X2,Y2,'k.')                                   % plot list-mode data
for j=1:N2                                          % convert list-mode data to array-mode
    for i=1:N2
        if X2(i,j) > 100 && X2(i,j) < 1001 && Y2(i,j) > 1 && Y2(i,j) < 1001;
            NN(floor(Y2(i,j)),floor(X2(i,j))) = NN(floor(Y2(i,j)),floor(X2(i,j))) + 1;
        end
    end
end
%
% *** Noise (mostly nonspecific binding) ***
%
r3=0.5;sx3=150;sy3=150;mx3=400;my3=350;N3=30;       % noise parameters
Z=randn(N3);                                       % N3xN3 measurements
X3=sx3*Z+mx3;                                       % noise along x axis...
Y3=sy3*(r3*Z+sqrt(1-r3^2)*randn(N3))+my3;          % ...along y axis
plot(X3,Y3,'k.');hold off                           % plot list-mode noise
for j=1:N3                                          % convert list-mode noise to array-mode
    for i=1:N3
        if X3(i,j) > 100 && X3(i,j) < 1001 && Y3(i,j) > 1 && Y3(i,j) < 1001;
            NN(floor(Y3(i,j)),floor(X3(i,j))) = NN(floor(Y3(i,j)),floor(X3(i,j))) + 1;
        end
    end
end;mN=max(max(NN));                                % find largest array element
%
%
figure;imagesc(flip(NN),[0 0.3*mN]);colormap gray   % plot array mode data. See note at end!!
% figure;plot(1:1000,sum(NN,2),'k')                  % projection of array mode data,
%                                                    % only valid when simulating FL1 data
comp1a=2*(1-r1);comp1b=2*(2-r1);                     % these are the contour criteria
comp2a=2*(1-r2);comp2b=2*(2-r2);
el=zeros(1000);                                     % ellipses as contours
for yj=1:1000                                        % find 1sigma and 2sigma contours
    for xk=1:1000                                    % (crude! ellipse1 could be more elegant)
        p1=((xk-mx1)/sx1)^2+((yj-my1)/sy1)^2 - 2*r1*(xk-mx1)*(yj-my1)/(sx1*sy1);
        p2=((xk-mx2)/sx2)^2+((yj-my2)/sy2)^2 - 2*r2*(xk-mx2)*(yj-my2)/(sx2*sy2);
        if p1 > 0.98*comp1a && p1 < 1.02*comp1a; el(yj,xk)=1; end;
        if p1 > 0.98*comp1b && p1 < 1.02*comp1b; el(yj,xk)=1; end;
        if p2 > 0.98*comp2a && p2 < 1.02*comp2a; el(yj,xk)=1; end;
```

```
        if p2 > 0.98*comp2b && p2 < 1.02*comp2b; el(yj,xk)=1; end;
    end
end
figure;imagesc(flip(el));colormap(gray);  % See following note to see why data are flipped.
% NOTES
% Since MATLAB plots image data with the origin in the top left corner, I flipped
% the vertical axis for display, effectively placing the origin in the bottom left.
% To generate Figs 9.3 b-d, values for mean, std, and corr coeff need to change.
%                   ***** Data parameters *****
% theta_1: mx1 = 500, my1 = 170, sx1 = 120, sy1 = 40,  N1 = 10^4, and r1 = 0.3
% theta_2: mx2 = 150, my2 = 400, sx2 = 120, sy2 = 250, N2 =10^4,  and r2 = 0.7
% theta_3: mx3 = 400, my3 = 350, sx3 = 150, sy3 = 150, N3 =   900, and r3 = 0.5
```

9.2   Covariance Diagonalization and Whitening Transformations

Partitioning the measurement space $\mathbf{X} = (\{x_1\}, \{x_2\}, \{x_3\}) = \{FSC, SSC, FL1\} \in \mathbb{R}^{N \times 3}$ into cell types enables classification of the nth measurement $\mathbf{x}_n$ as belonging to either class θ_1 or class θ_2. Ignoring the small noise contribution as a distinct class, we have $\Pr(\theta_1) + \Pr(\theta_2) \simeq \Pr(S) = 1$. Figure 9.3a shows that correlations between measurements x_1 and x_2 vary among the classes, making it challenging to carve up the event space in a way that minimizes classification errors for these measurements. We seek measures of class separability and the means to exploit class distinctions to correctly classify measurements when labeled data are available for classifier training, and we cluster data into groups when labeled data are not available. As shown in (3.57) and Figure 9.3e, applying a linear transformation to $\mathbf{X}$ does not change the Mahalanobis distance but does change the coordinate axes. This avenue has promise.

First, we take the time to describe *parametric classifiers*, some ad hoc and others rigorous. Each is straightforward to apply to the flow cytometry data discussed earlier. We begin with supervised methods that involve the use of *training data* having known class membership. I provide code at each step and related end-of-chapter problems that I hope prompts readers to try these methods. Principal component analysis (Section 5.1) and linear discriminant analyses (Section 9.3) are classic techniques, well understood in the data science community, but only truly understood by those who get their hands dirty with data. PCA and LDA are bricks in the foundation of modern machine learning approaches.

9.2.1   Orthonormal Transformation

Ideally, a similarity transformation (Appendix A) can be found that diagonalizes covariance matrix $\mathbf{K}$ without altering data properties (i.e., the eigenvalues are unchanged). We can achieve this using an *orthonormal transformation*. To illustrate the process, consider the 2-D data sample $\mathbf{X} = (\mathbf{x}_1, \mathbf{x}_2) \in \mathbb{R}^{N \times 2}$ drawn from population $\mathbf{X} = (\{x_1\}, \{x_2\})$ and shown diagrammatically in Figure 9.3e. One Mahalanobis isocontour Δ_x represents the small control-cell data in Example 9.1.1 labeled θ_2.

First, we shift the coordinate system to the origin for $\mathbf{X}' = (\mathbf{x}_1', \mathbf{x}_2')$, where $\mathbf{x}_1' = \mathbf{x}_1 - \mu_1$ and $\mathbf{x}_2' = \mathbf{x}_2 - \mu_2$ assuming a wide-sense stationary (WSS) process.

We are searching for vector $\mathbf{x}'_j$ that maximizes $\Delta_{x'}$ subject to $\mathbf{x}'^T_j \mathbf{x}'_j = 1$ to avoid scaling the eigenvalues. This is a *constrained optimization* problem (see Appendix E),[1]

$$L(\mathbf{x}'_j, \beta) = f(\mathbf{x}'_j) + \beta q(\mathbf{x}'_j) = \Delta_z(\mathbf{x}'_j) + \beta(\mathbf{x}'^T_j \mathbf{x}'_j - 1)$$

$$\nabla_{\mathbf{x}'_j, \beta} L = \frac{\partial}{\partial \mathbf{x}'_j, \beta} \left(\mathbf{x}'^T_j \mathbf{K}^{-1} \mathbf{x}'_j - \beta(\mathbf{x}'^T_j \mathbf{x}'_j - 1) \right) = \begin{cases} 2\mathbf{K}^{-1}\mathbf{x}'_j - 2\beta \mathbf{x}'_j = 0 \\ \mathbf{x}'^T_j \mathbf{x}'_j = 1 \end{cases},$$

where β is a Lagrange multiplier. Rearranging terms, we find

$$\mathbf{K}\mathbf{x}'_j = \frac{1}{\beta}\mathbf{x}'_j \quad \text{and} \quad \det\left(\mathbf{K} - \frac{1}{\beta}\mathbf{I}\right) = 0, \tag{9.1}$$

which is the *characteristic equation* for $\mathbf{K}$. The vector $\mathbf{x}'_j$ that satisfies this equation must be an eigenvector of the data $\mathbf{u}$, and $1/\beta$ is the corresponding eigenvalue λ. If $\mathbf{U}$ is the eigenvector matrix (eigenvectors oriented as columns) and Λ is the diagonal eigenvalue matrix, the *canonical form* of (9.1) that includes orthonormal eigenvectors from (6.2) is

$$\mathbf{K}\mathbf{U} = \mathbf{U}\Lambda, \quad \text{where} \quad \mathbf{U}^T\mathbf{U} = \mathbf{I}, \quad \mathbf{U} = \begin{pmatrix} U_{11} & U_{12} \\ U_{21} & U_{22} \end{pmatrix} \quad \text{and} \quad \Lambda = \begin{pmatrix} \lambda_1 & 0 \\ 0 & \lambda_2 \end{pmatrix}. \tag{9.2}$$

The indices in element U_{jk} denote data dimension j and eigenvector k. Covariance matrix symmetry ensures the eigenvalues are real, and if $\lambda_1 \neq \lambda_2$, the eigenvectors are orthogonal. Great! The *principal component transformation (PCT)*, from (5.3) achieves our goal when the $\mathbf{X} \in \mathbb{R}^{N \times 2}$ data matrix is transformed by $\mathbf{U} \in \mathbb{R}^{2 \times 2}$ to give $\mathbf{Y} \in \mathbb{R}^{N \times 2}$:

$$\mathbf{Y} = \mathbf{X}\mathbf{U} = \begin{pmatrix} x_{11}u_{11} + x_{12}u_{21} & x_{11}u_{12} + x_{12}u_{22} \\ x_{21}u_{11} + x_{22}u_{21} & x_{21}u_{12} + x_{22}u_{22} \\ x_{31}u_{11} + x_{32}u_{21} & x_{31}u_{12} + x_{32}u_{22} \\ \vdots & \vdots \\ x_{N1}u_{11} + x_{N2}u_{21} & x_{N1}u_{12} + x_{N2}u_{22} \end{pmatrix} = \begin{pmatrix} \mathbf{y}_1 & \mathbf{y}_2 \end{pmatrix}. \tag{9.3}$$

The transformed data matrix $\mathbf{Y}$ has a diagonal covariance matrix with unequal variances. The coordinates of $\mathbf{Y}$ are rotated to the eigenvectors $\mathbf{u}_1$ and $\mathbf{u}_2$ as shown in Figure 9.3e; i.e., $\mathbf{y}_j = \mathbf{X}\mathbf{u}_j$. The first principal component $\mathbf{u}_1$ corresponds to the largest eigenvalue λ_1, and therefore is oriented in the plane along the direction of the largest variability in $\mathbf{X}$. Because the eigenvectors of covariance matrices are orthogonal, we know that $\mathbf{u}_2$ is perpendicular to $\mathbf{u}_1$. Increasing the number of measurement dimensions $J > 2$, we would find that $\mathbf{u}_2$ is oriented along an axis orthogonal to $\mathbf{u}_1$ with the second greatest variability.

The columns of $\mathbf{X}$ become infinitely long for continuous measurements. In turn, the matrix product in (9.3) becomes an array of integrals, and the orthonormal expansion is called a *Karhunen–Loève transformation* or *Hotelling transformation* [42].

---

[1] Methods for completing matrix derivatives are summarized in Appendix A.

9.2.2   Whitening Transformation

PCT is an affine transformation of the data; i.e., it preserves points, lines, and planes. Thus, it comes as no surprise that, ignoring coordinate translations, $\Delta_x = \Delta_{x'} = \Delta_y$, as seen in Figure 9.3e. A *whitening transformation* further converts the diagonal covariance matrix of the transformed data to an identity matrix by dividing each of the J semi-major-minor axes by the square root of its corresponding eigenvalue (see Figure 9.3e and 9.3f). The whitening process is coupled to diagonalization using the *whitened principal components transform (WPCT)*,

$$\mathbf{Z} = \mathbf{XU\Lambda}^{-1/2}. \qquad \text{(WPCT)} \qquad\qquad (9.4)$$

WPCT combines PCA with a whitening transformation. It is straightforward to show that sample variances are scaled to 1. The sample covariance matrix of $\mathbf{Z}$ is

$$\mathbf{K}_z = \frac{1}{(N-1)}\mathbf{Z}^\top\mathbf{Z} = \mathbf{\Lambda}^{-1/2}\mathbf{U}^\top\left(\frac{1}{(N-1)}\mathbf{X}^\top\mathbf{X}\right)\mathbf{U\Lambda}^{-1/2}$$
$$= \mathbf{\Lambda}^{-1/2}\mathbf{\Lambda}\mathbf{\Lambda}^{-1/2} = \mathbf{I},$$

showing $\mathbf{Z} \sim \mathcal{N}(\mathbf{0}, \mathbf{I})$. By itself, whitening is not an orthonormal transformation, $\mathbf{\Lambda}^{-1/2}\mathbf{U}^\top\mathbf{U\Lambda}^{-1/2} = \mathbf{\Lambda}^{-1} \neq \mathbf{I}$, so properties of the data change as the eigenvalues are individually scaled. However, properties of all data in the space change in the same way. We may choose to whiten the multivariate data of one class (equalize the variances) to then decorrelate another class also in Ω, without recorrelating the first.

9.2.3   Data Simulation

Transformations that diagonalize and whiten multivariate data may also be used to simulate data with known properties [37]. Suppose we wish to simulate N samples having J measurements each, $\mathbf{X} \sim \mathcal{N}(\boldsymbol{\mu}, \mathbf{K})$ where $\mathbf{X} \in \mathbb{R}^{N \times J}$. Specifying $\boldsymbol{\mu}$ and $\mathbf{K}$, we are able to compute the eigenvalue $\mathbf{\Lambda}$ and eigenvector $\mathbf{U}$ matrices for $\mathbf{K}$. Simulating random matrix $\mathbf{Z} \in \mathbb{R}^{N \times J}$ using a standard normal generator, e.g., z=randn(N,J), we find the data matrix $\mathbf{X}$ using

$$\mathbf{X} = \mathbf{Z\Lambda}^{1/2}\mathbf{U}^\dagger + \boldsymbol{\mu}. \qquad\qquad (9.5)$$

MATLAB supplies a MVN random number generator mvnrnd that is very easy to use.

Example 9.2.1. *Demonstrate WPCT: Generate the bivariate data matrix* $\mathbf{X} = (\{FSC\},$ $\{SSC\}) \sim p(\mathbf{x}|\theta_2)$ *displayed in Figure 9.3a using* mvnrnd. *For convenience, subtract the mean vector throughout:* $\mathbf{X}' = (\mathbf{x}_1 - \boldsymbol{\mu}_1, \mathbf{x}_2 - \boldsymbol{\mu}_2)$. *Apply PCT and WPCT and display the results to show the diagonalizing and whitening effects of the transformations. Compute the sample covariances at each step and discuss the findings.*

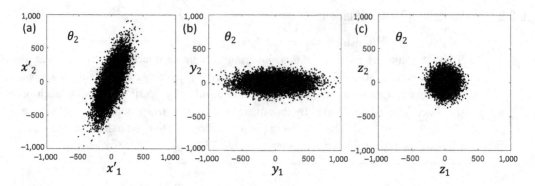

Figure 9.4 (a) Data from class θ_2 in Figure 9.3a are reproduced on the (x_1', x_2') axes. (b) The result of applying the principal component transform (PCT) (9.3) to $\mathbf{X}'$ are plotted on the (y_1, y_2) axes. (c) The result of applying the whitened principal component transform (WPCT) (9.4) to $\mathbf{X}'$. The results in (c) are scaled by 100 to keep the plot ranges comparable.

Solution

The solution code is given here and the results are reported in Figure 9.4. I suggested using a zero mean data vector for convenience to keep the pdf centered at the origin. Note that the sample covariances of the transformed data are diagonal and whitened within about 5%, depending on sample size. Remembering that sample covariance matrices are random variables, we find the covariances are less than 5% of the variances and the variances for the WPCT outputs are within a few percent of each other. Try it! You will need to adjust the plot axes for fair visual comparisons. The WPCT results in Figure 9.4c were multiplied by 100 to keep the scale for Z *roughly equivalent to that for* X *and* Y*.*

```
%%%%%%%%%%%% Solutions to Example 9.2.1.  Results are plotted in Fig. 9.4 %%%%%%%%%%%%
N=10^4;mx1=150;my1=400;sx1=120;sy1=250;r1=0.7;      % initialize para's for theta 2
% m=ones(N,2);m(:,1)=m(:,1)*mx1;m(:,2)=m(:,2)*my1;  % mean vector Nx2: not used here
m=zeros(N,2);                                        % use this for the mean vector
K=[sx1^2 r1*sy1*sx1;r1*sx1*sy1 sy1^2];               % 2x2 population cov matrix
X=mvnrnd(m,K);                                        % random data generated
Kx=cov(X);                                            % 2x2 sample cov matrix for X'
plot(X(:,1),X(:,2),'k.');axis square
[U,D]=eig(K);                                         % choose population cov to find U
figure;Y=X*U;plot(Y(:,2),Y(:,1),'k.');axis square    % invert axis to match eigenvalue order!!
Ky=cov(Y);                                            % sample cov after PCT
figure;Z=100*X*U/sqrt(D);plot(Z(:,2),Z(:,1),'k.');axis square
Kz=cov(Z);                                            % sample cov after WPCT
```

9.2.4   Simultaneous Diagonalization

Fukunaga [37] describes a method for simultaneously diagonalizing two arbitrary distributions. Consider again the two-class bivariate densities $p(\mathbf{x}|\theta_1)$ and $p(\mathbf{x}|\theta_2)$ from Figure 9.3a that are reproduced in Figure 9.5a. Fukunaga shows that diagonalizing the

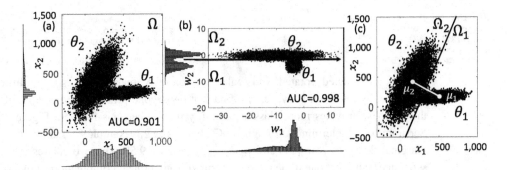

Figure 9.5 (a) Noise-free scatter plots of **X** from Figure 9.3a. (b) The same distributions after simultaneous diagonalization via (9.6). The operator whitens $\mathbf{K}_1$ and diagonalizes $\mathbf{K}_2$, which improves separability along the w_2 axis compared to x_1 or x_2. The histograms approximate marginal densities. Areas under the ROC curves (AUC) were generated from marginal densities along the x_2 and w_2 axes. Identifying $w_2 = -2$ as the minimum error point (the point where the marginal densities cross), a line is drawn along $(w_1, -2)$ and the data transformed back in (c) to show how data classes θ_1 and θ_2 are partitioned into decision regions Ω_1 and Ω_2 within (x_1, x_2). The white line illustrates d, the distance between mean vectors (see Example 9.2.2).

ratio of covariances $\mathbf{K}_1^{-1}\mathbf{K}_2$ from these distributions and properly scaling the results effectively applies a WPCT to $\mathbf{K}_1$ and PCT to $\mathbf{K}_2$.

The procedure is to find eigenvector $\mathbf{U}_s$ and eigenvalue $\mathbf{\Lambda}_s$ matrices via $\mathbf{K}_1^{-1}\mathbf{K}_2\mathbf{U}_s = \mathbf{U}_s\mathbf{\Lambda}_s$ [see (9.2)]. The scaling factor $\mathbf{S} = \mathbf{U}_s^{\dagger}\mathbf{K}_1\mathbf{U}_s$ is applied to ensure the eigenvectors are orthogonal, since we have no reason to expect $\mathbf{K}_1^{-1}\mathbf{K}_2$ to be symmetric. The simultaneous transformation converting $\mathbf{X}$ into $\mathbf{W}$ is

$$\mathbf{W} = \mathbf{X}\mathbf{U}_s\mathbf{S}^{-1/2}, \qquad \text{(Simultaneous diagonalization)} \qquad (9.6)$$

where $\mathbf{W}$ is the transformed data matrix. The code for applying the transformation is shown here, and the results from implementing the code in this example are found in Figure 9.5.

```
%%%%%%%%%%%%%%% Implementation of (9.6) with results displayed in Fig. 9.5 %%%%%%%%%%%%%%%%%%%%%%%%
close all;N=10^4;                                  % 10^4 measurements for both dist's
mx1=500;my1=170;sx1=120;sy1=40;r1=0.3;             % initialize para's for theta 1
mx2=150;my2=400;sx2=120;sy2=250;r2=0.7;            % initialize para's for theta 2
m1=ones(N,2);m1(:,1)=m1(:,1)*mx1;m1(:,2)=m1(:,2)*my1;  % mean vector for theta 1
m2=ones(N,2);m2(:,1)=m2(:,1)*mx2;m2(:,2)=m2(:,2)*my2;  % mean vector for theta 1
K1=[sx1^2 r1*sy1*sx1;r1*sx1*sy1 sy1^2];            % 2x2 pop cov matrix for theta 1
K2=[sx2^2 r2*sy2*sx2;r2*sx2*sy2 sy2^2];            % 2x2 pop cov matrix for theta 2
X1=mvnrnd(m1,K1);X2=mvnrnd(m2,K2);                 % random data for theta 1 and 2
Kx1=cov(X1);Kx2=cov(X2);                           % both sample cov matrices
figure;plot(X1(:,1),X1(:,2),'r.');axis square;hold on
plot(X2(:,1),X2(:,2),'k.');hold off
K12=K1\K2;                                         % compute cov ratio: inv(Kx1)*Kx2
[Us,Ds]=eig(K12);                                  % compute eigen-system
S=abs(sqrt(Us'*K1*Us));                            % compute scaling value
figure;W1=X1*Us/S;plot(W1(:,2),W1(:,1),'r.');      % transform and scale theta 1
axis equal;axis square;hold on
```

```
W2=X2*Us/S;plot(W2(:,2),W2(:,1),'k.');hold off          % transform and scale theta 2
KW=[cov(W1) cov(W2)];                                   % sample cov in case you are curious
X=[X1;X2];W=[W1;W2];                                    % assemble for scatter histogram
figure;scatterhist(X(:,1),X(:,2));
figure;scatterhist(W(:,2),W(:,1));                      % need to rescale axes to be equal
```

$\mathbf{W}$ is a translated, rotated, and scaled version of $\mathbf{X}$, making values more difficult to physically interpret. However, we find that the second principal component, along the w_2 axis, provides more class separability than either axis for $\mathbf{X}$ in Figure 9.5a. Separability is determined through ROC analysis (e.g., Example 8.9) applied to the marginal densities $p(w_2|\theta_1)$ and $p(w_2|\theta_2)$ in Figure 9.5b. The two classes are well separated, with just one dimension w_2 formed from a linear combination of measurements (x_1, x_2). The minimum-total-error classification strategy using training data is to set the threshold $w_t \triangleq w_2 = -2$. Then assign data for which $w_2 > w_t$ to class θ_2 and those for which $w_2 \leq w_t$ to θ_1.

We can transform the data, along with the classification line at $w_2 = -2$, back to the original coordinate axes by applying the inverse operator to (9.6) using $\mathbf{X} = \mathbf{W}(\mathbf{U}_s \mathbf{S}^{-1/2})^{-1} = \mathbf{W}\mathbf{S}^{1/2}\mathbf{U}_s^{-1}$. Since eigenvector matrix $\mathbf{U}_s$ is not orthogonal, it is important to apply $\mathbf{U}_s^{-1}$, rather than $\mathbf{U}_s^{\top}$. The code for this inverse operation is shown here, and the result appears in Figure 9.5c.

```
%%%%%%%%%%%%%%%%%%%%%%%%%%%%%%%  Continuing from the preceding code....  %%%%%%%%%%%%%%%%%%%%%%%%%%%%%%%%
y=[-40:0.1:20]';N=length(y);          % define y = w2 axis
XW1=W1*S'/Us;XW2=W2*S'/Us;              % inverse operation applied to theta 1 & theta 2 data separately
figure;plot(XW1(:,1),XW1(:,2),'r.');hold on;axis square
plot(XW2(:,1),XW2(:,2),'k.')
q=[-2*ones(N,1) y];                    % generate threshold line at w1=-2 for all w2 as points
b=q*S'/Us;                             % apply inverse operation to the line
plot(b(:,1),b(:,2),'b.');hold off
```

The inverse operator preserves the linearity of the classification line. We examine formal approaches to linear classifiers next. First, let's examine quantitative measures of class separability.

9.2.5   Quantifying Class Separability

The inherent separability of C classes distributed in a J-variate data space (J measurements per sample) is expressed by the scalar-valued *Hotelling trace* [10, 26, 56],

$$J_h = tr(\mathbf{S}_w^{-1}\mathbf{S}_b) = \sum_{i=1}^{C}(\mathbf{S}_w^{-1}\mathbf{S}_b)_{ii}, \qquad \text{(Hotelling trace)} \qquad (9.7)$$

where $\mathbf{S}_w$ is the average within-class *scatter matrix*, $\mathbf{S}_w = \sum_{i=1}^{C}\mathbf{K}_i \Pr(\theta_i)$, and $\mathbf{S}_b$ is the between-class scatter matrix, $\mathbf{S}_b = \sum_{i=1}^{C}(\boldsymbol{\mu}_i - \bar{\boldsymbol{\mu}})(\boldsymbol{\mu}_i - \bar{\boldsymbol{\mu}})^{\top} \Pr(\theta_i)$. The quantity $\bar{\boldsymbol{\mu}} = \sum_{i=1}^{C}\boldsymbol{\mu}_i \Pr(\theta_i)$ is the average of the class mean vectors.

J_h measures class separability as the square of distribution means per common covariance for any multivariate distribution. It is the C-class, J-dimensional generalization of detectability indices d'^2 and d_a^2 from (8.11). The latter indices measure separability for two class, univariate measurements (see Figure 8.7c). J_h parallels d'^2

when population parameters are used to compute the mean and covariance matrices; it parallels d_a^2 when $\mathbf{S}_w$ and $\mathbf{S}_b$ are estimated from the sample statistics. The relationship between J_h and the area under the ROC curve (AUC) is [10, 56]

$$\text{AUC} = \frac{1}{\sqrt{2\pi}} \int_{-\infty}^{\sqrt{2J_h}} dt \, e^{-t^2/2}, \qquad (9.8)$$

which is calculated in MATLAB using $0.5*(1+\text{erf}(\text{sqrt}(2*J)))$.[2]

For the *population statistics* of $\mathbf{X}$ in Example 9.2.1, where $C = J = 2$, I found $J_h = 4.83$. Using the *sample statistics* of $\mathbf{X}$, the result is $J_h = 4.77$. Finally, $J_h = 4.86$ when analyzing the sample statistics of $\mathbf{W}$, demonstrating the invariance of the Hotelling trace to linear transformations [26]. With J_h near 5, the equivalent d' value is approximately 2.2, so it is no surprise that AUC $\simeq 1$. As shown in Figure 9.5b, AUC is indeed much closer to 1 than the same data in the original measurement coordinates, shown in Figure 9.5a. Principal component analysis makes it easier to achieve classification performance near the intrinsic separability of the data than is possible using either of the original data axes. While J_h applies to multidimensional data and any number of classes $C \geq 2$, it is important to note that the comparison between J_h and AUC makes sense only for two-class data.

Example 9.2.2. *Comparing J_h and AUC: Show how J_h via (9.7) and AUC via (9.8) change as class mean separation increases from zero to the values displayed in Figure 9.5a and 9.5c. The distance between means is $d = \sqrt{(\mu_{x1} - \mu_{x2})^2 + (\mu_{y1} - \mu_{y2})^2}$.*

Solution

The solution is given in the code shown here, and the results are plotted in Figure 9.6. The data in this example show that AUC $\simeq 1$ for $J_h > 1$. The covariance matrices for each class can have very different variances and arbitrary correlation coefficients. Consequently, the results in Figure 9.6 are specific to this example and generally do not provide the intuition that d'^2 does for univariate data.

```
%%%%%%%%%%% Solution to the problem from Example 9.2.2.  Results are plotted in Fig. 9.6 %%%%%%%%%%%%
d=zeros(1,34);slope=(150-400)/(500-170);J=d;AUC=d;     % initialization for these particular data
mx1=500;my1=150;sx1=120;sy1=40;r1=0.3;                 % initialize para's for theta 1
mx2=170;my2=400;sx2=120;sy2=250;r2=0.7;                % initialize para's for theta 2
K1=[sx1^2 r1*sy1*sx1;r1*sx1*sy1 sy1^2];                % 2x2 pop cov matrix for theta 1
K2=[sx2^2 r2*sy2*sx2;r2*sx2*sy2 sy2^2];                % 2x2 pop cov matrix for theta 2
for j=1:34                             % use population statistics and vary means over a range
    mp1=mx2+(j-1)*10;mq1=my2+(mp1-mx2)*slope;          % vary mu1 from mu2
    d(j)=sqrt((mx2-mp1)^2+(my2-mq1)^2);                % compute distance between 2D mean vectors
    mb=[(mp1+mx2)/2;(mq1+my2)/2];                      % find average of class mean vectors
    mm1=([mp1;mq1]-mb)*([mp1;mq1]-mb)';
    mm2=([mx2;my2]-mb)*([mx2;my2]-mb)';
    Sb=(mm1+mm2)/2;                                    % between class scatter matrix
    Sw=(K1+K2)/2;                                      % within class scatter matrix
    J(j)=trace(Sw\Sb);                                 % Hotelling trace
    AUC(j)=0.5*(1 + erf(sqrt(2*J(j))));                % area under ROC curve
end
figure;plot(d,J,d,AUC)
```

---

[2] The error function in MATLAB is $\text{erf}(x) = (2/\sqrt{\pi}) \int_0^x dt \, \exp(-t^2)$, which necessitates that the constants appear in the code.

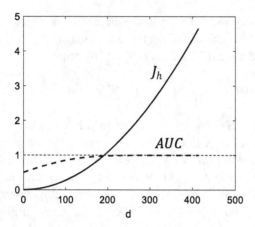

Figure 9.6 Plot of Hotelling trace J_h from (9.7) versus distance between class-mean vectors d for the data in Figure 9.5a. J_h is computed at various points as the mean for θ_1 data, μ_1, is moved along the white line in Figure 9.5c from μ_2, where $d = 0$, to μ_1, seen in the scatter plots at $d = 414.0$. Since J_h is invariant under nonsingular linear transformations, these plotted values apply to all three plots in Figure 9.5a, 9.5b, and 9.5c. Using (9.8), AUC values are found for the same range of d.

The *Bhattacharyya distance* [37] is another important measure of class separability:

$$B = \frac{1}{8}(\mu_2 - \mu_1)^\top S_w^{-1}(\mu_2 - \mu_1) + \frac{1}{2} \ln \left(\frac{\det(S_w)}{\sqrt{\det(K_1)\det(K_2)}} \right) = B(\mu) + B(K),$$

(9.9)

where S_w is the within-class scatter matrix. Unlike the Hotelling trace, the Bhattacharyya distance is based on a normal assumption and applies only to two-class problems. Its utility lies in its two terms that can distinguish MVN class separability caused by differences in means versus differences in covariance. For the data in Figure 9.5a, $B = 2.94$ with $B(\mu) = 2.32$ attributed to differences between class means and $B(K) = 0.62$ to class covariances.

9.3   Discriminant Analysis

Once we simultaneously diagonalized both covariances of 2-D, two-class data θ_1 and θ_2 in Figure 9.5, we found a way to partition the measurement plane into decision regions Ω_1 and Ω_2. We reduced the two data dimensions to one, and then moved the threshold along the w_2 axis. Insofar as these test data are representative of future experiments, we have an ad hoc method for classifying data. This partitioning of the measurement space achieves a classification performance AUC $= 0.998$ that is close to the best possible value predicted by the Hotelling trace $J_h = 4.8$ having an equivalent AUC of 1.0. Figure 9.5b showed that separating these 2-D data was intrinsically a 1-D task. So, by setting decision threshold w_t at a point where approximate marginal densities $p(w_2|\theta_1) = p(w_2|\theta_2)$, we minimized the net error (see Problem 8.3), and minimizing total error is a valid criteria. However, we may need to consider other criteria when the risks associated with decision errors are unequal. This section discusses *discriminant analysis* as a statistical method for partitioning measurement spaces into decision regions Ω_i based on error rates and costs. The approach described here borrows from the classic treatment by van Trees [110].

9.3.1   Quantifying Classification Errors and Risk

The net *classification error probability* for a binary decision is given by

$$\varepsilon = \Pr(\theta_1)\Pr(D_2|\theta_1) + \Pr(\theta_2)\Pr(D_1|\theta_2). \tag{9.10}$$

$\Pr(\theta_i)$ are prior probabilities of data from class θ_i and $\Pr(D_j|\theta_i)$ are decision probabilities defined in (8.10). Errors occur when $j \neq i$.

There are *costs* associated with making decisions, both correct and incorrect. Let $\mathcal{C}_{ji}$ be the cost of deciding data belong to the jth class when, in fact, the data were obtained from the ith class of an object. All medical decisions incur costs, such as lab and physician-services, but additional costs are associated with incorrect decisions, such as additional patient testing and malpractice insurance. It is reasonable to assume that the cost of diagnostic errors is never less than the cost of correct diagnoses; viz.,

$$\mathcal{C}_{ji}(j \neq i) \geq \mathcal{C}_{ji}(j = i). \tag{9.11}$$

Bayes' risk $\mathfrak{r}$ assumed by the decision maker may be defined by a combination of error probabilities and costs [110]:

$$\mathfrak{r} = \Pr(\theta_1)\Big(\mathcal{C}_{11}\Pr(D_1|\theta_1) + \mathcal{C}_{21}\Pr(D_2|\theta_1)\Big)$$
$$+ \Pr(\theta_2)\Big(\mathcal{C}_{12}\Pr(D_1|\theta_2) + \mathcal{C}_{22}\Pr(D_2|\theta_2)\Big). \tag{9.12}$$

Let the total event space in our flow cytometry example be $\Omega = \Omega_1 + \Omega_2$, which is the 2-D measurement plane (x_1, x_2) in Figure 9.5a and 9.5c and (w_1, w_2) in Figure 9.5b. Noting that

$$\Pr(D_j|\theta_i) = \int_{\Omega_j} d\mathbf{x}\, p(\mathbf{x}|\theta_i) \quad \text{and} \quad \int_{\Omega} d\mathbf{x}\, p(\mathbf{x}|\theta_i) = 1,$$

the risk equation in (9.12) becomes

$$\mathfrak{r} = \Pr(\theta_1)\left(\mathcal{C}_{11}\int_{\Omega_1} d\mathbf{x}\, p(\mathbf{x}|\theta_1) + \mathcal{C}_{21}\int_{\Omega_2} d\mathbf{x}\, p(\mathbf{x}|\theta_1)\right)$$
$$+ \Pr(\theta_2)\left(\mathcal{C}_{12}\int_{\Omega_1} d\mathbf{x}\, p(\mathbf{x}|\theta_2) + \mathcal{C}_{22}\int_{\Omega_2} d\mathbf{x}\, p(\mathbf{x}|\theta_2)\right). \tag{9.13}$$

The influence of the decision is seen by selecting the integration region, Ω_j. Eliminating Ω_2 from the equation, we have

$$\tau = \Pr(\theta_1)\left(\mathfrak{C}_{11}\int_{\Omega_1} d\mathbf{x}\, p(\mathbf{x}|\theta_1) + \mathfrak{C}_{21}\int_{\Omega-\Omega_1} d\mathbf{x}\, p(\mathbf{x}|\theta_1)\right)$$

$$+ \Pr(\theta_2)\left(\mathfrak{C}_{12}\int_{\Omega_1} d\mathbf{x}\, p(\mathbf{x}|\theta_2) + \mathfrak{C}_{22}\int_{\Omega-\Omega_1} d\mathbf{x}\, p(\mathbf{x}|\theta_2)\right)$$

$$= \Pr(\theta_1)\mathfrak{C}_{21}\int_{\Omega} d\mathbf{x}\, p(\mathbf{x}|\theta_1) - \int_{\Omega_1} d\mathbf{x}\, \Pr(\theta_1)(\mathfrak{C}_{21} - \mathfrak{C}_{11})\, p(\mathbf{x}|\theta_1)$$

$$+ \Pr(\theta_2)\mathfrak{C}_{22}\int_{\Omega} d\mathbf{x}\, p(\mathbf{x}|\theta_2) - \int_{\Omega_1} d\mathbf{x}\, \Pr(\theta_2)(\mathfrak{C}_{22} - \mathfrak{C}_{12})\, p(\mathbf{x}|\theta_2)$$

$$= \Pr(\theta_1)\mathfrak{C}_{21} + \Pr(\theta_2)\mathfrak{C}_{22} \tag{9.14}$$

$$+ \int_{\Omega_1} d\mathbf{x}\left(\left[\Pr(\theta_2)(\mathfrak{C}_{12} - \mathfrak{C}_{22})\, p(\mathbf{x}|\theta_2)\right] - \left[\Pr(\theta_1)(\mathfrak{C}_{21} - \mathfrak{C}_{11})\, p(\mathbf{x}|\theta_1)\right]\right).$$

Let's consider $p(\mathbf{x}|\theta_1)$ to be the pdf for "null-state" or on-target data. Then $p(\mathbf{x}|\theta_2)$ is the pdf for "positive-state" or off-target data so that, e.g., $\Pr(D_2|\theta_1)$ is a false-positive decision probability. The first two terms in (9.14) are therefore the fixed costs associated with false-positive and true-positive decisions, respectively.

The remaining integral is over region Ω_1 only. We can be sure both of the two terms in the integrand are positive because of (9.11). At measurement points where

$$\left[\Pr(\theta_2)(\mathfrak{C}_{12} - \mathfrak{C}_{22})\, p(\mathbf{x}|\theta_2)\right] \underset{\theta_1}{<} \left[\Pr(\theta_1)(\mathfrak{C}_{21} - \mathfrak{C}_{11})\, p(\mathbf{x}|\theta_1)\right], \tag{9.15}$$

the integral (and therefore the risk) is reduced because those points belong to θ_1.

Conversely, at measurement points where

$$\left[\Pr(\theta_2)(\mathfrak{C}_{12} - \mathfrak{C}_{22})\, p(\mathbf{x}|\theta_2)\right] \overset{\theta_2}{>} \left[\Pr(\theta_1)(\mathfrak{C}_{21} - \mathfrak{C}_{11})\, p(\mathbf{x}|\theta_1)\right], \tag{9.16}$$

the integral increases. Because risk increases, we must conclude that those points belong to class θ_2. The decision is indicated by the class label positioned above or below the inequality.

9.3.2 Bayes' Decision Rule

Bayes' test for minimum error is given by a comparison of posterior probabilities [110],

$$\Pr(\theta_2|\mathbf{x}) \underset{\theta_1}{\overset{\theta_2}{\gtrless}} \Pr(\theta_1|\mathbf{x}). \tag{9.17}$$

For example, if the probability of θ_1 given $\mathbf{x}$ is larger than that for θ_2, decide $\mathbf{x}$ is a member of class θ_1. Applying Bayes' theorem from Section 2.5 to continuous MVN variables, we can combine the probabilities for the jth class with the *likelihood function* $p(\mathbf{x}|\theta_i)$ and the *mixture density* $p(\mathbf{x})$ function (3.48) as follows:

$$\Pr(\theta_i|\mathbf{x})\, p(\mathbf{x}) = p(\mathbf{x}|\theta_i)\, \Pr(\theta_i). \tag{9.18}$$

9.3.3   Likelihood Ratio Test

Equations 9.17 and 9.18 can be combined to form the *likelihood ratio test* for making classification decisions with minimum error

$$\frac{p(\mathbf{x}|\theta_2)}{p(\mathbf{x}|\theta_1)} \overset{\theta_2}{\underset{\theta_1}{\gtrless}} \frac{\Pr(\theta_1)}{\Pr(\theta_2)}. \qquad \text{(Minimum error)} \qquad (9.19)$$

Here we assume mixture densities are the same for each class. Equation (9.19) is an example of a *Bayes' classifier* [37].

Further, if the prior probabilities for each class are also equal, then *minimum error* is achieved along a line (for bivariate normal densities) where $p(\mathbf{x}|\theta_1) = p(\mathbf{x}|\theta_2)$. For example, recall that after simultaneously diagonalizing the covariance matrices in Figure 9.5b where the priors were equal, and finding that the first principal component was most discriminating, we selected a minimum-error boundary line to separate Ω_1 and Ω_2 at $w_t = w_2$, where $p(w_2|\theta_1) = p(w_2|\theta_2) \ \forall \ w_1$. Because simultaneous diagonalization modifies all of the data eigenvalues uniformly, the error probability remains unchanged even as we transform back to the original measurement space. We did just that in Figure 9.5c to regain physical intuition regarding the measurement values and retained a linear discriminant function.

A more general likelihood ratio minimizes total risk by considering costs. Combining (9.15) and (9.16), we find

$$\frac{p(\mathbf{x}|\theta_2)}{p(\mathbf{x}|\theta_1)} \overset{\theta_2}{\underset{\theta_1}{\gtrless}} \frac{\Pr(\theta_1)(\mathfrak{C}_{21} - \mathfrak{C}_{11})}{\Pr(\theta_2)(\mathfrak{C}_{12} - \mathfrak{C}_{22})}, \qquad \text{(Minimum risk)} \qquad (9.20)$$

which reduces to the minimum error condition in (9.19) when the costs associated with decision errors are equal. For bivariate normal data, (9.20) becomes

$$\left(\det\left(\frac{\mathbf{K}_1}{\mathbf{K}_2}\right)\right)^{1/2} e^{-\frac{1}{2}\left[(\mathbf{x}-\mu_2)^\top \mathbf{K}_2^{-1}(\mathbf{x}-\mu_2)-(\mathbf{x}-\mu_1)^\top \mathbf{K}_1^{-1}(\mathbf{x}-\mu_1)\right]} \overset{\theta_2}{\underset{\theta_1}{\gtrless}} \frac{\Pr(\theta_1)(\mathfrak{C}_{21} - \mathfrak{C}_{11})}{\Pr(\theta_2)(\mathfrak{C}_{12} - \mathfrak{C}_{22})}.$$

The right side of (9.20), which is a constant with respect to $\mathbf{x}$, defines a new decision boundary between regions Ω_1 and Ω_2, that minimizes decision risks.

9.3.4   Quadratic Discriminant

Implementing this rule exactly requires detailed information about class priors and costs. Absent that information, we can set constant threshold t, which is systematically varied over some range as Type I and Type II errors are estimated, so that an ROC curve can be generated. Also, any monotonic transformation of (9.20) is another likelihood ratio with equal performance [110]. Hence, we consider the *log-likelihood ratio* $L(\mathbf{x})$ to simplify the analysis. From (9.20),

$$\mathsf{L}(\mathbf{x}) = \ln \frac{p(\mathbf{x}|\theta_2)}{p(\mathbf{x}|\theta_1)} \underset{\theta_1}{\overset{\theta_2}{\gtrless}} t \tag{9.21}$$

$$= (\mathbf{x} - \boldsymbol{\mu}_1)^\top \mathbf{K}_1^{-1}(\mathbf{x} - \boldsymbol{\mu}_1) - (\mathbf{x} - \boldsymbol{\mu}_2)^\top \mathbf{K}_2^{-1}(\mathbf{x} - \boldsymbol{\mu}_2) + \ln \frac{\det \mathbf{K}_1}{\det \mathbf{K}_2} \underset{\theta_1}{\overset{\theta_2}{\gtrless}} t,$$

where the constants are included in t. This expression is a *quadratic discriminant function* [56] because each data-dependent term depends on the square of $\mathbf{x}$. It is not easy to solve (9.20) for $\mathbf{x}$ without assumptions. Numerical approaches [105], including optimization methods, offer solutions. A more useful discriminant function is linear in the data.

9.3.5   Linear Discriminant

If class distributions have "similar" covariance matrices, we may simplify (9.21) by substituting the within-class scatter matrix from (9.7) in place of the covariance matrices, where $\mathbf{S}_w = \Pr(\theta_1)\mathbf{K}_1 + \Pr(\theta_2)\mathbf{K}_2$. This gives (see Problem 9.2)

$$\mathsf{L}(\mathbf{x}) \simeq \mathbf{a}^\top \mathbf{x} \underset{\theta_1}{\overset{\theta_2}{\gtrless}} b, \quad \text{and at the threshold,} \quad \sum_{j=1}^{J} a_j x_j = b, \tag{9.22}$$

where $\mathbf{a}^\top = 2(\boldsymbol{\mu}_2 - \boldsymbol{\mu}_1)^\top \mathbf{S}_w^{-1}, \quad b = t + \boldsymbol{\mu}_2^\top \mathbf{S}_w^{-1} \boldsymbol{\mu}_2 - \boldsymbol{\mu}_1^\top \mathbf{S}_w^{-1} \boldsymbol{\mu}_1,$

and J is the dimension of measurement space Ω. The discriminant line is a function of the threshold constant t, $x_2(t) = (b(t) - a_1 x_1)/a_2$, which is plotted in Figure 9.7 for three values: $t = -20, 0, 20$. The minimum error line at $t = 0$ is different from my ad hoc effort in Figure 9.5c, yet the AUC value is equivalent. To generate an ROC curve,

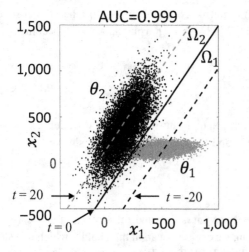

Figure 9.7 Applying the linear discriminant analysis (LDA) from (9.22) to the data in Figure 9.5, we find the solid line at $t = 0$. Comparing this statistical result to the linear discriminant in Figure 9.5c, we see much similarity. See the code in this section for details regarding the AUC estimate and alternative values of threshold t.

the range $-60 \leq t \leq 20$ was selected to span the true-positive and false-positive ranges of $[0, 1]$. (TPF,FPF) pairs are computed in the following code.

```
%%%%%%%%%%%%%%%%%%%% The linear discriminant analysis (LDA) function  %%%%%%%%%%%%%%%%%%%%%%%
close all;N=10^4;                                         % 10^4 measurements for both
mx1=500;my1=150;sx1=120;sy1=40;r1=0.3;                    % initialize para's for theta 1
mx2=170;my2=400;sx2=120;sy2=250;r2=0.7;                   % initialize para's for theta 2
m1=ones(N,2);m1(:,1)=m1(:,1)*mx1;m1(:,2)=m1(:,2)*my1;     % mean vector for simulation
m2=ones(N,2);m2(:,1)=m2(:,1)*mx2;m2(:,2)=m2(:,2)*my2;     % same for theta 2 data
M1=[mx1;my1];M2=[mx2;my2];                                % mean vectors for discriminant analysis
K1=[sx1^2 r1*sy1*sx1;r1*sx1*sy1 sy1^2];                   % 2x2 pop cov matrix for theta 1
K2=[sx2^2 r2*sy2*sx2;r2*sx2*sy2 sy2^2];                   % same for theta 2
Swi=2*eye(2)/(K1+K2);
X1=mvnrnd(m1,K1);X2=mvnrnd(m2,K2);                        % simulate data for theta 1 and 2
figure;plot(X1(:,1),X1(:,2),'r.');axis square;hold on     % theta 1 data
plot(X2(:,1),X2(:,2),'k.');                               % theta 2 data
t=-60:2:20;Nt=length(t);p=-400:1000;TPF=zeros(1,Nt);FPF=TPF; % initialize
for i=1:Nt                                                % range threshold t from -60 to 20
    a=2*(M2-M1)'*Swi;                                     % compute a vector
    b=t(i)+M2'*Swi*M2-M1'*Swi*M1;                         % compute b scalar
    q=(b-a(1)*p)/a(2);                                    % compute discriminant line
    C1=0;C2=0;                                            % initialize counters
    for j=1:N
        if(X1(j,2)>(b-a(1)*X1(j,1))/a(2))                 % find false positives
            C1=C1+1;
        end
        if(X2(j,2)>(b-a(1)*X2(j,1))/a(2))                 % find true positives
            C2=C2+1;
        end
    end
    TPF(i)=1e-4*C2;FPF(i)=1e-4*C1;                        % compute TPF and FPF
    if(t(i)==-20);plot(p,q,'r--');end                     % plot 3 discriminate functions
    if(t(i)==0);plot(p,q,'k--','linewidth',2);end
    if(t(i)==20);plot(p,q,'b--');hold off;end
end
figure;plot(FPF,TPF,'k','linewidth',2)
AUC=trapz(FPF,TPF)
```

9.3.6   Summary

These flow-cytometry simulation data are highly separable, as shown by the measures of intrinsic separability J_h, B, and AUC. Highly separable data were simulated intentionally to illustrate the methods. Interestingly, despite noticeable differences between the variance and covariance components of $\mathbf{K}_1$ and $\mathbf{K}_2$, they are similar enough for LDA to apply. LDA is often equivalent to quadratic discriminant analysis (QDA) for MVN data. This technique is an example of a supervised parametric classifier because it relies on training data; i.e., data are annotated to label class membership and associated probability densities. Before implementing data-driven methods, such as deep learning, it is always a good idea to first evaluate LDA methods provided you have sufficient training data to estimate the distribution parameters. High-quality training data are essential in supervised methods so that the classifier is able to achieve classification performance at levels commensurate with class separability.

Statistical pattern classifiers require a large number of testing samples per class to avoid problems associated with the *curse of dimensionality* [27].[3] Dimensionality

---

[3] The curse of dimensionality refers to situations where the number of measurement variables grows such that the sample size per dimension is too sparse to reach statistically significant decisions.

reduction, e.g., via PCA, can be vital when available data are limited. The number needed to achieve statistical significance in terms of classifier performance depends on many factors [31]. This topic is explored in Problem 9.3.

Many other supervised methods exist, each with its own strengths and weaknesses. One *nonparametric classifier* is the *support vector machine* (SVM) [103]. Like LDA, SVM is supervised by training on annotated, J-dimensional data. The result is a $J-1$ hyperplane (a line for 2-D data) that separates data based on known class membership. Because it is nonparametric, the SVM can achieve higher classification performance when the data classes are interspersed. For more information on SVM software, search for fitcsvm in MATLAB.

9.4   Clustering

Clustering techniques are generally unsupervised. They analyze N unlabeled points in data matrix $\mathbf{X} \in \mathbb{R}^{N \times J}$ that occupy measurement space Ω. The aim is to form K sets of data points among the total N to define distinct regions Ω_k in Ω. Points within the kth cluster minimize their squared distance to the cluster center vector $\mathbf{c}_k$ while maximizing distances to other cluster centers. Because the data are unlabeled, the assumption is that within-cluster data are associated with unique properties that define a class of data θ_i.

The rich literature on clustering [2, 17, 37] can account for a large range of prior information that might be available to generate meaningful results. We will discuss the *K-means clustering* algorithm as an example of an unsupervised, nonparametric technique where the operator provides only the data and K, the number of clusters. Here we do not think of sample vectors $\mathbf{x}$ as realizations of a random process, because each datum has a fixed and known location in Ω.

The algorithm inputs $\mathbf{X} = (\mathbf{x}_1, \mathbf{x}_2, \ldots, \mathbf{x}_N)^\top$ to minimize an objective function given by the following sum-squared difference,

$$\Theta = \sum_{k=1}^{K} \sum_{n=1}^{N} \|\mathbf{x}_n - \mathbf{c}_k\|^2. \tag{9.23}$$

Initially, data point $\mathbf{x}_n$ is randomly assigned to one of K clusters and then the center of each cluster is found as a sample mean, $\mathbf{c}_k = \sum_{n \in \Omega_k} \mathbf{x}_n / N_k$. N_k is the number of data points in the kth cluster within region Ω_k, where $N = \sum_k N_k$ and $\Omega = \sum_k \Omega_k$. Points are reassigned to the closest cluster center, and $\mathbf{c}_k$ are recomputed iteratively until further changes in cluster assignments fall below a tolerance level. The MATLAB function kmeans works well for this purpose. It uses the improvements of Arthur and Vassilvitskii [6] on the standard Lloyd algorithm [66].

Example 9.4.1. *Demonstrate a clustering method: While K-means clustering does not depend on any form of statistical properties, it is convenient to use MVN distributions to simulate the data to be clustered. We generate three distributions of bivariate normal data using the following parameters:*

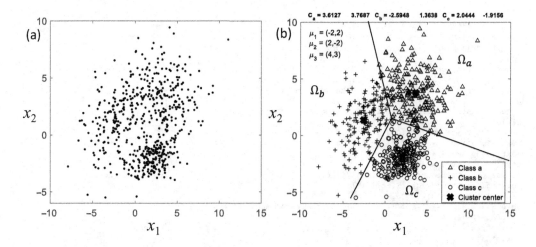

Figure 9.8 (a) An unlabeled 600-pt 2-D data vector was presented to a K-means clustering algorithm, where we asked it to find three classes of data. (b) This MATLAB algorithm generated clusters a–c, where we formed three regions: Region Ω_a is identified with data class θ_3, Ω_b with θ_1, and Ω_c with θ_2. Note that I added the discriminate lines to illustrate the clustering results. No lines are computed.

$$\mu_1 = \begin{pmatrix} -2 \\ 2 \end{pmatrix} \quad \mu_2 = \begin{pmatrix} 2 \\ -2 \end{pmatrix} \quad \mu_3 = \begin{pmatrix} 4 \\ 3 \end{pmatrix}$$

$$\mathbf{K}_1 : \sigma_1 = 2, \quad \sigma_2 = 2.5, \quad \rho = 0.5 \quad N_1 = 200$$
$$\mathbf{K}_2 : \sigma_1 = 1.5, \quad \sigma_2 = 1, \quad \rho = 0.2 \quad N_2 = 200$$
$$\mathbf{K}_3 : \sigma_1 = 2, \quad \sigma_2 = 2, \quad \rho = 0 \quad N_3 = 200$$
$$\Pr(\theta_1) = \Pr(\theta_2) = \Pr(\theta_3) = N_i/N.$$

A total of $N = 600$ unlabeled samples are displayed in Figure 9.8a. These data are entered into kmeans *while requesting three clusters and three replicates to examine reproducibility, and only the final results are displayed. The code that generates the results in Figure 9.8b is shown here.*

```
%%%%%%%%%%%%%%%%%  Script from Example 9.4.1 yields results in Fig. 9.8  %%%%%%%%%%%%%%%%%%%%%
close all;rng('default');                                     % fix the random number generator seed
M1=[-2;2];M2=[2;-2];N=200;M3=[4;3];                           % initialize mean vectors
m1=ones(2,N);m1(1,:)=m1(1,:)*M1(1);m1(2,:)=m1(2,:)*M1(2);     % set up means to simulate data
m2=ones(2,N);m2(1,:)=m2(1,:)*M2(1);m2(2,:)=m2(2,:)*M2(2);
m3=ones(2,N);m3(1,:)=m3(1,:)*M3(1);m3(2,:)=m3(2,:)*M3(2);
K1=[2^2 2*2.5*0.5;2*2.5*0.5 2.5^2];                           % initialize covariance matrices
K2=[1.5^2 1*1.5*0.2;1*1.5*0.2 1^2];
K3=[2^2 0;0 2^2];
%
X1=mvnrnd(m1',K1);X2=mvnrnd(m2',K2);                          % generate data for theta 1, 2, 3
X3=mvnrnd(m3',K3);
Swi=eye(2)/((K1+K2+K3)/3);                                    % divide by 3 for equal P(theta_i)
plot(X1(:,1),X1(:,2),'k.','MarkerSize',10);hold on
plot(X2(:,1),X2(:,2),'k.','MarkerSize',10);
plot(X3(:,1),X3(:,2),'k.','MarkerSize',10);hold off
%
X=[X1;X2;X3];                                                 % unlabeled data vector
options = statset('Display','final');
[idx,C] = kmeans(X,3,'Replicates',3,'Options',options);      % cluster the data
ic=0;                                                         % error counter
```

```
for j=1:N                                        % count misclassified samples
    if idx(j)        ~= 2; ic=ic+1;end
    if idx(j+N)      ~= 3; ic=ic+1;end
    if idx(j+2*N)    ~= 1; ic=ic+1;end
end
figure;plot(X(idx==1,1),X(idx==1,2),'k^','MarkerSize',6);hold on
plot(X(idx==2,1),X(idx==2,2),'k+','MarkerSize',6)
plot(X(idx==3,1),X(idx==3,2),'ko','MarkerSize',6)
plot(C(:,1),C(:,2),'kx','MarkerSize',12,'LineWidth',3);hold off
legend('Class a','Class b','Class c','Means','Location','SE')
title(['C_a = ',num2str(C(1,:)), ...
    '    C_b = ',num2str(C(2,:)),'   C_c = ',num2str(C(3,:))])
text(-9,9,'\mu_1 = (-2,2)')
text(-9,8,'\mu_2 = (2,-2)')
text(-9,7,'\mu_3 = (4,3)')
formatSpec = 'The error rate is %i/%i or %4.3f \n';      % report errors wrt theta_i
fprintf(formatSpec,ic,3*N,ic/(3*N))
mmm = (M1+M2+M3)/3;
Sb=0.5*((M1-mmm)*(M1-mmm)'+(M2-mmm)*(M2-mmm)'+(M3-mmm)*(M3-mmm)');
Jh=trace(Swi*Sb);
formatSpec = 'Hotelling trace is %5.2f \n';              % print to screen
fprintf(formatSpec,Jw)
```

Even though we are not classifying data, $J_h = 4.93$ suggest these classes are intrinsically highly separable. I also drew lines defining regions Ω_k, but keep in mind that classification errors are present, so the lines are suboptimal. MATLAB will parse Ω too, but there is some work involved. Also notice that because we are clustering and not classifying, source classes 1–3 are not directly assigned to regions a–c. Figure 9.8b shows that the cluster centers $\mathbf{c}_k$ do roughly approximate the population means $\boldsymbol{\mu}_i$.

Because these data were generated using parametric models, class membership of each point is known. The error rate is

$$\varepsilon = \frac{1}{N} \sum_{i \neq j}^{C} N_{i,j} = 0.112, \qquad (9.24)$$

where N_{ij} is the number of samples from class i that are labeled j. Errors were assessed to provide a measure of classification performance even though K-means clustering is not a classifier.

9.5   Problems

9.1 Write out the general expression for bivariate normal $p(\mathbf{x}|\theta_i)$, where $\mathbf{x} = (x_1, x_2)^{\mathsf{T}}$, in terms of $\mu_1, \mu_2, \sigma_1, \sigma_2, \rho$. How does the result change if measurements x_1 and x_2 are statistically independent?

9.2 Beginning with the quadratic discriminant function in (9.21), derive the result for the linear discriminant in (9.22).

9.3 Figure 9.9 shows contour lines describing two bivariate normal parent populations where class means are $\boldsymbol{\mu}_1 = \begin{pmatrix} 0 \\ 0 \end{pmatrix}$, $\boldsymbol{\mu}_2 = \begin{pmatrix} 0 \\ 40 \end{pmatrix}$; class priors are $\Pr(\theta_1) = \Pr(\theta_2) = 0.5$; and class covariance matrices are $\mathbf{K}_1 = \mathbf{K}_2$, where $\sigma_{x_1} = 40$, $\sigma_{x_2} = 25$,

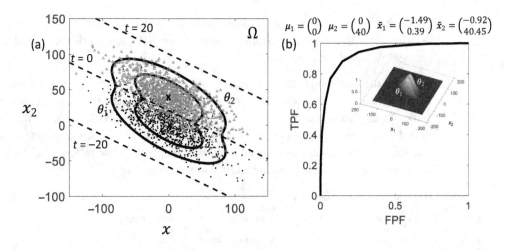

Figure 9.9 (a) Contour lines illustrating two MVN parent distributions for classes θ_1 and θ_2 in measurement space $\mathbf{x} = (x_1, x_2)$. In addition, 1,000 samples drawn from $p(\mathbf{x}|\theta_1)$ are indicated by $(\cdot)$ and those from $p(\mathbf{x}|\theta_2)$ by $(\triangle)$. Mean vectors are indicated with $\times$. Lines result from applying LDA via (9.22) over the range $-40 \leq t \leq 40$ (only three lines shown). (b) The ROC curve is computed from the discriminant lines. The parent distributions are shown as an insert. Population and sample means (for $N = 1,000$) are also shown.

and $\rho = -0.7$. (Don't confuse class subscripts μ_1, μ_2 with measurement dimension subscripts x_1, x_2.) In addition, 1,000 measurements for each class were simulated to train and evaluate a classifier.

(a) Adapt the code from Section 9.3.5 to obtain the results shown in Figure 9.9, including estimation of the ROC curve. Be sure to use sample statistics, unlike the code in the text. Compute the three class-separability metrics J_h, B, and AUC, showing the two classes are readily separable.

(b) How do the results of (a) change if the mean for θ_2 is changed to $\mu_2 = \begin{pmatrix} 0 \\ 20 \end{pmatrix}$?

(c) How do the results of (a) change if $\rho_1 = -0.7$ remains the same but $\rho_2 = 0.7$?

(d) How do the results of (a) change if the number of samples per class, N, changes from 10^1 to 10^2 to 10^3 to 10^4?

9.4 Apply (9.22) to find the linear discriminant for the following 2-class, 3-D linear problem. The two classes have the parameters

$$\mu_1 = \begin{pmatrix} 0 \\ 0 \\ -10 \end{pmatrix}, \ \mu_2 = \begin{pmatrix} 0 \\ 40 \\ 15 \end{pmatrix}, \ \text{and } \mathbf{K}_1 = \mathbf{K}_2,$$

where $\sigma_1 = 40$, $\sigma_2 = 40$, $\sigma_3 = 10$, $\rho_{12} = 0$, $\rho_{13} = \rho_{23} = 0.4$ for $N = 1,000$ samples/class. You are looking for a plane in $\mathbf{x} = (x_1, x_2, x_3)$ that yields the minimum classification error. After you generate the data, plot it using `scatter3`. Compute vector $\mathbf{a}$ and scalar b in (9.22) from class parameters and then solve for x_3 at four points within the (x_1, x_2) plane. With those points, you may use `patch` to generate the decision plane in the same `scatter3` output display. Also, what do the separability metrics tell you about these data?

9.5 Use the following code to generate two classes of data. However, you cannot use this information as is to solve the problem, because these data come to you unlabeled. Without class labels, there is no training data for LDA. You decide to cluster the data to find what groupings naturally exist so that later, if data labels became available, you can see how the clusters correspond to classes. Apply the code in Example 9.4.1 to show how unsupervised clustering might correspond to classes.

```
%%%%%%%%%%%%%%%%%%%%%%%%%  Data used for the clusering in Problem 9.5  %%%%%%%%%%%%%%%%%%%%%%%%%%%%%%%
rng('default');                             % fix the distribution
a=4;N=200;X1p=3*randn(N,2)+2*ones(N,2);     % parameters for class theta 1
X2=2*randn(20,2)-ones(20,2);                % class theta 2 data
N1=0;X1=zeros(length(N),2);                 % initialize parameters
for j=1:N                                   % here, remove theta 1 data in a circular pattern
    r=sqrt(X1p(j,1)^2+X1p(j,2)^2);          % radius of exclusion
    if r>a                                  % if outside radius, then use that point
        N1=N1+1;                            % count to keep track of how many of N are included
        X1(N1,:)=X1p(j,:);                  % collected included points
    end
end
X=[X1;X2];plot(X(:,1),X(:,2),'ko','MarkerSize',10);hold on
plot(X2(:,1),X2(:,2),'r.','MarkerSize',10);hold off        % mark X2 for illustration
xlabel('x_1');ylabel('x_2')
```

10   ODE Models I

Biological Systems

Previous chapters have focused on analyzing biomedical measurement tools and the data they produce. Since specifying the medical task is the first step in measurement design, it makes sense to also analyze and model the biological systems we measure. The tool of choice for these problems is *ordinary differential equations* (ODEs).

To model biological systems, we first briefly discuss complex-system properties by considering examples. Contraction force is an example of an *emergent property* of skeletal muscle behaving as a centimeter-scale *dynamical system*. Force generation is one whole-muscle property that emerges from the actions of interconnected and cooperating myocytes. To walk normally, many muscle groups and other vital subsystems cooperate to form a meter-scale dynamical complex system. Sustained normal gait involves cyclic movements of body systems at many scales, producing a steady forward translation of the body that we identify as a *dynamic equilibrium state* [40].[1] Vigorous walking may include arms and legs that are cycling furiously, yet the state variable "velocity" for the organism is constant when averaged over a few muscle cycles. To model a walking human exactly, we could include many differential equations describing state variables for all the subsystems that contribute to the activity and still never get it right. A good first attempt at modeling gait is to abstract some aspects of the complex system and form equations describing simple subsystems. We will show that this simplified approach predicts essential behavior trends even if the details are wrong or missing. This is acceptable because our modeling goal in this

---

[1] There are different types of equilibria. A *simple system*, such as air molecules in a container, is in *thermodynamic equilibrium* when there is no change in the *macroscopic* flow of its mass or energy, despite its *microscopic* features being in constant flux. A system is simple when all components behave similarly so there are few distinct functional modes. *State variables*, such as temperature, volume, and pressure, indicate the emergent properties of this simple system related to energy. In thermodynamic equilibrium, state variables are constant over the spatial extent of the system when averaged over a time corresponding to many mean free paths of molecular motion.

    Dynamical complex systems, including organisms, also express properties when in *dynamic equilibrium*, a state we identify from global features such as body temperature, blood pressure, and pH being constant at relevant scales. A system is complex when it expresses many functional modes operating at different spatiotemporal scales, and therefore can express a rich palette of emergent properties. A system must be in an equilibrium state for its properties to be defined; otherwise, it is transitioning between states. Healthy organisms are in dynamic equilibrium; only dead organisms can be in thermodynamic equilibrium.

treatment is to understand engineering parameters that regulate and predict *overall* system behavior in space and time.

System properties are essential to know because they predict the responses to environmental stimuli: A system must be in *equilibrium* for it to have well-defined properties. For example, if a healthy (the equilibrium state) adult (the system) has a finger poked unexpectedly with a sharp object (the stimulus), the hand is quickly withdrawn (the response), demonstrating a healthy reflex (the property). To model simplified aspects of biological systems, we first define state variables and determine relevant equilibrium states so that system properties can be defined and measured.

10.1   Mathematical Representations of Systems

State variables, such as hand location and speed of movement during reflex testing, are functions of independent variables such as position $\mathbf{x}$ and time t. Differential equations, which are expressions of derivatives of state variable with respect to independent variables, such as $dy/dt = ay$, are used to describe the *state of a system*.

If state variables depend on one independent variable, such as time, we model the system using ODEs. If state variables depend on more than one independent variable, such as position and time, we formulate the problem as a partial differential equation (PDE). This chapter describes systems of ODEs and how solutions offer fundamental concepts such as trajectories through state space, equilibrium points, and predictions of overall system stability.

The temporal behavior of a dynamical system described by a system of J state variables $y_j(t)$ and their temporal derivatives $\dot{y}_j(t) \triangleq dy_j/dt$, each a continuous, deterministic function of time t, may be expressed by a system of J ODEs:

$$\dot{y}_1(t) = f_1(y_1, \ldots, y_j, \ldots y_J, t; \theta_1, \ldots, \theta_j, \ldots, \theta_{J'})$$
$$\dot{y}_2(t) = f_2(y_1, \ldots, y_j, \ldots y_J, t; \theta_1, \ldots, \theta_j, \ldots, \theta_{J'})$$
$$\vdots \tag{10.1}$$
$$\dot{y}_J(t) = f_J(y_1, \ldots, y_j, \ldots y_J, t; \theta_1, \ldots, \theta_j, \ldots, \theta_{J'}).$$

Here, $f_j(\cdot)$ is a function of the variables and θ_j is one of J' model parameters, e.g., rate constants, that do not need to match the number of state variables J. In compact vector notation, (10.1) becomes

$$\dot{\mathbf{y}}(t) = \mathbf{f}(\mathbf{y}, t; \boldsymbol{\theta}). \tag{10.2}$$

Considering state variables and their first derivatives only, (10.1) and (10.2) are systems of equations given as *first-order* ODEs. Two essential roles for differential equations in biological sciences are to (1) generate models that offer parameters to explain observed behaviors and (2) predict future behavior of the system.

Example 10.1.1. *Radioactive decay: A mass of a single radioactive isotope is an example of a simple first-order system. Although there may be billions of atoms, this system is simple because element properties are identical. The property of radioactivity is completely characterized by a single state variable – the number of radioactive atoms, N.*

Radioisotopes are used in medicine as reporters when they are attached to biologically active molecules and distributed in the bloodstream of organisms during an imaging procedure. They report on movements of essential biomolecules in space and time. Initially, we restrict our model to the physical decay of the isotope.

Suppose instantaneous state variable $N(t)$ is the number of radioactive atoms in the mass at an instant of time. Observation shows that the change in N with time is proportional to the current number and the time increment, $\Delta N \propto N\Delta t$. Dividing both sides by Δt, declaring $-\lambda$ to be the proportionality constant (constant λ is always positive), and taking the limit $\Delta t \to 0$, we have

$$\dot{N}(t) = -\lambda N(t) \qquad \text{for } N, \lambda > 0. \tag{10.3}$$

The minus sign on the right side indicates that radioactivity (the change in the number of radioactive atoms) is decreasing over time. Expressing (10.3) as $\dot{N} + \lambda N = 0$, where λ has units of 1/time, we find a homogeneous first-order ODE *with constant coefficients that describes group behavior of identical atoms. Rearranging (10.3),*

$$\frac{dN}{N} = -\lambda \, dt$$

and applying indefinite integration

$$\ln N = -\lambda t + \ln C,$$

we find for integration constant $C > 0$

$$N(t) = Ce^{-\lambda t}. \tag{10.4}$$

Equation (10.4) is the general solution *to (10.3). Mathematically, the term "homogeneous" refers to the absence of terms in the ODE that are independent of state variable N. Physically, the homogeneous label tells us there are no independent sources of isotope being added or subtracted over time. Atoms can only be lost through decay.*

To find a particular solution, *we need to measure our state variable at two times, e.g., $N_0(t = t_0)$ and $N_1(t = t_1)$, and solve for unknown constants, C and λ, in terms of these measurements. Applying definite integration[2] and the two measurements, we solve for λ:*

---

[2] Definite integration variables disappear after integration and are replaced by limit values. To avoid confusing integration variables and limit values, we change to dummy variable names N' and t'.

$$\int_{N_0}^{N_1} \frac{dN'}{N'} = -\lambda \int_{t_0}^{t_1} dt'$$

$$\ln \frac{N_1}{N_0} = -\lambda \, (t_1 - t_0)$$

$$\lambda = \frac{1}{t_1 - t_0} \ln \frac{N_0}{N_1}. \tag{10.5}$$

With λ now defined in terms of measurements, we can combine (10.4) and (10.5) while applying the initial state condition $N(0) = N_0$ to find

$$\int_{N_0}^{N} \frac{dN'}{N'} = -\lambda \int_{0}^{t} dt'$$

$$N(t) = N_0 e^{-\lambda t} \tag{10.6}$$

$$= N_0 \exp \left(-\frac{t}{t_1} \ln \frac{N_0}{N_1} \right).$$

Equation (10.6) shows that we expect the isotope to decay exponentially at a rate defined by its characteristic constant λ, which we determine from theory or experiments.

Example 10.1.2. *Half-life:* How much time must we wait for a mass of Tc-99m to physically decay to 1/20th of its original mass?

Solution
From (10.6), we select $N(t_1) = N_1 = N_0/20$, in which case,

$$N_1 = N_0 \exp \left(-\frac{t}{t_1} \ln \frac{N_0}{N_1} \right)$$

$$\ln \frac{N_1}{N_0} = -\frac{t}{t_1} \ln \frac{N_0}{N_1}$$

$$t = t_1.$$

Of course! But that is not an answer.

In terms of the physical half-life $t_{1/2p}$ of the isotope, we note that $N/N_0 = 0.5 = \exp(-\lambda t_{1/2p})$. Consequently, $t_{1/2p} = \ln 2/\lambda$. Activity is defined as $A \triangleq \lambda N = (\ln 2/t_{1/2})N$. Applying (10.5), $t_1 = \frac{\ln 20}{\ln 2} t_{1/2p} = 4.322 t_{1/2p}$. Even if we're unsure of the half-life of Tc-99m (6.0 hours), we know that waiting 4.322 physical half-lives reduces the radioactive mass to 5% of the original amount. That is, $N(t_1) = 2^{-4.322} N_0 = 0.05 \, N_0$ after $4.332 \times 6.0 = 26.0$ hours. The decay constant serves as a "clock" for the experiment.

Example 10.1.3. *One-compartment dilution: Molecular imaging agents called* imaging probes *include a biologically active carrier molecule attached to a reporter, in this case a Tc-99m atom. Assume we inject[3] 210 MBq of Tc-99m–labeled MAA carrier (macroaggregated albumin: molecular weight = 66.43 kg/mole) into a patient for a lung perfusion study. The MAA molecule has an average size that causes it to lodge in the lung capillaries long enough to measure blood perfusion. The carrier molecule is cleared from blood by the liver and kidneys with a* biological half-life *equal to (but independent of) its* physical half-life, *6.0 hours. That is, $t_{1/2p} = t_{1/2b} = 6$ hours for this agent. The effective half-life considers both the biological and physical half-lives according to $1/t_{1/2e} = 1/t_{1/2p} + 1/t_{1/2b}$; in this example, $t_{1/2e} = 3$ hours.*
(a) Compute the carrier molecule concentration in the urine of an 89-kg man as a function of time after injection, assuming the carrier is cleared only by the kidneys with a negligible fraction captured by the lungs.
(b) Estimate the blood concentration over time.
(c) Estimate the total volume of blood that must pass through both kidneys to filter half of the isotope. Remember that we assume kidneys are the primary clearance mechanism.

Solution
(a) Assume each of the N MMA molecules injected has one Tc-99m atom attached. First, compute the concentration of carrier in the blood over time, which we begin by computing the mass injected:[4]

$$m_0 = N \frac{M_W}{N_A} = \frac{t_{1/2b}}{\ln 2} A \frac{M_W}{N_A}$$

$$= \frac{6\,hr \times 3,600\,s/hr}{0.693} (2.10 \times 10^8\,atoms/s) \frac{66,430\,g/mole}{6.022 \times 10^{23}\,atoms/mole}$$

$$= 7.22 \times 10^{-7}\,g = 722\,ng.$$

Assume the agent instantly mixes into the patient's $V = 6.7$ L blood volume (I looked up the average blood volume for a known body weight) after injection to give an initial concentration of $c(0) = c_0 = m_0/V = 107.8$ ng/L. Because we are measuring the concentration of MMA in urine, we will ignore the physical decay of Tc-99m because the decayed compound is not detected in urine. We consider only the biological half-life via kidney clearance.[5] Two kidneys receive 25% of the cardiac output (again, look it up), and therefore the flow through the kidneys is $q = 0.25CO$, where it is

---

[3] MBq (10^6 becquerel) of activity indicates 1 million Tc-99m atoms will decay to Tc-99 each second and each decay yields one 140-keV gamma ray that "reports" the event.
[4] A is isotope activity, N_A is Avogadro's number, and M_W is the molecular weight of the carrier molecule.
[5] This is a simplification. The liver and kidneys both clear this carrier.

reasonable to assume $CO = 6.7\ L/min$. For an injection at $t = 0$, the loss of carrier mass Δm over time interval Δt is

$$\Delta m(t) = \begin{cases} 0 & t < 0 \\ -\Delta t\ q\ c(t) & t \geq 0 \end{cases},$$

where $c(t) = m(t)/V$ is the instantaneous concentration. $\dot{c}(t)$ is found by dividing Δm by V and Δt while allowing $\Delta t \to 0$,

$$\frac{dc}{dt} = \lim_{\Delta t \to 0} \frac{\Delta c}{\Delta t} = -\frac{q}{V} c(t)$$

$$\dot{c}(t) + \frac{q}{V} c(t) = 0 \qquad \text{for } c, q, V > 0, \tag{10.7}$$

yielding a homogeneous, first-order ODE that holds at all times because $c(t < 0) = 0$. The units are c [g/L], $\dot{c}$ [g/L-min], q [L/min], and V [L]. From (10.6), the solution is

$$c(t) = c_0 e^{-qt/V} \text{step}(t) = \begin{cases} 0 & t < 0 \\ c_0 e^{-qt/V} & t \geq 0 \end{cases}. \tag{10.8}$$

ODE parameters q, V describe the physics of this simple system (flow and volume), while the ratio q/V is the time constant that serves the role as the "engineering parameter," meaning the ratio controls the rate at which the agent is cleared by the kidneys.

Equation (10.8) is the curve labeled * in Figure 10.1a but not yet the solution to part (a). The solution is $c_0 - c(t)$, describing the concentration of carrier found in urine accumulating in the bladder over time (Figure 10.1a).

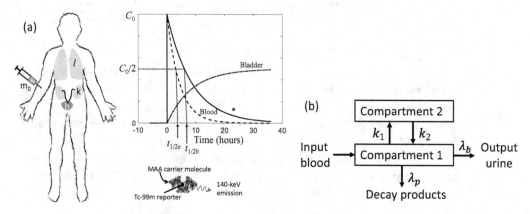

Figure 10.1 (a) Illustration of one-compartment dilution model from Example 10.1.3. The curve marked * is the result from part (a) of the example. The curve labeled "Blood" gives the results of part (b). The curve labeled "Bladder" is the liver/bladder measurement (bladder only shown). The liver and bladder are involved with biological clearance of MAA. The exact MAA molecular structure (bottom) is unknown except the molecular size ranges from 10 to 90 μm so that a large portion lodges in the pulmonary capillaries. (b) Diagram of a multicompartment model.

(b) Since $t_{1/2e} = 3$ hours, the result of (10.7) still applies, except that now we use $q/V = \ln 2/t_{1/2e}$. The curve of blood concentration over time is plotted in Figure 10.1a. This is the value found by drawing and assaying blood samples.

(c) We are assuming a resting CO of 6.7 L/min so that $q = 0.25(6.7)$ L/min. Hence, $q/V = \ln 2/t_{1/2e}$ and $V = (0.25 \times 6.7 \, L/min)(3 \, hr \times 60 \, min/hr)/0.693 = 435 \, L$. This is the net blood volume passing through the kidneys to filter half of the isotope.

The model in this example was simplified substantially. Realistically, humans have more compartments and clearance paths than those considered here, and the rates of transfer between any two compartments varies (see Figure 10.1b). Compartments temporarily store blood components, thus delaying clearance. What is realistic is that this patient has no significant activity after about 24 hours – approximately eight half lives. This determination is used for radiation safety purposes following a medical procedure. Although this model is too simple to accurately predict patient properties, it is a starting framework for adding realistic details.

Example 10.1.4. *Cellular uptake: Uptake of a nutrient in cell culture from media surrounding the cells is proportional to the difference in the concentration of the nutrient inside the cell, $c(t)$, and that in media outside the cell, c_m. The state variable of interest is $c(t)$. If the initial cell concentration is $c(0) = 0$ and media concentration c_m does not change during the time of the experiment, find t such that $c(t) = 0.9 \, c_m$.*

Solution
The expression from the problem statement is $\Delta c \propto (c_m - c)\,\Delta t$, where the concentration difference is passively driving cell uptake and $c_m \geq c$. The functional state variable for this problem is really $c_m - c$, even though we are interested in c. Dividing by Δt, adopting k as the constant of proportionality, and taking the limit as $\Delta t \to 0$, we find an inhomogeneous ODE

$$\dot{c} = k(c_m - c)$$

$$\dot{c}(t) + kc(t) = kc_m.$$

The ODE is inhomogeneous in variable c because the term kc_m does not depend on c, and it is autonomous *because kc_m is time independent. Let's change state variable c to $c' = c - c_m$ to solve a homogeneous ODE, and later change back to get the answer we want. The first change shifts the equilibrium concentration from c_m to zero as $t \to \infty$. Then, realizing $\dot{c}' = \dot{c}$ and $c'_0 \triangleq c'(0) = c(0) - c_m = -c_m$, the problem may be expressed in homogeneous form using*

$$\dot{c}' = -kc'.$$

As a consequence of (10.4) and the initial condition $c(0) = 0$,

$$c'(t) = c'_0 e^{-kt} = -c_m e^{-kt}.$$

For $c(t) = c'(t) + c_m$,

$$c(t) = c_m \left(1 - e^{-kt}\right). \tag{10.9}$$

*Nutrient concentration in the cell grows exponentially over time. Applying $c(t) =$
$0.9\,c_m$ to (10.9) and solving for t, we find $t = (\ln 10)/k = 2.303/k$. Let's make
sure these units make sense. Since transcendental functions like $\exp(-kt)$ must have
arguments with no units, rate constant k must have units of 1/time. Yes, our estimate
of t has units of time. Without knowing the uptake constant k, we must stop here. Note
that assuming k is a constant means that we assume every cell has exactly the same
nutrient dynamics given by (10.9).*

10.1.1   Summary

The preceding examples illustrate the time evolution of systems described by first-
order ODEs. There, each state variable was a function of the independent variable,
time. We were required to make measurements at a number of time points equal to
the number of unknown parameters, and, when we did, we could solve the ODE and
predict the state variable at all times. Anytime a change in state variable is proportional
to the variable itself, the solution to that ODE is an exponential function. The set of all
possible states that this system can assume is referred to as its *state space*. The path
taken by the variable through state space over time is its *trajectory*, which is revealing for second- and higher-order systems, as explained later in the chapter. If some
terms in the ODE are independent of the state variable, that ODE is *inhomogeneous*.
Further, if the inhomogeneous term is not a function of the independent variable, that
inhomogeneous ODE is *autonomous*. Non-autonomous inhomogeneous ODEs have
source and sink terms that are functions of the independent variable. Finally, first-
order ODEs most often describe simple components; complex systems require more
diversity in state variables obtained with higher-order derivatives and/or systems of
equations.

10.1.2   Generalizations

The general form of a homogeneous linear ODE for $J = 1$ (one equation variable)
is $\dot{g} = Ag$, where $g(0) = g_0$. It has the *ansatz* (general initial guess at a solution)
$g(t) = ue^{\lambda t}$ for constants u, λ that we must determine. Applying the ansatz to the
ODE, we find $u\lambda e^{\lambda t} = uAe^{\lambda t}$, showing that $\lambda = A$ when $u \neq 0$. Hence, the *general
solution* for this problem is $g(t) = ue^{At}$. Applying the initial condition to the general
solution, we find $u = g_0$ and the *particular solution* is $g(t) = g_0 e^{At}$. The general
constants u, λ from the ansatz are to be replaced by those of the specific problem – in
this case g_0, A.

The general form of an inhomogeneous, autonomous, linear ODE for $J = 1$ is
$\dot{g} = Ag + b$. At equilibrium, we have $\dot{g} = 0 = Ag_s + b$ and $g_s = -b/A$, where
g_s is the solution specific to the steady-state or equilibrium conditions. When we can

identify equilibrium points for the state variable, it is convenient to change the state variable to $g' = g - g_s$. Since $\dot{g}' = \dot{g}$, we can solve the homogeneous ODE, $\dot{g}' = Ag'$, for g' and convert back to g, as we did in Example 10.1.4. These methods can help us solve higher-dimensional systems of ODEs, i.e., for $J > 1$.

10.2   Linear Systems of Equations

A second-order ($J = 2$), inhomogeneous ($b_j \neq 0$), autonomous (b_j are constant), linear system with constant coefficients (A_{ji} are constant) has the form

$$\dot{g}_1(t) = A_{11}g_1(t) + A_{12}g_2(t) + b_1$$

$$\dot{g}_2(t) = A_{21}g_1(t) + A_{22}g_2(t) + b_2. \qquad (10.10)$$

In matrix form, the system of equations is $\dot{\mathbf{g}} = \mathbf{Ag} + \mathbf{b}$, where the matrix elements might contain modeling parameters, $A_{ji} = A_{ji}(\theta)$, and inhomogeneities $\mathbf{b} = \begin{pmatrix} b_1 \\ b_2 \end{pmatrix}$ are constant. Equation (10.10) describes a *second-order system* because state-space vector $\mathbf{g} = \begin{pmatrix} g_1 \\ g_2 \end{pmatrix}$ contains two variables; this system is linear because each term has one power of the variables or is constant. This pedantic description is intended to illustrate term definitions.

The ansatz for homogenous linear system, $\dot{\mathbf{g}} = \mathbf{Ag}$, is $\mathbf{g} = e^{\lambda t}\mathbf{u}$. We need to find two scalar constants $\lambda \rightarrow \lambda_1, \lambda_2$ and two vectors $\mathbf{u} \rightarrow \mathbf{u}_1, \mathbf{u}_2$ to complete the solution. Applying the ansatz, the derivative is $\dot{\mathbf{g}} = e^{\lambda t}\mathbf{Au} = e^{\lambda t}\lambda\mathbf{u}$. When scalar $e^{\lambda t} \neq 0$, we find $\mathbf{Au} = \lambda\mathbf{u}$. This is the *characteristic equation* for matrix $\mathbf{A}$ (see Appendix A) expressed as

$$(\mathbf{A} - \lambda\mathbf{I})\mathbf{u} = \mathbf{0}.$$

Solving for ansatz unknowns in $J \times J$ matrix $\mathbf{A}$ is the same as finding eigenvalues λ and eigenvectors $\mathbf{u}$ for that matrix.

Example 10.2.1. *Solving systems of equations: Find the general solution to the following system of equations:*

$$\dot{g}_1(t) = g_1(t) - 3g_2(t)$$

$$\dot{g}_2(t) = -2g_1(t) + 2g_2(t). \qquad (10.11)$$

Solution
Expressing (10.11) as $\dot{\mathbf{g}} = \mathbf{Ag}$, we find that

$$\mathbf{A} = \begin{pmatrix} 1 & -3 \\ -2 & 2 \end{pmatrix} \text{ has eigenstates } \mathbf{\Lambda} = \begin{pmatrix} -1 & 0 \\ 0 & 4 \end{pmatrix} \text{ and } \mathbf{U} = (\mathbf{u}_1, \mathbf{u}_2) = \begin{pmatrix} 3 & 1 \\ 2 & -1 \end{pmatrix}.$$

Solutions[6] are expressed as 2×2 matrix $\mathbf{Y}$ written as a row vector of column vectors: $\mathbf{Y} = (\mathbf{y}_1, \mathbf{y}_2) = (\mathbf{u}_1 e^{\lambda_1 t}, \mathbf{u}_2 e^{\lambda_2 t})$, where $\lambda_1 = -1$ and $\lambda_2 = 4$. Substituting the eigenstates,

$$\mathbf{Y}(t) = \begin{pmatrix} 3e^{-t} & e^{4t} \\ 2e^{-t} & -e^{4t} \end{pmatrix}, \quad \mathbf{y}_1(t) = e^{-t} \begin{pmatrix} 3 \\ 2 \end{pmatrix}, \quad \text{and} \quad \mathbf{y}_2(t) = e^{4t} \begin{pmatrix} 1 \\ -1 \end{pmatrix}.$$

$$(10.12)$$

To find a general solution, we invoke the superposition principle for homogeneous linear ODEs, which states that a linear combination of solutions is also a solution. For constant vector $\mathbf{c}$, the general solution is $\mathbf{g} = \mathbf{Y}\mathbf{c} = c_1 \mathbf{y}_1 + c_2 \mathbf{y}_2$, or

$$g_1(t) = 3c_1 e^{-t} + c_2 e^{4t}$$

$$g_2(t) = 2c_1 e^{-t} - c_2 e^{4t}.$$

$$(10.13)$$

Entering (10.13) into (10.11) shows that these solutions work for arbitrary $\mathbf{c}$. Initial conditions are required to determine the particular solution.

This method works well when the coefficients are numerical values, but is less helpful when the coefficients are expressions, as in Section 10.7. See Problem 10.6 for example.

10.3   Differential Operators for Linear Systems

Earlier chapters described linear systems using integral operators of the form $g(t) = \mathcal{H} f(t) = \int dt'\, f(t') h(t, t')$. That is, input function $f(t)$ is transformed into data function $g(t)$ by integral operator $\mathcal{H}$ that describes a linear measurement process. We now show this same method can help us solve nonautonomous ODEs.

An input–output relationship may also be expressed by a linear *differential operator, $\mathcal{D}$*. Applying $\mathcal{D}$ to g establishes its relationship to f via

$$\mathcal{D} g(t) = f(t) \tag{10.14}$$

$$\text{where} \quad \mathcal{D} = \frac{d^n}{dt^n} + k_1(t) \frac{d^{n-1}}{dt^{n-1}} + \cdots + k_{n-1}(t) \frac{d}{dt} + k_n(t).$$

We can think of time-dependent inhomogeneity f as a forcing *function* driving system $\mathcal{D}$ to generate output data g. Since $\mathcal{H}$ and $\mathcal{D}$ are both linear operators, we idealistically hope that $\mathcal{D} = \mathcal{H}^{-1}$.

Equation (10.14) expresses the general equation as an nth order ODE with coefficients k_n that depend on independent variable t. Fortunately, many natural

---

[6] It is smart to check the eigenanalysis. I might expect $\mathbf{U}\Lambda\mathbf{U}^\top = \mathbf{A}$, which is not true even if the eigenvectors are normalized. The reason is that $\mathbf{A}$ is asymmetric and hence its eigenvector matrix $\mathbf{U}$ (eigenvectors are columns of $\mathbf{U}$) is not unitary; i.e., $\mathbf{U}^\dagger \neq \mathbf{U}^{-1}$. Regardless, $\mathbf{U}\Lambda\mathbf{U}^{-1} = \mathbf{A}$ is true. Using $\mathbf{U}^{-1}$, factors eliminated by normalization are automatically cancelled in the synthesis equation $\mathbf{A} = \mathbf{U}\Lambda\mathbf{U}^{-1}$. Thus, eigenvector normalization is unnecessary.

and engineered systems are well represented by first- or second-order equations with constant coefficients.

10.3.1 First-Order Nonautonomous Systems

Consider the inhomogeneous, linear, first-order equation from (10.14),

$$\mathcal{D} g(t) = \dot{g}(t) + k\, g(t) = f(t). \tag{10.15}$$

This equation differs from those in previous sections of this chapter by having a time-varying inhomogeneity, $f(t)$. When the inhomogeneity is a function of the independent variable, the ODE is considered *nonautonomous*.

If the operator that maps input f to output g is linear, time-invariant (LTI), then solutions to (10.15) have the form

$$g(t) = \int_{-\infty}^{\infty} dt'\, h(t - t')\, f(t'), \tag{10.16}$$

suggesting a general solution can be found if we can identify the impulse response, $h(t)$. We know from Chapter 4 that $h(t)$ may be measured from the output of the LTI system when the input is $f(t) = A\,\delta(t)$.

This idea is implemented using the *causal*[7] ansatz, $A'e^{-kt}$ step(t). To verify the ansatz equals $h(t)$, we compute its derivative,

$$\dot{g}(t) = -kA'e^{-kt}\,\text{step}(t) + A'e^{-k}\,\delta(t) = -k\,g(t) + A'e^{-kt}\delta(t)$$

$$\dot{g}(t) + k\,g(t) = A'e^{-kt}\delta(t). \tag{10.17}$$

Combining (10.17) and (10.15), we find

$$f(t) = A'e^{-kt}\delta(t) = A\delta(t) \quad \text{if } A = A'.$$

Hence

$$h(t) \triangleq g_\delta(t) = Ae^{-kt}\,\text{step}(t). \tag{10.18}$$

The general solution to (10.15) is found by interpreting constants A and k for the problem and then combining (10.16) and (10.18).

Example 10.3.1. *One-compartment dilution with source: Continuing the discussion in Example 10.1.3, the problem is modified so that instead of an initial bolus (impulse) injection, we slowly infuse the molecular agent into the patient beginning at $t = 0$, where initially $c(t < 0) = 0$. We have a time-varying inhomogeneity (source term) balancing physical and biological clearance losses. Specifically, (10.7) becomes*

$$\dot{c}(t) = -\frac{q}{V}c(t) + \frac{r}{V}\,\text{step}(t),$$

where the source term is $f(t) = \frac{r}{V}$ step(t) and r is the infusion rate [g/min]. The first term on the right side is negative, representing the losses. We have combined physical

---

[7] The output of a causal system depends on past and current inputs, but not on future inputs.

and biological clearances by applying the effective half-life via $q/V = \ln 2/t_{1/2e}$. The second term on the right is positive, representing the sources – in this case, the infusion.

There is no reason to doubt this system is LTI. From (10.18), where $A = 1$ and $k = q/V$, $h(t) = c\delta(t) = e^{-qt/V}$ step(t) and

$$c(t) = \int_{-\infty}^{\infty} dt'\, h(t')\, f(t - t') = \frac{r}{V} \int_{-\infty}^{\infty} dt'\, e^{-qt'/V}\, \text{step}(t')\, \text{step}(t - t')$$

$$= \frac{r}{V} \int_0^t dt'\, e^{-qt'/V} = -\frac{r}{V}\frac{V}{q}\left[e^{-qt'/V}\right]_0^t = \frac{r}{q}\left[1 - e^{-qt/V}\right].$$

This solution is consistent with the initial condition, $c(0) = 0$, and gives the equilibrium state, $c(\infty) = r/q$. Physical parameters from the ODE r, q, and V combine to form engineering parameters q/V and r/q describing the state of the overall system.

Example 10.3.2. *Isotopes in radioactive equilibrium:* Tc-99m, $t_{1/2p} = 6$ hours, is formed from the decay of its parent isotope Mo-99, $t_{1/2p} = 66$ hours. The relatively large difference between the half-lives of the parent and the daughter produces a state in which the two are in equilibrium.[8] While the physical half-life of Tc-99m does not change, its production and decay rates form an equilibrium state that give it a half-life equivalent to its longer-lived parent Mo-99, which offers significant practical and economic advantages. The situation is mathematically the same as the classic two-compartment series dilution models described in Problem 10.2.

Let $N_M(t)$ be the number of radioactive molybdenum atoms in a sample. The activity of Mo-99 is defined as $A_M(t) \triangleq -\dot{N}_M(t)$. Also, $N_T(t)$ is the number of radioactive technetium atoms in the same sample, and $A_T(t) = -\dot{N}_T(t)$ is technetium's activity. Assume we initially have N_{0M} atoms of molybdenum in a sample container at measurement time $t = 0$, and $N_{0T} \triangleq N_T(0) = 0$. The problem is to find the daughter activity $A_T(t)$ from the parent activity $A_M(t)$, each as a function of time.

The ODE describing $\dot{N}_M(t)$ has a loss term but no source term,

$$\dot{N}_M(t) = -\lambda_M N_M(t) = -\lambda_M N_{0M}\, e^{-\lambda_M t}.$$

In contrast, $\dot{N}_T(t)$ has both source and loss terms. Expressing the ODE and identifying terms representing $h(t)$ and $f(t)$, we can use (10.16) to find $N_T(t)$:

$$\dot{N}_T(t) = \lambda_M N_M(t) - \lambda_T N_T(t) \quad \text{or} \quad \dot{N}_T + \lambda_T N_T = \lambda_M N_M, \quad t > 0$$

$$h(t) = N_T\delta(t) = e^{-\lambda_T t}\, \text{step}(t) \quad \text{and} \quad f(t) = \lambda_M N_{0M}\, e^{-\lambda_M t}\, \text{step}(t)$$

---

[8] Transient equilibrium occurs when the parent half-life is 2–20 times greater than that of the daughter isotope. Secular equilibrium occurs when the parent half-life is more than 50 times that of the daughter.

The loss of Mo-99 is a gain for Tc-99m, so that $f(t)$ is a positive quantity.

$$N_T(t) = \int_{-\infty}^{\infty} dt'\, h(t')\, f(t - t')$$

$$= \lambda_M N_{0M} \int_{-\infty}^{\infty} dt'\, e^{-\lambda_T t'}\, \text{step}(t')e^{-\lambda_M(t-t')}\, \text{step}(t - t')$$

$$= \lambda_M N_{0M}\, e^{-\lambda_M t} \int_0^t dt'\, e^{-(\lambda_T - \lambda_M)t'} = -\frac{\lambda_M N_{0M}}{\lambda_T - \lambda_M} e^{-\lambda_M t} \left(e^{-(\lambda_T - \lambda_M)t} - 1\right)$$

$$= \frac{\lambda_M N_{0M}}{\lambda_T - \lambda_M} \left(e^{-\lambda_M t} - e^{-\lambda_T t}\right).$$

The technetium activity as a function of the molybdenum activity is

$$A_T(t) = -\lambda_T N_T(t)$$

$$= -\frac{\lambda_T \lambda_M N_{0M}}{\lambda_T - \lambda_M} \left(e^{-\lambda_M t} - e^{-\lambda_T t}\right)$$

$$= \frac{\lambda_T A_{0M}}{\lambda_T - \lambda_M} \left(e^{-\lambda_M t} - e^{-\lambda_T t}\right)$$

$$= A_M(t)\frac{\lambda_T}{\lambda_T - \lambda_M} \left(1 - e^{-(\lambda_T - \lambda_M)t}\right).$$

An application of transient equilibrium is given in Problem 10.4.

10.4   Modeling Cell Growth

We continue the discussion of applying ODEs to model simple systems by examining cell growth in culture. The cell population is limited to one or two distinct types, each with identical properties, existing in a flat, i.e., uniformly resourced, environment. Classic problems in mathematical biology involve modeling the growth and decline of simple populations that are challenged by limited, flat resources and predation. Whether a population is human, animal, or microbial, there are many common features at the heart of population modeling that makes simple models worth discussing. Readers should be aware that realistic population models are multifactoral, such that system complexity is increased by introducing numerous state variables. Nevertheless, simple models can describe essential population features of realistic ecosystems, as we will show (see Problem 10.5). The only way to truly validate a model is through comparisons of predictions with experimental data.

Substantial insights may be gleaned from bacterial cultures. If you have ever inoculated a cell culture plate with bacteria, perhaps even by mistake from nonsterile lab practices, you have witnessed exponential population growth. The multiphase life cycle of common bacteria is well known from microbiology. We may use the mitotic cycle as an experimental clock to determine the age of cultures. When the life-cycle period is about the same for all individuals in the population and synchronized, then straightforward predictions of population sizes often match measurements over some number of cycles. Considering all healthy humans have bacteria living on or in our bodies that number more than 10 times the number of cells in our bodies (approximately 10^{14}), it becomes clear that bacterial growth is an important health matter from many perspectives.

10.4.1   Discrete and Continuous Linear Models

Assume 10^6 cells per milliliter are introduced into a standard culture medium at $t = 0$. Subsequently, we use cell counting methods to measure cell density $\rho(t)$ [cells/mL] each hour for 8 hours. The nine measurements are given in Table 10.1.

Measurement time is sampled: $t = mT$ for $T = 1$ hour and $0 \leq m \leq M - 1$ for $M = 9$. We ignore the effects of the cell's lifetime on the model by assuming the lifetime is much longer than the 8-hour measurement time.

Difference equations.
Cell density $\rho(t) = N(t)/V$ is the number of cells N found in a fixed-size media volume V [mL]. If we know the cell density at $t = mT$ and if the *cell growth factor r* [unitless] is constant, we can predict the density at the next measurement time, $t + T$:

$$r = \frac{\rho[m + 1]}{\rho[m]} \quad \text{or} \quad N[m + 1] = rN[m], \quad \text{for } r > 0. \tag{10.19}$$

Looking at the data in Table 10.1, we see that cell numbers double in a little more than 2 hours. That is, $N[m + nT] \simeq 2N[m]$ when $n = 2$, and therefore $r \simeq 2^{1/n} = \sqrt{2}$, and probably a little less. From the initial value $N[0]$, we can write the solution to the *difference equation* of (10.19) as $N[m] = r^m N[0]$. r includes all factors influencing growth of the cells, and in this simplified example r is a constant. When $r > 1$, the population grows. At $r = 1$, growth stops. When $0 < r < 1$, the population declines with time.

It is difficult to precisely observe cell doubling times (which are nominally equal to the mitotic cycle period) when cell numbers are sampled as coarsely as $T = 1$ hour. So let's change the cell-number model to be a continuous function of time and define

Table 10.1 Data from a cell growth experiment.

$t = mT$ (hr)	Cell density, $\rho(t)$ ($\times 10^6$ cells/mL)
0	1.0
1	1.4
2	1.9
3	2.7
4	3.6
5	5.2
6	6.9
7	10
8	13

a cell *growth rate*, k/hour. If bacterial growth is very small during an incremental interval between measurements $T \to 0$, then we will show $kT \simeq r - 1$. From (10.19),

$$N((m+1)T) = rN(mT), \quad \text{for } t = mT$$

$$N(t+T) = rN(t) = (kT+1)N(t)$$

$$\lim_{T \to 0} \left[\frac{N(t+T) - N(t)}{T} \right] = \frac{dN(t)}{dt} = kN(t).$$

From (10.6), we know $N(t) = N_0 e^{kt}$, which is exponential growth.

We can see that a difference equation or a differential equation results, depending whether the independent variable is discrete or continuous. The difference equation for cell density is $\rho[m+1] = r\rho[m]$, which has the solution $\rho[m] = \rho_0 r^m$, while the differential equation for cell number is $\dot{N}(t) = kN(t)$, with solution $N(t) = N_0 e^{kt}$. In both models, cell growth is exponential. These equivalent representations can be related by equating the growth factor constant r and growth rate constant k, where $\rho_0 = N_0/V$:

$$N(mT) = V\rho[m]$$

$$N_0 e^{kmT} = V\rho_0 r^m$$

$$kmT = m \ln r$$

$$k = (\ln r)/T. \tag{10.20}$$

This works dimensionally because r is unitless and T has the units of time. The last line of (10.20) is general and agrees with the earlier estimate $kT = r - 1$ for slow population growth conditions, i.e., when $\ln r \simeq r - 1$.[9]

Example 10.4.1. *Doubling time estimation: Find expressions for the cell population doubling time, t_2, using continuous and discrete-time models. Applying (10.20),*

$$\frac{N(t)}{N_0} = 2 = e^{kt_2}$$

$$t_2 = \frac{\ln 2}{k} = T \frac{\ln 2}{\ln r}.$$

These are the doubling times for the two models in terms of k and r. Writing k in terms of t_2, we also find the alternative expression,

$$N(t) = N_0 e^{(t \ln 2)/t_2} = N_0 2^{t/t_2}.$$

The population doubles at times $t = nt_2$, where $n = 1, 2, \ldots$.

---

[9] The first-order Taylor series of $\ln(1+x) \sim x$ for $x \in (-1, 1]$. Then letting $r = 1 + x$, we get $\ln(r) \sim r - 1$, but that approximation is lousy! Try it in MATLAB and you'll see. However, it helps the derivations of continuous results. A better approximation to $\ln(r)$ is the first term of the power series $\ln(r) = 2(r-1)/(r+1)$, but then the derivation doesn't work.

10.4.2   Doubling Time Computation from Measurements

We can apply least-square methods (see Section 3.12) to estimate model parameters from measurement data. The objective function to be minimized is

$$\arg \min_{\theta} \sum_m (\tilde{x}[m] - x[m; \theta])^2,$$

where $\{\tilde{x}\}$ are the time-series measurements, $\{x\}$ are the parametric model values at the same time points, and parameter vector $\theta = (\theta_1, \theta_2)^\top$.

If we can expect exponential growth conditions, we may take the logarithm of the cell-growth equation and apply the MATLAB function `polyfit`, which fits data to a first-order polynomial equation. Our continuous model applied to discrete measurements becomes $x[m; \theta] \triangleq \ln N[m] = kmT + \ln N_0$, with general linear form $x[m; \theta] = m\theta_1 + \theta_2$, where $\theta_1 = kT$ and $\theta_2 = \ln N_0$. For $\tilde{x}[m]$, we use the cell count numbers $\tilde{N}[m]$ from Table 10.1.

The function `P = polyfit(t,N',1)`, where $t = (T, \ldots, mT, \ldots, MT)^\top$, $N' = (\tilde{N}[1], \ldots, \tilde{N}[m], \ldots, \tilde{N}[M])^\top$, and the 1 in the third argument indicates a first-order polynomial (linear) regression. Also, $P = (\theta_1, \theta_2)^\top$, which are the parameters that minimize the objective function. Running this line for the data in Table 10.1 gives $\theta_1 = 0.3232$ and $\theta_2 = 13.8227$. Therefore, $k = 0.3232/\text{hour}$ and $N_0 = e^{13.8227} \simeq 1.01 \times 10^6$ cells, with the latter giving the initial number of cells within 1%. Now we can estimate the doubling time: $t_2 = \ln 2/k = 2.14$ hours. We find the more precise estimate $r = e^{kT} = 1.38$ is a bit smaller than our rough estimate of $\sqrt{2} = 1.41$. There are usually several ways to estimate model parameters, and it is generally a good idea to estimate parameters as many different ways as you can to gain confidence in your results.

10.4.3   More Realistic Discrete Models: Accounting for Limited Resources

In reality, cell numbers cannot grow exponentially for very many mitotic cycles because the available resource and waste removal limits would also need to grow exponentially. Since growth factor r in the discrete-time model represents all factors influencing growth, it is not realistic that it should remain constant as conditions change over time. Instead, this growth factor should reflect temporal changes in population density; e.g., $\rho[m + 1] = r(\rho[m]) \times \rho[m]$.[10] Hoppensteadt and Peskin [53] describe a version of the Verhulst model proposed in 1845 to predict human populations. Their growth factor is time-dependent to account for cell growth in a *limited-resource environment* via

$$r(\rho[m]; \theta) = \frac{r_0}{1 + \rho[m]/\rho'}, \quad \text{where } \theta = \begin{pmatrix} r_0 \\ \rho' \end{pmatrix} \text{ and } \rho(t) \geq 0 \; \forall \, t. \tag{10.21}$$

---

[10] To be clear, $r(\rho[m])$ states r is a continuous function of ρ that is a discrete function of integer m.

Parameters r_0, ρ' are constants; specifically, ρ' is a "stabilization constant" as described later in this chapter. Equation (10.21) is a first-order difference equation with variable coefficients.[11]

If the initial population is small compared to ρ', i.e., $\rho_0 \ll \rho'$, the model anticipates exponential growth with $r(\rho) \simeq r_0$. Conversely, at $\rho_0 \gg \rho', r(\rho_0) \simeq r_0 \rho'/\rho_0 < 1$ and we see a rapid decline in the population. For any initial value, the population stabilizes, i.e., $r = 1$, at $\rho(\infty) = (r_0 - 1)\rho'$. Assuming population growth is limited primarily by the available resources, then parameters r_0, ρ' can be combined to describe resources. However, the rate at which the population changes also depends on initial condition ρ_0. To see this graphically, we plot the populations estimated from the difference equation. In this simplified case, we can also solve the equation in closed form and then use the solution to predict time-varying populations. Let's look at this problem more closely.

Example 10.4.2. *Discrete models: Suppose we know the initial cell density ρ_0 and $\theta = (r_0, \rho')^{\top}$. Find a closed-form solution for cell density at an arbitrary time following inoculation, $\rho[m + 1]$, given (10.21). We find the solution $\rho[m]$ by applying an analytic iteration method [53],*

$$\rho[m + 1] = r(\rho[m]) \times \rho[m] = \frac{r_0 \rho[m]}{1 + \rho[m]/\rho'}.$$

Defining $X = \rho^{-1}$, we have

$$X[m + 1] = \frac{X[m]}{r_0}\left(1 + \frac{1}{\rho' X[m]}\right) = \frac{X[m]}{r_0} + \frac{1}{r_0 \rho'}.$$

Since

$$X[m] = \frac{X[m - 1]}{r_0} + \frac{1}{r_0 \rho'},$$

then

$$X[m + 1] = \frac{X[m - 1]}{r_0^2} + \frac{1}{r_0^2 \rho'} + \frac{1}{r_0 \rho'}$$

and generally

$$X[m + 1] = \frac{X_0}{r_0^{m+1}} + \frac{1}{r_0 \rho'}\left(1 + \frac{1}{r_0} + \frac{1}{r_0^2} + \ldots + \frac{1}{r_0^m}\right). \qquad (10.22)$$

The geometric series (see Section 5.7) in the parentheses of (10.22) equals

$$\sum_{n=0}^{m} r_0^{-n} = \frac{1 - r_0^{-m-1}}{1 - r_0^{-1}}.$$

---

[11] Modeling note: $r(\rho) = r_0/(1 + \rho/\rho')$ is given in (10.21). It is important that $r, r_0, \rho' > 0$ even though $\rho(t) \geq 0$. That is, the only time $r(\rho(t)) = 0$ is when $\rho(t) = 0$. To see this, let $r(\rho) = \epsilon \gtrsim 0$. Then $\epsilon(1 + \rho/\rho') = r_0$ or $\rho = \rho'(r_0/\epsilon - 1)$. Since $\rho \geq 0$ and $\rho' > 0, r_0/\epsilon - 1 \geq 0$ and so $r_0 > \epsilon$. This is all common sense. However, such details are important for stability when performing numerical modeling.

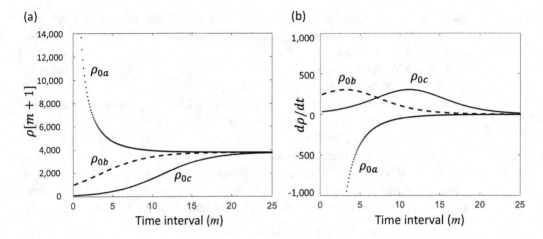

Figure 10.2 (a) Plots of cell population densities $\rho[m + 1]$ at discrete time intervals $t = mT$ from (10.23). The three lines are the results for initial populations: $\rho_{0a} = 10^6$, $\rho_{0b} = 10^3$, and $\rho_{0a} = 10^2$. In each case, $\rho' = 10^4$ and $r_0 = 1.38$, and consequently, $\rho(\infty) = 3,800$. (b) The derivatives of the population densities versus time are plotted. Notice that for the smallest initial population, ρ_{0c}, the initial growth is exponential.

Therefore, (10.22) becomes

$$X[m + 1] = X_0 r_0^{-m-1} + \frac{1 - r_0^{-m-1}}{(r_0 - 1)\rho'}$$

$$\frac{1}{\rho[m + 1]} = \frac{r_0^{-m-1}}{\rho_0} + \frac{1 - r_0^{-m-1}}{(r_0 - 1)\rho'}$$

$$\rho[m + 1] = \frac{(r_0 - 1)\rho_0 \rho'}{(r_0 - 1)\rho' r_0^{-m-1} + (1 - r_0^{-m-1})\rho_0}. \tag{10.23}$$

This method exploits the recursive nature of sequential density samples to solve the difference equation. Like any good model, the solution depends only on parameter vector θ and initial condition ρ_0. The independent variable is time via $m = t/T$.

Equation (10.23) is plotted in Figure 10.2a for three populations: The initial population A is larger than ρ', while populations B and C are smaller. Figure 10.2b shows the derivatives of the same three population density curves. As predicted, the cell population stabilizes at $\rho(\infty) = (r_0 - 1)\rho' = 3,800$, since $r_0 = 1.38$ and $\rho' = 10^4$. In Figure 10.2a, population C has $\rho_0/\rho' = 0.01$. This initial population density is very small, so initial cell growth is exponential. We see the same initial growth in the derivative for plot C in Figure 10.2b. In population B, $\rho_0/\rho' = 0.1$ and initially there is growth but it is not exponential. Population A is initially very dense, $\rho_0/\rho' = 100$, so it rapidly declines because $r(\rho_0) \simeq 0.01 \ll 1$.

10.4.4 Additional Discrete-Model Influences: Predation

You probably see the trend by now. Begin simply and gradually add influences to the model that match your experimental problem until you predict the measured behavior within an acceptable error level. Let's examine the situation of cells growing in limited resources that are also being actively eliminated by predation. For example, for bacteria in a mammalian system, one predator is the immune system. The Verhulst growth-factor model can be modified to predict *cell growth in limited resources with predation* [53]:

$$r(\rho; \boldsymbol{\theta}) = \frac{r'\rho}{1 + (\rho/\rho')^2}, \qquad (10.24)$$

where $\rho = \rho[m]$ [cell/mL] at time $t = mT$. The influence of the immune response is to multiply the limited-resource growth factor in (10.21) by an additional factor.[12] The result converts the difference equation $\rho[m + 1] = r(\rho'; \boldsymbol{\theta})\, \rho[m]$ to a nonlinear form that is difficult to solve in closed form. However, population dynamics can still be predicted numerically. $\boldsymbol{\theta} = (r', \rho')^\top$ are the constant model parameters. r' has units of [mL/cell] so that $r(\rho; \boldsymbol{\theta})$ is unitless.

Numerical methods for parametric model building involve proposing equations based on an understanding of the physics and biology of the situation, fitting the numerical predictions to experimental data, stopping when the regression yields residuals within the predetermined error bounds, and evaluating the parameters for their ability to describe, predict, and control the system under investigation. Least-squares techniques (Section 10.4.2) will come in handy here.

In situations where we cannot solve the difference equation, we can often find points in the plotted results where all the influences balance to give stable populations. Of course, you may always plot the equations to predict behavior. However, situations that yield stable populations (equilibrium points) are special because they may define eigenvalues of the system, as we will describe in Section 10.7. The difference equation,

$$\rho[m + 1] = \frac{r'\rho^2[m]}{1 + (\rho[m]/\rho')^2},$$

has two nontrivial roots, or stable points where $\rho[m + 1] = \rho[m] \ \forall m$. Substituting this into the preceding expression and solving for $\rho[m]$ yields $\rho = r'\rho'[1 \pm \sqrt{1 - 4/(r'\rho')^2}]/2$. These give the two population equilibrium densities.

10.5 Linearizing Equations via Taylor Series

When first encountering a nonlinear system of equations, a good approach is to linearize the system about equilibrium points. We found a solution to a linear equation

---

[12] From (10.21), $r = r_0/(1 + \rho/\rho')$. Multiplying r by predation factor $[(r'\rho/r_0)/(1 - \rho/\rho')]$ results in (10.24).

about its equilibrium point in Example 10.1.4, and now we describe a general method for linearizing nonlinear equations about their equilibria.

Let's begin with the general ODE system expression (10.2); that is, $\dot{\mathbf{y}} = \mathbf{f}(\mathbf{y})$, where here we suppress t and θ dependencies in the notation until necessary for clarity. The symbol $\mathbf{f}$ represents a vector of equations, each of which is nonlinear in state variables $\mathbf{y}$. If a steady-state or equilibrium point is given by $\mathbf{y}_s = (y_{s1}, \ldots, y_{sJ})^\top$, then the deviation about equilibrium is $\xi(t) = \mathbf{y}(t) - \mathbf{y}_s$. Because $\mathbf{y}_s$ is time independent, then as we saw in Example 10.1.4, $\dot{\mathbf{y}} = \dot{\xi}$. Consequently, writing out (10.2) initially as one equation with J variables, we find for $y_n = y_{sn} + \xi_n$ that

$$\dot{\mathbf{y}} = f(\mathbf{y}) = f(y_1, \ldots, y_n, \ldots, y_J) = f(y_{s1} + \xi_1, \ldots, y_{sn} + \xi_n, \ldots, y_{sJ} + \xi_J) = \dot{\xi}.$$

The second-order Taylor series expansion of $\dot{\xi}(t)$ is expressed in two equivalent notations:

$$\dot{\xi} = f(y_{s1}, \ldots, y_{sJ}) + \sum_{n=1}^{J} \xi_n \left.\frac{\partial f(y_n)}{\partial y_n}\right|_{y_n = y_{sn}}$$

$$+ \frac{1}{2}\sum_{n=1}^{J}\sum_{m=1}^{J} \xi_n \xi_m \left.\frac{\partial^2 f(y_n, y_m)}{\partial y_n \partial y_m}\right|_{\substack{y_n = y_{sn}\\ y_m = y_{sm}}} + O(3)$$

$$= f(\mathbf{y}_s) + \sum_{n=1}^{J} \xi_n f^{(1)}(y_{sn})$$

$$+ \frac{1}{2}\sum_{n=1}^{J}\sum_{m=1}^{J} \xi_n \xi_m f^{(2)}(y_{sn}, y_{sm}) + O(3). \quad \text{(Taylor series)} \quad (10.25)$$

The first three terms on the right side of both forms are the zeroth, first, and second derivatives for the J variables evaluated at the equilibrium point in J-dimensional state space.[13] The exact expansion includes all orders of derivatives. The last term, $O(3)$, indicates we are adding derivative terms of third and higher order. In practice, high-order derivatives can be neglected as small. The zeroth-order derivative at the equilibrium values is constant, so that $f(\mathbf{y}_s) = 0$. The linearized approximation is

$$\dot{\xi}(t) = \sum_{n=1}^{J} \xi_n(t)\frac{\partial f(y_{sn})}{\partial y_n} = \sum_{n=1}^{J} \xi_n(t) f^{(1)}(y_{sn}). \quad (10.26)$$

We can use an equal sign because it is understood that (10.26) is the linearized version of the nonlinear ODE. Specifically, it is the first-order Taylor series expansion of $\dot{y} = \dot{\xi}$ about state-space equilibrium point $\mathbf{y}_s$. We see that (10.26) is linear in the

[13] Common compact notations for the first and second partial derivatives evaluated at equilibrium points y_{sn} and y_{sm} are $f^{(1)}(y_{sn})$ and $f^{(2)}(y_{sn}, y_{sm})$, respectively. The differentiation variables are implicitly indicated, but are clear when comparing the two forms of (10.25).

shifted state variable $\xi(t)$ because the derivatives are all constants. I added time depen-dence to remind us which factors are time varying in the final form.

A truncated Taylor-series expansion is a basis decomposition that facilitates lin-earization. By studying functions near equilibrium, we are able to convert nonlinear, inhomogeneous, autonomous forms to linear, homogeneous equations with closed-form solutions. These solutions yield accurate predictions provided we do not venture too far from equilibrium.

The ith of J equations, each with J variables, is given by

$$\dot{\xi}_i(t) = \sum_{n=1}^{J} \frac{\partial f_i(y_{sn})}{\partial y_n} \xi_n(t), \quad \text{for } 1 \le i, n \le J. \tag{10.27}$$

By expressing all J equations in $J \times 1$ vector $\dot{\xi}$, (10.27) becomes

$$\dot{\xi}(t) = \mathbf{A}\xi(t) = \begin{pmatrix} \frac{\partial f_1}{\partial y_1} & \cdots & \frac{\partial f_1}{\partial y_n} & \cdots & \frac{\partial f_1}{\partial y_J} \\ & \vdots & & & \\ \vdots & & \frac{\partial f_i}{\partial y_n} & & \vdots \\ & & \vdots & & \\ \frac{\partial f_J}{\partial y_1} & \cdots & \frac{\partial f_J}{\partial y_n} & \cdots & \frac{\partial f_J}{\partial y_J} \end{pmatrix} \begin{pmatrix} \xi_1(t) \\ \vdots \\ \xi_n(t) \\ \vdots \\ \xi_J(t) \end{pmatrix}. \quad \text{(Linearized system)}$$

$$\tag{10.28}$$

$\mathbf{A}$ is a Jacobian matrix showing us how function $\mathbf{f}$ linearly maps ξ into $\dot{\xi}$. In sum-mary, the linearized version of $\dot{\mathbf{y}}(t) = \mathbf{f}(\mathbf{y}(t))$ is $\dot{\mathbf{y}}(t) = \mathbf{A}\mathbf{y}(t)$, and (10.26) is one row of (10.28). Equation (10.28) is a linearized approximation of the general nonlinear system of equations $\dot{\mathbf{y}} = \mathbf{f}(\mathbf{y})$, where $\xi(t) = \mathbf{y}(t) - \mathbf{y}_s$.

10.6   Nonlinear Continuous Models: Logistic Equation

The following classic approach to population modeling treats the state variable describing a large *single population* system as a continuous function of time, $N(t)$. Positive (negative) terms are added to the ODE to represent influences that increase (decrease) $N(t)$, which replaces modifications to the single growth factor r as discussed in Section 10.4 for discrete-time state variables.

The growth of a population with identical individuals is ultimately held in check by birth and death rates and resource limitations, such as the supply of nutrients or oxygen or by a buildup of toxins, ionic imbalances seen as a change in pH (especially for cell cultures), or some combination of influences. The change in population number over time $\dot{N}$ is modeled as being equal to one positive term that increases in proportion to

N and a second negative term that increases in proportion to N^2. The resulting ODE is nonlinear in N:

$$\dot{N}(t) = kN(t) - \alpha N^2(t), \qquad \text{where } k, \alpha, N \geq 0. \qquad (10.29)$$

The quadratic term ensures that ultimately growth will be limited as that term dominates at large N, making $\dot{N}$ negative.

Constant parameter k is the *net* growth rate of the population assuming no resource limitations, while α is the rate of decline due specifically to resource limitations. A quadratic decline is expected for identical individuals consuming resources and excreting waste at roughly the same rate, assuming nutrients and waste are able to freely move throughout the environment.

10.6.1   Solution to the Linearized ODE

This is a reasonably straightforward ODE to solve, but one that affords us an opportunity to explore and compare the exact nonlinear solution with the linear approximation.

Considering the linearized problem where (10.29) is near equilibrium, i.e., at $\dot{N} \simeq 0$, we find the trivial point $N_s = 0$ and the more interesting point at $N_s = k/\alpha$. Since the trivial equilibrium point occurs when the population is zero, we only consider the nontrivial equilibrium point. Combining (10.26) and (10.29) where $y = N$, $J = 1$, $\xi = (N - N_s)$, and $f(N) = kN - \alpha N^2$, we find

$$\dot{N} = \dot{\xi} = f^{(1)}(N_s)\,\xi = (k - 2\alpha N_s)\xi$$

$$= \left(k - 2\alpha \left(\frac{k}{\alpha} \right) \right) \xi = -k\xi.$$

This is a problem we have already solved several times. For initial condition, $\xi_0 = N_0 - N_s$, we can write the solution directly:

$$\xi(t) = N(t) - N_s = \xi_0\, e^{-kt}$$

$$N(t) = N_s + (N_0 - N_s)\, e^{-kt}. \qquad \text{(Linear Approximation)} \qquad (10.30)$$

Because the exponent is negative, the solution is stable (more on this later). This solution is plotted in Figure 10.3 for several parameters. We will return to (10.30) after examining the exact solution to the nonlinear problem.

10.6.2   Solution to the Nonlinear ODE

Often nonlinear ODEs are challenging to solve exactly in closed form, which is why we first linearized them to gain a limited understanding of the overall system behavior near N_s. We now solve this particular nonlinear problem by applying the method of *separation of variables.* It remains convenient to interpret parameters in terms of the equilibrium point at $N_s = k/\alpha$. From (10.29),

$$\frac{dN}{dt} = kN\left(1 - \frac{\alpha}{k}N\right) = kN\left(1 - \frac{N}{N_s}\right) = kN\left(\frac{N_s - N}{N_s}\right).$$

For the initial condition $N(0) = N_0$, we can separate the variables and apply partial fractions:

$$\int_{N_0}^{N} dN' \frac{N_s}{N'(N_s - N')} = \int_0^t k \, dt'$$

$$\int_{N_0}^{N} \frac{dN'}{N'} + \int_{N_0}^{N} \frac{dN'}{N_s - N'} = kt$$

$$\ln\frac{N}{N_0} - \ln\frac{N_s - N}{N_s - N_0} = \ln\frac{N(N_s - N_0)}{(N_s - N)N_0} = kt \tag{10.31}$$

$$N(t) = \frac{N_0 N_s e^{kt}}{N_s - N_0 + N_0 e^{kt}}. \quad \text{(Exact solution)}$$

Equation (10.31) is an example of the *logistic law of growth*. As with the linear approximation of (10.30), the exact solution of (10.31) satisfies the initial condition, and we see that $N(\infty) = N_s$. When $N_s \gg N_0$, then $N(t) \simeq N_0 e^{kt}$ and the population grows exponentially, as seen near the origin of Figure 10.3b for $N_0 = 1$. However, when $N_s \ll N_0$, then $N(t)$ declines "exponentially," as also shown in Figure 10.3 for $N_0 = 200$. The linear approximation holds true only when $N(t) \simeq N_s$. For example, see the plots for $N_0 = 80$. Notice the similarities between the logistic equation of

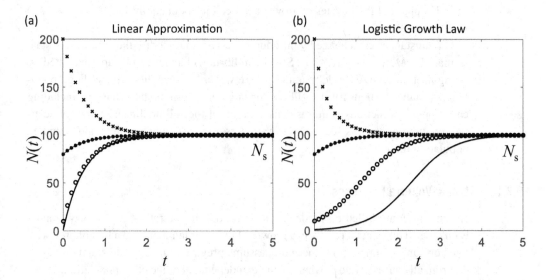

Figure 10.3 Plots of the linear approximation from (10.30) (a) and the exact logistic equation from (10.31) (b) for different initial conditions. In all cases, $k = 2$ and $N_s = 100$. However, N_0 was varied as shown. As expected, the approximation is reasonably accurate only when $N_0 \simeq N_s$.

(10.31) in continuous time that is plotted in Figure 10.3 and the plots from the discrete Verhulst model in Figure 10.2.

10.6.3   Summary

In this section, we reviewed single population models that are discrete and continuous functions of time. We also examined models with infinite resources, limited resources, and predation. This led to classic problems involving linear and nonlinear ODEs, equations with constant and variable coefficients, and the logistic equation. Next, we examine the interactions within a system of two coupled populations with predation.

10.7   The Predator–Prey Problem

Consider a healthy ecosystem where a population of lynx (also known as bobcats) preys on a population of rabbits. They need each other to maintain equilibrium populations given the limited environmental resources. Depending on birth, death, and interaction rates, and other factors, the instantaneous population of one of these animals may fluctuate. If it does, so will the other to form a codependent system of *coupled oscillators*.

An increase in prey numbers $N(t)$ results in increasing predator numbers $M(t)$ a short time later as the predator food supply increases. As the rate of predation approaches the birth rate of the prey, N declines, followed by a decrease in M some time later. Without additional influences, these two simple population models remain linked in phase-shifted cycles of growth and decline about equilibrium value (M_s, N_s) in 2-D state space.

Circumstances can change by introducing factors that modify the equilibrium state so that $(N_s, M_s)_1 \longrightarrow (N_s, M_s)_2$. Small equilibrium changes can be approximated by changing linear ODE model parameters. However, to accurately model large population variations about fixed equilibria or large equilibrium-state changes, including catastrophic extinctions, requires consideration of the full nonlinear model for accurate predictions. In this section, we examine some variations on this widely studied population model.

10.7.1   Lotka–Volterra Equations

Assume prey are killed by predators in proportion to the rate at which they contact each other. This contact results in a growth of predator numbers and a loss of prey with rate constant α [time]$^{-1}$. Furthermore, assume prey reproduce at rate k and predators die from all causes at rate β, where both are independent of the other population. These assumptions give rise to the *Lotka–Volterra equations*:

$$\dot{M}(t) = \gamma \alpha N(t)M(t) - \beta M(t) \qquad \alpha \geq 0, \ \beta > 0, \qquad \text{(Predator)}$$
$$\dot{N}(t) = kN(t) - \alpha N(t)M(t), \qquad\qquad\qquad\qquad \text{(Prey)} \qquad (10.32)$$

Here, γ is a unitless ratio of predator births that survive to maturity relative to the prey birth rate. The model makes a strong assumption that the availability of prey is the only limit to predator birth and that the natural lifetime of prey is much longer than the time between interactions with predators. As in (10.29), the right sides of the equations contain terms with two powers of population variables, making them nonlinear ODEs. Also, all parameters are positive so that positive terms on the right side describe increases in populations and negative terms are losses.

10.7.2 Solutions to Linearized Equations

To linearize the Lotka–Volterra equations, we first identify the equilibrium conditions by setting (10.32) to zero and solving for $M(t)$ and $N(t)$:

$$\dot{M}(t) = 0 = \gamma \alpha N(t)M(t) - \beta M(t)$$
$$\dot{N}(t) = 0 = kN(t) - \alpha N(t)M(t)$$

yield only one nontrivial solution,

$$\mathbf{N}_s = \begin{pmatrix} M_s \\ N_s \end{pmatrix} = \begin{pmatrix} k/\alpha \\ \beta/\alpha\gamma \end{pmatrix}. \tag{10.33}$$

Elements of vector $\mathbf{N}_s$ contain the equilibrium points for both populations, which is the point (M_s, N_s) in this 2-D state space. Applying (10.33) to (10.28), where

$$\mathbf{N}(t) = \begin{pmatrix} M(t) \\ N(t) \end{pmatrix}, \quad \xi(t) = \mathbf{N}(t) - \mathbf{N}_s, \quad \dot{\mathbf{N}}(t) = \begin{pmatrix} \dot{M}(t) \\ \dot{N}(t) \end{pmatrix} = \dot{\xi}(t),$$

$$\text{and } \mathbf{A} = \begin{pmatrix} \frac{\partial \dot{M}}{\partial M} & \frac{\partial \dot{M}}{\partial N} \\[6pt] \frac{\partial \dot{N}}{\partial M} & \frac{\partial \dot{N}}{\partial N} \end{pmatrix}_{M_s, N_s} = \begin{pmatrix} \gamma \alpha N_s - \beta & \gamma \alpha M_s \\ -\alpha N_s & k - \alpha M_s \end{pmatrix}, \tag{10.34}$$

yields the linearized solution

$$\dot{\mathbf{N}}(t) = \dot{\xi}(t) = \mathbf{A}\xi(t)$$

$$= \begin{pmatrix} (\gamma \alpha N_s - \beta)\xi_1(t) + \gamma \alpha M_s \xi_2(t) \\ -\alpha N_s \xi_1(t) + (k - \alpha M_s)\xi_2(t) \end{pmatrix} = \begin{pmatrix} (\frac{\gamma \alpha \beta}{\gamma \alpha} - \beta)\xi_1(t) + \frac{\gamma \alpha k}{\alpha}\xi_2(t) \\ \frac{-\alpha \beta}{\alpha \gamma}\xi_1(t) + (k - \frac{\alpha k}{\alpha})\xi_2(t) \end{pmatrix}$$

$$= \begin{pmatrix} \gamma k \xi_2(t) \\ -\frac{\beta}{\gamma}\xi_1(t) \end{pmatrix} = \begin{pmatrix} \gamma \alpha M_s \xi_2(t) \\ -\alpha N_s \xi_1(t) \end{pmatrix} = \begin{pmatrix} \gamma k(N(t) - N_s) \\ -\frac{\beta}{\gamma}(M(t) - M_s) \end{pmatrix}. \tag{10.35}$$

The three forms of the solution shown on the right side of (10.35) are based on the equilibrium point from (10.33). Interestingly, $\dot{\xi}_1 \propto \xi_2$ and $\dot{\xi}_2 \propto \xi_1$.

The problem includes two state variables M, N and one independent variable t. We can plot the two time series $M(t)$ and $N(t)$ and compare them, or we can eliminate time from the equations and plot $N(M)$ to view the behavior of both populations in state space.

10.7.3   Trajectories in State Space

To eliminate time from the solution, rewrite (10.35) so that both equations equal $\alpha\, dt$,

$$\frac{d\xi_1}{\gamma M_s \xi_2} = \alpha\, dt = -\frac{d\xi_2}{N_s \xi_1}$$

$$\frac{\xi_1 \, d\xi_1}{\gamma M_s} = -\frac{\xi_2 \, d\xi_2}{N_s} \quad \dots \text{ integrating both sides } \dots$$

$$\frac{\xi_1^2}{\gamma M_s} + \frac{\xi_2^2}{N_s} = a, \tag{10.36}$$

where a is a constant combining the two integration constants to reveal an ellipse. Unitless a is determined from the initial conditions: When $\xi_2(t) = N(t) - N_s = 0$, we have $\xi_1^2(t)/\gamma M_s = a$ so that $\xi_{1min,max} = \pm\sqrt{\gamma a M_s} = \pm\sqrt{\gamma a k/\alpha}$. The two values are symmetric about M_s. Similarly, when $\xi_1(t) = M(t) - M_s = 0$, then $\xi_{2min,max} = \pm\sqrt{a N_s} = \pm\sqrt{a\beta/\alpha\gamma}$. Equation (10.36) provides the expressions

$$\xi_2(\xi_1) = \left(N_s \left(a - \frac{\xi_1^2}{\gamma M_s} \right) \right)^{1/2} \quad \text{and} \quad N(M) = N_s + \left(N_s \left(a - \frac{(M - M_s)^2}{\gamma M_s} \right) \right)^{1/2}.$$

$$\tag{10.37}$$

The second form is plotted in Figure 10.4c.

10.7.4   Time Series

Expressions for $\xi_1(t)$ and $\xi_2(t)$ are found by noting $a = \xi_{1max}^2/\gamma M_s$ and then solving (10.36) for $1/\xi_2$:

$$\frac{\xi_1^2}{\gamma M_s} + \frac{\xi_2^2}{N_s} = \frac{\xi_{1max}^2}{\gamma M_s}$$

$$\frac{\xi_{1max}^2 - \xi_1^2}{\gamma M_s} = \frac{\xi_2^2}{N_s}$$

$$\frac{1}{\xi_2} = \left(\frac{\gamma M_s}{N_s(\xi_{1max}^2 - \xi_1^2)} \right)^{1/2}. \tag{10.38}$$

Substituting (10.38) into the first line of (10.36) and integrating the left side from ξ_{1max} to $\xi_1(t)$ and the right side from 0 to t, we find

$$\frac{d\xi_1}{\gamma \alpha M_s} \left(\frac{\gamma M_s}{N_s(\xi_{1max}^2 - \xi_1^2)} \right)^{1/2} = \frac{1}{\alpha\sqrt{\gamma M_s N_s}} \frac{d\xi_1}{(\xi_{1max}^2 - \xi_1^2)^{1/2}} = dt$$

$$\int_{\xi_{1max}}^{\xi_1} \frac{d\xi'_1}{(\xi_{1max}^2 - \xi'^2_1)^{1/2}} = \alpha\sqrt{\gamma M_s N_s} \int_0^t dt'$$

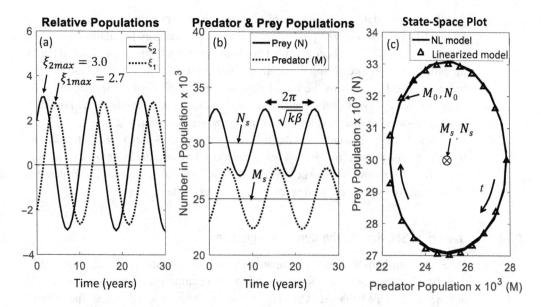

Figure 10.4 Solutions to the linearized Lotka–Volterra equations (10.32) for $M_0, N_0 \simeq M_s, N_s$. (a) Time series for the relative populations (in thousands), $\xi_1(t), \xi_2(t)$. (b) Time series for the populations, $M(t), N(t)$. (c) Trajectories through state space for the linear solution via (10.37) and the nonlinear solution (computed numerically for $M_0, N_0 \simeq M_s, N_s$). The initial populations and equilibrium populations are indicated. Arrows show the rotation with time. Since the maximum deviation of the population oscillations is within approximately 10% of the equilibrium population values, the linearized and nonlinear solutions are nearly equal. For this model, $k = 0.5/\text{year}, \alpha = 0.02/\text{year}, \gamma = 1.0, \beta = 0.6/\text{year}, a = 0.3, M_0 = 2.3 \times 10^4$, $N_0 = 3.2 \times 10^4, M_s = 2.5 \times 10^4, N_s = 3.0 \times 10^4$. The temporal oscillation frequency is $\sqrt{k\beta}/2\pi$.

$$\cos^{-1}\left[\frac{\xi_1'}{\xi_{1max}}\right]_{\xi_{1max}}^{\xi_1} = \cos^{-1}\left(\frac{\xi_1}{\xi_{1max}}\right) - \cos^{-1}(1) = \alpha\sqrt{\gamma M_s N_s}\, t$$

$$\xi_1(t) = \xi_{1max}\cos\left(2\pi\frac{\alpha\sqrt{\gamma M_s N_s}}{2\pi}t\right) \text{ or } M(t) = \sqrt{\frac{\gamma ak}{\alpha}}\cos\left(\Omega_0 t\right) + M_s. \quad (10.39)$$

The solution to the predator population as determined from the linearized system of ODEs shows the predator number oscillates in time about equilibrium value M_s with temporal frequency $u_0 = \alpha\sqrt{\gamma M_s N_s}/2\pi = \sqrt{k\beta}/2\pi$ and radial frequency $\Omega_0 = \sqrt{k\beta}$. We can determine $\xi_{1max} = \xi_1(0)$ from an experiment or use $\xi_{1max} = \sqrt{\gamma a M_s} = \sqrt{\gamma ak/\alpha}$ when numerical modeling, as we did earlier.

It is now straightforward to find

$$\xi_2(t) = \xi_{2\,max}\cos\left(2\pi\frac{\alpha\sqrt{\gamma M_s N_s}}{2\pi}t + \pi/2\right) \text{ or } N(t) = -\sqrt{\frac{a\beta}{\alpha\gamma}}\sin\left(\Omega_0 t\right) + N_s,$$

$$(10.40)$$

where $\xi_{2max} = \sqrt{a N_s} = \sqrt{a\beta/\alpha\gamma}$.

There is a fixed 90-degree phase shift between the two population cycles, as seen in Figure 10.4a and 10.4b. The predator population leads the prey population in time by a quarter cycle. The frequency of the population cycles are equal and determined by the birth rate of the prey k and the death rate of the predator β. The population amplitudes are also functions of the model parameters. The plots of $N(M)$ from (10.37) in Figure 10.4c show that the linearized solution closely approximates the nonlinear equations solved numerically in Section 10.7.5 when $\xi_1(t), \xi_2(t) \ll M_s, N_s \; \forall \, t$.

Rubinow [91] points out in his discussion of the predator–prey problem that during times after both populations are significantly reduced by an outside factor, it is the prey population that recovers first. The reason is that the birth rate of the prey is independent of the predator population, so it recovers at its natural rate k.

10.7.5   Numerical Solutions to the Nonlinear Equations

MATLAB can help us directly compare the linearized solutions of (10.37), (10.39), and (10.40) to the nonlinear solution found by direct numerical computation of (10.32) in function `lotka1` via `ode45`. The code shown here is set up to give the plots of Figure 10.4. Increasing the initial populations M_0, N_0 away from the equilibrium point gives the results of Figure 10.5. Notice that solutions to the nonlinear equations under the small-cycle-amplitude approximation give numerical results for the population amplitude and frequency that are equal to the values computed in closed form for the linearized equations.

```
%%%%%%% This nonlinear solver has parameters set to give linear response since N0 ~ Ns & M0 ~ Ms %%%%%%%%
% Observe NL numerical solutions by changing N0 and commenting out linearized solution section %%%%%%%%%%%
%%%%%% Careful, I use N0, Ns, and N to indicate both populations, e.g., N = [M;N]   %%%%%%%%%%%%%%%%%%%%%%%%
k=0.5;alpha = 0.02;                          % k=prey birth rate;alpha=interaction rate or pred efficiency
gamma = 1.0;beta = 0.6;                       % gamma=ratio of predator to prey birth; beta=pred death rate
theta=[k alpha;gamma beta];                   % parameter matrix used on ode45
Ns=[k/alpha;beta/(alpha*gamma)];              % equilib point from model parameters [Ms;Ns]
N0=[23;32];                                    % initial population vector: N0(1)=predators, N0(2)=prey
trange=[0 31];                                 % time range (years) for modeling in ODE45
[tt,N]=ode45(@(tt,N) lotka1(tt,N,theta),trange,N0);    % solve 2-D nonlinear system of eqs
xi=[N(:,1)-Ns(1),N(:,2)-Ns(2)];               % relative populations [M-Ms;N-Ns]
subplot(1,3,1);plot(tt,xi);title('Relative Populations');
xlabel('time (yrs)');legend('\xi_1','\xi_2');
subplot(1,3,2);plot(tt,N)                       % time series for (M,N) populations
title('Predator & Prey Populations');xlabel('time (yrs)');
ylabel('Number in Population x 10^3');legend('Predator (M)','Prey (N)','Location','North')
subplot(1,3,3);plot(N(:,1),N(:,2));hold on  % state-space trajectory for (M,N) output
title('Phase Plane Plot');xlabel('Predator Population x 10^3');ylabel('Prey Population x 10^3')
hold on;plot(Ns(1),Ns(2),'+');                % add equilibrium point to plot
%
% Compute and plot the closed-form linearized solution
%
x1min=min(N(:,1)-Ns(1));x1max=max(N(:,1)-Ns(1));
dx1=(x1max-x1min)/10;a=3^2/Ns(2);x1=x1min:dx1:x1max;        % find \xi_1 range...
x2=real(sqrt((a-x1.^2*alpha/(gamma*k))*beta/(alpha*gamma)));   % ...then compute (10.37)
plot(x1+Ns(1),x2+Ns(2),'k^');hold on;plot(x1+Ns(1),-x2+Ns(2),'k^');hold off;
legend('NL model','Linearized Model')                        % plot result on same graph
%
% Lotka-Volterra equations as external MATLAB function for use in ODE45
%
function yp=lotka1(t,y,p)
%    modified MATLAB's lotka function to accept parameters
    yp=diag([-p(2,2) + p(2,1)*p(1,2)*y(2), p(1,1) - p(1,2)*y(1)])*y; % must be column vector
end                                           % to understand this, understand what diag(.) does!
```

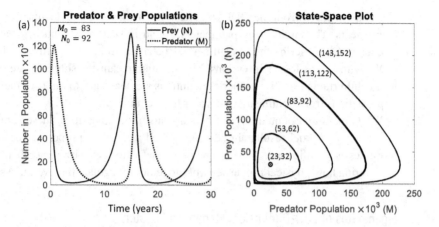

Figure 10.5 Exact numerical solutions to the Lotka–Volterra equations. (a) The time series for a model with the same parameters as that in Figure 10.4 except now the initial populations are large compared to the equilibrium populations; i.e., $M_0 > M_s, N_0 > N_s$. Under these conditions, the solutions are highly nonlinear to the point where minimum populations become perilously close to zero. If they reach zero, the system collapses, which is less likely to occur in linearized solutions. (b) State-space trajectories for the initial population values (M_0, N_0) indicated; other parameters are the same. The equilibrium point and the smallest initial populations are the same as those in Figure 10.4. One of these trajectories is plotted in the time series in (a).

With this script, you have an example of how to generate numerical solutions. The `lotka1` function at the end is called by `ode45`. Typing "help" in MATLAB will not give this form of `ode45`. The *handle* indicated by the @ sign in the argument of `ode45` was modified to allow a matrix of parameters to be passed into `lotka1`. You are asked to adapt this code in Problem 10.5 to model measurements published a century ago.

Figure 10.5 shows the effects of selecting initial populations far from equilibrium populations. We see the populations remain cyclic with somewhat longer periods. However, the time series are no longer sinusoidal and the state-space plots are no longer elliptical.

Summary.
When modeling compartmental flows in Example 10.3.1, we noticed that the physical parameters of flow q and volume V combined to form engineering parameter q/V with units of a rate constant. In that example, the ratio described how quickly the concentration of a molecular agent in the patient's circulation changed over time, a systemic property. Systemic engineering parameters like q/V emerge from closed-form solutions to linearized ODEs.

Similarly the predator–prey system has *physical parameters* that regulate individual components and *engineering parameters* for controlling the overall linearized system behavior. The physical parameters in this basic predator–prey problem are rate constants α, β, γ, k from the ODEs that describe the growth and decline of individual components. The engineering parameters, found in vectors $\mathbf{N}_s$ and ξ_{max} and scalar frequency Ω_0, describe the overall system behavior from the solution of linearized ODEs with closed-form expressions. Linearization means the solutions are predictive near equilibrium point $\mathbf{N}_s$. Numerical solutions are necessary to improve the accuracy of predictions from nonlinear systems far from equilibrium.

While measured population numbers are stochastic quantities, these solutions are deterministic. Model results are best interpreted as average properties for stationary processes at equilibrium. Models are most useful for revealing overall trends and discovering how physical properties work together to regulate the overall system.

10.7.6   Eigenstates of Low-Dimensional Linearized Systems

In Section 10.2, we found that applying the ansatz to an autonomous linear system of equations expressed in matrix form, i.e., $\dot{\xi}(t) = \mathbf{A}\xi(t)$, gave a characteristic equation. Solutions are found from *eigenstates* of Jacobian matrix $\mathbf{A}$. From (10.34),

$$\det(\mathbf{A} - \lambda\mathbf{I}) = \begin{vmatrix} \gamma\alpha N_s - \beta - \lambda & \gamma\alpha M_s \\ -\alpha N_s & k - \alpha M_s - \lambda \end{vmatrix} = \begin{vmatrix} -\lambda & \gamma k \\ -\beta/\gamma & -\lambda \end{vmatrix} = 0, \quad (10.41)$$

yielding $\lambda^2 + k\beta = 0$ with imaginary roots at $\lambda_{1,2} = \pm i\sqrt{k\beta}$ that are eigenvalues of the linearized population system.

The two eigenvalues form a purely imaginary conjugate pair, which tells us that the two state variables oscillate with constant amplitude and equal frequency about $\mathbf{N}_s$ for as long as the system can maintain the equilibrium state. The magnitude of the eigenvalues is the radial frequency of oscillation, $|\lambda| = \Omega_0 = \sqrt{k\beta}$.

Equations (10.39) and (10.40) may be expressed in terms of their eigenvalues,

$$\xi_1(t) = \xi_{1max}\cos(\Omega_0 t) = \xi_{1max}\Re\{e^{\lambda_1 t}\} \tag{10.42}$$
$$\xi_2(t) = \xi_{2max}\cos(\Omega_0 t + \pi/2) = \xi_{2max}\Re\{e^{\lambda_2 t + \pi/2}\} = \xi_{2max}\Im\{e^{\lambda_2 t}\},$$

where $\Re$ and $\Im$ indicate the real and imaginary parts of the complex arguments, respectively.

This linearized system has two linearly independent variables M, N or, equivalently, ξ_1, ξ_2, which is the rank of $\mathbf{A}$; this occurs despite thousands of individuals in each subpopulation being modeled as having identical properties. Thus, it is not surprising that this *simple system* has just a few engineering parameters regulating overall behavior. For each eigenvalue, there is an eigenvector describing temporal dynamics of the subpopulations – those are *functional modes* of the system. Each mode has a characteristic time scale given by $2\pi/\Omega_0$ that is spatially uniform (featureless over the region of interest). We would need to model a higher-order system of equations with additional independent variables that include spatial variations within interacting subgroups. The richness of properties wrought by large and diverse sets of interacting subgroups generates a *complex system* expressing *emergent properties* like those found in nature. The linearized model of (10.42) is just a first step, but a very important one because the "transparency" provided by eigenanalysis reveals the overall system properties.

10.7.7 Transitions

Returning to the 2-D Lotka–Volterra equations of (10.32), let's examine the response of these populations to a sudden change in equilibrium state at $t = t_0$. When $t \leq t_0$, we will assume the system is as described earlier and the parameters given in Figure 10.4 hold. When $t > t_0$, however, a new loss term (hunter harvest) appears in the predator equation that reduces their numbers at constant rate κ/year, where $\kappa < 0$. To account for this change, we add the term $\kappa M(t) \, \text{step}(t - t_0)$ to the predator equation to find

$$\dot{\mathbf{N}}(t) = \begin{pmatrix} \gamma \alpha N(t)M(t) + \kappa \, M(t) \, \text{step}(t - t_0) - \beta M(t) \\ kN(t) - \alpha N(t)M(t) \end{pmatrix} \qquad \text{(Exact)}$$

$$\dot{\xi}(t) = \mathbf{A}\xi(t) = \begin{pmatrix} 0 & \gamma \alpha M_s \\ -\alpha N_s & 0 \end{pmatrix} \xi(t), \qquad \text{(Linearized)} \quad (10.43)$$

where now $N_s(t) = (\beta - \kappa \, \text{step}(t - t_0))/\alpha\gamma$ is a function of time. To maintain linearity, we model the time dependence as piecewise constant: At $t = t_0$, the equilibrium-state vector suddenly shifts from $\mathbf{N}_s = \begin{pmatrix} M_s \\ N_s \end{pmatrix} = \begin{pmatrix} k/\alpha \\ \beta/\alpha\gamma \end{pmatrix}$ to $\begin{pmatrix} k/\alpha \\ (\beta - \kappa)/\alpha\gamma \end{pmatrix}$. That's practical, but not rigorous.

The equilibrium state, and therefore system properties, change when the eigenstates change. At $t \leq t_0$, $\lambda_{1,2} = \pm i\sqrt{k\beta}$ as found from (10.41); birth and death rates must remain positive, $k, \beta > 0$. At $t > t_0$, $\lambda_{1,2} = \pm\sqrt{-k(\beta - \kappa)}$, where κ can be positive or negative. When the imposed change in predator population κ is much less than the natural death rate of predators, $|\kappa| \ll \beta$, the system responds linearly to the small change, the eigenvalues remain imaginary, and so the populations continue to oscillate but with new engineering parameters. If $\kappa \lesssim \beta$, the larger change maintains imaginary eigenvalues, but the change is unlikely to be modeled accurately using linearized equations. If $\kappa \geq \beta$, the eigenvalues become real, and the model fails to represent observations.

To illustrate, assume $\kappa = -0.025 \ll \beta = 0.6$ year. At $t_0 = 30$ year, we suddenly reduce $M(t)$, with equilibrium value $M_s = 25,000$, by just 25 animals each year. Adapting the code given earlier, we found the results visualized in Figure 10.6. The oscillation period increases from $2\pi/|\sqrt{k\beta}| = 11.47$ years to $2\pi/|\sqrt{k(\beta - \kappa)}| = 11.72$ years. Also, the amplitudes of both populations decrease. The step reduction at $M(t_0)$ produces a step reduction in N_s at $t = t_0$ but not M_s. Specifically, N_s is reduced from 30,000 to 28,750, while M_s remains the same. When substantial changes in results occur from small changes in physical parameters, the system is responding nonlinearly.

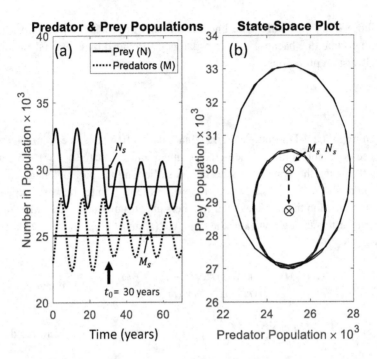

Figure 10.6 Solutions to the exact LV system with small-amplitude parameters listed in Figure 10.4 for $0 \leq t \leq t_0 = 30$ years. For $t_0 \geq 30$ years, the predator population is suddenly reduced by 25/year; $\kappa = -0.025$ from (10.43). The system responds by reducing the prey equilibrium value N_s by 1,250 animals and changing the amplitude and period of both populations. In addition, the equilibrium value in state space shifts as shown. This numerical solution to the NL-LV equations matches the linearized closed-form predictions because $\kappa \ll \beta$.

10.8   Linear Stability, Model Limitations, and Generalizations

Systems in stable equilibrium respond to input with state variables that remain near the equilibrium point. System output tends to wander away from unstable equilibria. *Linear stability* is contingent upon only small inputs applied to the system that do not push the state variable far from equilibrium. For example, imagine a marble in a hemispherical bowl that is being carried around. As the carrier walks slowly about, a small force moves the marble around the bowl, but the bottom of the bowl remains a linearly stable equilibrium point for the marble-bowl-carrier system because the marble is very likely to return once input forces are removed. However, if the carrier runs and jumps while carrying the bowl, the input forces are large and the marble is likely to fall out as that system is driven to respond nonlinearly. This system is linearly stable, but it becomes unstable when driven to respond nonlinearly.

How do we identify linearly stable equilibrium points? As we saw in (10.42), time-varying state variables have the general form $\xi \propto \exp(\lambda t)$, where λ is an eigenvalue of the system. Eigenvalues are generally complex, $\lambda = \sigma + i\Omega$ for real σ, Ω. Hence,

$\exp(\lambda t) = e^{\sigma t} e^{i\Omega t}$. The exponential factor with the imaginary part oscillates and has magnitude $|e^{i\Omega t}| = 1$, so Ω does not affect overall stability. However, the factor containing the real part of the eigenvalue diverges over time if $\sigma > 0$, so *the criterion for linear stability is that the real parts of all eigenvalues must be less than zero*, $\sigma < 0$. In Chapter 11, we use linear stability analysis to evaluate the performance of sensors initially at rest that respond linearly to input and return linearly to rest. Linear stability analysis works well for sensors and similar low-dimensional systems.

The Lotka–Volterra system is different. We abstracted core features of real populations to form this 2-D model and found purely imaginary eigenvalues. The linear stability of the model is unclear from the criterion just given. If the real parts of λ become nonzero, the model predicts that both populations will spiral to zero if negative, or grow unsustainably if positive. In either case, the population model collapses over time. Section 10.7 showed the model could transition to a new equilibrium state for small changes provided the eigenvalues remain imaginary.

A quasi-stable Lotka–Volterra system seems to be awaiting disaster at the slightest hint of a nonzero real part of an eigenvalue. This suggests the model has major limitations regarding its ability to represent realistic populations. Linearized models are wonderful choices for revealing engineering parameters that help us understand overall function, but they accurately predict system responses only near equilibria where we pretty much already know the population values, as shown in Figure 10.3. The stability of an equilibrium point can be predicted for nonlinear models by applying *Lyapunov functions* [112]. They apply over a much larger neighborhood of state space than linearized models and therefore can address nonlinear responses.

More fundamentally, we must question the use of low-dimensional models for population prediction or for modeling any complex system. Organisms and ecosystems express a rich palette of properties that require models with many state variables for comprehensive accuracy. Diversity and adaptability are built into genetic evolution to maximize the survival of organisms in a variable environment [40]. Healthy economies benefit similarly. Linearized models of complex systems (if they could be found) would have hundreds or thousands of functional modes operating over many spatiotemporal scales, enabling the system to remain in a dynamic equilibrium despite being subjected to large environmental stimuli. Robust equilibria tell us that system properties remain fixed in the short term, which allows time for incremental evolutionary change. The simple models illustrated in this chapter cannot reflect nature's true robustness. They falsely predict that life is very fragile because they don't have the robustness of high-dimensional hierarchical complexity [40]. Nevertheless, a simple model of abstracted features is the best place to start, and it may be enough if the goal of the model is to understand and not predict. George Box [19] was right: All models are wrong and sometimes that is okay!

10.8.1 Toward Higher-Dimensional Models

To briefly touch on approaches to higher-dimensional modeling, consider a system with L subpopulations (L is an even integer). We rename the population variables for convenience to $\mathbf{z} \in \mathbb{R}^{L \times 1}$, where at $L = 2$, Section 10.7 told us that $M(t) = z_1(t)$, $N(t) = z_2(t)$, and $z_i \geq 0$. The general L-dimensional Lotka–Volterra equation is [82]

$$\dot{\mathbf{z}} = \mathbf{z} \circ (\mathbf{b} + \mathbf{C}\mathbf{z}), \qquad \text{where} \quad \dot{z}_i = z_i \left(b_i + \sum_{j=1}^{L} C_{ij} z_j \right)$$

and $\mathbf{z} \circ \mathbf{b}$ is a Hadamard product between the vectors. The 2-D system is

$$\dot{z}_1 = z_1(b_1 + C_{11}z_1 + C_{12}z_2) \qquad \text{and} \qquad \dot{z}_2 = z_2(b_2 + C_{21}z_1 + C_{22}z_2).$$

Terms linear in z_i describe changes in that subpopulation without interactions. Populations may grow or decline, depending on the sign of b_i. Terms proportional to $z_i z_j$ describe spatiotemporally homogenous interactions between subpopulations, where the sign of C_{ij} determines who is preying on who.

This 2-D system is similar to other second-order dynamical systems described in the fields of mechanics, chemical kinetics, and electromagnetics, and any system with coupled oscillators. Solutions to generalized L-dimensional models can be found using a *Hamiltonian system* [60, 61, 82] with model equations

$$\dot{\mathbf{z}} = \mathbf{J} \nabla H_a. \tag{10.44}$$

$\mathbf{J} = \begin{pmatrix} 0 & \mathbf{I} \\ -\mathbf{I} & 0 \end{pmatrix}$ is *skew-symmetric* and $\mathbf{I}$ is an $L/2 \times L/2$ identity matrix. $H_a(\mathbf{z}, t)$ is a smooth scalar Hamiltonian function of state-space variables and time. Problem 10.9 asks you to find the Hamiltonian function that combines with (10.44) to give the linearized solution of (10.35). Interested readers are encouraged to search the growing literature on this approach to modeling high-dimensional populations.

10.9   Modeling Infectious Disease in Populations

Another important problem in biological modeling is predicting the dynamics of infectious disease. We will describe a basic model of viral or bacterial epidemics to examine the engineering parameters predicting the spread of infection in the population, so that a mitigation strategy can be developed. An *epidemic* is a widespread simultaneous occurrence of illness in a local population. It has temporary prevalence, with a duration that is short compared with a healthy lifespan. A *pandemic* is a global-scale epidemic.

The most basic time-varying models have very stringent assumptions that make them low dimensional, simple to express, and limited in detail. Efforts to build realistic models add features intended to relax many of the following assumptions. First, births, unrelated deaths, and migration during the epidemic are excluded because the recovery period is assumed to be relatively short. Second, there is no latent infectious period, meaning infected individuals become infectious themselves immediately. Third, all infected individuals recover, and recovery confers lifetime immunity against reinfection. Fourth, the population is assumed to mix thoroughly and continuously, so that changes in the number of infected individuals over a time increment are proportional to the number of interactions between infected and susceptible sub-populations. Sequestered groups necessitate higher-dimensional models with more interacting state variables. Finally, a contagious individual can infect any member of the susceptible population with equal probability.

10.9.1 SIR Model

These limiting conditions are properties of the basic *SIR model* [52]. A vulnerable population of N consists of three disjoint sets: the number of individuals susceptible to infection S, those infected I, and those who have recovered from the infection R, where

$$N = S + I + R \quad \text{and} \quad \dot{N} = \dot{S} + \dot{I} + \dot{R} = 0.$$

Model equations for the three subpopulations include the efficacy of infectious trans-mission to susceptible individuals $\alpha > 0$ and the rate of recovery $\gamma > 0$, which are given by time derivatives of the subpopulations,

$$\begin{aligned} \dot{S} &= -\alpha SI \\ \dot{I} &= \alpha SI - \gamma I \qquad \text{(SIR model)} \\ \dot{R} &= \gamma I. \end{aligned} \qquad (10.45)$$

If the relevant unit of time is days, then αS is the number of new infections each day from each infectious individual. We can consider $1/\gamma$ to be the time during which an individual is infectious.

Example 10.9.1. *Applying the SIR model: Consider data from a flu epidemic at an English boarding school with $N = 763$ students [59, 71]. Initially, the number of new infections per day per infected individual was $\alpha S = 1.66/day$, and the recovery rate was $\gamma = 0.455/day$. The initial subpopulations selected have at least one member: $S \simeq N - 2$, $I = 1$, and $R = 1$. The daily numbers of infected and model curves predicting the three populations are shown in Figure 10.7. The code that solves this nonlinear system is structured similar to that of the predator–prey models described in Section 10.7.*

```
%%%%% Solver for the SIR model of (10.45).  N is a column vector with subpopulations S,I, and R %%%%%%%%%
%
N0=[761;1;1];                                      % initial subpopulation numbers
alpha=0.0022;gamma=0.455;                          % model parameters
theta=[alpha;gamma];                               % parameter vector for use in function SIR
trange=[0 21];                                     % duration of the model in days, used in ode45
[tt,N]=ode45(@(tt,N) SIR(tt,N,theta),trange,N0);   % N = (S;I;R)
plot(tt,N)
title('SIR Sub-Populations');xlabel('time (days)');
%
% Function for computing the system of equations solved in ode45
%
function z=SIR(t,y,p)
    z = ([-p(1)*y(2),0,0;p(1)*y(2),-p(2),0;0,0,p(2)*y(2)/y(3)])*y;   % system of equations for SIR model
end
```

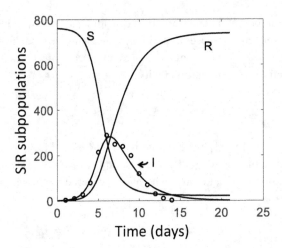

Figure 10.7 SIR model in Example 10.9.1 fit to infected numbers from flu epidemic data (o) [59].

Initially, we see the numbers of infected students rise exponentially $I \propto \exp(\alpha St)$ as the numbers susceptible to infection fall. Soon thereafter, we see an exponential rise in the number recovered but at a rate different from those becoming infected, all as predicted by (10.45) near $t = 0$.

Figure 10.7 shows that the number of infected students peaks at just over 6 days, at $\dot{I} = 0 = \alpha SI - \gamma I$, or $S = \gamma/\alpha$. Infections die out when $\dot{I} < 0$, which sets a lower bound for the initial susceptible population at $S(0) \simeq N > \gamma/\alpha$. The threshold for an epidemic is given by

$$R_0 = \frac{N\alpha}{\gamma} > 1. \qquad \text{(Epidemic threshold)} \qquad (10.46)$$

R_0 is the *basic reproductive ratio* [51] describing the number of new infections per day multiplied by the infectious period. This unitless engineering parameter for the population is composed of the physical parameters of the ODE model. R_0 is also the mean number of individuals infected by a single infected individual during the entire infectious period. In this example, $R_0 = 3.6$, showing an epidemic was possible.

The results show that everyone who is susceptible does not become infected. To investigate, divide $\dot{S}$ by $\dot{R}$ and find

$$\frac{dS}{dR} = -\frac{\alpha I S}{\gamma I} = -\frac{R_0}{N} S.$$

Separating variables and integrating S from N to $S(t)$ and R from 0 to $R(t)$,

$$\ln S(t) - \ln N \simeq -\frac{R_0}{N} R(t)$$

$$\frac{S(t)}{N} \simeq \exp\left(-R_0 \frac{R(t)}{N}\right).$$

As $t \to \infty$ and $R(\infty)/N \simeq 1$, we find that $S(\infty)/N \simeq e^{-R_0} \simeq 0.027 > 0$. Hence, approximately 3% of S never becomes infected. This model does not run unless $I(0), R(0) > 0$, but the conclusion remains despite nonzero initial subpopulations as long as $N \gg 1$.

Fortunately, not everyone becomes sick at the same time. We can also predict the fraction of the population expected to become infected at the peak of the epidemic by expressing I_{max}, where $S = \gamma/\alpha$. Dividing $\dot{I}$ by $\dot{S}$,

$$\frac{dI}{dS} = -1 + \frac{\gamma}{\alpha S} = \frac{N}{R_0 S} - 1$$

$$\int_0^{I(t)} dI' = \int_N^{S(t)} dS' \left(\frac{N}{R_0 S'} - 1 \right)$$

$$I(t) = \frac{N}{R_0} \ln \left(\frac{S(t)}{N} \right) - S(t) + N$$

$$I_{max} = N \left(1 - \frac{1 + \ln R_0}{R_0} \right). \tag{10.47}$$

Since $S = \gamma/\alpha$ at I_{max}, (10.46) gives $S = N/R_0$, which was applied in (10.47). Equation (10.47) predicts $I_{max}/N = 0.37$; that is, 37% of the boarding school population was infected at its peak. Model data are useful for resource planning during an outbreak.

Summary.
Equation (10.46) is critical when formulating a strategy to prevent epidemics:

- $1/\gamma$ is small when treatment of infected people is effective.
- $S \ll N$ when there is an effective vaccine or herd immunity.
- α is small when personal protective equipment, distancing, and hand washing are maintained to keep transmission rates low.

Even a very simple model makes the risks and potential mitigation strategies clear.

10.9.2   Stability of Public Health

We now examine the linearized model for insights. From (10.33), we know the equilibrium points of a system are found when that system is at rest. The equilibrium points are

$$\mathbf{N}_{s1} = \begin{pmatrix} S_s \\ I_s \\ R_s \end{pmatrix} = \begin{pmatrix} N \\ 0 \\ 0 \end{pmatrix} \quad \text{or} \quad \mathbf{N}_{s2} = \begin{pmatrix} \epsilon N \\ 0 \\ (1 - \epsilon)N \end{pmatrix},$$

which occur before and after an epidemic, respectively. The fraction of S that do not become ill during an epidemic is $\epsilon < 1$. The linear stability of this system (Section 10.8) is evaluated from Jacobian matrix $\mathbf{A}$ (10.28) and its eigenvalues:

$$A = \begin{pmatrix} \partial \dot{S}/\partial S & \partial \dot{S}/\partial I & \partial \dot{S}/\partial R \\ \partial \dot{I}/\partial S & \partial \dot{I}/\partial I & \partial \dot{I}/\partial R \\ \partial \dot{R}/\partial S & \partial \dot{R}/\partial I & \partial \dot{R}/\partial R \end{pmatrix} = \begin{pmatrix} -\alpha I_s & -\alpha S_s & 0 \\ \alpha I_s & \alpha S_s - \gamma & 0 \\ 0 & \gamma & 0 \end{pmatrix}$$

(10.48)

$$\det(A - \lambda I)_{N=N_s} = \begin{cases} N_{s1}: \; \lambda^2(\alpha N - \gamma - \lambda) = 0; \; \lambda_1 = \lambda_2 = 0, \lambda_3 = \alpha N - \gamma \\ N_{s2}: \; \lambda^2(\alpha \epsilon N - \gamma - \lambda) = 0; \; \lambda_1 = \lambda_2 = 0, \lambda_3 = \alpha \epsilon N - \gamma \end{cases}.$$

This two-variable, rank 1, linearized system has one real eigenvalue < 0 at the first equilibrium point provided $R_0 = N\alpha/\gamma < 1$. Similarly, at the second equilibrium point, $\lambda_3 < 0$ and real when $\epsilon R_0 < 1$. This is the same result as in (10.46). If $R_0 < 1$, then $\epsilon R_0 < 1$ because $\epsilon < 1$. The model predicts both equilibrium points are stable as long as the basic reproductive ratio, a core engineering parameter, remains less than 1.

Interested readers should read the extensive literature on infectious disease models that go far to relax the stringent assumptions as needed to create more realistic models. One extension to SIR is explored in Problem 10.8.

10.10   Problems

10.1   At time $t = 0$, you have a 10 pM (pico-molar) solution of Tc-99m (i.e., 10 pmoles per liter of solution) with a half-life of 6.0 hours. Compute the initial activity of a one-L solution in units of [Bq].

10.2   Two-compartment series dilution.[14] Consider an extension of Example 10.3.1, where we have two compartments in series, as illustrated in Figure 10.8. Imagine there is a way to lock a molecular substance within cells of an organism so that the substance does not appear in the blood until suddenly, at $t = 0$, we trigger cells to slowly release that substance into the bloodstream. Once in the blood, the substance is cleared by the kidneys via the urine, and we infuse a solute-free solution into the organism at a rate equal to urine production so volumes of each compartment are unchanged.

Compartment 1 consists of cells at constant volume V_1 and initial substance concentration $c_1(0) = c_0 > 0$. Compartment 2 comprises the circulating blood, with constant volume V_2 and initial solute concentration $c_2(0) \simeq 0$. The flow between compartments is constant q. The time courses for state variables c_1 and c_2 are shown in Figure 10.8 for physical parameters q, V_1, V_2. Since we don't know the compartment volumes exactly, we decided to compute the results for different ratios $a = V_1/V_2$. Find expressions for $c_1(1)$ and $c_2(t)$ and verify the results plotted in Figure 10.8. What are the engineering parameters of this system?

10.3   One-compartment metabolic turnover. Returning to the single-compartment situation, consider a metabolite at equilibrium; that is, the metabolite is created and consumed at equal rates, r [mg/min]. Estimate r from measurements of a labeled tracer injected in a blood pool that is eliminated in the same way as the unlabeled

[14] The problem is contrived to keep it relatively straightforward.

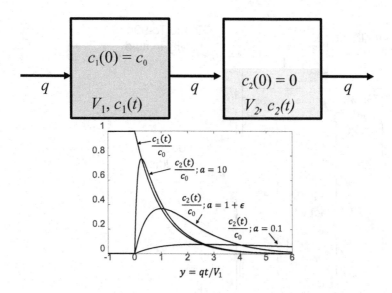

Figure 10.8 Illustration of the two-compartment series dilution in Problem 10.2.

metabolite. I did not provide enough information for you to find a numerical answer, but you can adapt the discussion on one-compartment dilution to develop equations that offer solutions if numerical values of parameters were measured.

10.4 In Example 10.3.2, we showed that Tc-99m is found in transient equilibrium with its parent isotope Mo-99. This provides a practical source for the short-lived Tc-99m for medical applications. Given the radioisotope generator diagrammed in Figure 10.9 and the equations for Tc-99m activity in equilibrium from Example 10.3.2, devise a schedule for "milking" the generator that maximizes the recovery of Tc-99m activity. Assume the eluting solvent is 90% efficient at extracting Tc-99m activity and that Mo-99 breakthrough is negligible.
(a) Plot the resulting generator activity of both Mo-99 and Tc-99m as you execute your strategy for recovering Tc-99m.
(b) Compute the total activity of Tc-99m produced during one week's time. Then give your recovery output as a fraction (percentage) of the total.

10.5 The following data are from records made by the Hudson Bay Company on populations ($\times 1,000$) of lynx and rabbit pelts harvested over two decades [76].

Year, Lynx, Rabbits	Year, Lynx, Rabbits	Year, Lynx, Rabbits	Year, Lynx, Rabbits
1900, 4.0, 30.0	1905, 41.7, 20.6	1910, 7.4, 27.1	1915, 51.1, 19.5
1901, 6.1, 47.2	1906, 19.0, 18.1	1911, 8.0, 40.3	1916, 29.7, 11.2
1902, 9.8, 70.2	1907, 13.0, 21.4	1912, 12.3, 57.0	1917, 15.8, 7.6
1903, 35.2, 77.4	1908, 8.3, 22.0	1913, 19.5, 76.6	1918, 9.7, 14.6
1904, 59.4, 36.3	1909, 9.1, 25.4	1914, 45.7, 52.3	1919, 10.1, 16.2
			1920, 8.6, 24.7

Adapt the code by adjusting parameters from Section 10.7.5 to select k, α, γ, β that do a reasonable job of modeling the data. Discuss any caveats you may have regarding your solution.

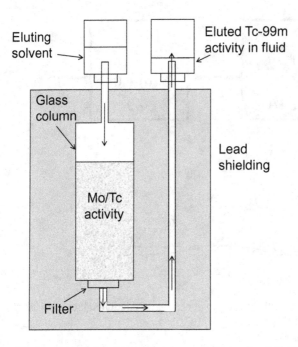

Eluting solvent

Eluted Tc-99m activity in fluid

Glass column

Lead shielding

Mo/Tc activity

Filter

Figure 10.9 Diagram of a Tc-99m generator for use in medical nuclear imaging clinics. A vacuum in the collecting vial draws fluid from the source vial through the glass column.

10.6 Apply the eigenanalysis methods of Section 10.2 and Example 10.2.1 to solve the 2-D linearized Lotka–Volterra system. Specifically, show the statements made leading to (10.42) result from those methods. Assume the model parameters listed in Figure 10.4.

10.7 Modify the code providing numerical solutions to the exact Lotka–Volterra equations to give the results in Figure 10.6 in Section 10.7.7. That is, introduce $\kappa = -0.025$/year at $t = t_0 = 30$ years to reproduce Figure 10.6.

10.8 Describe how the SIR model of Section 10.9 changes if permanent post-recovery immunity is compromised; i.e., kR recovered individuals return to the susceptible group each day. This is the SIRS model, where the last "S" refers to the return of recovered individuals to the susceptible subpopulation.
(a) Write the system of equations describing the SIRS model.
(b) Adapt the code in Section 10.9 to implement this system of equations.
(c) Find the equilibrium points.
(d) Determine the stability of the equilibrium points at $k = 0.03, 0.30$/day. Will this epidemic become *endemic*?

10.9 Propose a Hamiltonian function $H_a(\mathbf{N}, t)$ in state variables $\mathbf{N} = \begin{pmatrix} M \\ N \end{pmatrix}$ at equilibrium point $\mathbf{N}_s = \begin{pmatrix} \frac{k}{\alpha} \\ \frac{\beta}{\alpha\gamma} \end{pmatrix}$ that when it is applied to (10.44) results in (10.35).

11   ODE Models II

Sensors

11.1   Second-Order Systems

In Chapter 10, coupled first-order ODEs describing two state variables were given as an example of a simple second-order system. In this section, we consider another second-order system – one suitable for modeling sensors responses. This system equation contains a second derivative applied to one state variable, and is given by this linear, inhomogeneous, nonautonomous equation:[1]

$$\mathcal{D}_2 \, g(t) = \ddot{g}(t) + k_1 \dot{g}(t) + k_2 g(t) = k_0 f(t). \tag{11.1}$$

The physics described by (11.1) is much different from that of the Lotka–Volterra (LV) equations of Section 10.7. A second derivative term describes responses that are distinct from first derivatives: inertial effects versus energy-loss effects.[2]

11.1.1   Homogeneous Solution

Despite different physics, we can treat the *homogeneous form* of (11.1) mathematically in the same manner as the LV system by making a simple change of variables. Later, we will address the particular solution by including the inhomogeneity, $k_0 f(t)$.

Defining $x_1 = g$ and $x_2 = \dot{g}$ and applying them to (11.1) gives the following two first-order equations:

$$
\begin{aligned}
\dot{g} &= \dot{g} \longrightarrow \dot{x}_1 = x_2 \\
\ddot{g} &= -k_2 g - k_1 \dot{g} \longrightarrow \dot{x}_2 = -k_2 x_1 - k_1 x_2
\end{aligned}
\quad \text{or} \quad
\begin{pmatrix} \dot{x}_1 \\ \dot{x}_2 \end{pmatrix} = \begin{pmatrix} 0 & 1 \\ -k_2 & -k_1 \end{pmatrix} \begin{pmatrix} x_1 \\ x_2 \end{pmatrix}
$$

$$\text{or} \quad \dot{\mathbf{x}} = \mathbf{A}\mathbf{x}. \tag{11.2}$$

It is straightforward to see that the Jacobian matrix $\mathbf{A}$ is full rank with eigenvalues,

$$\lambda_{1,2} = \frac{-k_1 \pm \sqrt{k_1^2 - 4k_2}}{2}. \tag{11.3}$$

---

[1] If units of f are [obj] and those of g are [data], then other terms have the following units: $\dot{g} \triangleq dg/dt$, [data/time]; $\ddot{g} \triangleq d^2 g/dt^2$, [data/time^2]; k_0, [data/obj-time^2]; k_1, [1/time]; and k_2, [1/time^2].

[2] Equation (11.1) is the harmonic oscillator equation. If it represents a mechanical system, k_1 is the ratio of mechanical damping to mass and k_2 is the ratio of spring constant to mass. The equation can represent many types of sensors, mechanical and otherwise.

As discussed in Section 10.2, solutions to homogeneous second-order ODEs have the general form (*ansatz*), $g_h(t) = c_1 \exp(\lambda_1 t) + c_2 \exp(\lambda_2 t)$, where c_1, c_2 are nonzero constants. Entering this form into the $\ddot{g}$ in (11.2), we find

$$c_1 e^{\lambda_1 t}\left[\lambda_1^2 + k_1\lambda_1 + k_2\right] + c_2 e^{\lambda_2 t}\left[\lambda_2^2 + k_1\lambda_2 + k_2\right] = 0.$$

Since $c_j \exp(\lambda_j t) \neq 0$ for $j = 1, 2$, then it must be true that $\lambda_j^2 + k_1\lambda_j + k_2 = 0$ and we confirm the findings in (11.3). We do not yet have enough information to find c_1, c_2.

11.1.2   Particular Solution Input

To find a particular solution g_p, we follow the discussion in Section 10.3 and compute the response to a step input, $f(t) = f_p \text{ step}(t)$. In the steady state at $t \to \infty$, $\dot{g} = \ddot{g} = 0$ and (11.1) reduces to[3]

$$k_2 g_p(t) = k_0 f_p \text{ step}(t) \quad \text{and} \quad g_p(t > 0) = k_0/k_2,$$

where for simplicity we set $f_p = 1$. The output response of this system that is initially at rest, i.e., $g(t \leq 0) = 0$, is the sum of the homogeneous and particular responses:

$$g(t) = g_h(t) + g_p(t)$$
$$= \begin{cases} 0 & t \leq 0 \\ c_1 e^{\lambda_1 t} + c_2 e^{\lambda_2 t} + k_0/k_2 & t > 0 \end{cases}. \tag{11.4}$$

The impulse response is the derivative of $g(t)$ from (11.4), since[4]

$$g(t) = \int_{-\infty}^{\infty} dt'\, h(t') \text{ step}(t - t')$$
$$\dot{g}(t) = \begin{cases} \int_{-\infty}^{\infty} dt'\, h(t') \delta(t - t'), & t > 0 \\ 0, & \text{otherwise} \end{cases} = h(t)$$
$$h(t) = \left(c_1\lambda_1 e^{\lambda_1 t} + c_2\lambda_2 e^{\lambda_2 t}\right) \text{ step}(t). \quad \text{(Impulse response)} \tag{11.5}$$

Applying the initial conditions $\dot{g}(0) = g(0) = 0$ to (11.4) and (11.5), we can solve for unknowns c_1, c_2,

$$\begin{array}{l} c_1 + c_2 + k_0/k_2 = 0 \\ c_1\lambda_1 + c_2\lambda_2 = 0 \end{array} \quad \text{resulting in} \quad \begin{array}{l} c_1 = -\frac{k_0}{k_2}\frac{\lambda_2}{\lambda_2 - \lambda_1} \\ c_2 = \frac{k_0}{k_2}\frac{\lambda_1}{\lambda_2 - \lambda_1} \end{array}. \tag{11.6}$$

When we examine (11.1) and (11.3)–(11.6), it becomes clear that output $g(t)$ responds to any input function $f(t)$ that preserves linearity via the noise-free LTI system equation,

---

[3] The ratio $k_0 f_p/k_2$ has the units of [data/(obj-time^2)][obj]/[1/time^2] = [data]. The units are consistent.
[4] From Appendix D, $\dot{g}(t) = [\dot{f} * h](t) = [f * \dot{h}](t)$.

$$g(t) = [h * f](t) = \int_0^\infty dt' \left(c_1 \lambda_1 e^{\lambda_1 t'} + c_2 \lambda_2 e^{\lambda_2 t'} \right) f(t' - t), \qquad (11.7)$$

which depends only on physical parameters k_0, k_1, k_2 through engineering parameters $c_1 \lambda_1$ and $c_2 \lambda_2$. Note that the products $c_j \lambda_j$ set the correct units for $h(t)$ to [data/obj-time]. Equation (11.7) is the solution to the second-order, nonautonomous ODE of (11.1).

Let's examine the elements of $\mathbf{A}$ in (11.2). Unlike the linearized LV equations in (10.43), one of these diagonal elements is nonzero, assuming $k_1 \neq 0$. Consequently, the eigenvalues can be complex and their real part determines the *gain*[5] of the sensor response that, as we will see later, also regulates the bandwidth and sensitivity of its response to input signals. A second-order system is *stable*[6] if the real part of its eigenvalues are negative. This is consistent with $\text{tr}(\mathbf{A}) < 0$ and $\det(\mathbf{A}) > 0$, as discussed in Chapter 10. Unstable outputs grow exponentially until the sensor cannot meet the energy demand and the signal "saturates" or "clips," which is a nonlinear response from a linear device.

11.2 Sensor Impulse Response

11.2.1 Transition Point and Damping Factor

The response of the sensor, $g(t)$, is the sum of its response from being driven by $f(t)$, which we label $g_p(t)$ in (11.4), and its natural resonant response after the input is removed, which we label $g_h(t)$. Modeling a linear sensor response is really just modeling its impulse response, $h(t)$, by selecting k_0, k_1, k_2 that give the engineering properties we desire. The nature of the sensor response is modified by changing the eigenvalues from real to complex. From (11.3), we see this occurs at $k_1^2 - 4k_2 = 0$. To quantify how close we are to this transition point, we write $k_1^2 - 4k_2\zeta^2 = 0$ and solve for the *damping factor* ζ,

$$\zeta = \frac{k_1}{2\sqrt{k_2}}. \qquad \text{(Damping factor)} \qquad (11.8)$$

For example, let $k_0 = 0.5, k_1 = 2$, and $k_2 = 1$. Then, $k_1^2 - 4k_2 = 0$, $\zeta = 1$, and the eigenvectors are equal, real, and negative; from (11.3), $\lambda_1 = \lambda_2 = -1$. However, c_1 and c_2 from (11.6) are undefined. At the transition point $\zeta = 1$, the impulse response $h(t)$ is not a function; it is approximated by a Dirac delta.

---

[5] $0 \leq$ gain $< \infty$. When the gain is greater than 1, the signal amplitude is amplified. When the gain is less than 1, the signal amplitude is attenuated.

[6] A system is BIBO stable if for every bounded input there is a bounded output. For continuous-time LTI systems represented by $g(t) = [h * f](t)$, the system is BIBO stable if it is absolutely integrable, $\int_{-\infty}^{\infty} dt \, |h(t)| = \|h\|_1 < \infty$.

11.2.2   Real Eigenvalues, $\zeta > 1$

The impulse responses of systems with real eigenvalues are composed of real-valued exponentials, meaning h rises in time from zero, it peaks, and then it falls back to zero. These are characteristics of second-order systems with real eigenvalues that are initially at rest.

We first illustrate real eigenvalues near the transition point, where we set $k_0 = 0.5, k_1 = 2$, and $k_2 = 0.9975$ (k_2 is a little less than 1). Consequently, $k_1^2 - 4k_2 = 0.01$, $\zeta = 1.00125$, $\lambda_1 = -0.95$, and $\lambda_2 = -1.05$. From (11.6), we find the constants, $c_1 = -10.5$ and $c_2 = 9.5$. Entering these results into (11.5) gives h with minimal damping,

$$h(t) = c_1 \lambda_1 \left(e^{\lambda_1 t} - e^{\lambda_2 t} \right) \qquad \text{(Impulse response, real eigenvalues)}$$

$$(11.9)$$

$$= 9.975 \left(e^{-0.95t} - e^{-1.05t} \right) \quad \text{for } \zeta = 1.00125.$$

Equation (11.9) follows from (11.5) because $c_1 \lambda_1 = -c_2 \lambda_2$. Assuming a causal system, we made the step function in (11.9) implicit.

Let's continue to set $k_0 = 0.5, k_1 = 2$, but now $k_2 = 0.5$ (less than 1). We find $k_1^2 - 4k_2 = 2$ and $\zeta = \sqrt{2}$ to indicate an increase in damping. Also, $\lambda_1 = -1 + 1/\sqrt{2} = -0.2929$, $\lambda_2 = -1 - 1/\sqrt{2} = -1.7071, c_1 = -1.207$, and $c_2 = 0.207$. Consequently, (11.9) gives

$$h(t) = 0.3535 \left(e^{(-1+1/\sqrt{2})t} - e^{(-1-1/\sqrt{2})t} \right) \quad \text{for } \zeta = \sqrt{2}.$$

Both impulse responses $h(t)$ are plotted at the top of Figure 11.1 as dashed lines. For fixed k_1, the k_2 parameter works through the eigenvalues to influence both the response duration (bandwidth) and the overall sensitivity of the device. More on this later.

11.2.3   Complex Eigenvalues, $0 < \zeta < 1$

In the $0 < \zeta < 1$ range, (11.8) shows that $k_1^2 - 4k_2 < 0$, and (11.3) shows the eigenvalues are complex constants. They are conveniently expressed using the notation

$$\lambda_1 = \sigma_1 + i\Omega_1, \quad \lambda_2 = \sigma_1 - i\Omega_1, \quad \text{where} \quad \sigma_1 = -\frac{k_1}{2} \quad \text{and} \quad \Omega_1 = \frac{\sqrt{4k_2 - k_1^2}}{2}$$

$$(11.10)$$

are real constants. Note that $\lambda_1^* = \lambda_2$.

Combining (11.5), (11.6), and (11.10), we find

$$h(t) = c_1 \lambda_1 e^{\sigma_1 t} \left(e^{i\Omega_1 t} - e^{-i\Omega_1 t} \right) \text{step}(t)$$

$$= v e^{\sigma_1 t} \sin(\Omega_1 t) \text{step}(t),$$

$$(11.11)$$

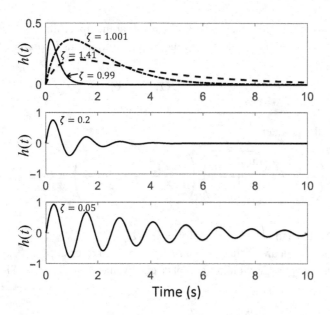

Figure 11.1 (top, dashed lines) Impulse response functions from (11.9) for real eigenvalues and two damping factors, $\zeta > 1$. (solid lines in all three plots) Impulse responses from (11.15) for complex eigenvalues where $0 < \zeta < 1$ and $\Omega_0 = 5$ Hz. Notice the vertical axes are scaled differently.

where $v = 2ic_1\lambda_1$. The exponential factor from the real part of the eigenvalues shows the oscillating amplitude can grow or decline depending on the sign of σ_1. A constant-amplitude oscillating response was found for state variables of the second-order LV system in Section 10.7.

11.2.4 Engineering Parameters

Currently, the engineering parameters are v, σ_1, Ω_1, which are not clearly interpretable. We can fix that. Applying (11.11) to $g(t) = [h * f](t)$ gives the particular solution $g_p(\infty)$ for a special input $f(t) = \text{step}(t)$ at $t \to \infty$. We already found in (11.4) that $g_p(\infty) = k_0/k_2$, but let's look from this view to find v in terms of other parameters. The result is[7]

$$g_p(t \to \infty) = \int_{-\infty}^{\infty} dt'\, h(t')\,\text{step}(t - t')$$

$$= \int_{-\infty}^{t \to \infty} dt'\, h(t') = v \int_0^{t \to \infty} dt'\, e^{\sigma_1 t'} \sin(\Omega_1 t')$$

$$= \frac{v\sigma_1}{\sigma_1^2 + \Omega_1^2}\, e^{\sigma_1 t'} \left(\sin(\Omega_1 t') - \frac{\Omega_1}{\sigma_1} \cos(\Omega_1 t') \right)\Bigg|_0^{t \to \infty}.$$

---

[7] The integral in the last line of the following equation was found by integrating by parts twice. To see if it is a correct solution, begin with that last-line solution and differentiate; you will find $\exp(\sigma t') \sin \Omega t'$.

Since stable systems require that $\sigma_1 < 0$, the result at the upper limit is zero because the factor $\exp(\sigma_1 t) \to 0$ as $t \to \infty$. At the lower limit, the factor in parentheses is Ω_1/σ_1, and

$$g_p(\infty) = \frac{\nu \Omega_1}{\sigma_1^2 + \Omega_1^2} = \frac{k_0}{k_2}. \tag{11.12}$$

From (11.10), the product

$$\lambda_1 \lambda_2 = \left(-\frac{k_1}{2} + i \frac{\sqrt{4k_2 - k_1^2}}{2}\right)\left(-\frac{k_1}{2} - i \frac{\sqrt{4k_2 - k_1^2}}{2}\right) = k_2 = \sigma_1^2 + \Omega_1^2. \tag{11.13}$$

Combining (11.12) and (11.13), we find $\nu = k_0/\Omega_1$.

Next, define the square of the *resonant frequency* of the sensor, $\Omega_0^2 \triangleq k_2 = \sigma_1^2 + \Omega_1^2$, in the damping range yielding complex eigenvalues $0 < \zeta < 1$. Since $\sigma_1 = -k_1/2$,

$$\zeta = \frac{k_1}{2\sqrt{k_2}} = -\frac{\sigma_1}{\Omega_0} \qquad \text{or} \qquad \sigma_1 = -\zeta \Omega_0, \tag{11.14}$$

and so $\quad \Omega_1 = \sqrt{\Omega_0^2 - \sigma_1^2} = \Omega_0 \sqrt{1 - \zeta^2}.$

Parameters ν, σ_1, Ω_1 are each functions of physical parameters k_0, k_1, k_2. They become more interpretable when converted into input gain k_0, resonant frequency Ω_0, and damping factor ζ. These are not yet the engineering properties we seek. Returning to (11.11), we have for $0 < \zeta < 1$,

$$h(t) = \frac{k_0 e^{-\Omega_0 \zeta t}}{\Omega_0 \sqrt{1 - \zeta^2}} \sin\left(\Omega_0 \sqrt{1 - \zeta^2}\, t\right) \text{step}(t) \qquad \text{(IR, complex eigenvalues)} \tag{11.15}$$

with units of [data/(obj-time)] from the ratio k_0/Ω_0. Note that gain k_0 must be set in the linear range of the device. We might also write (11.15) as $h(t; \boldsymbol{\theta})$, where $\boldsymbol{\theta} = (k_0, \Omega_0, \zeta)^\top$ is a vector of constants that parameterize the impulse response for $0 < \zeta < 1$.

The *engineering parameters* we seek are those that influence task performance – in this case, measurement quality. They are pulse frequency $\Omega_0 \sqrt{1 - \zeta^2}$ and pulse bandwidth $2\zeta \Omega_0$, as explained later. Both depend on the damping factor and resonant frequency, which themselves depend on physical parameters k_1 and k_2.

11.2.5   Summary and Interpretation

Recall that the two plots at the top of Figure 11.1, where $\zeta > 1$ (dashed lines), have real eigenvalues and are given by (11.9). Examples of $h(t)$ with complex eigenvalues from (11.15) are shown as solid lines in Figure 11.1 for three damping factors, $\zeta = 0.99, 0.2, 0.05$. In each case, $\Omega_0/2\pi = 5$ Hz. Note that vertical-axis scaling on the top plot is magnified four times relative to the bottom two plots. Plots from ζ values on either side of the $\zeta = 1$ transition singularity are very different.

Comparing the results of (11.9) and (11.15), we see that the shortest-duration impulse response occurs near the $\zeta = 1$ transition, where $h(t) \to \delta(t)$. Increasing damping factor ζ always decreases the response amplitude, but the influence on pulse length depends on which side of $\zeta = 1$ the damping parameter is set. Applying input to an *overdamped system* ($\zeta > 1$) initially at rest generates an output peak that exponentially decays back to zero without oscillation. Perturbing an *underdamped systems* ($0 < \zeta < 1$) produces an oscillating response. Whether we wish the response to oscillate or not depends on the application. However, at no time should we set $\zeta < 0$, because (11.15) shows the system becomes unstable over time as the response grows exponentially.

A second transition point occurs at $\zeta = 0$, where the system is *critically damped*. It has purely imaginary eigenvalues, and the sensor's response oscillates continuously and sinusoidally over time. At $\zeta = 1$, the impulse response is the shortest possible; at $\zeta = 0$, it is infinitely long. However, the system is unstable at either transition point. At $\zeta = 0$, the sensor is in a *quasi-stable state*, where the system becomes stable if ζ increases just a small amount and unstable if ζ decreases at all (see the discussion in Section 10.8).

11.2.6   Acoustic Sensors

The preceding discussion described "sensors" as a general concept that can be adapted to specific optical, electrical, and mechanical devices. Piezoelectric crystals used in ultrasonic transducers [72] are examples of resonant electromechanical devices described by (11.15).

Lead zirconate titanate (PZT) crystals are stimulated to change shape by applying a voltage potential across opposing surfaces. Because PZT crystals have mass and are elastic, applying and removing a voltage quickly causes the element to deform and snap back, but its inertia causes the rebound to overshoot so the crystal oscillates about its resting shape. The forced, damped, harmonic oscillator equation of (11.1) is perfect to model these dynamics. The resonant frequency, Ω_0, depends on the size of the element. The thinnest crystals produce the highest resonant frequencies. Crystal oscillations can be electrically and mechanically damped, which sets $\zeta > 0$ to control the sound pulse duration, as seen in the middle and lower pulse profiles in Figure 11.1. Ω_0, ζ are the engineering parameters that determine the amplitude and duration of the sound pulses generated. These sensor properties ultimately determine clinical sonographic task performance via SNR, contrast, and spatial resolution features (Chapter 7).

The Fourier transform of the impulse response in (11.15) gives the scaled power spectral density $|H(u)|^2$ (see Section 5.12). It describes the frequency distribution of

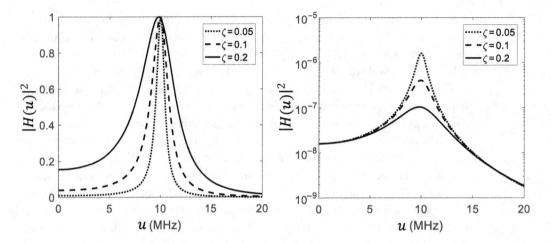

Figure 11.2 (a) Normalized spectral density curves from (11.16) for a $u_0 = \Omega_0/2\pi = 10$ MHz ultrasound transducer element with damping factors $\zeta = 0.05, 0.1, 0.2$. The *fractional bandwidth* for each ζ value is FBW $= 2\zeta/\sqrt{1-\zeta^2} \simeq$ FWHM$/u_0$, where FWHM is the full-width at half-maximum spectral value. (b) Without normalization, the sensitivity of each device becomes apparent.

transmitted power for the device. The Fourier transform pairs listed in Appendix D give[8]

$$H(\Omega; \boldsymbol{\theta}) = \mathcal{F}h(t; \boldsymbol{\theta}) = \frac{k_0}{\Omega_0^2(1-\zeta^2) + \left(\zeta\Omega_0 + i\Omega\sqrt{1-\zeta^2}\right)^2}. \qquad (11.16)$$

Here, we explicitly indicate that $H(\Omega)$ is parameterized by $\boldsymbol{\theta} = (\Omega_0, \zeta)^\top$. Examples of $|H(\Omega)|^2$ are plotted in Figure 11.2 for three values of ζ and a fixed resonant frequency of $\Omega_0/2\pi = u_0 = 10$ MHz. By normalizing the three spectral density functions in Figure 11.2a to have the same peak value, we can visualize relative bandwidths.

From (11.15), the pulse frequency is $\Omega_0\sqrt{1-\zeta^2}$ and the pulse bandwidth is $BW = 2\zeta\Omega_0$; both have units of rad$/\mu$s. The fractional bandwidth is the ratio, $FBW = 2\zeta/\sqrt{1-\zeta^2}$. The narrowest-band pulse has the longest duration and the least damping, $\zeta = 0.05$. Its 10% bandwidth ($FBW \simeq 0.1$) gives a duration of roughly 10 cycles. Since the wavelength at 10 MHz is about 0.154 mm in most soft tissues, the axial resolution is approximately 1.5 mm.

By comparison, the 40% bandwidth pulse ($\zeta = 0.2$) has about 2.5 cycles for an axial resolution of about 0.4 mm. Narrow-band pulses have more energy, for the same k_0 gain. Figure 11.2b shows the greatest pulse energy occurs near Ω_0, where $\zeta \to 0$. Hence, narrow-band pulses offer the greatest sensitivity and the poorest temporal resolution. This discussion is intended to illustrate how to analyze the impulse-response

---

[8] The impulse response $h(t)$ [data/(obj-time)] has Fourier transform $H(u)$ [data/obj].

function and spectrum equations to select sensors based on their application features. Similar analyses may be applied to design and assess the performance of other sensors.

```
%%%%%%%%%%%%%%%%%%%%%%%%%%%%%%%%%%%%%%  Code to generate spectral plots in Fig. 11.2  %%%%%%%%%%%%%%%%%%%%%%%%%%%%%%%%%%%%
k0=0.5;u=0:0.01:20;N=length(u);om0=2*pi*10;H=zeros(3,N);       % initialize parameters and variables
zeta=[0.05,0.1,0.2];om=2*pi*u;den=zeros(1,N);                  % initialize parameters and variables
for j=1:3                                                      % repeat for 3 damping constants
    den=om0^2*(1-zeta(j)^2)+(zeta(j)*om0 + 1i*om*sqrt(1-zeta(j)^2)).^2; % compute denominator of (11.16)
    H(j,:)=(k0./den).*conj(k0./den);                          % compute |H(u)|^2
end
figure;plot(u,H(1,:),'k:');hold on;plot(u,H(2,:),'k--');plot(u,H(3,:),'k');hold off
figure;plot(u,H(1,:)/max(H(1,:)),'k:');hold on;
plot(u,H(2,:)/max(H(2,:)),'k--');plot(u,H(3,:)/max(H(3,:)),'k');hold off
```

11.2.7 Graphical Design of Acoustic Sensors

The graphical representations of sensor eigenvalues shown in Figure 11.3a can offer significant intuition about how changes in physical or engineering parameters can achieve desired pulse properties. We keep the eigenvalues complex for acoustic sensors so that pulses oscillate stably as required for medical imaging; specifically, we keep $0 < \zeta < 1$. In this range, the resonant frequency is Ω_0 but the pulse frequency is $\Omega_0\sqrt{1 - \zeta^2}$ and bandwidth is $2\zeta\Omega_0$. Damping influences both.

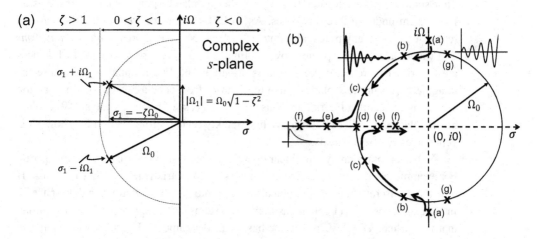

Figure 11.3 (a) Illustration of the complex s-plane where sensor properties are defined. $s = \sigma + i\Omega$ for continuous variables σ, Ω. Eigenvalues appear at conjugate-pair points called "poles" or Dirac deltas, marked by X: $\lambda_{1,2} = \sigma_1 \pm i\Omega_1 = -\zeta\Omega_0 \pm i\Omega_0\sqrt{1 - \zeta^2}$ via (11.14). The resonant frequency of the device is Ω_0 and the damping factor is $\zeta = -\sigma_1/\Omega_0$ for $0 < \zeta < 1$. In the same range, $\Omega_0\sqrt{1 - \zeta^2}$ is the pulse frequency and $2\zeta\Omega_0$ is the pulse bandwidth. (b) One pair of poles is shown moving together as k_1 is adjusted. Motion is constrained to the imaginary axis, a circle of radius Ω_0, and the real axis. The poles move symmetrically in opposite directions along the imaginary axis $i\Omega$ when the eigenvalues are imaginary and along the circle when they are complex. The poles move in opposite directions along σ for real eigenvalues such that $\lambda_1 \to 0$ as $\lambda_2 \to \infty$. The subplots illustrate temporal impulse response functions at points (b) and (g).

Eigenvalues appear in the s-plane as "poles" (Dirac deltas) at symmetric conjugate positions that are constrained to follow paths. For example, selecting k_2 determines Ω_0, which is the radius of the circles in Figure 11.3. Next, selecting k_1 changes ζ to symmetrically move poles along the circle, as shown in Figure 11.3b. Think of k_1 and k_2 as tuning knobs to set the pulse frequency and bandwidth. Operating outside these bounds is risky. For example, selecting $\zeta = 0$ places poles on the imaginary Fourier axis $i\Omega$, where $h(t) \propto \sin(\Omega_0 t)$ is quasi-stable. Tuning k_1 so that $\zeta < 0$ is a clear mistake. Now $\exp(+\Omega_0 \zeta t)$ in (11.15) grows the impulse response $h(t)$ to saturation for any input. Negative damping means the device is unstable.

Summary.
For the PZT sensor element in the preceding example, we adjusted the ODE parameters k_1, k_2 to give complex eigenvalues where the oscillating pulse has the frequency and bandwidth required for the task. We can follow the adjustments by tracking (11.15) or interpreting that equation graphically as in Figure 11.3.

11.3   Laplace Transforms

In Section 11.1.1, we found that the ansatz for a second-order ODE is $c_1 e^{\lambda_1 t} + c_2 e^{\lambda_2 t}$. That solution was used to compute $h(t)$, the impulse response for the sensor as an LTI system under different regimes. As an LTI system, we can predict the sensor's response $g(t)$ to many practical inputs $f(t)$. This section explores *Laplace transforms* as a way to access the s-plane shown in Figure 11.3, where properties of LTI devices are described more completely than is possible with Fourier analysis. Because the imaginary axis in the s-plane is also the Fourier frequency axis, it should not be too surprising to hear that Fourier analysis is a special case of Laplace analysis. To explain how Laplace generalizes Fourier, we first quickly review (from Section 6.5) how SVD generalizes Fourier analysis.

SVD is a tool for analyzing linear operators that exist in two distinct vector spaces. For example, SVD is designed for analyzing system matrix $\mathbf{H}$ in $\mathbf{g} = \mathbf{Hf} + \mathbf{e}$ because $\mathbf{H}$ transforms discrete functions in object space $\mathbb{U}$ into discrete functions in data space $\mathbb{V}$. In special cases of LTI/LSI measurement systems described by square, Hermitian matrices where $\mathbb{U} = \mathbb{V}$, a Fourier basis can decompose $\mathbf{H}$ just as well as it can decompose measurement data $\mathbf{g}$ or $\mathbf{e}$ noise. For this special case, SVD, eigenanalysis, and Fourier analyses all give the same result.

Laplace analysis also assumes the special case where $h(t)$ operates within one vector space. The Laplace "frequency" variable, $s = \sigma + i\Omega$, defines a 2-D plane that includes the 1-D Fourier frequency axis $i\Omega$ at $\sigma = 0$. Adding the real axis dimension to s helps us understand how increasing the damping in $h(t)$ affects frequency bandwidth. For example, in Figure 11.3a, we see that increasing damping moves the poles away from the $i\Omega$ axis for $0 < \zeta < 1$. As they move away, their influence on the imaginary axis becomes more diffuse, which widens the frequency bandwidth and lowers the amplitude of the frequency spectrum. This effect is illustrated later in this chapter in Figure 11.5. We will show that the Laplace integral may exist when the Fourier integral strictly does not. By specifying regions of convergence, Laplace analysis tells us about device stability. Let's look at the details.

Eigenfunctions of any LTI system have the general form e^{st}. Consequently, output $g(t)$ must satisfy $g(f(t)) = \lambda f(t)$ for input $f(t) = e^{st}$ and constant λ:

$$g(t) = [h * f](t) = \int_{-\infty}^{\infty} dt'\, h(t')\, e^{s(t-t')}$$

$$= e^{st} \int_{-\infty}^{\infty} dt'\, h(t')\, e^{-st'} = e^{st}\, H(s).$$

Because $H(s)$ is a constant with respect to t, this shows that $H(s)$ is an eigenvalue of $h(t)$ that we record in $g(t)$ when we input $f(t) = e^{st}$. More generally, $H(s)$ is the *system response* obtained from a *Laplace transform* of impulse response $h(t)$,

$$H(s) \triangleq \mathcal{L}h(t) = \int_{-\infty}^{\infty} dt\, h(t)\, e^{-st}. \quad \text{(Two-sided forward Laplace transform)}$$

$$(11.17)$$

The two-sided forward Laplace transform is closely related to the Fourier transform. Restricting our attention to the imaginary axis in the s-plane by setting $\sigma = 0$, we have the *continuous-time Fourier transform*, where $\Omega = 2\pi u$,

$$H(\Omega) = \mathcal{F}h(t) = \int_{-\infty}^{\infty} dt\, h(t)\, e^{-i\Omega t}, \quad (11.18)$$

that is recovered with the inverse continuous-time Fourier transform,[9]

$$h(t) = \mathcal{F}^{-1} H(\Omega) = \frac{1}{2\pi} \int_{-\infty}^{\infty} d\Omega\, H(\Omega)\, e^{i\Omega t}$$

$$= \frac{1}{2\pi} \int_{-\infty}^{\infty} d\Omega \left[\int_{-\infty}^{\infty} dt'\, h(t')\, e^{-i\Omega t'} \right] e^{i\Omega t}$$

$$= \int_{-\infty}^{\infty} dt'\, h(t') \left[\frac{1}{2\pi} \int_{-\infty}^{\infty} d\Omega\, e^{i\Omega(t-t')} \right]$$

$$= \int_{-\infty}^{\infty} dt'\, h(t')\, \delta(t - t') = h(t).$$

Let's try the same technique for computing the inverse Laplace transform. We first define $h'(t) \triangleq h(t)\, e^{-\sigma t}$. Then, from (11.17),

$$H(s) = \int_{-\infty}^{\infty} dt\, h(t)\, e^{-st}$$

$$H(\sigma + i\Omega) = \int_{-\infty}^{\infty} dt\, h'(t)\, e^{-i\Omega t} = H'(\Omega).$$

---

[9] In previous chapters, I used the notations $H(\Omega)$ and $H(u)$ to denote CT-FT in the radial and temporal frequency domains. Other authors use $H(i\Omega)$, which is quite natural in this context.

So the Laplace transform of $h(t)$ equals the Fourier transform of $h'(t)$: $H(s) = H(\sigma + i\Omega) = H'(\Omega)$. Therefore, applying a change of variable,

$$h(t) = e^{\sigma t} h'(t) = \frac{e^{\sigma t}}{2\pi} \int_{-\infty}^{\infty} d\Omega \, H'(\Omega) \, e^{i\Omega t}$$

$$= \frac{1}{2\pi} \int_{-\infty}^{\infty} d\Omega \, H(\sigma + i\Omega) \, e^{(\sigma + i\Omega)t}$$

$$h(t) = \frac{1}{2\pi} \int_{\sigma-i\infty}^{\sigma+i\infty} ds \, H(s) \, e^{st}. \quad \text{(Inverse Laplace transform)} \quad (11.19)$$

The integration variable change from Ω to s changes the integration limits from $\Omega = \pm\infty$ to $s = \sigma \pm i\infty$. Solutions exist in the *region of convergence* of the integral. Because of the factor $\exp(-\sigma t)$, the Laplace transform exists when the Fourier transform does not, which is a source of interest in Laplace transforms.[10] For causal systems, we use the one-sided Laplace transform (1SLT):

$$H(s) \triangleq \mathcal{L}h(t) = \int_0^{\infty} dt \, h(t) \, e^{-st}, \quad \text{(One-sided forward Laplace transform)}$$

$$(11.20)$$

which has the same inverse as (11.19).

Example 11.3.1. *Computing Laplace transforms: Find $H(s)$ for $h(t) = A \, e^{-at} \, \text{step}(t)$ and $a \in \mathbb{R} > 0$.*

Solution
The presence of the step function makes this a one-sided Laplace transform:

$$H(s) = \mathcal{L}h(t) = \int_{-\infty}^{\infty} dt \, A \, e^{-at} \, \text{step}(t) \, e^{-st}$$

$$= A \int_0^{\infty} dt \, e^{-(a+s)t} = -\left(\frac{A}{a+s}\right) e^{-(a+s)t} \Big|_0^{\infty}$$

$$= \frac{A}{a+s}, \quad \text{where ROC: } s > -a.$$

The integral exists when the argument of the exponent in the integrand $-(a+s) < 0$ or $\Re\{s\} > -a$, which is the region of convergence. *Outside the region of convergence, the integral diverges. There is one pole on the real axis at $s = -a$, as shown in Figure 11.4, and one zero at $s = \infty$. Since the eigenvalue is real, this 1-D system has an overdamped response. Importantly, this region of convergence includes the imaginary axis, so we are sure the Fourier transform exists in this example.*

---

[10] Laplace transforms are for continuous-time signals. For discrete-time signals, z transforms apply [77].

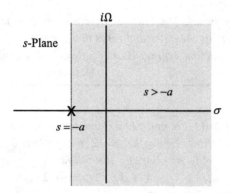

$i\Omega$

s-Plane

$s > -a$

σ

$s = -a$

Figure 11.4 Illustration of the region of convergence in Example 11.3.1.

- When $H(s)$ is expressed as a ratio of polynomials, $H(s) = \frac{(s+a)}{(s^2+(c-b)s-bc)}$, the roots of the numerator are the zeros of the system and the roots of the denominator are the poles of the system response. They define system properties in the Laplace domain.
- Because the polynomial coefficients are real, all poles and zeros must either be real or exist as conjugate pairs.
- There are equal numbers of zeros and poles in each equation. For example, $H(s) = \frac{(s+a)}{(s-b)(s+c)}$ has zeros at $s = -a, \infty, -\infty$ and poles at $s = b, -c, \infty$.
- We can ignore poles or zeros at $s = 0, \infty, -\infty$ as having no influence on the system properties. For example, $H(s) = 1/(s(s+e))$ has zeros at $s = \infty, -\infty$ and poles at $s = 0, -e$, but only the pole at $s = -e$ influences the system properties.
- Regions of convergence cannot contain any poles.
- Stable systems have all their poles in the negative half-plane at $\sigma < 0 \; \forall \, i\Omega$.

11.4 MATLAB Solutions via Symbolic Math

The symbolic math toolbox in MATLAB offers analytic solutions to the one-sided Laplace transforms. You may need to spend a little time reading up about *symbolic math*, although usage is straightforward. The next three sections of code use `laplace` and `ilaplace` to compute the forward and inverse transforms for (a) Example 11.3.1 and (b) the real and (c) the complex impulse responses in Section 11.2. I urge readers to try these.

Example 11.4.1.

```
%(a)********* MATLAB symbolic math solution to Example 11.2.1 ************
syms t s a A              % define variables and constants to begin
h=A*exp(-a*t);            % function to be transformed
H=laplace(h,t,s)          % computes 1SLT that I output here...

H = A/(a + s)

a=0.5;                    % defining constants, the pole-zero plot is obtained by...
HH=tf([0,0],[1,a]);       % finding transfer function (tf), include polynomial coeff's
pzmap(HH)                 % ... and then plotting the poles and zeros.
hp=ilaplace(H)            % find inverse LT to show return to h.  Reg of conv is up to you
```

This code is relatively straightforward. The Laplace transform and its inverse are computed. The function `pzmap` *plots the poles and zeros that are not at zero or* $\pm\infty$ *from* `tf`, *which is the transfer function expressed as a polynomial. I explain some details in the following discussion.*

Example 11.4.2. *We now return to (11.9) and (11.3) and revisit the second-order ODE with real eigenvalues. Selecting* $k_0 = k_1 = 2$ *and* $k_2 = 0.5$, *we find* $c_1\lambda_1 = \sqrt{2} = -c_2\lambda_2$. *We simplify by adopting the following notation:* $\lambda_1 = -1 - 1/\sqrt{2} = -1.7071 = -a$, $\lambda_2 = -1 + 1/\sqrt{2} = -0.2929 = -b$, *and* $A = c_1\lambda_1 = \sqrt{2}$ *according to the "k" parameters.*

```
%%*** MATLAB symbolic math solution to H(s) where h(t) has real eigenvalues (Eqs 11.3 and 11.9) *****
syms t s a b A                   % variables for the IR for 2nd-order ODE with real eigenvalues
h=A*(exp(-b*t) - exp(-a*t));     % similar to #11 of Appendix F
H=laplace(h);pretty(H)           % pretty makes it easier to read the following result...
%
A=0.5;k2=0.5;a=1+sqrt(1-k2);b=1-sqrt(1-k2); % define numerical parameters to plot poles
pzmap([0 0 0],[1 a+b a*b])                   % do this after obtaining the symbolic solution!
% The Laplace transform of h(t) is
        /   1         1    \
H = -A  | ----- - ----- |
        \ a + s    b + s /
```

It's not immediately clear that this answer is the same as that in Appendix F, number 11. We can rewrite the result as follows:

$$H(s) = A\left(\frac{1}{s+b} - \frac{1}{s+a}\right) = \frac{A(a-b)}{(s+a)(s+b)}.$$

Now we see the MATLAB result is the same as number 11 in Appendix F. Further, since $(a - b) = \sqrt{2}$, the product $A(a - b) = 2$.

The denominator of the transfer function $(s + a)(s + b)$, needs to be multiplied out to provide `pzmap` with polynomial coefficients. Specifically, to describe poles for the denominator, $(s + a)(s + b) \to s^2 + (a + b)s + ab \to$ `[1 a+b a*b]`. For zeros in the numerator, we just give `[0 0 0]`.

Example 11.4.3. *Finally, we compute the Laplace transform of (11.15) for complex eigenvalues. I renamed* $a = \Omega_0\sqrt{1 - \zeta^2}$ *and* $b = \zeta\Omega_0$ *so that* $h(t) = k_0(\exp(-bt)/a)\sin(at)$. *This is number 18 in Appendix F. The following code computes the transform, its inverse, and the pole–zero plot (not shown), assuming* $\Omega_0 = 2\pi\times(10\ MHz)$ *and* $\zeta = 0.2$. *The region of convergence is up to you to figure out.*

```
% (c) *** MATLAB symbolic math solution to h(t) with complex eigenvalues in Section 11.1.3 ***
>> syms t s a b k0                 % zeta = 0.2 and Omega_0 = 2\pi x 10 MHz
>> h=k0*exp(-b*t)*sin(a*t)/(a);    % IR 2nd-order ODE with complex eigenvalues
>> H=laplace(h);pretty(H)          % "pretty" makes it easier to read close-form result...


             k0
H = -------------
           2    2
      (b + s) + a
```

```
>> a=2*pi*10*sqrt(1-0.2^2);b=2*pi*10*0.2;      % provide numerical values to constants,
>> HH=tf([0 0 0],[1 2*b a^2+b^2]);             % compute the transfer function,
>> pzmap(HH)                                    % and output the pole-zero plot
>> hp=ilaplace(H);pretty(hp)                    % find the inverse transform to return to h

       k0 exp(-b t) sin(a t)
h = ---------------------
                a
```

The symbolic output in this example gives the results for H *and the recovered* h. *If you generate the pole–zero plot, you will find* $\lambda_{1,2} = 12.6 \pm 61.6i$. *Noting that* $\lambda_{1,2} = \Omega_0(-\zeta \pm i\sqrt{1-\zeta^2})$, *we see that the plot values match the analytic values. The Fourier transform in (11.16) is found from the* MATLAB-*generated Laplace transform,* $H(s) = k_0/((s+b)^2 + a^2)$, *after projecting s onto the imaginary axis:* $s \to i\Omega = \pm i\Omega_1 = \pm i\Omega_0\sqrt{1-\zeta^2}$ *(see Figure 11.3a.) Finally, the region of convergence is* $\Re\{s\} > -b = -\zeta\Omega_0$.

Figure 11.5a is a pole–zero graph for the parameters of Example 11.4.3. Figure 11.5b shows a portion of the s-plane for $|H(s)|$ rendered in grayscale to visualize the continuously varying system response resulting from complex poles. These poles are like tent poles that rise sharply at two bright points in the plane (see the plot in Figure 11.5c), while casting broad, low-amplitude shadows on the Fourier axis at

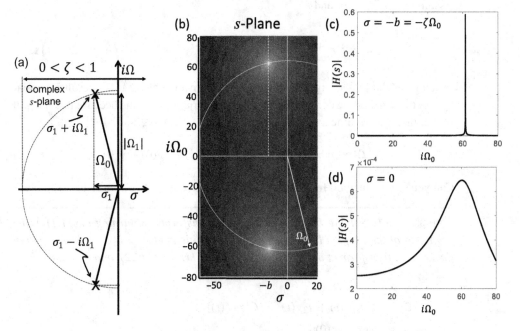

Figure 11.5 (a) Pole–zero diagram specifically for Example 11.4.3. (b) Grayscale version of the graph in (a). A line through one pole is plotted in (c) along the dotted line shown in (b) at $\sigma = -b$. The positive imaginary axis is plotted in (d). Off-axis poles influence values on the Fourier frequency axis at $\sigma = 0$.

$s = (0, i\Omega_0)$, as seen in Figure 11.5d. The grayscale amplitude is log compressed to display amplitude variations more clearly, while the plots have linear scales. Also compare Figure 11.5d with the Fourier spectrum for $\zeta = 0.2$ in Figure 11.2. While the axes are scaled differently, the shape is the same. In this way, the Laplace transform is related to the Fourier transform.

11.5   Laplace Transform Approach to Solving ODEs

Laplace transforms offer solutions to ODEs of the form $\mathcal{D}g(t) = k_0 f(t)$ as a way to describe those linear systems. The method begins with a Laplace transform of each term in the differential equation using the properties identified in Appendix F to obtain a polynomial equation in s. We then solve for $H(s) = G(s)/F(s)$ and inverse-transform the result to find $h(t)$. For any input $f(t)$, we can predict output $g(t)$ using $g(t) = [h * f](t)$.

1SLT of derivatives. Before we apply the table of 1SLT pairs, let's look at two important pairs. For this we need to integrate the following equation by parts:

$$\mathcal{L}\{\dot{g}(t)\} = \int_0^\infty dt \frac{dg}{dt} e^{-st}.$$

Assigning $dt \frac{dg}{dt} \to dv$ and $e^{-st} \to u$,

$$= g(t) e^{-st} \Big|_0^\infty + s \int_0^\infty dt\, g(t)\, e^{-st} = -g(0) + s G(s), \qquad (11.21)$$

which gives the Laplace transform of the first derivative of $g(t)$. The result is similar but not the same as the derivative theorem for Fourier transforms because this integral is over the positive time axis only. We can apply (11.21) to $\dot{g}$ to find $\mathcal{L}\{\ddot{g}(t)\}$,

$$\mathcal{L}\{\ddot{g}(t)\} = s\mathcal{L}\{\dot{g}\} - \dot{g}(0) = s^2 G(s) - sg(0) - \dot{g}(0). \qquad (11.22)$$

Can you see the method for computing the nth-order derivative? On to the examples.

Example 11.5.1. *Solving ODEs with Laplace transforms: Returning to (11.1) at the beginning of the chapter, let's assume initial conditions $g(0) = \dot{g}(0) = 0$. Computing the Laplace transform of both sides according to (11.21), (11.22), and Appendix F, we find*

$$\mathcal{L}\{\ddot{g}(t) + k_1\dot{g}(t) + k_2 g(t)\} = \mathcal{L}\{k_0 f(t)\}$$
$$s^2 G(s) - sg(0) - \dot{g}(0) +$$
$$k_1 s G(s) - k_1 g(0) + k_2 G(s) = k_0 F(s)$$
$$G(s)\left(s^2 + k_1 s + k_2\right) = k_0 F(s)$$

$$\frac{G(s)}{F(s)} = H(s) = \frac{k_0}{s^2 + k_1 s + k_2} \tag{11.23}$$

$$= k_0 \frac{1}{\left(s + \frac{k_1 - \sqrt{k_1^2 - 4k_2}}{2}\right)\left(s + \frac{k_1 + \sqrt{k_1^2 - 4k_2}}{2}\right)}$$

$$= \frac{k_0}{(s - \lambda_1)(s - \lambda_2)}.$$

The poles are eigenvalues of the system as found in (11.3). We may take the inverse Laplace transform to find

$$\mathcal{L}^{-1}\{H(s)\} = h(t) = \frac{k_0}{\lambda_1 - \lambda_2}\left(e^{\lambda_1 t} - e^{\lambda_2 t}\right).$$

Applying (11.6) and (11.13), we can show that $k_0/(\lambda_1 - \lambda_2) = c_1\lambda_1 = -c_2\lambda_2$. Consequently, we obtain the same result as (11.9), showing Laplace transforms offer another method for solving ODEs.

Example 11.5.2. One-compartment dilution with source revisited: *Let's return to Example 10.3.1, where we had a single compartment with a constant input of a tracer substance at rate r,*

$$v\,\dot{c}(t) + q\,c(t) = r \quad \text{for } t > 0.$$

Previously, we set the initial concentration $c(0) = 0$, but this time we leave it arbitrary. Applying the forward 1SLT properties from Appendix F,

$$v\left[s\,C(s) - c(0)\right] + q\,C(s) = \frac{r}{s}.$$

Solving for $C(s)$,

$$C(s) = \frac{c(0)}{s + q/v} + \frac{r/v}{s(s + q/v)},$$

and applying the inverse transforms from Appendix F, we find

$$c(t) = c(0)\,e^{-qt/v} + \frac{r}{q}(1 - e^{-qt/v}).$$

The second term on the right is what we found in Example 10.3.1. The solution obtained in this case is more general because here we allow the initial concentration to be arbitrary instead of zero.

Example 11.5.3. Two-compartment series dilution: *Let's return to Problem 10.2, which is the two-compartment series dilution problem. The differential equation is*

$$\dot{c}_2(t) + \frac{q}{v_2}c_2(t) = \frac{q}{v_2}c_1(0)\,e^{-qt/v_1}\,\text{step}(t).$$

Initially, we have $c_2(0) = 0$. Proceeding as described earlier,

$$s C_2(s) - c_2(0) + \frac{q}{v_2} C_2(s) = \frac{q\, c_1(0)}{v_2} \frac{1}{s + q/v_1}. \qquad \text{(Transform 5 in Appendix F)}$$

Solving for $C(s)$,

$$C_2(s) = \frac{q\, c_1(0)}{v_2} \left(\frac{1}{s + q/v_1} \right) \left(\frac{1}{s + q/v_2} \right).$$

To transform back, we use transform Appendix F, number 11:

$$c_2(t) = \frac{c_1(0)\, v_1}{v_2 - v_1} \left[e^{-qt/v_2} - e^{-qt/v_1} \right].$$

It is the same as the solution to Problem 10.2 if we multiply the numerator and denominator by -1.

Example 11.5.4. *Diffusion between two compartments. Let's try a different problem. Consider two compartments separated by a semi-permeable barrier that allows diffusion of a solute [99]. The diffusion rate from one compartment to the other is proportional to the difference in concentrations, c_1 and c_2, and will be driven toward the compartment with the lower concentration. The two first-order ODEs are*

$$v_1 \dot{c}_1(t) = k(c_2(t) - c_1(t)) = -v_2 \dot{c}_2(t),$$

where k is a constant that varies with solute properties, barrier properties, and perhaps other things. The total molar mass of solute is the sum of masses in both compartments. Initially, we have

$$m_{total} = v_1\, c_1(0) + v_2\, c_2(0). \qquad \text{(Initial condition)}$$

Once the two compartments have reached equilibrium, we have $c(\infty) \triangleq c_1(\infty) = c_2(\infty)$ and

$$c(\infty) = \frac{m_{total}}{v_1 + v_2} = \frac{c_1(0)v_1 + c_2(0)v_2}{v_1 + v_2},$$

and therefore

$$m_{total} = (v_1 + v_2)c(\infty). \qquad \text{(Equilibrium condition)}$$

Using the transform method,

$$v_1(s C_1(s) - c_1(0)) = k(C_2(s) - C_1(s))$$
$$v_2(s C_2(s) - c_2(0)) = k(C_1(s) - C_2(s)). \qquad (11.24)$$

From the second equation in (11.24), we find

$$C_2(s) = \frac{k C_1(s) + v_2\, c_2(0)}{s v_2 + k}.$$

Inserting this result back into the first of equations (11.24),

$$v_1(sC_1(s) - c_1(0)) = \frac{k^2 C_1(s) + kv_2\, c_2(0)}{sv_2 + k} - kC_1(s)$$

$$C_1(s) = \frac{sv_1v_2c_1(0) + v_1k\, c_1(0) + v_2k\, c_2(0)}{s^2v_1v_2 + kv_1s + kv_2s}$$

$$= \frac{c_1(0)}{s+a} + k\frac{c_1(0)/v_2 + c_2(0)/v_1}{s(s+a)},$$

where
$$a = k\frac{v_1 + v_2}{v_1v_2}.$$

Applying numbers 5 and 6 from Appendix F, the inversion gives

$$c_1(t) = c_1(0)\, e^{-at} + \frac{k}{a}\left(\frac{c_1(0)}{v_2} + \frac{c_2(0)}{v_1}\right)(1 - e^{-at})$$

$$= c_1(0)\, e^{-at} + \left(\frac{c_1(0)v_1 + c_2(0)v_2}{v_1 + v_2}\right)(1 - e^{-at})$$

$$= (c_1(0) - c(\infty))\, e^{-at} + c(\infty), \text{ where } c(\infty) = \frac{c_1(0)v_1 + c_2(0)v_2}{v_1 + v_2}.$$

The solution for $c_2(t)$ is entirely symmetric in the sense that repeating the procedure from (11.24) onward for $c_2(t)$ leads to

$$c_2(t) = (c_2(0) - c(\infty))\, e^{-at} + c(\infty).$$

11.6   Summary

The last two chapters showed that biological systems and the sensors that measure their properties can be analyzed with ODE models and related basic engineering techniques. There is value in keeping models as simple as possible and focused on specific aspects of system behavior when the goal is to understand how systems work. While linear models work well for sensors, they often miss many important details of biological systems. At the same time, they offer valuable clues about overall behavior via engineering parameters.

One key to obtaining essential insights into the measurement process is to use basis decompositions of modeled object functions, recorded data, and measurement devices. For example, singular values can reveal the uncoupled properties of a measurement device, while their corresponding singular vectors describe the functional modes by which input stimuli engage system components to generate output responses. These are core tools for understanding how data encode task information so that the costs and risks associated with a measurement pay off. Those of us who rely on well-designed measurement technologies to ensure patient health have a wealth of available engineering methods at our disposal. We just need to know where to find them and how they can help reveal the hidden information we seek.

11.7   Problems

11.1   Find impulse response $h(t)$ for the following system. Is the system stable? Solve for $g(t)$ assuming the particular solution $f(t) = f_p \, \text{step}(t)$ from Section 11.1.

$$\ddot{g}(t) - 4\dot{g}(t) + 3g(t) = f(t), \quad \text{where } g(0) = \dot{g}(0) = 0.$$

11.2   Compute output $g(t)$ for the following system, assuming we apply the specific driving function, $f(t) = \cos at$. Is the system stable?

$$\ddot{g}(t) + a^2 g(t) = f(t), \quad \text{where } g(0) = \dot{g}(0) = 0 \text{ and } a > 0$$

11.3   The mechanical sensor in Figure 11.6 is a second-order system, where position x is the state variable and t is the independent variable. This sensor requires maximum realistic sensitivity to applied forces. That is, you must select complex eigenvalues when possible. The nonautonomous ODE model is given by the driven, damped, harmonic oscillator equation,

$$m\ddot{x}(t) + b\dot{x}(t) + kx(t) = f(t), \quad \text{where } x(0) = \dot{x}(0) = 0.$$

Find the impulse response for this sensor. Be sure to reduce it to the simplest form.

11.4   Use Laplace transforms to solve for $g(t)$. Give the nontrivial poles and zeros for the system. Is the system stable?

$$\ddot{g} + 3\dot{g} + 2g = 0, \quad \text{where } g(0) = 1, \ \dot{g}(0) = 0$$

Validate your answer.

11.5   You are asked to model and compare the responses of two LTI sensors with temporal impulse responses that are indicated graphically in the pole–zero diagram of Figure 11.7. The resonant frequency of sensor 1 is given by $\Omega_0 = 2\pi \times 15$ rad/s

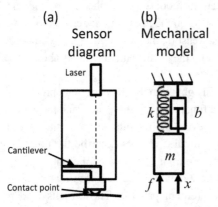

(a)   (b)

Sensor   Mechanical
diagram   model

Figure 11.6 Diagram of a mechanical sensor (a) based on a cantilever in a clear viscous fluid that is scanned by a laser interferometer. The equivalent mechanical model (b) shows mass m, damping constant b, and elastic coefficient k. Also indicated are input force f and output displacement x.

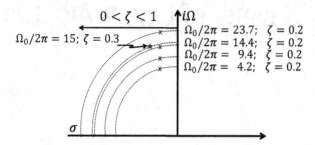

Figure 11.7 Pole–zero plot describing two LTI sensors related to Problem 11.5. Resonant frequencies $\Omega_0/2\pi$ are indicated for sensor 1 (*) on the left and for sensor 2 (×) on the right. Since the real parts of the poles (only one of each conjugate pair is shown) are negative, both sensor responses will be stable.

and $\zeta = 0.3$ so that pulse frequency$/2\pi = 14.3$ Hz, bandwidth BW$/2\pi = 9$ Hz, and FBW $= 0.63$. These engineering parameters are defined in the text after (11.16). One pole from the conjugate pair (indicted by *) for sensor 1 is shown in Figure 11.7. Sensor 1 is completely specified by that one pole; it has impulse response $h_1(t)$.

Sensor 2 is a composite of four sensors whose impulse responses are given by vector $\mathbf{h}^\top = (h_a(t), h_b(t), h_c(t), h_d(t))$ and indicated by ×'s on the right side of Figure 11.7. The net impulse response for the sensor is a weighted sum of the four components. The weight vector is $\mathbf{w}^\top = (1, 2, 4, 8)$, and the net impulse response is $h_4(t) = \mathbf{h}^\top \mathbf{w}$.

The assignment is to model both sensors and compare their properties by having them measure the following object function. This is a sequence of square wave oscillations with increasing frequency, whose sharp transitions emphasize the need for high temporal bandwidth.

```
%%%%%%% Input object function f(t) for testing the response of two sensors %%%%%%%
        f=ones(size(t));   % generate square-wave object over 10s
        f(51:100)=-1;f(151:200)=-1;f(251:300)=-1;f(351:400)=-1;
        f(426:450)=-1;f(476:500)=-1;f(526:550)=-1;f(576:600)=-1;
        f(612:625)=-1;f(637:650)=-1;f(662:675)=-1;f(687:700)=-1;
        f(712:725)=-1;f(737:750)=-1;f(762:775)=-1;f(787:800)=-1;
        f(806:812)=-1;f(818:824)=-1;f(830:836)=-1;f(842:848)=-1;
        f(854:860)=-1;f(866:872)=-1;f(878:884)=-1;f(890:896)=-1;
        f(902:908)=-1;f(914:920)=-1;f(926:932)=-1;f(938:944)=-1;
        f(950:956)=-1;f(962:968)=-1;f(974:980)=-1;f(986:992)=-1;
```

Other things to include in your model: (a) Each impulse response function should be individually normalized to have energy $= 1$ before being combined. Also, normalize $h_1(t)$ for sensor 1 and composite $h_4(t)$ for sensor 2 to have the same peak spectral value. (b) Assume SNR $= \int dt\ f^2(t)/\mathrm{var}(e) \sim 70$ dB, where aquisition noise is added after convolution for these LTI devices. (c) Compare the frequency spectra for the impulse responses with each other and with that of the object. This will guide your explanation of the results found. (d) Compare h_1 and h_4 as well as individual components to help explain how well each device is able to measure the object function. You are allowed to apply gain or attenuation to measurements $g_1(t)$ and $g_4(t)$ after noise is added to scale the results for comparisons with object $f(t)$.

If sensor 2 costs twice as much as sensor 1, would you recommend spending the money?

Appendix A Review of Linear Algebra

This extended appendix aims to review linear algebra concepts, define terms and properties, and provide examples that augment the discussions in the chapters. However, it is assumed that readers have had some exposure to linear algebra equivalent to an undergraduate course. I find the following book to be very accessible and a complete introduction: Gareth Williams, *Linear Algebra with Applications, 8/e* (Alternate), Jones & Bartlett Learning, Burlington, MA, 2014.

Notation. Define a column sequence of real numbers as column vector

$$\mathbf{u} = \begin{pmatrix} u_1 \\ \vdots \\ u_n \\ \vdots \\ u_N \end{pmatrix} \in \mathbb{R}^{N \times 1}. \text{ Alternatively, } \mathbf{u}^\top = (u_1, \ldots, u_n, \ldots, u_N) \in \mathbb{R}^{1 \times N}, \text{ where}$$

superscript $\top$ indicates *transpose*. If vector $\mathbf{u}$ contains complex numbers, then $\mathbf{u} \in \mathbb{C}^{N \times 1}$. Generally, boldface uppercase letters are matrices, e.g., $\mathbf{A}, \mathbf{B}, \mathbf{C} \in \mathbb{R}^{M \times N}$; boldface lowercase letters are vectors, e.g., $\mathbf{u}, \mathbf{v}, \mathbf{w} \in \mathbb{R}^{N \times 1}$; and italic lowercase letters are scalars, e.g., $c, d \in \mathbb{R}^{1 \times 1}$. The blackboard bold font $\mathbb{R}$ denotes the vector space of all real numbers.

Vector Space

$\mathbb{V}$ is a set of elements (vectors $\mathbb{R}^{N \times 1}$, matrices $\mathbb{R}^{M \times N}$, or functions $f(x)$) for which addition and scalar multiplication are defined. They satisfy *closure axioms* [e.g., the sum of any two vectors in $\mathbb{V}$, $\mathbf{u} + \mathbf{v}$ (Figure A.1), is an element of $\mathbb{V}$ and the scalar product $c\mathbf{u}$ is also an element of $\mathbb{V}$], *addition axioms* [e.g., commutative property $\mathbf{u} + \mathbf{v} = \mathbf{v} + \mathbf{u}$, associative property $(\mathbf{u} + \mathbf{v}) + \mathbf{w} = \mathbf{u} + (\mathbf{v} + \mathbf{w})$, and identity property $\mathbf{u} + \mathbf{0} = \mathbf{u}$], and *scalar multiplication axioms* [e.g., distributive properties $c(\mathbf{u} + \mathbf{v}) = c\mathbf{u} + c\mathbf{v}$; $(c + d)\mathbf{u} = c\mathbf{u} + d\mathbf{u}$; $c(d\mathbf{u}) = (cd)\mathbf{u}$], and identity property $(1\mathbf{u} = \mathbf{u})$.

The properties of a vector space capture the linear structure of data that is essential to linear models of the measurement process. Specifically, linear operations on vectors in $\mathbb{V}$ yield other vectors living in the same space. They also provide a geometric interpretation via lengths and angles that allows us to compare data.

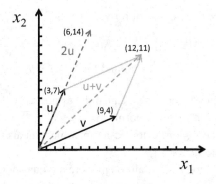

Figure A.1 A geometric illustration of vector addition, $\mathbf{u} + \mathbf{v}$, and scalar multiplication, $2\mathbf{u}$.

Subspaces

Let $\mathbb{S}$ be a nonempty subset of $\mathbb{V}$. If $\mathbb{S}$ is a vector space as defined earlier and it contains the zero vector $\mathbf{0}$ of $\mathbb{V}$, then it is a *subspace* of $\mathbb{V}$. Note that a subspace containing only the zero vector is not empty. That subspace has *dimension* zero.

Linear Independence

The set of vectors $\{\mathbf{v}_m\} = \mathbf{v}_1, \ldots \mathbf{v}_M$ in vector space $\mathbb{V}$ is *linearly dependent* if there exist scalars $\{c_m\}$ that are *not* all zero such that $c_1\mathbf{v}_1 + c_2\mathbf{v}_2 + \ldots + c_m\mathbf{v}_M = \mathbf{0}$. Vectors $\{\mathbf{v}_m\}$ in $\mathbb{V}$ are *linearly independent* if $c_1\mathbf{v}_1 + c_2\mathbf{v}_2 + \ldots + c_m\mathbf{v}_M = \mathbf{0}$ only if $c_m = 0 \ \forall \ m$.

Spanning Basis of a Vector Space

$\{\mathbf{v}_m\}$ *spans* vector space $\mathbb{V}$ if every vector in $\mathbb{V}$ can be expressed as a linear combination of vectors in $\{\mathbf{v}_m\}$. If $\{\mathbf{v}_m\}$ spans $\mathbb{V}$ and the elements of the set are *linearly independent*, $\{\mathbf{v}_m\}$ is a *basis* for $\mathbb{V}$.

Norms

The norms of a vector or a matrix, denoted by $\| \cdot \|$, are an important way to *quantify* their properties. For example, the *Euclidean norm* of vector $\mathbf{v}$, denoted $\|\mathbf{v}\|_2$, measures its length, while the vector ∞-norm quantifies the element with the largest magnitude. Also, the condition number of a matrix, which is based on norms, measures how close a matrix is to being singular. The many types of norms enable us to measure their many properties; each is a positive scalar. The exceptions are a zero vector and a zero matrix, which have zero norms. Among the simplest 2-norms is the *absolute value* of scalar a; i.e., $\|a\|_2 = |a|$. For example, if $a = 3$, $\|a\|_2 = \sqrt{3^2} = 3$; if $a = -2$, $\|a\|_2 = 2$; if $a = 2 - 3i$, $\|a\|_2 = \sqrt{a^*a} = \sqrt{13}$. Also, if $\mathbf{x}$ is a *unit vector*, then $\|\mathbf{x}\|_2 = 1$.

Vector p-Norms. For $p = [1, \infty]$, the ℓ_p norm for vector $\mathbf{v} \in \mathbb{C}^{N \times 1}$ is

$$\|\mathbf{v}\|_p = \begin{cases} \left(\sum_{j=1}^{N} |v_j|^p \right)^{1/p} & p = [1, \infty) \\ \max_j |v_j| & p = \infty \text{ and for } 1 \le j \le N \end{cases}.$$

The most important are (a) the vector 1-norm, also called the ℓ_1 norm, $\|\mathbf{v}\|_1 = |v_1| + \ldots + |v_N|$; (b) the ℓ_2, Euclidean, or vector 2-norm, $\|\mathbf{v}\|_2 = \sqrt{\langle \mathbf{v}, \mathbf{v} \rangle} = \sqrt{\mathbf{v}^\dagger \mathbf{v}} = (|v_1|^2 + \ldots + |v_N|^2)^{1/2}$; and (c) the ℓ_∞ or vector ∞-norm, which equals the largest absolute value of all elements in the vector.

The norms are strictly undefined for $p < 1$, although there are quasinorms in this range. For example, the ℓ_0 norm can be defined as $\|\mathbf{v}\|_0 \triangleq |\text{supp}(\mathbf{v})|$, which is the cardinality of the *support* for $\mathbf{v}$. In plain words, the ℓ_0 norm describes the number of nonzero elements in $\mathbf{v}$, which may be found in MATLAB using `length(find(v))`.

Errors and Significant Digits. Let elements of $\mathbf{v}'$ be stochastic with mean vector $\mathcal{E}\mathbf{v}'$. Then $\|\mathbf{v}' - \mathcal{E}\mathbf{v}'\|_2^2$ is proportional to *variance*, describing the energy in the stochastic variability of the recorded data. For any ℓ_p norm, $\|\mathbf{v} - \mathbf{v}'\|_p$ is a measure of *absolute error* and $\|\mathbf{v} - \mathbf{v}'\|_p / \|\mathbf{v}\|_p$ is a measure of *relative error*. For example, Golub and van Loan [41] show that the relative ℓ_∞-norm error of $\mathbf{v}'$ can be used to estimate the number of significant digits in the recorded data. They show that

$$\frac{\|\mathbf{v} - \mathbf{v}'\|_\infty}{\|\mathbf{v}\|_\infty} \simeq 10^{-q},$$

where q approximates the number of significant digits in the largest element of $\mathbf{v}'$.

For example, if $\mathbf{v} = \begin{pmatrix} 1.0451 \\ 2.1232 \end{pmatrix}$ and $\mathbf{v}' = \begin{pmatrix} 0.9591 \\ 2.1159 \end{pmatrix}$, then $\|\mathbf{v} - \mathbf{v}'\|_\infty / \|\mathbf{v}\|_\infty = 0.0860/2.1232 = 0.0405 \simeq 10^{-1.4}$, suggesting that, based on the relative error, $\mathbf{v}'$ may be specified with one or two significant digits. Note that the same result can by found using MATLAB and `A=log10(norm((v-vp),inf)/norm(v,inf))`. Generally, when the matrix elements are deviations of data about a mean value, then p-norms weight the largest deviations increasingly more than small deviations as the p value increases.

Matrix p-Norms

General statements about matrix norms are more complicated. Nevertheless, many useful specific statements are quite simple. The matrix p-norms for $\mathbf{A} \in \mathbb{R}^{M \times N}$ are defined in terms of vector p-norms [41]. Specifically, for $\mathbf{v} \in \mathbb{R}^{N \times 1}$,

$$\|\mathbf{A}\|_p = \sup_{\mathbf{v} \neq \mathbf{0}} \frac{\|\mathbf{A}\mathbf{v}\|_p}{\|\mathbf{v}\|_p} = \max_{\|\mathbf{v}\|_p = 1} = \|\mathbf{A}\mathbf{v}\|_p, \tag{A.1}$$

where sup indicates the *supremum* of the quantity, which is the "least upper bound."[1] For most situations involving finite size $\mathbf{v}$, the supremum is the largest absolute-value

---

[1] Note that "supp" and "sup," which are both defined in this section, are different operations.

element. For example, if $\mathbf{v} = (-2.3, \quad 0, \quad 4, \quad 2-i, \quad -4-2i)$, then $\sup(\mathbf{v}) = \sqrt{20} = 4.4721$. The second form of (A.1), which uses the max operator, may be a little clearer. However, this expression is not specific since there are many vectors $\mathbf{v}$ that might apply.

Fortunately, there are straightforward specific expressions for the most common matrix norms.

- For $\mathbf{A} \in \mathbb{R}^{M \times N}$, the *matrix 1-norm* is

$$\|\mathbf{A}\|_1 = \max_n \sum_{m=1}^{M} |A_{mn}|,$$

which sums the absolute values of elements in each *column* of $\mathbf{A}$ and selects the largest out of the N possible sums.

- The *matrix 2-norm* is

$$\|\mathbf{A}\|_2 = \max_j (\varsigma_j(\mathbf{A})) = \text{the largest singular value of matrix } \mathbf{A}.$$

Note that we find the singular values of $\mathbf{A}$ by taking the positive square roots of the eigenvalues of $\mathbf{A}^\dagger \mathbf{A}$.

- The *Frobenius norm* is the matrix analog to the *Euclidean norm* for vectors,

$$\|\mathbf{A}\|_f = \left(\sum_{n=1}^{N} \sum_{m=1}^{M} |A_{mn}|^2 \right)^{1/2}.$$

The relationship between the Frobenius norm and the matrix 2-norm is bounded by [41],

$$\|\mathbf{A}\|_2 \leq \|\mathbf{A}\|_f \leq \sqrt{N} \|\mathbf{A}\|_2.$$

- The *matrix ∞-norm* is

$$\|\mathbf{A}\|_\infty = \max_m \sum_{n=1}^{N} |A_{mn}|,$$

which sums the absolute values of elements in each *row* of $\mathbf{A}$ and selects the largest sum. The matrix 1-norm and matrix ∞-norm probe the column and row spaces of $\mathbf{A}$, respectively.

In the following example, assume $\mathbf{A} = \begin{pmatrix} 1 & -2 & 3 \\ -4 & 5 & -6 \\ 7 & -8 & 9 \end{pmatrix}$. Using MATLAB, we find $\|\mathbf{A}\|_1 = 18$, $\|\mathbf{A}\|_2 = 16.8481$, $\|\mathbf{A}\|_f = 16.8819$, $\|\mathbf{A}\|_\infty = 24$.

Two other norms have applications in modern measurement analysis.

- The mixed $\ell_{2,1}$ norm or *matrix 2,1-norm* is a variation on the Frobenius norm. Writing $M \times N$ matrix $\mathbf{A}$ as a row vector composed of column vectors $\mathbf{a}_n$, $\mathbf{A} = (\mathbf{a}_1, \ldots, \mathbf{a}_n, \ldots, \mathbf{a}_N)$, the matrix 2,1-norm is defined as

$$\|\mathbf{A}\|_{2,1} = \sum_{n=1}^{N}\left(\sum_{m=1}^{M}|A_{mn}|^2\right)^{1/2} = \sum_{n=1}^{N}\|\mathbf{a}_n\|_2.$$

This metric is more robust than ℓ_1 and ℓ_2 norms to data errors, such as from signal noise. It essentially takes the Euclidean norm of each column vector of $\mathbf{A}$, which is a vector 2-norm, and then sums the results, which is a vector 1-norm operation. Hence the name. In the recovery of compressively sampled data (Section 6.8), the 2,1-norm attempts to minimize the noise energy, via the 2-norm, and recover sparsity, via the 1-norm.

- The *nuclear norm*, denoted $\|\mathbf{A}\|_*$, is formed from the sum of singular values of $\mathbf{A}$,

$$\|\mathbf{A}\|_* = \sum_{j=1}^{\min(M,N)} \varsigma_j(\mathbf{A}) = \mathrm{tr}\left((\mathbf{A}^\dagger\mathbf{A})^{1/2}\right).$$

It is a sum of all singular values. The *trace* of square matrix $\mathbf{A}^\dagger\mathbf{A}$, denoted as $\mathrm{tr}(\mathbf{A}^\dagger\mathbf{A})$, sums values along the major diagonal. There is an important property that $\mathrm{tr}\mathbf{A}^\dagger\mathbf{A}$ equals the sum of the eigenvalues of $\mathbf{A}^\dagger\mathbf{A}$. Also, $\mathbf{A}^\dagger\mathbf{A}$ is positive semi-definite, so the square root of the matrix is well behaved. Further, the singular values of $\mathbf{A}$ are equal to the square root of the eigenvalues of $\mathbf{A}^\dagger\mathbf{A}$. However, since $\mathbf{A}$ is $M \times N$, we only sum singular values up to the smaller of the two dimensions. The nuclear norm emphasizes low-rank solutions in *constrained optimization* problems involving ℓ_1 constraints.

- To complete the example begun earlier, I find $\|\mathbf{A}\|_{2,1} = 28.9927$ and $\|\mathbf{A}\|_* = 17.9165$.
- Two useful properties are $\|\mathbf{A}\|_p\|\mathbf{B}\|_p \ge \|\mathbf{AB}\|_p$ and $\|\mathbf{A}\|_1\|\mathbf{A}\|_\infty \ge \|\mathbf{A}\|_2^2$.

Dimension

If vector space $\mathbb{V}$ has a basis consisting of M vector elements, then the *dimension* of $\mathbb{V}$ is M. While there are an infinite number of different bases for $\mathbb{V}$, they are all the same size. For example, the dimension of $\mathbb{R}^3$ is 3 and its *standard basis* has three elements:

$$\mu_1 = \begin{pmatrix} 1 \\ 0 \\ 0 \end{pmatrix}, \quad \mu_2 = \begin{pmatrix} 0 \\ 1 \\ 0 \end{pmatrix} \quad \text{and} \quad \mu_3 = \begin{pmatrix} 0 \\ 0 \\ 1 \end{pmatrix}.$$

Other mathematical structures form a vector space where basis and dimension are defined. The standard basis for a 2×2 matrix is

$$\begin{pmatrix} 1 & 0 \\ 0 & 0 \end{pmatrix}, \quad \begin{pmatrix} 0 & 0 \\ 1 & 0 \end{pmatrix}, \quad \begin{pmatrix} 0 & 1 \\ 0 & 0 \end{pmatrix}, \quad \begin{pmatrix} 0 & 0 \\ 0 & 1 \end{pmatrix}$$

because any 2×2 matrix can be written as

$$\begin{pmatrix} w & y \\ x & z \end{pmatrix} = w\begin{pmatrix} 1 & 0 \\ 0 & 0 \end{pmatrix} + x\begin{pmatrix} 0 & 0 \\ 1 & 0 \end{pmatrix} + y\begin{pmatrix} 0 & 1 \\ 0 & 0 \end{pmatrix} + z\begin{pmatrix} 0 & 0 \\ 0 & 1 \end{pmatrix}.$$

Can you show they are linearly independent? The concepts generalize to $M \times N$ matrices, where the dimension of the vector space is MN.

Finally, consider the vector space of an N-degree polynomial written as

$$k_{N+1}x^N + k_N x^{N-1} + \ldots + k_3 x^2 + k_2 x + k_1 = 0$$

for constants $k_1, \ldots, k_{N+1}$. The functions $(x^N, x^{N-1}, \ldots, x^2, x, 1)$ form a basis because (a) they span the vector space of the polynomial and (b) they are linearly independent: The only solution to the preceding equation is $k_1 = k_2 = \ldots = k_{N+1} = 0$ for arbitrary x. The dimension of this vector space is $N + 1$.

Vector Products

Let $\mathbf{u} = (u_1, u_2, \ldots, u_N)^\top$ and $\mathbf{v} = (v_1, v_2, \ldots, v_N)^\top$, where $\mathbf{u}, \mathbf{v} \in \mathbb{R}^{N \times 1}$. Denoting $N \times 1$ unit vectors as $\hat{\mathbf{e}}_1, \ldots, \hat{\mathbf{e}}_N$, we can also write $\mathbf{u} = u_1 \hat{\mathbf{e}}_1 + u_2 \hat{\mathbf{e}}_2 + \ldots + u_N \hat{\mathbf{e}}_N$ and $\mathbf{v} = v_1 \hat{\mathbf{e}}_1 + v_2 \hat{\mathbf{e}}_2 + \ldots + v_N \hat{\mathbf{e}}_N$. With these definitions, we can express four vector products.

$$\mathbf{u}^\top \mathbf{v} = \begin{pmatrix} u_1 & u_2 & \cdots & u_N \end{pmatrix} \begin{pmatrix} v_1 \\ v_2 \\ \vdots \\ v_N \end{pmatrix}$$

$$= u_1 v_1 + u_2 v_2 + \cdots + u_N v_N \quad \text{is the } inner \ product, \text{ a scalar.}$$

$$\mathbf{u}\mathbf{v}^\top = \begin{pmatrix} u_1 \\ u_2 \\ \vdots \\ u_N \end{pmatrix} \begin{pmatrix} v_1 & v_2 & \cdots & v_N \end{pmatrix}$$

$$= \begin{pmatrix} u_1 v_1 & u_1 v_2 & \cdots & u_1 v_N \\ u_2 v_1 & u_2 v_2 & \cdots & u_2 v_N \\ \vdots & \vdots & & \vdots \\ u_N v_1 & u_N v_2 & \cdots & u_N v_N \end{pmatrix} \quad \text{is the } outer \ product, \text{ an } N \times N \text{ matrix.}$$

$$\mathbf{u} \circ \mathbf{v} = (u_1 v_1, u_2 v_2, \ldots, u_N v_N)^\top \quad \text{is the } Hadamard \ product, \text{ an } N \times 1 \text{ vector.}$$

$$\mathbf{u}\mathbf{v} = \begin{pmatrix} u_1 v_1 \hat{\mathbf{e}}_1 \hat{\mathbf{e}}_1 & u_1 v_2 \hat{\mathbf{e}}_1 \hat{\mathbf{e}}_2 & \cdots & u_1 v_N \hat{\mathbf{e}}_1 \hat{\mathbf{e}}_N \\ u_2 v_1 \hat{\mathbf{e}}_2 \hat{\mathbf{e}}_1 & u_2 v_2 \hat{\mathbf{e}}_2 \hat{\mathbf{e}}_2 & \cdots & u_2 v_N \hat{\mathbf{e}}_2 \hat{\mathbf{e}}_N \\ \vdots & \vdots & & \vdots \\ c u_N v_1 \hat{\mathbf{e}}_N \hat{\mathbf{e}}_1 & u_N v_2 \hat{\mathbf{e}}_N \hat{\mathbf{e}}_2 & \cdots & u_N v_N \hat{\mathbf{e}}_N \hat{\mathbf{e}}_N \end{pmatrix}$$

is the *dyadic product*, a second-order tensor,

where, for example,

$$\hat{e}_1\hat{e}_1 = \begin{pmatrix} 1 & 0 & \cdots & 0 \\ 0 & 0 & \cdots & 0 \\ \vdots & \vdots & & \vdots \\ 0 & 0 & \cdots & 0 \end{pmatrix} \quad \text{and} \quad \hat{e}_1\hat{e}_N = \begin{pmatrix} 0 & 0 & \cdots & 1 \\ 0 & 0 & \cdots & 0 \\ \vdots & \vdots & & \vdots \\ 0 & 0 & \cdots & 0 \end{pmatrix},$$

each an $N \times N$ matrix. A *dyad* is one component of the dyadic product, such as

$$\begin{pmatrix} u_1 v_1 & 0 & \cdots & 0 \\ 0 & 0 & \cdots & 0 \\ \vdots & \vdots & & \vdots \\ 0 & 0 & \cdots & 0 \end{pmatrix}.$$ This dyadic product is an $N \times N$ matrix of $N \times N$ matrices.

Finally, a Hadamard product is computed in MATLAB using u.*v.

Frame of an Inner-Product Space

A *frame* for inner-product space $\mathbb{W}$ is a generalization of the basis concept for vector spaces. An *inner-product space* contains both a set of frame vectors $\{\mathbf{w}_n\}$ and any scalars resulting from inner products between measurement vector $\mathbf{g}$ and frame vector $\mathbf{w}_n$, viz., $c_n = \langle \mathbf{w}_n, \mathbf{g} \rangle = \mathbf{w}_n^\dagger \mathbf{g}$. Like a basis, $\{\mathbf{w}_n\}$ is a set of vectors that spans $\mathbb{W}$; however, unlike a basis, these vectors may be overdetermined and hence *linearly dependent*. Consequently, coefficients $\{c_n\}$ may not uniquely determine the function $\mathbf{g}$ they decompose. Frames offer the redundancy necessary for *wavelet analysis* using non-orthonormal wavelets. In this context, see the discussion of wavelets by Barrett and Myers [10].

The set of vectors $\{\mathbf{w}_n\}_{n=1}^N$, can be assembled into a frame matrix with N columns, each of length M, $\mathbf{W} = (\mathbf{w}_1, \mathbf{w}_2, \cdots, \mathbf{w}_N) \in \mathbb{C}^{M \times N}$, where $M < N$. This set is a frame if it satisfies the stability condition,

$$A \leq \frac{\sum_{n \in \mathbb{W}} |\langle \mathbf{w}_n, \mathbf{g} \rangle|^2}{\|\mathbf{g}\|_2^2} \leq B, \tag{A.2}$$

for positive real numbers A and B, such that $0 < A \leq B < \infty$, and for each $\mathbf{g}$ in $\mathbb{W}$. A and B are the bounds on the frame that are, respectively, the smallest and largest eigenvalues of $\mathbf{W}\mathbf{W}^\dagger$. Unlike bases, frames can be adjusted by adding or discarding $\mathbf{w}_n$ within the stability limits of (A.2). When A is as large as possible and B is as small as possible, then the frame $\mathbf{W}$ is optimal. In particular, when $A = B = 1$, we have a *Parseval frame*, which is a basis because the set are linearly independent. *All bases are special cases of frames.*

For continuous signals, the summation in (A.2) is replaced by the integral $\int_{n \in \mathbb{W}} dt\, |\langle \mathbf{w}_n(t), g(t) \rangle|^2$. Similar to the case with the basis decompositions in (5.5) and (5.6), the frame-decomposition analysis expression is $c_n = \langle \mathbf{w}_n, \mathbf{g} \rangle$ or $\mathbf{c} = \mathbf{W}^\dagger \mathbf{g}$, and the frame-synthesis expression is $\mathbf{g} = \sum_{n \in \mathbb{W}} c_n \mathbf{w}_n = \mathbf{W}\mathbf{c}$.

Matrix Rank

Let $\mathbf{A}$ be an $M \times N$ matrix of real numbers. Each row is a vector in row space $\mathbb{R}^M$ and each column is a vector in column space $\mathbb{R}^N$. The dimension of the row space equals that of the column space and is called the *rank* of matrix $\mathbf{A}$. The rank of $\mathbf{A}$ is the number linearly independent columns or rows, which must be equal. That is, if $M \neq N$, then $\text{rank}(\mathbf{A}) \leq \min(M, N)$. One way to find the rank of a matrix is from the number of nonzero rows of $\mathbf{A}$ once it is placed in reduced echelon form. MATLAB provides the reduced echelon form using $\text{B} = \text{rref}(\text{A})$. Likewise, $\text{rank}(\text{A})$ works.

Examples: Find the rank of the following. The reduced echelon forms are on the right.

$$\mathbf{A}_1 = \begin{pmatrix} 1 & 2 & 3 \\ 4 & 3 & 2 \\ 3 & 1 & -1 \end{pmatrix} \xrightarrow{\text{rref}} \begin{pmatrix} 1 & 0 & -1 \\ 0 & 1 & 2 \\ 0 & 0 & 0 \end{pmatrix} \quad \text{Note that R2} - \text{R1} = \text{R3} \qquad \mathbb{R}^{3 \times 3}$$

$$\mathbf{A}_2 = \begin{pmatrix} 1 & i & -i & 2 \\ -i & 3+i & 0 & 1 \\ i & 0 & 2i & -1 \end{pmatrix} \xrightarrow{\text{rref}} \begin{pmatrix} 1 & 0 & 0 & 0.94 - 1.24i \\ 0 & 1 & 0 & 0.76 + 0.06i \\ 0 & 0 & 1 & -0.47 + 1.12i \end{pmatrix} \qquad \mathbb{C}^{3 \times 4}$$

$$\mathbf{A}_3 = \begin{pmatrix} 1 & 1 & 1 & 1 \\ 1 & e^{i2\pi/4} & e^{i2\pi2/4} & e^{i2\pi3/4} \\ 1 & e^{i2\pi2/4} & e^{i2\pi4/4} & e^{i2\pi6/4} \\ 1 & e^{i2\pi3/4} & e^{i2\pi6/4} & e^{i2\pi9/4} \end{pmatrix}$$

$$= \begin{pmatrix} 1 & 1 & 1 & 1 \\ 1 & i & -1 & -i \\ 1 & -1 & 1 & -1 \\ 1 & -i & -1 & i \end{pmatrix} \xrightarrow{\text{rref}} \begin{pmatrix} 1 & 0 & 0 & 0 \\ 0 & 1 & 0 & 0 \\ 0 & 0 & 1 & 0 \\ 0 & 0 & 0 & 1 \end{pmatrix} \quad \mathbb{C}^{4 \times 4}.$$

$\mathbf{A}_1$ has rank 2 because there are only two linearly independent rows; the third row equals the difference of the first two. Also, the *reduced echelon form* of $\mathbf{A}_1$ has two nonzero rows. Using MATLAB to compute the eigenvalues of $\mathbf{A}_1$, $[\text{V}, \text{D}] = \text{eig}(\text{A1}); \text{diag}(\text{D})'$, we obtain $\lambda = 6.217, -3.217, 0$, which has one zero eigenvalue indicating a nontrivial *null space*. The size of a matrix $[\min(M, N) = 3]$ is equal to the rank (2) plus the size of its null space (1, where $\mathbf{Ax} = 0$). If this matrix describes a measurement transformation, then it is a mapping from a 3-D object to 2-D data. The reduced echelon form of matrix $\mathbf{A}$ is found in MATLAB using $\text{rref}(\text{A})$.

$\mathbf{A}_2$ has rank 3, which we find from the number of nonzero rows in the reduced echelon form.

$\mathbf{A}_3$ has columns that are unnormalized Fourier basis vectors. The matrix is *full rank*, $\text{rank}(\text{A3}) = 4$. The middle matrix shows the numerical values of the matrix on the left.

Determinant

We use two standard symbols for the determinant of matrix $\mathbf{A}$: det$\mathbf{A}$ and $|\mathbf{A}|$. The latter form should not be confused with matrix norms. In this book, I use the former when possible.

The determinant of 1×1 matrix $\mathbf{A} = a$ is the value of the single element, det$\mathbf{A} = a$.

The determinant of a 2 × 2 matrix is

$$\text{For } \mathbf{A} = \begin{pmatrix} a_{11} & a_{12} \\ a_{21} & a_{22} \end{pmatrix}, \quad \det\mathbf{A} = a_{11}a_{22} - a_{21}a_{21}.$$

The determinant of a 3 × 3 matrix is

$$\text{For } \mathbf{A} = \begin{pmatrix} a_{11} & a_{12} & a_{13} \\ a_{21} & a_{22} & a_{23} \\ a_{31} & a_{32} & a_{33} \end{pmatrix},$$

$$\det\mathbf{A} = a_{11}(a_{22}a_{33} - a_{23}a_{32}) - a_{12}(a_{21}a_{33} - a_{23}a_{31}) + a_{13}(a_{21}a_{32} - a_{22}a_{31}).$$

The determinant of a 4 × 4 matrix is

$$\det\mathbf{A} = a_{11}\begin{vmatrix} a_{22} & a_{23} & a_{24} \\ a_{32} & a_{33} & a_{34} \\ a_{42} & a_{43} & a_{44} \end{vmatrix} - a_{12}\begin{vmatrix} a_{21} & a_{23} & a_{24} \\ a_{31} & a_{33} & a_{34} \\ a_{41} & a_{43} & a_{44} \end{vmatrix}$$

$$+ a_{13}\begin{vmatrix} a_{21} & a_{22} & a_{24} \\ a_{31} & a_{32} & a_{34} \\ a_{41} & a_{42} & a_{44} \end{vmatrix} - a_{14}\begin{vmatrix} a_{21} & a_{22} & a_{23} \\ a_{31} & a_{32} & a_{33} \\ a_{41} & a_{42} & a_{43} \end{vmatrix},$$

where we have interchanged both notations. The result of taking a determinant is always a scalar.

Matrix Inverse

The inverse of a matrix, denoted as $\mathbf{A}^{-1}$, is defined only for square matrices where $\mathbf{A}^{-1}\mathbf{A} = \mathbf{I}$. If $\mathbf{A}$ has an inverse, it is said to be invertible or *nonsingular*. If $\mathbf{A}$ does not have an inverse, it is a *singular* matrix.

One way to compute an inverse by hand is to first find the *cofactor matrix* $\mathbf{A}^c$. If

$$\mathbf{A} = \begin{pmatrix} 3 & 1 & 2 \\ -2 & 5 & 4 \\ 1 & 3 & 6 \end{pmatrix},$$

then

$$\mathbf{A}^c = \begin{pmatrix} (-1)^{1+1}\begin{vmatrix} 5 & 4 \\ 3 & 6 \end{vmatrix} & (-1)^{1+2}\begin{vmatrix} -2 & 4 \\ 1 & 6 \end{vmatrix} & (-1)^{1+3}\begin{vmatrix} -2 & 5 \\ 1 & 3 \end{vmatrix} \\ (-1)^{2+1}\begin{vmatrix} 1 & 2 \\ 3 & 6 \end{vmatrix} & (-1)^{2+2}\begin{vmatrix} 3 & 2 \\ 1 & 6 \end{vmatrix} & (-1)^{2+3}\begin{vmatrix} 3 & 1 \\ 1 & 3 \end{vmatrix} \\ (-1)^{3+1}\begin{vmatrix} 1 & 2 \\ 5 & 4 \end{vmatrix} & (-1)^{3+2}\begin{vmatrix} 3 & 2 \\ -2 & 4 \end{vmatrix} & (-1)^{3+3}\begin{vmatrix} 3 & 1 \\ -2 & 5 \end{vmatrix} \end{pmatrix}$$

$$= \begin{pmatrix} 18 & 16 & -11 \\ 0 & 16 & -8 \\ -6 & -16 & 17 \end{pmatrix}.$$

The *adjoint* of $\mathbf{A}$, denoted as $\mathbf{A}^a$, is defined as the transpose of the cofactor matrix $\mathbf{A}^c$: $\mathbf{A}^a = (\mathbf{A}^c)^\top$. It can be shown that

$$\mathbf{A}^a\mathbf{A} = \mathbf{A}\mathbf{A}^a = \det(\mathbf{A})\mathbf{I}.$$

Provided that $\det\mathbf{A} \neq 0$, we can divide through by the determinant (remember, $\det\mathbf{A}$ is a scalar!):

$$\left(\frac{\mathbf{A}^a}{\det\mathbf{A}}\right)\mathbf{A} = \mathbf{A}\left(\frac{\mathbf{A}^a}{\det\mathbf{A}}\right) = \mathbf{I}.$$

Therefore, $\mathbf{A}^{-1} \triangleq \mathbf{A}^a/\det\mathbf{A}$. From the earlier example, $\det\mathbf{A} = 3(18) - 1(-16) + 2(-11) = 48$, and

$$\mathbf{A}^{-1} = \frac{\mathbf{A}^a}{\det\mathbf{A}} = \frac{\begin{pmatrix} 18 & 0 & -6 \\ 16 & 16 & -16 \\ -11 & -8 & 17 \end{pmatrix}}{48} = \begin{pmatrix} 3/8 & 0 & -1/8 \\ 1/3 & 1/3 & -1/3 \\ -11/48 & -1/6 & 17/48 \end{pmatrix}.$$

Notice a few things.

- If $\mathbf{A}$ is diagonal, then the inverse matrix is composed of the inverses of each nonzero element. For example,

$$\text{If } \mathbf{A} = \begin{pmatrix} a_{11} & 0 & 0 \\ 0 & a_{22} & 0 \\ 0 & 0 & a_{33} \end{pmatrix}, \text{ then } \mathbf{A}^{-1} = \begin{pmatrix} \frac{1}{a_{11}} & 0 & 0 \\ 0 & \frac{1}{a_{22}} & 0 \\ 0 & 0 & \frac{1}{a_{33}} \end{pmatrix}.$$

- The inverse of a 2×2 matrix

$$\mathbf{A} = \begin{pmatrix} a_{11} & a_{12} \\ a_{21} & a_{22} \end{pmatrix} \text{ is } \mathbf{A}^{-1} = \frac{1}{\det\mathbf{A}}\begin{pmatrix} a_{22} & -a_{12} \\ -a_{21} & a_{11} \end{pmatrix}.$$

- If $\mathbf{A}$ is an upper *triangular matrix* (only elements from the diagonal and above are nonzero) or a lower triangular matrix, then $\mathbf{A}^{-1}$ is also upper/lower triangular with diagonal elements given by inverses of diagonal elements of $\mathbf{A}$, but other elements must be found.

PROPERTY A.0.1   *Matrix inverses.*

1. *The inverse of a matrix is unique.*
2. *If the determinant of a matrix is zero, the matrix is singular (does not have an inverse).*
3. $(\mathbf{A}^{-1})^{-1} = \mathbf{A}.$
4. $(\mathbf{ABC})^{-1} = \mathbf{C}^{-1}\mathbf{B}^{-1}\mathbf{A}^{-1}$ *to any number. of matrices.*
5. $(\mathbf{A}^\top)^{-1} = (\mathbf{A}^{-1})^\top.$
6. $(\lambda\,\mathbf{A})^{-1} = (1/\lambda)\,\mathbf{A}^{-1}$ *for scalar λ.*
7. *The inverse of a nonsingular symmetric matrix is also symmetric.*

8. $\mathbf{A}^{-n} = (\mathbf{A}^{-1})^n$ *so that if* $\mathbf{A} = \begin{pmatrix} \frac{1}{3} & \frac{1}{2} \\ \frac{1}{2} & 1 \end{pmatrix}$, *then*

$$\mathbf{A}^{-2} = (\mathbf{A}^{-1})^2 = \begin{pmatrix} 12 & -6 \\ -6 & 4 \end{pmatrix}^2 = \begin{pmatrix} 180 & -96 \\ -96 & 52 \end{pmatrix}.$$

Eigenanalysis

Let $\mathbf{A}$ be an $N \times N$ matrix. The scalar λ is an eigenvalue of $\mathbf{A}$ if there exists a nonzero vector $\mathbf{u}$ in $\mathbb{R}^N$ such that

$$\mathbf{Au} = \lambda\mathbf{u}.$$

Vector $\mathbf{u} = (x, y, z)^\top$ is an eigenvector of $\mathbf{A}$ that corresponds to eigenvalue λ. Eigenvalues are found from roots of the *characteristic equation*

$$\det(\mathbf{A} - \lambda\mathbf{I}) = 0.$$

Example: Find the eigenspaces, basis, and dimensions of

$$\mathbf{A} = \begin{pmatrix} 15 & 7 & -7 \\ -1 & 1 & 1 \\ 13 & 7 & -5 \end{pmatrix}.$$

To find the eigenvalues,

$$\mathbf{A} = \begin{pmatrix} 15 & 7 & -7 \\ -1 & 1 & 1 \\ 13 & 7 & -5 \end{pmatrix} \xrightarrow{R1-R3} \begin{pmatrix} 2 & 0 & -2 \\ -1 & 1 & 1 \\ 13 & 7 & -5 \end{pmatrix} \xrightarrow{C3+C1} \begin{pmatrix} 2 & 0 & 0 \\ -1 & 1 & 0 \\ 13 & 7 & 8 \end{pmatrix}.$$

The characteristic equation, $\det(\mathbf{A} - \lambda\mathbf{I}) = 0$, gives $(2 - \lambda)(1 - \lambda)(8 - \lambda) = 0$. The eigenvalues of $\mathbf{A}$ are the roots, $\lambda_1 = 2$, $\lambda_2 = 1$ and $\lambda_3 = 8$.

To find the eigenvectors,

$$\text{For } \lambda_1 = 2, \quad (\mathbf{A} - \lambda_1\mathbf{I})\mathbf{u}_1 = \begin{pmatrix} 15 - 2 & 7 & -7 \\ -1 & 1 - 2 & 1 \\ 13 & 7 & -5 - 2 \end{pmatrix} \begin{pmatrix} x_1 \\ y_1 \\ z_1 \end{pmatrix} = 0$$

$$\begin{matrix} 13x_1 + 7y_1 - 7z_1 = 0 \\ -x_1 - y_1 + z_1 = 0 \end{matrix} \; ; \quad \mathbf{u}_1 = r \begin{pmatrix} 0 \\ 1 \\ 1 \end{pmatrix} \text{ for random constant } r.$$

The *eigenspace* is $\left\{ r \begin{pmatrix} 0 \\ 1 \\ 1 \end{pmatrix} \right\}$, an *orthonormal basis* is $\frac{1}{\sqrt{2}} \begin{pmatrix} 0 \\ 1 \\ 1 \end{pmatrix}$, and the *dimension* of this eigenspace is 1.

$$\text{For } \lambda_2 = 1, \quad (\mathbf{A} - \lambda_2 \mathbf{I})\mathbf{u}_2 = \begin{pmatrix} 15-1 & 7 & -7 \\ -1 & 1-1 & 1 \\ 13 & 7 & -5-1 \end{pmatrix} \begin{pmatrix} x_2 \\ y_2 \\ z_2 \end{pmatrix} = \mathbf{0}$$

$$\begin{array}{l} 14x_2 + 7y_2 - 7z_2 = 0 \\ -x_2 + z_2 = 0 \; ; \\ 13x_2 + 7y_2 - 6z_2 = 0 \end{array} \quad \mathbf{u}_2 = s \begin{pmatrix} 1 \\ -1 \\ 1 \end{pmatrix} \text{ for random constant } s.$$

The eigenspace is $\left\{ s \begin{pmatrix} 1 \\ -1 \\ 1 \end{pmatrix} \right\}$, an orthonormal basis is $\frac{1}{\sqrt{3}} \begin{pmatrix} 1 \\ -1 \\ 1 \end{pmatrix}$, and its dimension is 1.

$$\text{For } \lambda_3 = 8, \quad (\mathbf{A} - \lambda_3 \mathbf{I})\mathbf{u}_3 = \begin{pmatrix} 15-8 & 7 & -7 \\ -1 & 1-8 & 1 \\ 13 & 7 & -5-8 \end{pmatrix} \begin{pmatrix} x_3 \\ y_3 \\ z_3 \end{pmatrix} = \mathbf{0}$$

$$\begin{array}{l} 7x_3 + 7y_3 - 7z_3 = 0 \\ -x_3 - 7y_3 + z_3 = 0 \; ; \\ 13x_3 + 7y_3 - 13z_3 = 0 \end{array} \quad \mathbf{u}_3 = p \begin{pmatrix} 1 \\ 0 \\ 1 \end{pmatrix} \text{ for random constant } p.$$

The eigenspace is $\left\{ p \begin{pmatrix} 1 \\ 0 \\ 1 \end{pmatrix} \right\}$, an orthonormal basis is $\frac{1}{\sqrt{2}} \begin{pmatrix} 1 \\ 0 \\ 1 \end{pmatrix}$, and its dimension is 1.

Eigenvalue Multiplicity

The characteristic equation of $\mathbf{A} = \begin{pmatrix} 5 & 4 & 2 \\ 4 & 5 & 2 \\ 2 & 2 & 2 \end{pmatrix}$ is $(10 - \lambda)(1 - \lambda)^2 = 0$. The

eigenspace for $\lambda_1 = 10$ is $\left\{ r \begin{pmatrix} 2 \\ 2 \\ 1 \end{pmatrix} \right\}$, and there is a one-dimensional basis $\frac{1}{3} \begin{pmatrix} 2 \\ 2 \\ 1 \end{pmatrix}$.

The eigenspace for $\lambda_{2,3} = 1$ is $\left\{ \begin{pmatrix} s-p \\ -s \\ 2p \end{pmatrix} \right\}$ with two *linearly independent*

eigenvectors, $s \begin{pmatrix} 1 \\ -1 \\ 0 \end{pmatrix}$ and $p \begin{pmatrix} -1 \\ 0 \\ 2 \end{pmatrix}$. A 2-D eigenspace with orthogonal basis is

$\left\{ s \begin{pmatrix} 1 \\ -1 \\ 0 \end{pmatrix}, p \begin{pmatrix} -1 \\ 0 \\ 2 \end{pmatrix} \right\}$.

The characteristic equation for $\mathbf{A}$ has the root $\lambda = 1$ with an *algebraic multiplicity* (multiplicity of eigenvalues) of 2. Since the eigenvectors of this eigenspace are linearly independent, the *geometric multiplicity* (dimension of corresponding eigenspaces) is also 2 and a subspace is formed. If the two eigenvectors in this eigenspace were linearly dependent, the algebraic multiplicity would still be 2 but the geometric multiplicity would be 1 and the eigenspace would not form a subspace.

Dimensions of Eigenspaces

The dimension of an eigenspace of $\mathbf{A}$ equals the geometric multiplicity and is less than or equal to the algebraic multiplicity. Matrix $\mathbf{A}$ is diagonalizable if and only if the algebraic multiplicity of every eigenspace equals the geometric multiplicity. Diagonalizability of a matrix is discussed later in this appendix.

Eigenvalue Properties

Here are a couple of properties of matrices regarding eigenvalues.

- The eigenvalues of an upper or lower triangular matrix of any size are given by the diagonal elements.
- For 2×2 matrix $\mathbf{A} = \begin{pmatrix} a & b \\ c & d \end{pmatrix}$, the characteristic equation gives

$$\lambda^2 - (a+d)\lambda + ad - bc = \lambda^2 - \text{tr}(\mathbf{A}) + \det(\mathbf{A}).$$ Therefore,

$$\lambda_{1,2} = \frac{a+d}{2} \pm \frac{\sqrt{(a+d)^2 - 4(ad-bc)}}{2} = \frac{a+d}{2} \pm \frac{\sqrt{(a-d)^2 - 4bc)}}{2}.$$

- The concept of linear stability discussed in Section 10.8 is key to understanding the stability of solutions for models composed of systems of ODEs. It is true that linearized solutions near an equilibrium point are linearly stable when the real parts of the eigenvalues are negative. For the 2×2 Jacobian matrices of (10.28),

$$2\lambda_{1,2} = \text{tr}(\mathbf{A}) \pm \sqrt{(\text{tr}(\mathbf{A}))^2 - 4\det(\mathbf{A})}.$$

Consequently, the 2×2 system is linearly stable if $\text{tr}(\mathbf{A}) < 0$ and $\det(\mathbf{A}) > 0$.

Derivatives

Being able to compute derivatives of matrix quantities is an important part of optimization problems and maximum likelihood estimation. A few properties are reviewed here.

- If $N \times 1$ vector $\mathbf{y} = \mathbf{f}(\mathbf{x}) = (y_1(x_1, x_2, \ldots, x_L), \ldots, y_n(x_1, x_2, \ldots, x_L), \ldots, y_N(x_1, x_2, \ldots, x_L))^\top$, then

$$\frac{\partial \mathbf{y}}{\partial \mathbf{x}} = \begin{pmatrix} \nabla y_1 \\ \vdots \\ \nabla y_N \end{pmatrix} \triangleq \begin{pmatrix} \frac{\partial y_1}{\partial x_1} & \frac{\partial y_1}{\partial x_2} & \cdots & \frac{\partial y_1}{\partial x_L} \\ \vdots & \vdots & & \vdots \\ \frac{\partial y_N}{\partial x_1} & \frac{\partial y_N}{\partial x_2} & \cdots & \frac{\partial y_N}{\partial x_L} \end{pmatrix}. \tag{A.3}$$

∇ is the gradient operator, defined as $\nabla_{\mathbf{x}} \triangleq (\partial/\partial x_1, \partial/\partial x_2, \ldots, \partial/\partial x_L)$. Vector notation can be a little confusing, so take a few minutes to look closely at the indices. For example, the net derivative of vector $\mathbf{y} \in \mathbb{R}^{N \times 1}$, each of whose elements is a function of another vector $\mathbf{x} \in \mathbb{R}^{L \times 1}$, generates an $N \times L$ matrix of values.

- Let $\mathbf{g}(\mathbf{x}) = \mathbf{H}\mathbf{f}(\mathbf{x}')$ be the noise-free convolution equation for an LSI measurement. $\mathbf{H}$ is not a function of $\mathbf{f}$, but $\mathbf{f}$ is a function of position. We are interested in how data $\mathbf{g}$ vary with position. We have $\mathbf{g} \in \mathbb{R}^{M \times 1}$, $\mathbf{f} \in \mathbb{R}^{N \times 1}$, and $\mathbf{x}, \mathbf{x}' \in \mathbb{R}^{L \times 1}$, so that $\mathbf{H}$ is $M \times N$. Then $\partial \mathbf{g}/\partial \mathbf{f} = \mathbf{H}$, and, using the chain rule,

$$\frac{\partial \mathbf{g}}{\partial \mathbf{x}} = \frac{\partial \mathbf{g}}{\partial \mathbf{f}} \frac{\partial \mathbf{f}}{\partial \mathbf{x}'} = \mathbf{H} \frac{\partial \mathbf{f}}{\partial \mathbf{x}'}.$$

Another way to see this is by examining the components:

$$g_m = \sum_{n=1}^{N} h_{mn} f_n$$

$$\frac{\partial g_m}{\partial x_\ell} = \sum_{n=1}^{N} h_{mn} \frac{\partial f_n}{\partial x'_\ell}, \qquad \text{where } \ell = 1, 2, \dots, L.$$

- Scalar $\beta(\mathbf{y}) = \mathbf{y}^\top \mathbf{B} \mathbf{y}$ for $N \times 1$ vector $\mathbf{y}$. Then, by the chain rule, we have

$$\frac{\partial \beta}{\partial y_\ell} = \frac{\partial}{\partial y_\ell} \left(\sum_{n'=1}^{N} \sum_{n=1}^{N} b_{nn'} y_n y_{n'} \right)$$

$$= \sum_{n'=1}^{N} b_{\ell n'} y_{n'} + \sum_{n=1}^{N} b_{n\ell} y_n$$

and therefore, $\nabla \beta = \dfrac{\partial \beta(\mathbf{y})}{\partial \mathbf{y}} = \mathbf{B}\mathbf{y} + \mathbf{B}^\top \mathbf{y} = (\mathbf{B} + \mathbf{B}^\top)\mathbf{y}$.

Since the gradient of a scalar is a vector, there are unit vectors associated with each element that are only implied in the preceding expressions. If, in addition, $\mathbf{B}$ is a symmetric matrix, i.e., $\mathbf{B} = \mathbf{B}^\top$, then $\partial \beta / \partial \mathbf{y} = 2\mathbf{B}\mathbf{y}$. Notice the similarity with the analogous scalar expression $d\beta'/dy = d(By^2)/dy = 2By$.

- A similar function associated with MVN pdfs, $\beta(\mathbf{B}) = \mathbf{y}^\top \mathbf{B}^{-1} \mathbf{y} = \text{tr}(\mathbf{B}^{-1}\mathbf{y}\mathbf{y}^t)$ [37].

$$\frac{\partial \beta}{\partial \mathbf{B}} = -\mathbf{B}^{-1} \left(\mathbf{y}\mathbf{y}^\top + (\mathbf{y}\mathbf{y}^\top)^\top \right) \mathbf{B}^{-1} + \text{diag}(\mathbf{B}^{-1}\mathbf{y}\mathbf{y}^\top \mathbf{B}^{-1})$$

$$= -2\mathbf{B}^{-1}\mathbf{y}\mathbf{y}^\top \mathbf{B}^{-1} + \text{diag}(\mathbf{B}^{-1}\mathbf{y}\mathbf{y}^\top \mathbf{B}^{-1}).$$

-

$$\frac{\partial \det \mathbf{B}}{\partial \mathbf{B}} = \det \mathbf{B}(2\mathbf{B}^{-1} - \text{diag}(\mathbf{B}^{-1})).$$

- If scalar $\beta^2(\mathbf{x}) = \|\mathbf{f}(\mathbf{x})\|_2^2 = \mathbf{f}^\top \mathbf{f}$ for $\mathbf{f} \in \mathbb{R}^{N \times 1}$ (square of the Euclidean norm of $\mathbf{f}$ that is a function of position $\mathbf{x}$), then

$$\beta^2(\mathbf{x}) = \sum_{n=1}^{N} f_n^2(\mathbf{x})$$

$$\frac{\partial \beta^2(\mathbf{x})}{\partial x_\ell} = 2 \sum_{n=1}^{N} f_n(\mathbf{x}) \frac{\partial f_n(\mathbf{x})}{\partial x_\ell}$$

$$\nabla \beta^2(\mathbf{x}) = \frac{\partial \beta^2(\mathbf{x})}{\partial \mathbf{x}} = 2\mathbf{f}^\top(\mathbf{x}) \frac{\partial \mathbf{f}(\mathbf{x})}{\partial \mathbf{x}}.$$

In this case, the gradient of scalar β^2 is a row vector. Be sure to follow the indices!

- Similarly, for $\beta(\mathbf{x}) = \|\mathbf{f}(\mathbf{x})\|_2$, the derivative of the Euclidean norm (not its square) is

$$\nabla \beta(\mathbf{x}) = \nabla \|\mathbf{f}(\mathbf{x})\|_2 = \nabla \left(\sum_{n=1}^{N} f_n^2(\mathbf{x}) \right)^{1/2}$$

$$= \frac{1}{2} \left(\sum_{n=1}^{N} f_n^2(\mathbf{x}) \right)^{-1/2} \left(2 \sum_{n=1}^{N} f_n(\mathbf{x}) \nabla f_n(\mathbf{x}) \right)$$

$$= \frac{\sum_{n=1}^{N} f_n(\mathbf{x}) \nabla f_n(\mathbf{x})}{\|\mathbf{f}(\mathbf{x})\|_2} = \frac{\mathbf{f}^\top(\mathbf{x})(\partial \mathbf{f}(\mathbf{x})/\partial \mathbf{x})}{\|\mathbf{f}(\mathbf{x})\|_2}.$$

- The following result will be helpful in Section 7.6.1:

$$\Theta = \frac{1}{2}\|\mathbf{g} - \mathbf{Hf}\|_2^2 + \frac{\beta_1}{2}\|\mathbf{Af}\|_2^2 + \beta_1\|\mathbf{f}\|_1$$

$$\frac{d\Theta}{d\mathbf{f}} = -\mathbf{H}^\top(\mathbf{g} - \mathbf{Hf}) + \beta_1\mathbf{A}^\top\mathbf{Af} + \beta_2. \tag{A.4}$$

- For $M \times N$ matrix $\mathbf{H}(\theta)$,

$$\frac{\partial \mathbf{H}}{\partial \theta} \triangleq \begin{pmatrix} \frac{\partial h_{11}}{\partial \theta} & \cdots & \frac{\partial h_{1N}}{\partial \theta} \\ \vdots & & \vdots \\ \frac{\partial h_{M1}}{\partial \theta} & \cdots & \frac{\partial h_{MN}}{\partial \theta} \end{pmatrix}.$$

- Finally, we can find $\partial \mathbf{H}^{-1}/\partial \theta$ as follows:

$$\frac{\partial(\mathbf{H}^{-1}\mathbf{H})}{\partial \theta} = \frac{\partial \mathbf{I}}{\partial \theta} = \mathbf{0}.$$

$$\frac{\partial \mathbf{H}^{-1}}{\partial \theta}\mathbf{H} + \mathbf{H}^{-1}\frac{\partial \mathbf{H}}{\partial \theta} = \mathbf{0}$$

$$\frac{\partial \mathbf{H}^{-1}}{\partial \theta} = -\mathbf{H}^{-1}\frac{\partial \mathbf{H}}{\partial \theta}\mathbf{H}^{-1}.$$

Normal Matrix

If $\mathbf{A}$ is a square matrix with N^2 complex elements, then $\mathbf{A}$ is a normal matrix if $\mathbf{A}^\dagger \mathbf{A} = \mathbf{A}\mathbf{A}^\dagger$, where $\mathbf{A}^\dagger$ is the *conjugate transpose* of $\mathbf{A}$. (If we consider operator $\mathcal{A}$ instead of a matrix $\mathbf{A}$, then $\mathcal{A}^\dagger$ is the *adjoint operator*, as discussed at the end of this appendix.) If the elements are all real, then the equivalent statement is $\mathbf{A}^\top \mathbf{A} = \mathbf{A}\mathbf{A}^\top$, where $\mathbf{A}^\top$ is matrix transpose. The trace of $\mathbf{A}^\dagger \mathbf{A}$ is $\text{tr}\{\mathbf{A}^\dagger \mathbf{A}\} = \sum_{n=1}^{N} |\lambda_n|^2$, where λ_n is the nth eigenvalue of $\mathbf{A}$.

Example: For the real 3×3 normal matrix

$$\mathbf{A} = \begin{pmatrix} 1 & 1 & 2 \\ 1 & 1 & 0 \\ 2 & 0 & 1 \end{pmatrix}, \quad \text{we find } \mathbf{A}^\top \mathbf{A} = \mathbf{A}\mathbf{A}^\top = \begin{pmatrix} 6 & 2 & 4 \\ 2 & 2 & 2 \\ 4 & 2 & 5 \end{pmatrix}$$

and $\lambda_{1,2,3} = 3.2361, 1.0000, -1.2361$. Finally, $\text{tr}\{\mathbf{A}^\top \mathbf{A}\} = 13.0 = \sum_{n=1}^{3} |\lambda_n|^2$.

Symmetric Matrix

For square real $\mathbf{B} \in \mathbb{R}^{N \times N}$, $\mathbf{B}$ is a *symmetric matrix* if $\mathbf{B} = \mathbf{B}^\top$. That is, $B_{mn} = B_{nm}$. In the preceding example, $\mathbf{A}$ is both normal and symmetric. An example of a symmetric matrix frequently encountered is a stationary covariance matrix.

For *skew-symmetric matrices*, $\mathbf{B}^\top = -\mathbf{B}$. Consequently, $B_{nm} = -B_{mn}$.

Hermitian Matrix

For square complex $\mathbf{C} \in \mathbb{C}^{N \times N}$, $\mathbf{C}$ is a Hermitian matrix if $\mathbf{C} = \mathbf{C}^\dagger$. That is, $C_{mn} = C_{nm}^*$. A system matrix is Hermitian only if it maps input vectors from one vector space into output vectors in the same vector space, i.e., $\mathbb{U} = \mathbb{V}$. Note that diagonal elements of a Hermitian matrix must be real, and therefore the matrix must have real eigenvalues. Its determinant is also real.

Example: For the 3×3 Hermitian matrix,

$$\mathbf{C} = \begin{pmatrix} 1+0i & 3+2i & 4-1i \\ 3-2i & 1+0i & 1+3i \\ 4+1i & 1-3i & 1+0i \end{pmatrix}, \quad \mathbf{C}^\dagger \mathbf{C} = \mathbf{C}\mathbf{C}^\dagger,$$

$$\text{tr}\{\mathbf{C}^\dagger \mathbf{C}\} = \sum_{n=1}^{3} |\lambda_n|^2 = 83.0, \quad \det(\mathbf{C}) = -0.85.$$

A Hermitian matrix of rank N can always be diagonalized by a unitary matrix. The resulting diagonal elements are all real. Consequently, the N eigenvalues are real and the N eigenvectors are linearly independent. If the eigenvalues are distinct, the eigenvectors are also orthogonal. The inverse of a Hermitian matrix (if it exists) is also Hermitian.

In MATLAB, the command C′ computes $\mathbf{C}^\dagger$, the conjugate transpose of complex $\mathbf{C}$. If you wish to take the transpose of $\mathbf{C}$ without also computing the conjugate, the MATLAB command is C.′.

Positive-Definite Matrix

If Hermitian matrix $\mathbf{C}$ is positive definite, then $\mathbf{x}^\dagger \mathbf{C} \mathbf{x} > 0$ and is real for all nonzero vectors $\mathbf{x}$ having real or complex values. If $\mathbf{C}$ is positive semi-definite, the criteria are the same except that $\mathbf{x}^\dagger \mathbf{C} \mathbf{x} \geq 0$.

Example:

For $\mathbf{x} = \begin{pmatrix} x \\ y \\ z \end{pmatrix}$, then $\mathbf{x}^\dagger \mathbf{C} \mathbf{x} = \begin{pmatrix} x^* & y^* & z^* \end{pmatrix} \begin{pmatrix} 1 & i & 0 \\ -i & 2 & -1 \\ 0 & -1 & 1 \end{pmatrix} \begin{pmatrix} x \\ y \\ z \end{pmatrix}$

$$= x^*(x + iy) + y^*(-ix + 2y - z) + z^*(-y + z)$$
$$= x^*x + ix^*y - iy^*x + 2yy^* - y^*z - yz^* + zz^*$$
$$= (x + iy)(x^* - iy^*) + (y - z)(y^* - z^*)$$
$$= |x + iy|^2 + |y - z|^2 \geq 0.$$

Because $|\mathbf{x}|^2 = \mathbf{x}^\dagger \mathbf{x}$ is a scalar that is greater than zero if $\mathbf{x} \neq 0$, we say $\mathbf{C}$ is positive definite. If a matrix is Hermitian and positive definite, then

- Its eigenvalues are real and positive;
- The matrix is invertible and $\mathbf{C}^{-1}$ exists; and
- The rank of $\mathbf{C}$ equals the rank of $\mathbf{C}^\dagger \mathbf{C}$.

It is also true that $\mathbf{x}^\dagger \mathbf{C} \mathbf{x} = \text{tr}\{\mathbf{C}\mathbf{x}\mathbf{x}^\dagger\}$. To illustrate using Hermitian matrix $\mathbf{C}$ from the preceding example,

$$\text{tr}\{\mathbf{C}\mathbf{x}\mathbf{x}^\dagger\} = \text{tr}\left\{ \begin{pmatrix} 1 & i & 0 \\ -i & 2 & -1 \\ 0 & -1 & 1 \end{pmatrix} \begin{pmatrix} xx^* & xy^* & xz^* \\ yx^* & yy^* & yz^* \\ zx^* & zy^* & zz^* \end{pmatrix} \right\}$$

$$= \text{tr} \begin{pmatrix} xx^* + iyx^* & xy^* + iyy^* & xz^* + iyz^* \\ -ixx^* + 2yx^* - zx^* & -ixy^* + 2yy^* - zy^* & ixz^* + 2yz^* - zz^* \\ -yx^* + zx^* & -yy^* + zy^* & -yz^* + zz^* \end{pmatrix}$$

$$= xx^* + iyx^* - ixy^* + 2yy^* - zy^* - yz^* + zz^*$$

$$= (x + iy)(x^* - iy^*) + (y - z)(y^* - z^*) = |x + iy|^2 + |y - z|^2 = \mathbf{x}^\dagger \mathbf{C} \mathbf{x}.$$

Toeplitz Matrix

For $\mathbf{H} \in \mathbb{R}^{N \times N}$, $\mathbf{H}$ is a Toeplitz matrix if elements along each of the diagonals have the same value. More specifically, if the first column of $\mathbf{H}$ is $[h_0, h_1, h_2, \ldots, h_{N-1}]^\top$ and the first row is $[h_0, h_{N-1}, h_{N-2}, \ldots, h_1]$, then $H_{mn} = h_{m-n}$ for a Toeplitz matrix. Specifically, for $N = 5$,

$$\mathbf{H} = \begin{pmatrix} H_{00} & H_{01} & H_{02} & H_{03} & H_{04} \\ H_{10} & H_{11} & H_{12} & H_{13} & H_{14} \\ H_{20} & H_{21} & H_{22} & H_{23} & H_{24} \\ H_{30} & H_{31} & H_{32} & H_{33} & H_{34} \\ H_{40} & H_{41} & H_{42} & H_{43} & H_{44} \end{pmatrix} = \begin{pmatrix} h_0 & h_4 & h_3 & h_2 & h_1 \\ h_1 & h_0 & h_4 & h_3 & h_2 \\ h_2 & h_1 & h_0 & h_4 & h_3 \\ h_3 & h_2 & h_1 & h_0 & h_4 \\ h_4 & h_3 & h_2 & h_1 & h_0 \end{pmatrix}.$$

However, a Toeplitz matrix does not need to be square. For example, we can implement the linear convolution $g[m] = (h * f)[m]$ using matrix multiplication $\mathbf{g} = \mathbf{H}\mathbf{f}$, where we set $\mathbf{g}$ to be a 10×1 column vector and $\mathbf{f}$ to be an 8×1 vector. The convolution kernel, **h**, has just three nonzero elements in this simple example, but it must be used appropriately to build a 10×8 Toeplitz matrix. Let $\mathbf{h}^\top = [h_0 \; h_1 \; h_2 \; 0 \; 0 \; 0 \; 0 \; 0] = [0.25 \; 0.50 \; 0.25 \; 0 \; 0 \; 0 \; 0 \; 0]$. The length of **h** must equal that of **f**. Then the equation for the Ath measurement is

$$\mathbf{g}_A = \begin{pmatrix} g_1 \\ g_2 \\ g_3 \\ g_4 \\ g_5 \\ g_6 \\ g_7 \\ g_8 \\ g_9 \\ g_{10} \end{pmatrix} = \begin{pmatrix} h_0 & 0 & 0 & 0 & 0 & 0 & 0 & 0 \\ h_1 & h_0 & 0 & 0 & 0 & 0 & 0 & 0 \\ h_2 & h_1 & h_0 & 0 & 0 & 0 & 0 & 0 \\ 0 & h_2 & h_1 & h_0 & 0 & 0 & 0 & 0 \\ 0 & 0 & h_2 & h_1 & h_0 & 0 & 0 & 0 \\ 0 & 0 & 0 & h_2 & h_1 & h_0 & 0 & 0 \\ 0 & 0 & 0 & 0 & h_2 & h_1 & h_0 & 0 \\ 0 & 0 & 0 & 0 & 0 & h_2 & h_1 & h_0 \\ 0 & 0 & 0 & 0 & 0 & 0 & h_2 & h_1 \\ 0 & 0 & 0 & 0 & 0 & 0 & 0 & h_2 \end{pmatrix} \begin{pmatrix} f_1 \\ f_2 \\ f_3 \\ f_4 \\ f_5 \\ f_6 \\ f_7 \\ f_8 \end{pmatrix} . \tag{A.5}$$

There is one column in **H** for each element in **f**, and one row in **H** for each element in **g**. If $\mathbf{g}_A$ is $M \times 1$ and **f** is $N \times 1$, then **H** is $M \times N$. The end effects (in this case, the top-two and bottom-two rows of **H**) yield biased results because not all nonzero elements are present. One way to deal with these end effects is to conduct a circular convolution, which is appropriate when **f** is also periodic. Since DFT is a discrete-to-discrete mapping from time to frequency, **f** is periodic in both domains, whether or not the presampled $f(t)$ is periodic. In that case, the circulant approximation to the Toeplitz matrix is used.

Circulant Matrix
A circulant matrix is a special type of Toeplitz matrix in which each row is rotated toward the right by one element relative to the row above, while wrapping elements that leave the matrix on the right. Using the convolution example given earlier, we implement a *circular convolution* $g_B[m] = (h * f)[m]$ by forming **H** as a square circulant matrix,

$$\mathbf{g}_B = \begin{pmatrix} g_1 \\ g_2 \\ g_3 \\ g_4 \\ g_5 \\ g_6 \\ g_7 \\ g_8 \end{pmatrix} = \begin{pmatrix} h_1 & h_0 & 0 & 0 & 0 & 0 & 0 & h_2 \\ h_2 & h_1 & h_0 & 0 & 0 & 0 & 0 & 0 \\ 0 & h_2 & h_1 & h_0 & 0 & 0 & 0 & 0 \\ 0 & 0 & h_2 & h_1 & h_0 & 0 & 0 & 0 \\ 0 & 0 & 0 & h_2 & h_1 & h_0 & 0 & 0 \\ 0 & 0 & 0 & 0 & h_2 & h_1 & h_0 & 0 \\ 0 & 0 & 0 & 0 & 0 & h_2 & h_1 & h_0 \\ h_0 & 0 & 0 & 0 & 0 & 0 & h_2 & h_1 \end{pmatrix} \begin{pmatrix} f_1 \\ f_2 \\ f_3 \\ f_4 \\ f_5 \\ f_6 \\ f_7 \\ f_8 \end{pmatrix} . \tag{A.6}$$

In (A.6), we wrapped the elements and eliminated the top and bottom rows from (A.5) because they became redundant. Make no mistake – there are still end effects and bias when you consider that realistic objects $f(t)$ are rarely, if ever, periodic. However, computational conveniences are available when using Fourier methods to implement a convolution that are highly desirable in terms of ease and speed of computation.

If the size of **f** greatly exceeds the support of **h**, i.e., the length of the nonzero elements in **h**, then circulant measurement matrices are very useful provided that data near the boundaries of **g** can be ignored. MATLAB code to generate a circulant matrix for a 1-D impulse response appears in Example 4.2.1, and that to generate a 2-D impulse response can be found in Example 4.7.1.

Fourier Implementation of Circular Convolution

There is an easier way. Given the Fourier convolution theorem of Section 5.6, we have $g_C = \mathcal{F}^{-1}\{\mathcal{F}\{\mathbf{h}\} \times \mathcal{F}\{\mathbf{f}\}\}$. This is implemented in MATLAB using

```
g=fftshift(ifft(fft(h).*fft(f)));
```

For example, if we throw away the first and last points in g_A, we can compare **f** with all three ways of computing **g**. The three ways are linear convolution g_A, circular convolution g_B, and Fourier implementation of circular convolution g_C. I used $\mathbf{f}^\top = [0\ 0.5\ 1.0\ 0.5\ 0\ 1.0\ 0\ 1.0]$ and $\mathbf{h}^\top = [0\ 0\ 0\ 0.25\ 0.50\ 0.25\ 0\ 0]$, and the results are plotted in Figure A.2. The results are the same except near the left end. It is recommended to use Fourier techniques to compute convolution whenever it is possible to assume the measurement can be represented by a circulant approximation to a Toeplitz system matrix **H**.

Unitary and Orthogonal Matrices

A square matrix with complex elements is a unitary matrix if **U** is nonsingular and $\mathbf{U}^\dagger = \mathbf{U}^{-1}$. Note that nonsingular matrices have inverses and nonzero determinants. Since $\mathbf{A}^{-1}\mathbf{A} = \mathbf{I}$ for any nonsingular matrix, then $\mathbf{U}^\dagger\mathbf{U} = \mathbf{U}\mathbf{U}^\dagger = \mathbf{I}$ when **U** is unitary. If the elements of **U** are real, then we have an orthogonal matrix if $\mathbf{U}^\top = \mathbf{U}^{-1}$. In either case, the columns and rows of **U** form orthonormal bases.

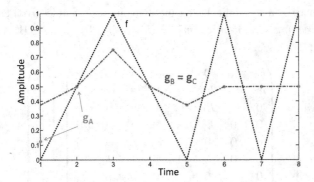

Figure A.2 The input vector **f** is shown (dotted black line) along with the results of a matrix implementation of a linear convolution g_A (red dots), circular convolution g_B, and Fourier-based convolution g_C (both of the latter are shown by the blue dot-dashed line). Only the points at integer time values are computed. Clearly, the blurring caused by this impulse response is more severe as one moves toward the right. We find $g_B = g_C$ exactly at all points, as expected from the Fourier convolution theorem. The linear and circular convolutions are the same except possibly at the ends.

To illustrate, consider the 2×2 matrix $\mathbf{Z}$:

$$\mathbf{Z} = \frac{1}{2}\begin{pmatrix} 1+i & 1-i \\ 1-i & 1+i \end{pmatrix} \text{ and } \mathbf{ZZ}^{\dagger} = \frac{1}{2}\begin{pmatrix} 1+i & 1-i \\ 1-i & 1+i \end{pmatrix} \frac{1}{2}\begin{pmatrix} 1-i & 1+i \\ 1+i & 1-i \end{pmatrix}$$

$$= \frac{1}{4}\begin{pmatrix} 4 & 0 \\ 0 & 4 \end{pmatrix} = \mathbf{I}.$$

Since $\mathbf{Z}$ is unitary, its columns must be orthonormal. That is, the norm of each column equals $\sqrt{(1+i)(1-i)+(1-i)(1+i)}/2 = 1$. Expressing the matrix as a row vector of column vectors, $\mathbf{Z} = [\mathbf{z}_1 \ \mathbf{z}_2]$, it can be shown that their inner products $\langle \mathbf{z}_1, \mathbf{z}_2 \rangle = \sum_{n=1}^{2} z_{1n}^* z_{2n} = \delta_{12}$, which is true for unitary matrices.

Another property of unitary matrices is that $|\det(\mathbf{U})| = [\det^*(\mathbf{U}) \times \det(\mathbf{U})]^{1/2} = 1$. In the example, $\det(\mathbf{Z}) = [(0.5+0.5i)^2 - (0.5-0.5i)^2] = i$, so indeed the *magnitude* $|\det(\mathbf{Z})| = 1$. Unitary matrices are wonderful to run into because they often greatly simply calculations.

Diagonalization and Eigenanalysis

The processes of diagonalizing a matrix and computing its eigenvalues are closely related. This section provides some examples that show the relationships and illustrate other properties discussed in this appendix.

First, let $\mathbf{A}$ and $\mathbf{B}$ be $N \times N$ matrices with real or complex elements. We say that $\mathbf{B}$ is *similar* to $\mathbf{A}$ if there exists a nonsingular matrix $\mathbf{Q}$ such that $\mathbf{B} = \mathbf{Q}^{-1}\mathbf{A}\mathbf{Q}$.

Second, similar matrices have the *same eigenvalues* with the *same geometric multiplicity* (same number of linearly independent eigenvectors in each eigenspace). We can prove the part about eigenvalues as follows:

$$\det(\mathbf{B} - \lambda\mathbf{I}) = \det(\mathbf{Q}^{-1}\mathbf{A}\mathbf{Q} - \lambda\mathbf{Q}^{-1}\mathbf{Q}) = \det(\mathbf{Q}^{-1}(\mathbf{A} - \lambda\mathbf{I})\mathbf{Q})$$

$$= \det(\mathbf{Q}^{-1})\det(\mathbf{A} - \lambda\mathbf{I})\det(\mathbf{Q}) = \det(\mathbf{A} - \lambda\mathbf{I})\det(\mathbf{Q}^{-1}\mathbf{Q}) = \det(\mathbf{A} - \lambda\mathbf{I}).$$

This result uses the determinant relation $\det(\mathbf{AB}) = \det(\mathbf{A})\det(\mathbf{B})$. If $\mathbf{B}$ is similar to $\mathbf{A}$ and $\mathbf{B}$ is a *diagonal matrix*, then we can diagonalize $\mathbf{A}$ by identifying $\mathbf{Q}$.

Example 1: Show that $\mathbf{Q}$ diagonalizes $\mathbf{A}$, where

$$\mathbf{A} = \begin{pmatrix} 7 & -10 \\ 3 & -4 \end{pmatrix} \text{ and } \mathbf{Q} = \begin{pmatrix} 2 & 5 \\ 1 & 3 \end{pmatrix}. \text{ We find that } \mathbf{Q}^{-1} = \frac{\text{adj}(\mathbf{Q})}{\det(\mathbf{Q})} = \begin{pmatrix} 3 & -5 \\ -1 & 2 \end{pmatrix},$$

where $\text{adj}(\mathbf{Q})$ is the adjoint of the matrix and in this example, the determinant is 1. It is easy to verify that $\mathbf{QQ}^{-1} = \mathbf{I}$. Assembling the parts, we find that $\mathbf{B}$ is a diagonal matrix,

$$\mathbf{B} = \mathbf{Q}^{-1}\mathbf{A}\mathbf{Q} = \begin{pmatrix} 3 & -5 \\ -1 & 2 \end{pmatrix}\begin{pmatrix} 7 & -10 \\ 3 & -4 \end{pmatrix}\begin{pmatrix} 2 & 5 \\ 1 & 3 \end{pmatrix} = \begin{pmatrix} 2 & 0 \\ 0 & 1 \end{pmatrix}. \quad \text{(A.7)}$$

Diagonalizing $\mathbf{A}$ gives us its eigenvalues. Furthermore, the columns of $\mathbf{Q}$ are the corresponding eigenvectors.

Let's show this is true is by finding the eigenspaces of $\mathbf{A}$.

Eigenvalues: From the characteristic equation,

$$\det(\mathbf{A} - \lambda \mathbf{I}) = 0 = \det \begin{pmatrix} 7 - \lambda & -10 \\ 3 & -4 - \lambda \end{pmatrix}$$

$$= (7 - \lambda)(-4 - \lambda) + 30 = \lambda^2 - 3\lambda + 2.$$

The eigenvalues are the roots of the polynomial $(\lambda - 1)(\lambda - 2)$, which are $\lambda_1 = 2$ and $\lambda_2 = 1$, and also the diagonal elements of $\mathbf{B}$ in (A.7).

Eigenvectors: The two eigenvectors $\mathbf{u}_1 = \begin{pmatrix} x_1 \\ y_1 \end{pmatrix}$ and $\mathbf{u}_2 = \begin{pmatrix} x_2 \\ y_2 \end{pmatrix}$ are found by substituting eigenvalues back into the characteristic equation, one at a time, and following the procedure described in the "Eigenanalysis" section. For $\mathbf{u}_1$,

$$(\mathbf{A} - \lambda_1 \mathbf{I})\mathbf{u}_1 = \mathbf{0} = \begin{pmatrix} 7 - 2 & -10 \\ 3 & -4 - 2 \end{pmatrix} \begin{pmatrix} x_1 \\ y_1 \end{pmatrix} : \quad \begin{matrix} 5x_1 - 10y_1 = 0 \\ 3x_1 - 6y_1 = 0 \end{matrix}.$$

This gives

$$\mathbf{u}_1 = r \begin{pmatrix} 2 \\ 1 \end{pmatrix},$$

where r is an arbitrary constant. Repeating the process for the second eigenvector,

$$\mathbf{u}_2 = s \begin{pmatrix} 5 \\ 3 \end{pmatrix}.$$

We find that diagonal matrix $\mathbf{B}$ does, indeed, have the eigenvalues of $\mathbf{A}$ for its diagonal elements, while the columns of $\mathbf{Q}$ are the eigenvectors:

$$\mathbf{B} = \Lambda = \begin{pmatrix} \lambda_1 & 0 \\ 0 & \lambda_2 \end{pmatrix} \quad \text{and} \quad \mathbf{Q} = \mathbf{U} = [\mathbf{u}_1 \ \mathbf{u}_2] = \begin{pmatrix} x_1 & x_2 \\ y_1 & y_2 \end{pmatrix} = \begin{pmatrix} 2r & 5s \\ r & 3s \end{pmatrix}.$$

Can you think of a way to compute constants r and s in this example? (They will be computed in the next example.) The decomposition of $\mathbf{A}$ is

$$\mathbf{A} = \mathbf{U}\Lambda\mathbf{U}^{-1}, \tag{A.8}$$

where $\mathbf{B} = \Lambda$ and $\mathbf{Q} = \mathbf{U}$, which returns us to the original statement $\mathbf{B} = \mathbf{Q}^{-1}\mathbf{A}\mathbf{Q}$ in (A.7).

Example 2: Hermitian matrix

$$\mathbf{A} = \begin{pmatrix} 1 & -2i \\ 2i & 1 \end{pmatrix}$$

is diagonalized by a unitary matrix $\mathbf{Q}$. Because $\mathbf{A}$ is Hermitian, its eigenvalues must be real.

We can find $\mathbf{Q}$ by computing the eigenvector matrix $\mathbf{U}$. First, the eigenvalues are

$$\det(\mathbf{A} - \lambda \mathbf{I}) = \det \begin{pmatrix} 1 - \lambda & -2i \\ 2i & 1 - \lambda \end{pmatrix}, \quad \lambda^2 - 2\lambda - 3 = 0 \quad \text{with roots} \quad \lambda_1 = 3, -1.$$

The eigenvectors are complex and found, as in the last example, by solving for $\mathbf{v}_1$ and $\mathbf{v}_2$:

$$
\begin{aligned}
-2x_1 - 2iy_1 &= 0 \\
2ix_1 - 2y_1 &= 0
\end{aligned}, \quad \mathbf{u}_1 = r\begin{pmatrix} 1 \\ i \end{pmatrix} \quad \text{and} \quad
\begin{aligned}
2x_2 - 2iy_2 &= 0 \\
2ix_2 + 2y_2 &= 0
\end{aligned}, \quad \mathbf{u}_2 = s\begin{pmatrix} i \\ 1 \end{pmatrix}.
$$

The diagonal elements of $\mathbf{U} = (\mathbf{u}_1\ \mathbf{u}_2)$ are real, which we know they must be.

We can find r and s by ensuring the eigenvectors are orthonormal. To have a magnitude of 1, the inner product $\langle \mathbf{u}_1, \mathbf{u}_1 \rangle = 1 = r\sqrt{(1^2 + i(-i))} = r\sqrt{2}$. Hence $r = 1/\sqrt{2}$. Similarly, $s = 1/\sqrt{2}$. Now that we know the constants, we see that the eigenvectors are orthogonal, $\langle \mathbf{u}_1, \mathbf{u}_2 \rangle = 0$. Since the eigenvalues are different, we say there are two eigenstates each of multiplicity 1. We know $\mathbf{A}$ is full rank because it is diagonalizable.

The decomposition of Hermitian matrix $\mathbf{A}$ is achieved though eigenanalysis, where Λ is a diagonal matrix of eigenvalues and $\mathbf{U}$ is an orthonormal matrix whose columns are the eigenvectors of $\mathbf{A}$:

$$
\begin{pmatrix} 1 & -2i \\ 2i & 1 \end{pmatrix} = \mathbf{A} = \mathbf{U}\Lambda\mathbf{U}^\dagger = \frac{1}{2}\begin{pmatrix} 1 & i \\ i & 1 \end{pmatrix}\begin{pmatrix} 3 & 0 \\ 0 & -1 \end{pmatrix}\begin{pmatrix} 1 & -i \\ -i & 1 \end{pmatrix}. \tag{A.9}
$$

Checking, we see that $\mathbf{A} = \mathbf{A}^\dagger \neq \mathbf{A}^{-1}$ is Hermitian and $\mathbf{U} \neq \mathbf{U}^\dagger = \mathbf{U}^{-1}$ is unitary. There are an infinite number of basis sets that can restore $\mathbf{A}$ using (A.8) because $\mathbf{U}^{-1}$ normalizes the result correctly. However, (A.9) holds only if $\mathbf{U}^\dagger$ has orthonormal eigenvectors. Otherwise, the 1/2 factor on the right side of (A.9) is not present.

Eigenvectors of Symmetric Matrices

Covariance matrices $\mathbf{K}$ are symmetric. Symmetric matrices not only have real eigenvalues, but *eigenvectors corresponding to distinct eigenvalues are orthogonal*. To see this feature, select any two eigen-pairs of $\mathbf{K}$ – say $\lambda_m, \mathbf{u}_m$ and $\lambda_n, \mathbf{u}_n$, where $\lambda_m \neq \lambda_n$. Then

$$
\mathbf{K}\mathbf{u}_m = \lambda_m\mathbf{u}_m \quad \text{and} \quad \mathbf{K}\mathbf{u}_n = \lambda_n\mathbf{u}_n
$$

$$
\mathbf{u}_n^\top\mathbf{K}\mathbf{u}_m - \lambda_m\mathbf{u}_n^\top\mathbf{u}_m = \mathbf{u}_m^\top\mathbf{K}\mathbf{u}_n - \lambda_n\mathbf{u}_m^\top\mathbf{u}_n
$$

$$
\mathbf{u}_n^\top\mathbf{K}\mathbf{u}_m - \mathbf{u}_m^\top\mathbf{K}\mathbf{u}_n + (\lambda_m - \lambda_n)\mathbf{u}_n^\top\mathbf{u}_m = 0.
$$

Because $\mathbf{K}$ is symmetric, $(\mathbf{u}_n^\top\mathbf{K}\mathbf{u}_m)^\top = \mathbf{u}_m^\top\mathbf{K}^\top\mathbf{u}_n = \mathbf{u}_m^\top\mathbf{K}\mathbf{u}_n$. Therefore, $\mathbf{u}_n^\top\mathbf{K}\mathbf{u}_m - \mathbf{u}_m^\top\mathbf{K}\mathbf{u}_n = 0$. Also, since $\lambda_m \neq \lambda_n$, it must be true that $\mathbf{u}_n^\top\mathbf{u}_m = 0$.

Sparse and Low-Rank Matrices

Generally, sparse and low-rank matrices are very different. At least half of the elements of a sparse matrix are zero, while at least half of the eigenvalues (or singular values) of a low-rank matrix are zero. To illustrate, consider $\mathbf{A}, \mathbf{B}, \mathbf{C}, \mathbf{D} \in \mathbb{R}^{5 \times 5}$, where

$$A = \begin{pmatrix} 1 & 0 & 0 & 0 & 0 \\ 0 & 2 & 0 & 0 & 0 \\ 0 & 0 & 3 & 0 & 0 \\ 0 & 0 & 0 & 4 & 0 \\ 0 & 0 & 0 & 0 & 5 \end{pmatrix} \quad \text{and} \quad B = \begin{pmatrix} 1 & 2 & 3 & 4 & 5 \\ 0 & 0 & 0 & 0 & 0 \\ 0 & 0 & 0 & 0 & 0 \\ 0 & 0 & 0 & 0 & 0 \\ 0 & 0 & 0 & 0 & 0 \end{pmatrix}$$

are both sparse matrices. However, A is full rank while B has rank 1. In contrast,

$$C = \begin{pmatrix} 1 & 2 & 3 & 4 & 5 \\ 2 & 4 & 6 & 8 & 10 \\ 5 & 10 & 15 & 20 & 25 \\ 3 & 6 & 9 & 12 & 15 \\ 4 & 8 & 12 & 16 & 20 \end{pmatrix} \quad \text{and} \quad D = \begin{pmatrix} 4 & 5 & 2 & 3 & 1 \\ 5 & 1 & 2 & 4 & 3 \\ 1 & 4 & 2 & 5 & 3 \\ 1 & 2 & 5 & 3 & 4 \\ 3 & 2 & 5 & 4 & 1 \end{pmatrix}$$

are both dense matrices, although C is rank 1 and D is full rank. Because C is low rank and square, it can be made sparse through an eigendecomposition. Specifically, $\Lambda = Q^{\dagger} C Q$ is both sparse and low rank, while C is dense and low rank – but because they are *similar matrices*, they share identical properties.

Finally, consider $X = C + R$, where $R \in \mathbb{R}^{5 \times 5}$ has random elements drawn from $\mathcal{N}(0, (0.1)^2)$. X is a noisy version of deterministic C but with relatively high SNR ($= 10\texttt{*log10(var(reshape(C,1,25))}/0.1\texttt{\^{}\{2\}})$ = 36 dB). Strictly speaking, the noise makes X full rank. However, sometimes X is well represented by a *low-rank approximation*, where eigenvalues $\lambda_2 \dots \lambda_5$ are set to zero. The idea is that the *signal subspace* of X, where $\lambda_1 \gg \lambda_2 \gg \dots \gg \lambda_5$, can be represented by the one nonzero eigenvalue of C. Setting the smaller eigenvalues to zero reduces, but does not eliminate, the noise power.

Condition Number

Consider the linear equation $g = Hf$, where $H \in \mathbb{R}^{N \times N}$ and $g, f \in \mathbb{R}^{N \times 1}$. We are often interested in solving the equation for f given g and H, i.e., $\hat{f} = H^{-1}g$ assuming H is nonsingular. Regardless of the algorithm used, the *stability* of the solution will depend on *the condition of* H [41]. The condition of a matrix is a measure of how close it is to being singular: A well-conditioned matrix has a condition number near 1, while a poorly conditioned matrix has a larger condition number.

In a well-conditioned matrix, a small change in g produces a small change in $\hat{f}$. The condition number of a matrix can be found numerically from the ratio of the largest to smallest singular values (or the square root of eigenvalues if H is square and nonsingular, as in this example). To see this, apply SVD analysis (see Section 6.5) to H, which reveals that

$$\hat{f} = H^{-1}g = (V\Sigma^{1/2}U^{\dagger})^{-1}g = U\Sigma^{-1/2}V^{\dagger}g = \sum_{k=1}^{N} \frac{1}{\sqrt{\varsigma_k}} \mu_k v_k^{\dagger} g. \tag{A.10}$$

Thus, small changes in H or g are amplified when forming values of $\hat{f}$ when the kth singular value $\sqrt{\varsigma_k}$ is small.

There is always uncertainty (noise) in acquired data $\mathbf{g}$, which we model as $\mathbf{g} = \mathbf{H}\mathbf{f} + \mathbf{e}$. In turn, stability is a concern. Solving for $\mathbf{f}$, we estimate $\hat{\mathbf{f}} = \mathbf{H}^{-1}(\mathbf{g} - \mathbf{e})$. This solution is stable when the relative ℓ_2 norm of $\mathbf{H}^{-1}\mathbf{e}$ is on the order of the relative norm of $\mathbf{H}^{-1}\mathbf{g}$, i.e., $\|\mathbf{H}^{-1}\mathbf{e}\|_2/\|\mathbf{e}\|_2 \sim \|\mathbf{H}^{-1}\mathbf{g}\|_2/\|\mathbf{g}\|_2$. Norms were discussed at the beginning of this appendix. Expressed as a ratio, we have, with the help of (A.1) and the properties of matrix norms,

$$\frac{\frac{\|\mathbf{H}^{-1}\mathbf{e}\|_2}{\|\mathbf{H}^{-1}\mathbf{g}\|_2}}{\frac{\|\mathbf{e}\|_2}{\|\mathbf{g}\|_2}} = \frac{\|\mathbf{H}^{-1}\mathbf{e}\|_2}{\|\mathbf{e}\|_2} \frac{\|\mathbf{g}\|_2}{\|\mathbf{H}^{-1}\mathbf{g}\|_2} \to \max \frac{\|\mathbf{H}^{-1}\mathbf{e}\|_2}{\|\mathbf{e}\|_2} \frac{\|\mathbf{H}\mathbf{f}\|_2}{\|\mathbf{f}\|_2}$$

$$= \|\mathbf{H}^{-1}\|_2 \|\mathbf{H}\|_2 \geq \|\mathbf{H}^{-1}\mathbf{H}\|_2 = 1,$$

provided $\mathbf{e} \neq \mathbf{0}$ and $\mathbf{f} \neq \mathbf{0}$. When the product of the matrix norm and that of its inverse are on the order of unity, the matrix is well conditioned. The condition number assuming ℓ_2 norms apply is

$$\kappa(\mathbf{H}) = \|\mathbf{H}^{-1}\|_2 \|\mathbf{H}\|_2 = \sqrt{\frac{\varsigma_{max}}{\varsigma_{min}}}, \tag{A.11}$$

which has a form that is the square root of the ratio of the largest to smallest singular values.

Example. Find the condition numbers of the following matrices:

$$\mathbf{H}_1 = \begin{pmatrix} 1 & 2 & 3 \\ 3 & 2 & 1 \\ 4 & 2 & 3 \end{pmatrix} \qquad \mathbf{H}_2 = \begin{pmatrix} 1 & 2 & 3 \\ 3 & 2 & 1 \\ -1 & -2.0033 & -3 \end{pmatrix}.$$

MATLAB gives cond(H1) = 8.9490 and cond(H2) = 3.1523e+03. Both are full rank so both inverses exist, but you can see in $\mathbf{H}_2$ that row 3 is *almost* -1 times row 1. The large value of $cond(\mathbf{H}_2)$ means $\mathbf{H}_2$ is poorly conditioned, while the smaller number for $cond(\mathbf{H}_1)$ shows that $\mathbf{H}_1$ is well conditioned. Computing SVD in MATLAB via [U S V]=svd(H2), the ratio of the largest to smallest singular values is S(1,1)/S(3,3)=6.0658/0.0019, which gives the same value as the cond(H2) command.

Let's examine both inverse solutions in a little detail. Letting $\mathbf{g}_1 = (1\ 3\ 2)^\top$, we find fh1=H1\g1 = $(0.3333\ 1.3333\ -0.6667)^\top$. For $\mathbf{g}_2 = (1.1\ 2.8\ 2.2)^\top$, we find fh2=H1\g2 = $(0.3667\ 1.0917\ -0.4833)^\top$, which shows that $\|\mathbf{g}_1 - \mathbf{g}_2\|_2/\|\mathbf{g}_1\|_2 = 0.08$. Thus, an 8% perturbation in $\mathbf{g}_1$ yields roughly a $\|\mathbf{f}_1 - \mathbf{f}_2\|_2/\|\mathbf{f}_1\|_2 = 0.19$ or 19% perturbation in $\hat{\mathbf{f}}$.

For the same $\mathbf{g}_1$ and $\mathbf{g}_2$, fh1=H2\g1 = $(451.0000\ -900.0000\ 450.0000)^\top$ and fh2=H2\g2 = $(495.9125\ -990.0000\ 495.0625)^\top$. The 8% variation in $\mathbf{g}$ values produces a 10% variation in output, but the solution errors are all highly amplified. Better-conditioned $\mathbf{H}_2$ generates much smaller (and more reasonable) inverse solutions.

Moore–Penrose Pseudoinverse

Consider system matrix $\mathbf{H} \in \mathbb{R}^{M \times N}$ in the noiseless linear acquisition equation $\mathbf{g} = \mathbf{Hf}$. If $M \neq N$ and/or $\mathbf{H}$ is not full rank, we know that $\mathbf{H}^{-1}$ does not exist. The Moore–Penrose pseudoinverse of $\mathbf{H}$, denoted as $\mathbf{H}^{+}$, offers the unique solution $\hat{\mathbf{f}} = \mathbf{H}^{+}\mathbf{g}$ that minimizes the ℓ_2 norm $\|\mathbf{g} - \mathbf{Hf}\|_2$.

$\mathbf{H}^{+}$ is found from the following equations [41]:

$$\mathbf{H}^{+} = \lim_{\epsilon \to 0} (\mathbf{H}^{\top}\mathbf{H} + \epsilon^2 \mathbf{I})^{-1} \mathbf{H}^{\top}, \qquad \text{(Left invertible)} \qquad \text{(A.12)}$$

so that if $\mathbf{H}$ has linearly independent columns, then $\mathbf{H}$ is left invertible and $\mathbf{H}^{+} = (\mathbf{H}^{\top}\mathbf{H})^{-1}\mathbf{H}^{\top}$. However,

$$\mathbf{H}^{+} = \lim_{\epsilon \to 0} \mathbf{H}^{\top}(\mathbf{H}^{\top}\mathbf{H} + \epsilon^2 \mathbf{I})^{-1}, \qquad \text{(Right invertible)} \qquad \text{(A.13)}$$

so that if $\mathbf{H}$ has linearly independent rows, then $\mathbf{H}$ is right invertible and $\mathbf{H}^{+} = \mathbf{H}^{\top}(\mathbf{H}^{\top}\mathbf{H})^{-1}$.

Penrose [81] showed that $\mathbf{H}^{+} \in \mathbb{R}^{N \times M}$ satisfies the following four conditions:

$$\mathbf{H} = \mathbf{HH}^{+}\mathbf{H}$$
$$\mathbf{H}^{+} = \mathbf{H}^{+}\mathbf{HH}^{+}$$
$$\mathbf{HH}^{+} = (\mathbf{HH}^{+})^{\top}$$
$$\mathbf{H}^{+}\mathbf{H} = (\mathbf{H}^{+}\mathbf{H})^{\top}.$$

The uniqueness property is a result of the last two statements regarding the symmetry of products $\mathbf{HH}^{+}$ and $\mathbf{H}^{+}\mathbf{H}$. Readers are urged to select small, medium, and large square matrices and an appropriately dimensioned vector to try H\g, g\H, inv(H)*g, and pinv(H)*g and discover what works and how long MATLAB takes to find each result. Chapter Problems 7.3, 7.4, and 7.6 apply the Moore–Penrose pseudoinverse for image restoration problems.

Operators

Mappings associated with various types of operators are diagrammed in Figure A.3. A wonderfully detailed description of measurement operators is given in [10]. Assume a continuous-to-discrete operator where objects are "measured" according to the mapping $\mathbf{g} = \mathcal{A}f(\mathbf{x})$. $\mathcal{A}$ is the linear operator that describes measurements as a transformation of continuous object functions of position, $f(\mathbf{x})$ in vector space $\mathbb{U}$, to data vectors $\mathbf{g}$ in vector space $\mathbb{V}$. We can map back to $\mathbb{U}$ using the adjoint operator $\mathcal{A}^{\dagger}$. If $\mathcal{A}$ is Hermitian, then $\mathcal{A} = \mathcal{A}^{\dagger}$, $\mathbb{U} = \mathbb{V}$, and we say operator $\mathcal{A}$ is *self-adjoint*. If $\mathcal{A}$ is unitary, then $\mathcal{A}^{\dagger} = \mathcal{A}^{-1}$. $\mathcal{A}^{\dagger}$ is defined [10] using inner products involving vectors in different spaces,

$$\langle \mathbf{g}, \mathcal{A}\mathbf{f} \rangle_{\mathbb{V}} = \langle \mathcal{A}^{\dagger}\mathbf{g}, \mathbf{f} \rangle_{\mathbb{U}}. \qquad \text{(A.14)}$$

$\mathbf{g}$ and $\mathbf{f}$ can each be either continuous (an $\infty \times 1$ vector) or discrete (an $N \times 1$ vector). Depending on the forms of $\mathbf{g}$ and $\mathbf{f}$, $\mathcal{A}$ will change from an integral operator to a matrix multiplication, as illustrated in Figure A.3.

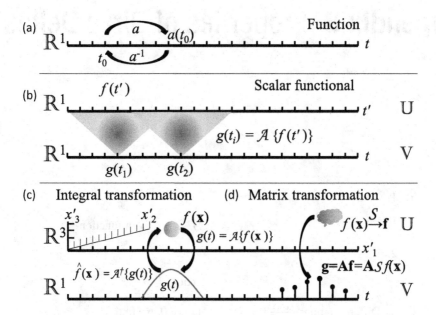

Figure A.3 Four types of operators are illustrated. (a) A *function* maps point t_0 on time axis t into a real-valued scalar function at t_0, labeled $a(t_0)$. We can map back using the inverse function $t_0 = a^{-1}\{a(t_0)\}$, assuming a^{-1} exists and is one-to-one. (b) A linear *functional* is a mapping of function $f(t')$ in vector space $\mathbb{U}$ into a scalar point possibly in another vector space, e.g., $g(t_1)$ and $g(t_2)$ on axis t. (c) A *continuous-to-continuous integral transformation* from 3-space to time and back is illustrated (e.g., ultrasonic acquisition). While operator $\mathcal{A}$ takes us from $\mathbb{U} \to \mathbb{V}$, adjoint operator $\mathcal{A}^\dagger$ maps back to the original space $\mathbb{V} \to \mathbb{U}$. If the operator is unitary, $\mathcal{A}^\dagger = \mathcal{A}^{-1}$, and the acquisition noise is minimal, then there are only trivial null spaces and estimates $\hat{f}(\mathbf{x}) \simeq f(\mathbf{x})$. If the operator is not unitary, the adjoint operator will still return data to the object space but with undesirable properties such as blur. (c) A two-step *continuous-to-discrete transformation*, $\mathbf{g} = \mathbf{A}\mathcal{S}\{f(\mathbf{x})\}$, is illustrated. First, continuous function $f(\mathbf{x})$ is sampled via continuous-to-discrete sampling operator $f[n] = \mathcal{S}f(t)$ to generate $\mathbf{f}$ [see (5.23)]. Then discrete-to-discrete matrix operator $\mathbf{g} = \mathbf{A}\mathbf{f}$ is applied. The measurement can be represented as a sequence of transformations, $\mathbf{g} = \mathbf{A}\mathcal{S}\{f(\mathbf{x})\}$.

Equation (A.14) shows us how to compare object and data functions that exist in different vector spaces. Of course, it does not necessarily mean that $\hat{\mathbf{f}} = \mathcal{A}^\dagger \mathbf{g}$ is a good estimate. In fact, that estimator is usually pretty bad unless $\mathcal{A}$ is a unitary operator and the null space resulting from the measurement is negligible. We explore equation (A.14) further in Problem 6.2.

Appendix B Properties of Dirac Deltas

1.

$$\delta(-x) = \delta(x) \qquad \text{(Even function)}$$

2.

$$x\delta(x) = 0$$

3. Differentiating $\frac{d[x\delta(x)]}{dx} = x\frac{d\delta}{dx} + \delta(x) = 0$, we find

$$\dot{\delta}(x) \triangleq \frac{d\delta}{dx} = -\frac{\delta(x)}{x}. \qquad \text{(Odd function)}$$

4. The step function step(x) is found from (notice integration limits)

$$\int_{-\infty}^{x} dy\, \delta(y) = \text{step}(x) = \begin{cases} 0 & x < 0 \\ 1 & x > 0 \end{cases} \qquad \text{and} \qquad \frac{d\text{step}}{dx} = \delta(x).$$

5. The ramp function ramp(x) is found from

$$\int_{-\infty}^{x} dy\, \text{step}(y) = \text{ramp}(x) = \begin{cases} 0 & x < 0 \\ x & x > 0 \end{cases} \qquad \text{and} \qquad \frac{d^2\text{ramp}}{dx^2} = \frac{d\text{step}}{dx} = \delta(x).$$

6.

$$\int_{-\infty}^{\infty} dx\, g(x)\dot{\delta}(x - a) = -\dot{g}(a) \qquad \text{(First derivative)}$$

7.

$$\int_{-\infty}^{\infty} dx\, g(x)\delta^{(n)}(x - a) = (-1)^n g^{(n)}(a) \qquad \text{(nth derivative)}$$

8.

$$\delta(ax) = |a|^{-1}\delta(x) \quad \text{and} \quad \delta(ax - x_0) = |a|^{-1}\delta(x - x_0/a)$$

$$\text{(Scaling and shifting)}$$

9.

$$\int_{-\infty}^{\infty} dx\, \delta(a - x)\delta(x - b) = \delta(a - b) \qquad \text{(Nonzero only at } a = b\text{)}$$

10.

$$\int_{-\infty}^{\infty} dx \int_{-\infty}^{\infty} dy\, g(x, y)\, \delta(x - a, y - b) = g(a, b) \qquad \text{(2-D sifting property)}$$

Appendix C Signal Modulation

One of the themes of this book is that we need to be prepared to look for task information in data wherever it might be found. At other times, we may purposefully attach information to wave energy that is reliably transmitted and received so the information can then be recovered. It is therefore important to understand how information is encoded in data, and to develop the tools that can decode the information.

Broadcast radio is a well-known example of encoding a carrier wave with information. Recorded music signals are routinely transmitted into the atmosphere after combining the music signal with an electromagnetic (EM) wave at *carrier frequency* u_0. These waves are selected for their ability to travel long distances from a transmitting antenna, be received by a radio receiver, and be quickly stripped of the carrier wave to access the music payload.

An AM radio transmitter uses audio signals to *modulate the amplitude* of a carrier EM wave in the 0.5–1.6 MHz carrier-frequency range (in the United States). Your car radio receives the modulated carrier and *demodulates* the signals, i.e., removes the carrier wave, before presenting the remaining signal as a voltage to the car's speakers. Similarly, FM radio works with carrier frequencies between 88 and 108 MHz, but it *modulates the frequency* of the EM carrier wave with the audio signal instead of the amplitude. In medical practice, magnetic resonance imaging (MRI), optical coherence tomography (OCT), and ultrasonic imaging (US) systems employ carrier waves to safely transmit energy into and out of the body.

Example C.0.1. *Signal mixing: Carrier wave $g_0(t) = \cos(2\pi u_0 t)$ at $u_0 = 1$ Hz is mixed with signal wave $g_s(t) = \cos(2\pi u_s t)$ at $u_s = 0.2$ Hz by multiplying them. The results are shown in Figure C.1a.*

$$g(t) = \cos(2\pi u_0 t)\cos(2\pi u_s t) = \frac{1}{2}\Big(\cos\big(2\pi(u_0 + u_s)t\big) + \cos\big(2\pi(u_0 - u_s)t\big)\Big).$$

It is difficult to interpret $g_s(t)$ properties directly from $g(t)$ in Figure C.1a. However, the frequency domain shows us the sum-and-difference frequency peaks, which reveal two interfering sinusoids and their amplitudes and frequencies. Figure C.1b shows sidebands at $\pm(u_0 - u_s) = \pm 0.8$ Hz and $\pm(u_0 + u_s) = \pm 1.2$ Hz. We can retrieve $g_s(t)$ from $g(t)$ as shown in Figure C.1c and C.1d.

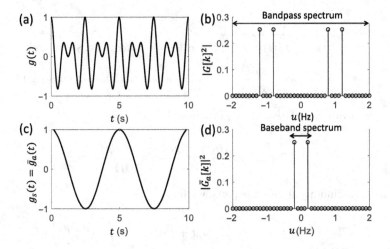

Figure C.1 (a) Mixed signal $g(t) = g_0(t)g_s(t) = \cos(2\pi u_0 t)\cos(2\pi u_s t)$ at carrier frequency $u_0 = 1$ Hz and signal frequency $u_s = 0.2$ Hz. (b) From its Fourier series, we find the modulus-squared transform $|G[k]|^2$ (see Example C.0.1). (d) The magnitude-squared spectrum of the complex amplitude $|\bar{G}_a[k]|^2$ and the corresponding time domain function $\bar{g}_a(t)$ in (c) gives us the demodulated signal $g_s(t) = \cos(2\pi u_s t)$.

Signal Properties

$|G[k]|^2$ in Figure C.1b is the magnitude-squared spectrum of a *bandpass signal*, which includes all nonzero frequency components in $g(t)$. The bandpass spectrum is ultimately limited by the sampling frequency, e.g., (5.28), but applying a Fourier series allows us to select a smaller frequency band within $\pm u_N$. It contains two "groupings," referred to as the *baseband signal* spectrum, centered about the carrier frequency at $\pm u_0$. Demodulated spectra remove the influence of the carrier frequency, here at $u_0 = 1$ Hz, so baseband spectra are centered at zero frequency, as shown in Figure C.1d.

Time series $g(t)$ may be expressed in terms of baseband and carrier-wave components using the following expression (see appendix 3A in [109]):

$$g(t) = g_i(t)\cos(2\pi u_0 t) - g_q(t)\sin(2\pi u_0 t). \tag{C.1}$$

g_i and g_q are the *in-phase-quadrature (IQ) signals* for $g(t)$. IQ signals describe baseband properties of $g(t)$, including all amplitude and phase information, but without the influence of the carrier frequency. This representation holds only for "narrowband" data, defined for bandpass signals with baseband frequency support $\leq 2u_0$. A simpler statement is that the percent bandwidth must be less than 200%. Any nonzero frequency coefficients outside that range must be removed by filtering.

This representation of $g(t)$ enables us to express any recorded signal as

$$g(t) = A(t)\cos(2\pi u_0 t + \theta(t)),$$

where

$$A(t) = \sqrt{g_i^2(t) + g_q^2(t)} \qquad \text{is the } \textit{envelope} \text{ of the signal and}$$

$$\theta(t) = \tan^{-1} \frac{g_q(t)}{g_i(t)} \qquad \text{is the } \textit{phase modulation} \text{ of the signal.}$$

The *frequency modulation* of $g(t)$ is defined by $\zeta(t)$,

$$2\pi\zeta(t) = \frac{d\theta(t)}{dt} = \frac{g_i(t)\frac{dg_q}{dt} - g_q(t)\frac{dg_i}{dt}}{A^2(t)}.$$

Finally, assembling all of these parts, we find

$$g(t) = A(t)\cos\left(2\pi u_0 t + \theta(0) + 2\pi \int_0^t dt' \zeta(t')\right).$$

$g(t)$ is now expressed in terms of its envelope signal $A(t)$, carrier frequency u_0, initial phase $\theta(0)$, and frequency modulation $\zeta(t)$, each of which is expressible in terms of IQ components, $g_i(t), g_q(t)$.

Analytic Signal and Complex Envelope

At the end of Appendix D, we define the Hilbert transform of signal $g(t)$ using the symbol $\breve{g}(t)$, as well as its Fourier transform $\mathcal{F}\breve{g}(t)$. Applying those definitions here, we introduce the *analytic signal* of bandpass time series $g(t)$ as

$$g_a(t) = g(t) + i\breve{g}(t) = (g_i(t) + ig_q(t))e^{i2\pi u_0 t}.$$

The first form shows that the real part of the analytic signal is just the recorded data itself, while the imaginary part is the Hilbert transform of $g(t)$. The second form of the equation is derived in [109]. It describes the analytic signal as being composed of the IQ signals and the carrier waveform $\exp(i2\pi u_0 t)$. In the previous section, we described real bandpass waveform $g(t)$ as a baseband signal modulated by a cosine carrier wave. Similarly, the analytic signal of any recorded waveform may be described as a complex envelope modulated via multiplication by a complex exponential. Let's see.

The Fourier transform of the analytic signal is (see Appendix D)

$$G_a(u) = G(u) + i\breve{G}(u) = G(u)(1 + \text{sgn}(u)) = \begin{cases} 2G(u) & u > 0 \\ G(u) & u = 0 \\ 0 & t < 0 \end{cases}.$$

"Why introduce $g_a(t)$?" you might ask. Recall that we are free to convert real signals recorded during an experiment to a complex form. We do this when we need access to both signal amplitude and phase. Those are the signal properties needed to ask "how much is there" and "where/when is it located." Analytic signal $g_a(t)$ provides this representation.

We can demodulate bandpass waveform $g(t)$ by multiplying its analytic form by $\exp(-i2\pi u_0 t)$. The result is the *complex envelope*, $\bar{g}_a(t)$,

$$\bar{g}_a(t) = g_a(t)e^{-i2\pi u_0 t} = (g(t) + i\breve{g}(t))e^{-i2\pi u_0 t} = g_i(t) + ig_q(t). \qquad \text{(C.2)}$$

Consequently, $g_i(t) = \Re\{\bar{g}_a(t)\}$ and $g_q(t) = \Im\{\bar{g}_a(t)\}$.

The complex envelope is a baseband representation of bandpass $g(t)$ that preserves all amplitude and phase information.

The Fourier transform of complex envelope $\bar{g}_a(t)$ is found using the Fourier shift theorem,

$$\bar{G}_a(u) = G_a(u + u_0) = \begin{cases} 2G(u + u_0) & u > 0 \\ G(u + u_0) & u = 0 \\ 0 & t < 0 \end{cases},$$

where $|\bar{G}_a(u)|^2$ is plotted in Figure C.1d. Computing the magnitude-squared transform discards the signal phase. The data presented in Figure C.1 were generated using the following code.

```
%%%%%%%%%%%%%%%%  Script generating data in Fig. C.1 from Example C.0.1  %%%%%%%%%%%%%%%%%%%%
T0=10;dt=0.01;t=0:dt:T0;N=length(t);               % initialize
u0=1;subplot(2,2,1);gp=zeros(size(t));             % u0 is carrier freq
g=cos(2*pi*1*t).*cos(2*pi*0.2*t);                  % mixed bandpass signal
plot(t,g,'k','linewidth',2);xlabel('t (s)');ylabel('g(t)')
u=-2:0.1:2;C=zeros(N,41);                           % select freq's displayed
for j=1:41                                          % Fourier series basis
    C(:,j)=cos(2*pi*u(j)*t);
end
G=2/T0*g*C*dt;G2=G.*conj(G);                        % G2=mag^2 spectrum
subplot(2,2,2);stem(u,G2,'k','linewidth',1);
xlabel('u (Hz)');ylabel('|G[k]|^2')
gt=hilbert(g).*exp(-2i*pi*u0*t);                    % compute complex envelope
Ga=1/T0*gt*C*dt;Ga2=Ga.*conj(Ga);                  % factor is 2/(2*T0) so 1/T0
subplot(2,2,4);stem(u,Ga2,'k','linewidth',1)
xlabel('u (Hz)');ylabel('|G_s[k]|^2')
for j=1:41                                          % Fourier series synthesis
    gp=gp+Ga(j)*exp(i*2*pi*u(j)*t);
end subplot(2,2,3);plot(t,gp/2,'k','linewidth',2)
xlabel('t (s)');ylabel('g_s(t)')
```

The analytic signal of $g(t)$ is found in MATLAB using `hilbert(g)`. This is not the Hilbert transform, as you might expect. $\breve{g}(t)$ is found using `imag(hilbert(g))`.

Example C.0.2. *Demodulation: Simulate and process one pulse-echo time series from an ultrasound experiment (see Section 4.4).*

(a) Generate a Gaussian-modulated ultrasound pulse $h(t)$, and a standard normal distribution to represent tissue scattering $f(t)$. Compute the noiseless recorded echo $g(t) = [h * f](t)$ and the envelope of $g(t)$, the latter being the B-mode image amplitude. Plot them in one figure.

(b) Compute and plot $|H(u)|$, the frequency spectrum of pulse $h(t)$ in a separate plot. Then compute the frequency spectrum of its analytic signal, $|H_a(u)|$ and complex envelope $|\bar{H}_a(u)|$ from components of $H(u)$. Plot all three together to compare them. Finally, compute the pulse envelope, $|h_a(t)|$ from $H_a(u)$ and plot it with the pulse in a separate plot.

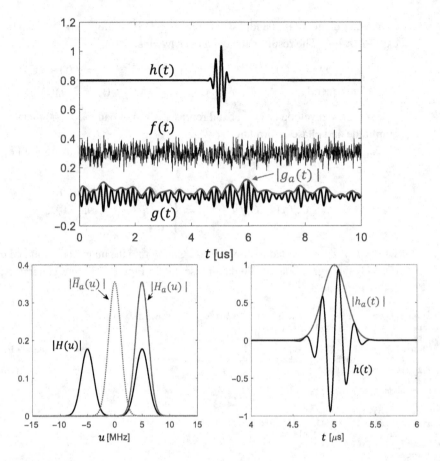

Figure C.2 Results from Example C.0.2.

```
%%%%%%%%%%%%%    Code to generate Fig. C.2 from Example C.0.2    %%%%%%%%%%%%%
```

```
%%%%%%%%%%  Part (a)  %%%%%%%%%%%%%%%%%%%%%%%%
T=0.01;t=0:T:10;sh=5;s2=0.02;u0=5;N=size(t);      % time axis and parameters
h=exp(-(t-sh).^2/(2*s2)).*sin(2*pi*u0*t);f=randn(N);  % simulate pulse and tissue scattering
gg=conv(f,h)*T;g=gg(501:1501);                    % simulate the RF echo signal
plot(t,g);hold on;plot(t,f/20+0.3);plot(t,h/4+0.8);   % plot g,f,h but shift so no overlap
env=abs(hilbert(g));plot(t,env,'r');hold off;     % find the envelope of g and plot on g
%%%%%%%%  Part (b)  %%%%%%%%%%%%%%%%%%%%%%
u=-1/(2*T):1/10:1/(2*T);
H=abs(fftshift(fft(h)))*T;figure;plot(u,H)        % Compute & plot H, FT of pulse h
Ha(1:500)=0;Ha(501)=H(501);Ha(502:1001)=2*H(502:1001);  % Compute Ha(u) using H(u)...
Hab=zeros(size(u));Hab(1:900)=Ha(51:950);hold on;  % ...& \bar Ha(u) from Ha(u)...
plot(u,Ha,'r');plot(u,Hab,'r:','linewidth',2);hold off  % ...& plot results together, compare
figure; ha=abs(fftshift(ifft(Ha)))/0.01;          % Use Ha(u) to find pulse envelope
plot(t,ha,'r');hold on;plot(t,h),hold off
```

I'm hoping that this code and the graphical results help you experiment with these tools for any associated problems you might encounter.

Relating the analytic signal to the IQ signal. The analytic signal representation $g_a(t)$ allows us to express real RF signal $g(t)$ measured during an experiment in terms of its magnitude and phase that may each be a function of time. The analytic signal is expressed in terms of the Hilbert transform $\breve{g}(t)$ as

$$g_a(t) = g(t) + i\breve{g}(t), \quad \text{and therefore} \quad g(t) = \frac{1}{2}\left(g_a(t) + g_a^*(t)\right).$$

We now take a circuitous path to show that

$$g_a(t) = (g_i(t) + ig_q(t))e^{i2\pi u_0 t}. \tag{C.3}$$

In-phase component, $g_i(t)$, is the baseband component of $2g(t)\cos(2\pi u_0 t)$ [109]. To show this, combine this expression with (C.1) to find:

$$
\begin{aligned}
2g(t)\cos(2\pi u_0 t) &= 2g_i(t)\cos^2(2\pi u_0 t) - 2g_q(t)\sin(2\pi u_0 t)\cos(2\pi u_0 t) \\
&= g_i(t) + g_i(t)\cos(4\pi u_0 t) - g_q(t)\sin(4\pi u_0 t).
\end{aligned}
$$

Therefore,

$$
\begin{aligned}
g_i(t) &= LowPass\{2g(t)\cos(2\pi u_0 t)\} = \frac{1}{2}LowPass\{(g_a(t) + g_a^*(t))(e^{i2\pi u_0 t} + e^{-i2\pi u_0 t})\} \\
&= \frac{1}{2}LowPass\left\{g_a(t)e^{i2\pi u_0 t} + g_a^*(t)e^{i2\pi u_0 t} + g_a(t)e^{-i2\pi u_0 t} + g_a^*(t)e^{-i2\pi u_0 t}\right\} \\
&= \frac{1}{2}\left(g_a(t)e^{-i2\pi u_0 t} + g_a^*(t)e^{i2\pi u_0 t}\right) \\
&= \Re\{g_a(t)e^{-i2\pi u_0 t}\} = \Re\{(g(t) + i\breve{g}(t))(\cos(2\pi u_0 t) - i\sin(2\pi u_0 t))\} \\
&= g(t)\cos(2\pi u_0 t) + \breve{g}(t)\sin(2\pi u_0 t). \tag{C.4}
\end{aligned}
$$

The use of the low-pass filter assumes $g(t)$ is "narrowband," e.g., the fractional signal bandwidth $BW/u_0 < 1$. This limit does not constrain ultrasound signals since in practice the fractional bandwidth is routinely less than 100%. The low-pass filter also provides anti-aliasing protection.

Similarly, $g_q(t)$ is the baseband component of $-2g(t)\sin(2\pi u_0 t)$ [109]:

$$
\begin{aligned}
-2g(t)\sin(2\pi u_0 t) &= -2g_i(t)\cos(2\pi u_0 t)\sin(2\pi u_0 t) + 2g_q(t)\sin^2(2\pi u_0 t) \\
&= -g_i(t)\sin(4\pi u_0 t) + g_q(t) - g_i(t)\cos(4\pi u_0 t).
\end{aligned}
$$

Therefore,

$$
\begin{aligned}
g_q(t) &= LowPass\{-2g(t)\sin(2\pi u_0 t)\} \;=\; \frac{i}{2} LowPass\{(g_a(t) \\
&\quad + g_a^*(t))(e^{i2\pi u_0 t} - e^{-i2\pi u_0 t})\} \\
&= \frac{i}{2} LowPass\left\{g_a(t)e^{i2\pi u_0 t} + g_a^*(t)e^{i2\pi u_0 t} - g_a(t)e^{-i2\pi u_0 t} - g_a^*(t)e^{-i2\pi u_0 t}\right\} \\
&= \frac{i}{2}\left(-g_a(t)e^{-i2\pi u_0 t} + g_a^*(t)e^{i2\pi u_0 t}\right) \\
&= \frac{i}{2}\{-(g(t)+i\breve{g}(t))(\cos(2\pi u_0 t) - i\sin(2\pi u_0 t)) + (g(t)-i\breve{g}(t)) \\
&\quad (\cos(2\pi u_0 t) + i\sin(2\pi u_0 t))\} \\
&= -g(t)\sin(2\pi u_0 t) + \breve{g}(t)\cos(2\pi u_0 t) \;=\; \Im\{g_a(t)e^{-i2\pi u_0 t}\}. \qquad \text{(C.5)}
\end{aligned}
$$

For narrowband RF signals, $g_i(t)$ and $g_q(t)$ are orthogonal. Combining (C.3) – (C.5),

$$
\begin{aligned}
g_a(t)e^{-i2\pi u_0 t} &= g_i(t) + ig_q(t) \\
&= [g(t)+i\breve{g}(t)]\cos(2\pi u_0 t) - i[g(t)+i\breve{g}(t)]\sin(2\pi u_0 t) \\
&= g_a(t)e^{-i2\pi u_0 t}.
\end{aligned}
$$

Appendix D Fourier Transform Theorems and Special Functions

Once you have computed a few transforms by hand, you'll notice patterns emerging. Some of the most useful patterns are listed in this appendix as theorems. Many insights come from knowledge of Fourier theorems, so much so that most experienced researchers keep lots of these theorems in their head so they can apply them quickly without using paper or laptop. The theorems presented here are mostly for CT-FT, but there are versions for each type of data.

Convolution Theorems

$$\mathcal{F}[h * f](x) = H(u)F(u) \quad \text{and} \quad \mathcal{F}\Big(h(x)f(x)\Big) = [H * F](u)$$

Correlation Theorem

$$\mathcal{F}\varphi(\tau) = \mathcal{F}\int_{-\infty}^{\infty} dx\, h(x - \tau)\, f^*(x) = H(u)F^*(u)$$

Shift Theorems

$$\mathcal{F}h(x - a) = H(u)e^{-i2\pi au} \quad \text{and} \quad \mathcal{F}\Big(e^{i2\pi u_0 x}h(x)\Big) = H(u - u_0)$$

Scaling Theorem

$$\mathcal{F}h(ax) = \frac{1}{|a|}H(u/a)$$

Separability

$$\mathcal{F}h(x, y) = \mathcal{F}_y\mathcal{F}_x h(x, y) = H(u, v)$$
$$= \int_{-\infty}^{\infty} dy\, e^{-i2\pi vy} \int_{-\infty}^{\infty} dx\, e^{-i2\pi ux}\, h(x, y)$$

Periodicity $H[k] = H[k + rN]$ for $N =$ points in h and integer r

Symmetry and Conjugate symmetry

$$\mathcal{F}h(-x) = H(-u), \quad H(u) = H^*(-u) \quad \text{and} \quad |H(u)| = |H(-u)|$$

Proof:

$$\mathcal{F}h(-x) = \int_{-\infty}^{\infty} dx\, h(-x)\, e^{-i2\pi ux}\,. \text{ Letting } x' = -x,$$

$$= \int_{-\infty}^{\infty} dx'\, h(x')\, e^{-i2\pi(-u)x'} \;=\; H(-u)\,. \text{ Also,}$$

$$H(u) = \Re H(u) + i\Im H(u)$$

$$H^*(-u) = (\Re H(u) - i\Im H(u))^* \;=\; \Re H(u) + i\Im H(u) \;=\; H(u)$$

Linearity

$$\mathcal{F}[ag(x) + bh(x)] = a\mathcal{F}g(x) + b\mathcal{F}h(x) \text{ for constants } a,b$$

Modulation-shift theorem

$$\mathcal{F}\big(h(x)\cos(2\pi u_0 x + \varphi)\big) = \frac{e^{i\varphi}}{2}\left[H(u) * \big(\delta(u + u_0) + \delta(u - u_0)\big)\right]$$

$$= \frac{e^{i\varphi}}{2}\big(H(u + u_0) + H(u - u_0)\big)$$

Conjugate theorem

$$\mathcal{F}h^*(x) = H^*(-u)$$

$$[\mathcal{F}h(x)]^* = H^*(u)$$

Proof:

$$\mathcal{F}h^*(x) \;=\; \int_{-\infty}^{\infty} dx\, h^*(x)\, e^{-i2\pi ux} = \left(\int_{-\infty}^{\infty} dx\, h(x)\, e^{i2\pi ux}\right)^*$$

$$\xrightarrow[u'=-u]{} \left(\int_{-\infty}^{\infty} dx\, h(x)\, e^{-i2\pi u'x}\right)^* = H^*(u') \;=\; H^*(-u)$$

Derivative theorem

$$\mathcal{F}\frac{d^n h(x)}{dx^n} = (i2\pi u)^n\, H(u)$$

Integral theorem

$$\mathcal{F}\left\{\int_{-\infty}^{x} dx'\, f(x')\right\} = \frac{F(u)}{i2\pi u} + a\delta(u) \text{ provided } \int_{-\infty}^{\infty} dx'\, (f(x') - a) = 0$$

Parseval's theorem (conservation of energy)

$$\int_{-\infty}^{\infty} dx\, g(x)h^*(x) = \int_{-\infty}^{\infty} du\, G(u)\, H^*(u)$$

Proof:

$$\int_{-\infty}^{\infty} dx\, g(x)\, h^*(x) = \int_{-\infty}^{\infty} dx\, g(x)\left[\int_{-\infty}^{\infty} du\, H(u)\, e^{i2\pi ux}\right]^*$$

$$= \int_{-\infty}^{\infty} dx\, g(x)\int_{-\infty}^{\infty} du\, H^*(u)\, e^{-i2\pi ux}$$

$$\int_{-\infty}^{\infty} du\, H^*(u)\int_{-\infty}^{\infty} dx\, g(x)\, e^{-i2\pi ux} = \int_{-\infty}^{\infty} du\, H^*(u)G(u)$$

Rayleigh's Theorem (Special Case of Parseval's Theorem)

$$\int_{-\infty}^{\infty} dx\, |h(x)|^2 = \int_{-\infty}^{\infty} du\, |H(u)|^2$$

Important Transforms and Properties

$$\mathcal{F}\text{rect}(x/X) = X\text{sinc}(Xu)$$

$$\mathcal{F}\sin(2\pi u_0 x) = \frac{i}{2}[\delta(u+u_0) - \delta(u-u_0)]$$

$$\mathcal{F}\cos(2\pi u_0 x) = \frac{1}{2}[\delta(u+u_0) + \delta(u-u_0)]$$

$$\mathcal{F}\{f(x)\cos(2\pi u_0 x)\} = \frac{1}{2}[F(u+u_0) + F(u-u_0)]$$

$$\mathcal{F}\{e^{-ax}\text{step}(x)\} = \frac{1}{a+i2\pi u} = \frac{1}{a+i\Omega}$$

$$\mathcal{F}\{e^{-a|x|}\} = \frac{2a}{a^2 + 4\pi^2 u^2} = \frac{2a}{a^2 + \Omega^2}$$

$$\mathcal{F}\left\{\frac{1}{\sigma\sqrt{2\pi}}e^{-(x-x_0)^2/2\sigma^2}\right\} = e^{-i2\pi u x_0}\, e^{-2\pi^2 u^2 \sigma^2}$$

$$\mathcal{F}\left\{e^{-x^2/2\sigma^2}\right\} = (\sigma\sqrt{2\pi})\, e^{-2\pi^2 u^2 \sigma^2}$$

$$\left(\mathcal{F}^{-1}\right)^* \{e^{a^2 u^2}\} = \frac{\sqrt{\pi}}{a} e^{-\pi^2 x^2/a^2}$$

$$\mathcal{F}\{e^{-at}\sin(\Omega_0 t)\text{step}(t)\} = \frac{\Omega_0}{\Omega_0^2 + (a+i\Omega)^2}$$

$$\mathcal{F}\{e^{-at}\cos(\Omega_0 t)\text{step}(t)\} = \frac{a+i\Omega}{\Omega_0^2 + (a+i\Omega)^2}$$

To find $\mathcal{F}\{\text{step}(t)\}$, note that $\displaystyle\int_{-\infty}^{\infty} dt\,(\text{step}(t) - 1/2) = 0$.

Applying the integral theorem where $a = 1/2$, we have

$$\mathcal{F}\{\text{step}(t)\} = \mathcal{F}\left\{\int_{-\infty}^{t} dt'\, \delta(t')\right\} = \frac{\mathcal{F}\delta(t)}{i2\pi u} + \frac{\delta(u)}{2} = \frac{1}{i2\pi u} + \frac{\delta(u)}{2}.$$

Sampling Operator and Its Fourier Transform (Section 5.7)

$$f_s(x) = \mathcal{S}^\dagger \mathcal{S} f(x) = \left[\frac{1}{X}\text{comb}(x/X)\right] \times f(x)$$

$$= f(x) \sum_{m=-\infty}^{\infty} \delta(x - mX) = f(mX) \quad [\text{Eq. (5.23)}]$$

$$\mathcal{F}\left\{\frac{1}{X}\text{comb}(x/X)\right\} = \mathcal{F}\left\{\sum_{m=-\infty}^{\infty} \delta(x - mX)\right\} = \frac{1}{X}\sum_{k=-\infty}^{\infty} \delta(u - k/X) = \text{comb}(Xu)$$

Discrete-Time Fourier Transform (Section 5.7)

$$G_s(u) = \mathcal{F}\mathcal{S}^\dagger \mathcal{S} g(t) = \mathcal{F}\left\{ \frac{1}{T}\mathrm{comb}(t/T)\, g(t) \right\} = \mathcal{F}\left\{ \sum_{n=-\infty}^{\infty} \delta(t-nT) \right\} * \mathcal{F}g(t)$$

$$= \int_{-\infty}^{\infty} du'\, G(u-u')\, \mathrm{comb}(Tu')$$

$$= \frac{1}{T}\sum_{k=-\infty}^{\infty} \int_{-\infty}^{\infty} du'\, G(u')\, \delta(u-k/T-u')$$

$$= \frac{1}{T}\sum_{k=-\infty}^{\infty} G(u-k/T)$$

Real, Imaginary, Even, Odd Properties (see 5.10)

Real, even functions transform to real, even functions.

Real, odd functions transform to imaginary, odd functions.

Gaussian Integral

$$\int_{-\infty}^{\infty} dx\, e^{-a^2 x^2 \pm bx} = e^{b^2/4a^2}\sqrt{\frac{\pi}{a^2}}$$

Functions with Dirac Deltas

$$\int_{-\infty}^{\infty} dx'\, \delta(x-x')f(x') = f(x)$$

$$\mathcal{F}a = a\delta(u)$$

$$\mathcal{F}\delta(x-x_0) = e^{-i2\pi u x_0}$$

$$\mathcal{F}e^{i(2\pi u_0 x + \varphi)} = e^{i\varphi}\delta(u-u_0)$$

Definitions and Properties

$$\mathrm{rect}\left(\frac{x-b}{a}\right) \triangleq \begin{cases} 1 & b-a/2 \le x \le b+a/2 \\ 0 & \text{otherwise} \end{cases}$$

$$\mathrm{step}(x-x_0) \triangleq \begin{cases} 1 & x \ge x_0 \\ 0 & x \le x_0 \end{cases}$$

$$\frac{d}{dx}\mathrm{step}(x-x_0) \triangleq \delta(x-x_0)$$

$$\mathrm{sinc}(x) \triangleq \frac{\sin(\pi x)}{\pi x}$$

$$\mathrm{circ}\left(\frac{r}{a}\right) \triangleq \begin{cases} 1 & r \le a \\ 0 & \text{otherwise} \end{cases}$$

$$\int_{-\infty}^{\infty} dx\, \delta(x) \triangleq 1$$

For $g(t) = [f * h](t)$, then $\dot{g}(t) = [\dot{f} * h](t) = [f * \dot{h}](t)$, where $\dot{g}(t) = dg/dt$.

Integration by Parts

The chain rule gives $d(uv) = du\,v + u\,dv$. Rearranging to $dv\,u = d(uv) - du\,v$ and integrating from a to b, we find $\int_a^b dv\,u = uv\big|_a^b - \int_a^b du\,v$.

Effective bandwidth of a baseband function applies its power spectral density $S(u)$:

$$\mathrm{BW}_{eff} = \frac{\int_{-\infty}^{\infty} du\, S(u)}{\max S}.$$

Hilbert transform of real signal $g(t)$, as indicated by $\breve{g}(t)$, is defined as

$$\breve{g}(t) = \int_{-\infty}^{\infty} dt'\, \frac{g(t')}{\pi(t - t')},$$

which is a convolution between $g(t)$ and $1/\pi t$. Actually, this form is not very interesting. Of greater interest is the Fourier transform of $\breve{g}(t)$. From the Fourier convolution theorem,

$$\mathcal{F}\breve{g}(t) = G(u)\left(\mathcal{F}\frac{1}{\pi t}\right) = G(u)\left(-i\,\mathrm{sgn}(u)\right).$$

Since

$$\mathrm{sgn}(t) = \begin{cases} 1 & t > 0 \\ 0 & t = 0 \\ -1 & t < 0 \end{cases},$$

$$\mathcal{F}\breve{g}(t) = \begin{cases} -i\,G(u) & t > 0 \\ 0 & t = 0 \\ i\,G(u) & t < 0 \end{cases}.$$

Some properties of Hilbert transforms:

- The Hilbert transform of a constant is zero.
- The Hilbert transform of the Hilbert transform of $g(t)$ equals $-g(t)$.
- The Hilbert transform of $[g_1 * g_2](t)$ equals $[\breve{g}_1 * g_2](t) = [g_1 * \breve{g}_2](t)$.
- $\int_{-\infty}^{\infty} dt\, g(t)\,\breve{g}(t) = 0$, which tells us a signal and its Hilbert transform are orthogonal.

Symbolic Math in MATLAB

The symbolic math elements will compute Fourier transforms for you analytically. The functions `fourier` and `ifourier` can be a comfort when you are computing by hand and want a cross-check, or when problems get complicated. To use these functions correctly, you need to know a little about the symbolic math toolbox and their assumed notations.

For parameters a, b and conjugate variables t, u (my notation), the assumed equation is,

$$F(u) = b \int_{-\infty}^{\infty} dt\, f(t)\, e^{iaut}.$$

If you do not assume variable names, MATLAB assumes x, w instead of t, u. If you do not specify the parameters, `fourier` assumes $a = -1, b = 1$, which is consistent with the use of radial frequency notation Ω instead of temporal frequency u. Specifically, $F(\Omega) = \int_{-\infty}^{\infty} dt\, f(t)\, e^{-i\Omega t}$, where $u \to \Omega$.

Let's try an example. We know that analytically,

$$f(t) = \exp(-t^2/2) \;\overset{\mathcal{F}}{\longleftrightarrow}\; F(u) = \sqrt{2\pi}\exp(-2\pi^2 u^2) = \sqrt{2\pi}\exp(-\Omega^2/2).$$

In MATLAB, we find:

```
>> syms t u                                    % define variables t and u
>> sympref('FourierParameters',[1, -2*pi]);    % change default parameters to b=1, a=-2*pi
>> F=fourier(exp(-t^2/2),t,u);                 % compute FT{exp(-t^2/2)}; define variables t,u
>> pretty(F)                                    % pretty makes results easier to read
                2   2
sqrt(2) sqrt(pi) exp(-2 u  pi )                % this is forward transform by MATLAB

>> ff=ifourier(F,t)                             % compute the inverse transform to...

ff = exp(-t^2/2)                                % ...return to original f(t)
```

Adjusting the Fourier parameters effectively changes u to temporal frequency. Also, it is better to remove the constants from the problem given to `fourier` to avoid the constants MATLAB gives in solutions. For example, dividing $f(t)$ by $\sqrt{2\pi}$ should cancel the $\sqrt{2\pi}$ in the numerator of the result. Here is what we find:

```
>> syms t u m s f(x)
>> sympref('FourierParameters',[1, -2*pi]);
>> F=fourier(exp(-t^2/2)/sqrt(2*pi),t,u);pretty(F)
                2   2
sqrt(2) sqrt(pi) exp(-2 u  pi ) 2251799813685248    % the not-so-pretty result
-------------------------------------------------
              5644425081792261
```

This result equals $\exp(-2\pi^2 u^2)$, which is reassuring, but the result given is messy.

Appendix E Constrained Optimization

There are times when we need to identify the value of functions at a minimum or maximum. This is an optimization problem. For example, Problems 3.4 and 3.5 asked you to find random-variable value x at the maximum value of likelihood function $p(x; \mu, \sigma^2)$ based on N measurement samples. The results gave ML expressions for sample mean and variance. Similarly, we can find optimum values for any differentiable function, $f(\mathbf{x})$, from the zeros of the first derivatives – the equilibria or *stationary points* – via gradient $\nabla f(\mathbf{x}) = (\partial f / \partial x_1, \ldots, \partial f / \partial x_N) = \mathbf{0}$. The second-order derivatives tell us whether a stationary point is a minimum, maximum, or saddle point.

Sometimes it is important to constrain the range of possible solutions to the optimization using an *equality constraint*. For example, we may wish to treat a patient's infection without exacerbating another condition in the same patient. In that situation, we could ask what is the maximum amount of antibiotic a patient can receive while avoiding known side effects.

We now introduce analytical methods through a simple example. Note that most practical problems cannot be solved analytically; they require numerical solutions.

Example E.0.1. *Maximize the rectangular area enclosed by a fence of length C. That is, we wish to maximize* $f(x, y) = xy$ *subject to the constraint that* $2x + 2y = C$, *which we express as* $q(x, y) = 2x + 2y - C$. *We may use the* method of Lagrange multipliers *to solve this problem.*

Here we want to find simultaneous stationary points for the function $f(x, y)$ *and its singular constraint* $q(x, y)$; *i.e.,* $\nabla_{x,y} f(x, y) = \beta \nabla_{x,y} q(x, y)$. *We added the real constant* β, *which is the* Lagrange multiplier, *to tell us how sensitive the maximum value of* $f(x, y)$ *is to changes in constraint* q. *Large values of* $|\beta|$ *tell us a small change in the constraint yields a large change in* $\max(f)$. *We incorporate both parts into the Lagrange equation,*

$$L(x, y, \beta) = f(x, y) + \beta \, q(x, y) = xy + \beta(2x + 2y - C). \tag{E.1}$$

Equation (E.1) is maximized by solving the system of equations for variables x, y, β:

$$\mathbf{0} = \nabla_{x, y, \beta} L(x, y, \beta) = \begin{cases} \frac{\partial L(x, y, \beta)}{\partial x} \\ \frac{\partial L(x, y, \beta)}{\partial y} \\ \frac{\partial L(x, y, \beta)}{\partial \beta} \end{cases} . \tag{E.2}$$

Combining (E.1) and (E.2), we have

$$\nabla_{x,y,\beta}L(x,y,\beta) = \begin{cases} y + 2\beta = 0 \\ x + 2\beta = 0 \\ 2x + 2y - C = 0 \end{cases}, \quad \text{which gives} \quad \begin{matrix} x = y = C/4 \\ \beta = -C/8 \end{matrix}.$$

It is obvious that this stationary point is a maximum and not a minimum or saddle point. We see the largest rectangular area for a fence of length C is a square of side C/4. Also, β changes in proportion to the length of available fencing C. Now suppose we doubled two opposite sides and halved the other two. The rectangular area is the same, but now the fence needs to be longer, 5C/4, so the constraint is violated.

This example was intuitive and simple. Let's try more general examples that become intuitive from graphical representations like those in Figure E.1.

Example E.0.2. *(a) Minimize $f(x,y) = x^2 + y^2$ for $-1 \le x, y \le 1$ and subject to (s.t.) $y = 0.4 \; \forall \; x$, i.e., $q(x,y) = y - 0.4$. Figure E.1a shows the functions and a projection of q onto f only for illustration. From (E.1),*

Solution

$$L(x,y,\beta) = f(x,y) + \beta\, q(x,y) = x^2 + y^2 + \beta(y - 0.4)$$

$$\nabla_{x,y,\beta}L(x,y,\beta) = \begin{cases} 2x = 0 \\ 2y + \beta = 0 \\ y - 0.4 = 0 \end{cases}, \quad \text{which gives} \quad \begin{matrix} x = 0 \\ y = 0.4 \\ \beta = -0.8 \end{matrix}.$$

Plugging these coordinates into $f(x,y)$, we find $f(0, 0.4) = 0.16$, which we see in Figure E.1. Note that the solution is not the minimum point for $f(x,y)$, but it is the minimum when the solution is constrained to values where $y = 0.4$.

(b) Minimize $f(x,y) = x^2 + y^2$ s.t. $y = 0.4$ and $y = 1 - x$ also for $-1 \le x, y \le 1$. The two constraints shown in Figure E.1b must both be considered as follows.

Solution

$$L(x,y,\beta_1,\beta_2) = f(x,y) + \sum_{j=1}^{J}\beta_j\, q_j(x,y) = x^2 + y^2 + \beta_1(y - 0.4) + \beta_2(x + y - 1)$$

$$\nabla_{x,y,\beta_1\beta_2}L(x,y,\beta_1,\beta_2) = \begin{cases} 2x + \beta_2 = 0 \\ 2y + \beta_1 + \beta_2 = 0 \\ y - 0.4 = 0 \\ x + y - 1 = 0 \end{cases}, \quad \text{which gives} \quad \begin{matrix} x = 0.6 \\ y = 0.4 \\ \beta_1 = 0.4 \\ \beta_2 = -1.2 \end{matrix}.$$

From the results, we find that $\min f(x,y)$ s.t. $q_1(x,y), q_2(x,y)$ is given by $f(0.6, 0.4) = 0.52$. The code for illustrating these functions follows.

(a)

(b)

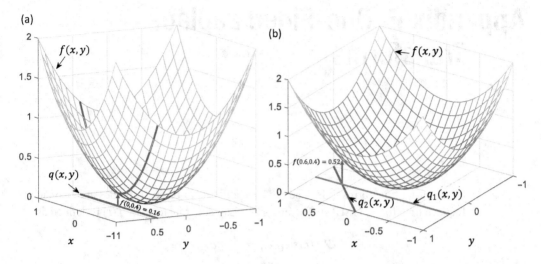

Figure E.1 The graphical representations of optimization using one (a) and two (b) constraints from Example E.0.2.

```
%%%%%%%%%%%%%   Exploring Lagrange multipliers in Example E.0.2 part b   %%%%%%%%%%%%%%%%%%%%%%%%%
close all;x=-1:0.1:1; [X,Y]=meshgrid(x,x);          % initialize spaces
f=X.^2+Y.^2;mesh(X,Y,f)                             % generate f(x,y)
%line(X(15,:),Y(15,:),f(15,:),'linewidth',3)        % draw lines on the axes
line(X(15,:),Y(15,:),zeros(1,21),'linewidth',3)
xx=0:0.1:1;line(xx,1-xx,zeros(1,11),'linewidth',3)  % after running this, rotate the results!
xlabel('x');ylabel('y')
```

One form of the general Lagrange equation for optimization using equality constraints is

$$L(\mathbf{x}, \boldsymbol{\beta}) = f(\mathbf{x}) + \sum_{j=1}^{J} \beta_j q_j(\mathbf{x}), \qquad (E.3)$$

which may be written as an objective function via

$$\underset{\mathbf{x}, \boldsymbol{\beta}}{\operatorname{argmin}} \left(f(\mathbf{x}) + \sum_{j=1}^{J} \beta_j q_j(\mathbf{x}) \right), \qquad (E.4)$$

telling us to find the variables for f, i.e., find $\mathbf{x}$, that maximize the objective function.

When the constraints are inequalities, the approach is different. Readers need to be aware that constrained optimization is a rich and broad area of analysis that requires further exploration before applying these solutions to research problems. See, for example, [16].

Appendix F One-Sided Laplace Transforms

Like Fourier transforms, it is handy to have a listing of basic properties and one-sided Laplace transforms (1SLTs). The one-sided Laplace transform of $f(t)$ is defined as

$$\mathcal{L}\{f(t)\,\text{step}(t)\} = \int_0^\infty dt\ f(t)\,e^{-st}.$$

In the following set of properties, we use the notation $\dot{f}(0) \triangleq df(t)/dt|_{t=0} = f^{(1)}(0)$ and $f^{(n)}(0) \triangleq d^n f(t)/dt^n|_{t=0}$. t and s are conjugate variables and a is a real constant.

PROPERTY F.0.1

$$\mathcal{L}\{a\,f(t)\} = a F(s) \qquad \text{(Scaling)}$$

$$\mathcal{L}\{f(at)\} = \frac{1}{a}F(s/a) \qquad \text{for } a > 0 \quad \text{(Scaling)}$$

$$\mathcal{L}\{\dot{f}(t)\} = sF(s) - f(0) \qquad \text{(First derivative)}$$

$$\mathcal{L}\{\ddot{f}(t)\} = s^2 F(s) - sf(0) - \dot{f}(0) \qquad \text{(Second derivative)}$$

$$\mathcal{L}\{f^{(n)}(t)\} = s^n F(s) - \sum_{j=1}^n s^{n-j} f^{(j-1)}(0) \qquad (n\text{th derivative})$$

$$\mathcal{L}\left\{\int_0^t dt'\ f(t')\right\} = \frac{F(s)}{s} \qquad \text{(Integral)}$$

$$\mathcal{L}\{f(t)e^{-at}\} = F(s+a) \qquad \text{(Shift)}$$

$$\mathcal{L}\{f(t-a)\,\text{step}(t-a)\} = F(s)e^{-as} \qquad \text{(Shift)}$$

Common One-Sided Laplace Transform Pairs
The time function is on the left and its 1SLT is on the right.

1. $\delta(t) \leftrightarrow 1$ and $\delta(t - a) \leftrightarrow e^{-as}$ (Impulse and shifted impulse)
2. $\text{step}(t)$ (constant for 1SLT) $\leftrightarrow s^{-1}$ (Step)
3. $t \leftrightarrow s^{-2}$ (Ramp)
4. $t^n \leftrightarrow \frac{n!}{s^{n+1}}$

5. $e^{-at} \leftrightarrow \frac{1}{s+a}$ for $a \neq 0$

6. $\frac{1}{a}\left(1 - e^{-at}\right) \leftrightarrow \frac{1}{s(s+a)}$ for $a \neq 0$

7. $\frac{1}{a}\left[t - \frac{1}{a}\left(1 - e^{-at}\right)\right] \leftrightarrow \frac{1}{s^2(s+a)}$ for $a \neq 0$

8. $t e^{-at} \leftrightarrow \frac{1}{(s+a)^2}$ for $a \neq 0$

9. $(1 - at)e^{-at} \leftrightarrow \frac{s}{(s+a)^2}$ for $a \neq 0$

10. $\frac{1}{a^2}\left[1 - (1 + at)e^{-at}\right] \leftrightarrow \frac{1}{s(s+a)^2}$ for $a \neq 0$

11. $\frac{1}{a-b}\left(e^{-bt} - e^{-at}\right) \leftrightarrow \frac{1}{(s+a)(s+b)}$ for $a \neq 0, b \neq 0, a \neq b$

12. $\frac{1}{2a}\left(e^{at} - e^{-at}\right) \leftrightarrow \frac{1}{s^2-a^2}$ for $a \neq 0$

13. $\frac{1}{2}\left(e^{at} + e^{-at}\right) \leftrightarrow \frac{s}{s^2-a^2}$ for $a \neq 0$

14. $\frac{1}{2a^2}\left(e^{at} + e^{-at} - 2\right) \leftrightarrow \frac{1}{s(s^2-a^2)}$ for $a \neq 0$

15. $\frac{1}{ab}\left(1 + \frac{be^{-at} - ae^{-bt}}{a-b}\right) \leftrightarrow \frac{1}{s(s+a)(s+b)}$

16. $\frac{ae^{-at} - be^{-bt}}{a-b} \leftrightarrow \frac{s}{(s+a)(s+b)}$

17. $\sin at \leftrightarrow \frac{a}{s^2+a^2}$

18. $\cos at \leftrightarrow \frac{s}{s^2+a^2}$

19. $Ae^{-bt}\sin at \leftrightarrow \frac{Aa}{(s+b)^2+a^2}$

20. $Be^{bt}\cos at \leftrightarrow \frac{B(s-b)}{(s-b)^2+a^2}$

21. $\sin(\Omega_0 t + \varphi) \leftrightarrow \frac{s \sin \varphi + \Omega_0 \cos \varphi}{s^2+\Omega_0^2}$

22. $\cos(\Omega_0 t + \varphi) \leftrightarrow \frac{s \cos \varphi - \Omega_0 \sin \varphi}{s^2+\Omega_0^2}$

23. $\frac{e^{at}}{(b-a)(c-a)} + \frac{e^{bt}}{(a-b)(c-b)} + \frac{e^{ct}}{(a-c)(b-c)} \leftrightarrow \frac{1}{(s+a)(s+b)(s+c)}$

24. $\frac{\sin at - at \cos at}{2a^3} \leftrightarrow \frac{1}{(s^2+a^2)^2}$

25. $\frac{t \sin at}{2a} \leftrightarrow \frac{s}{(s^2+a^2)^2}$

26. $\frac{\sin at + at \cos at}{2a} \leftrightarrow \frac{s^2}{(s^2+a^2)^2}$

27. $\cos at - \frac{1}{2}at \sin at \leftrightarrow \frac{s^3}{(s^2+a^2)^2}$

Appendix G Independent, Orthogonal, Uncorrelated

The title of this appendix lists three terms indicating a *lack of relationship* between variables expressed as functions or vectors. Our goal is to illustrate and contrast use of these terms as they relate to state variables and random variables. There is a nice two-page summary paper by Rodgers et al. [89] that summarizes the following three algebraic and geometric properties of data vectors describing state variables.

Linearly Independent

Two nonzero vectors[1] $\mathbf{u} \in \mathbb{C}^M$ and $\mathbf{v} \in \mathbb{C}^M$ are linearly independent if there is no scalar c for which $c\mathbf{u} + \mathbf{v} = \mathbf{0}$. This algebraic statement can be reinterpreted geometrically for $M = 3$: Vectors $\mathbf{u} = (u_1, u_2, u_3)^\top$ and $\mathbf{v} = (v_1, v_2, v_3)^\top$ are linearly independent if they do not lie along the same line (see the examples in Figure G.1). The concept can be extended to N vectors $\mathbf{u}_1 \ldots \mathbf{u}_N$, each of length M. By placing these vectors into columns of $M \times N$ matrix $\mathbf{U}$, we find they are linearly independent if the only solution to $\mathbf{Uc} = \mathbf{0}$ for $N \times 1$ vector $\mathbf{c}$ occurs when $c_1, c_2, \ldots, c_N$ are each zero. If $N > M$, then at most only M can be linearly independent. If $N = M$, the column vectors of $\mathbf{U} \in \mathbb{C}^{N \times N}$ are linearly independent if $\det \mathbf{U} \neq 0$. If $N < M$, then the N vectors are linearly independent if $\det \mathbf{U}^\dagger \mathbf{U} \neq 0$.

Orthogonal

Two nonzero vectors $\mathbf{u}$ and $\mathbf{v}$ are orthogonal if their inner product is zero; i.e., $\langle \mathbf{u}, \mathbf{v} \rangle = \mathbf{u}^\dagger \mathbf{v} = 0$ (see Appendix A). Orthogonality is a more limited criterion than linear independence because not only must the two vectors *not* lie along one line, but they must also be perpendicular to each other. Orthogonal vectors are always linearly independent.

Uncorrelated

In contrast, uncorrelatedness[2] is a property of the *central form* of vectors or functions. For example, consider real $M \times 1$ vectors $\mathbf{u}$ and $\mathbf{v}$, where $\bar{u} = \sum_{m=1}^{M} u_m / M$ and $\bar{v} = \sum_{m=1}^{M} v_m / M$. If $\mathbf{1}$ is an $M \times 1$ vector of 1s, then $\mathbf{u}$ and $\mathbf{v}$ are uncorrelated if inner product $(\mathbf{u} - \bar{u}\mathbf{1})^\top (\mathbf{v} - \bar{v}\mathbf{1}) = 0$. In other words, uncorrelated vectors have orthogonal

---

[1] N-dimensional zero vector $\mathbf{x} = \mathbf{0}$ has zero magnitude, $||\mathbf{x}|| = 0$.
[2] My apologies for making up words like *correlatedness* and *uncorrelatedness*. I hope you'll see my meaning as the properties characteristic of being correlated or uncorrelated, respectively.

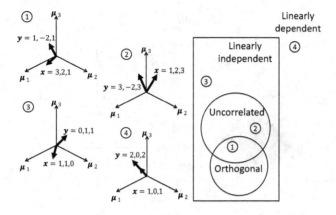

Figure G.1 Illustration of relationships between vector pairs representing various states of linear dependence, orthogonality, and correlatedness (from [89]). For example, situation (1) includes a pair of vectors that are linearly independent, orthogonal, and uncorrelated, while the pair in situation (3) are linearly independent but correlated and nonorthogonal. These situations do not correspond to the statistical assessments made in Figure G.2.

central-form vectors defined as $\mathbf{u} - \bar{u}\mathbf{1}$ and $\mathbf{v} - \bar{v}\mathbf{1}$. Geometrically, the original vectors $\mathbf{u}$ and $\mathbf{v}$ are not likely to have the same angle between them as their central-form vectors. The relationships are described in the Venn diagram shown in Figure G.1.

As shown on the right side of Figure G.1, all orthogonal vectors are linearly independent and all uncorrelated vectors are linearly independent. In this case, we examine the orthogonality and correlatedness of vector pairs. The degree to which only two vectors are correlated is rarely asked. Of much greater interest are the statistical equivalents of these three properties.

Statistical Independence, Orthogonality, and Correlations

We saw in Section 2.6 that if U and V are random variables, they are statistically independent if $p(u, v) = p(u)\, p(v)$. Knowing two random variables are statistically independent tells us nothing definitive about linear independence! As defined in Section 3.5, uncorrelated variables have $\mathrm{cov}(U, V) = \mathcal{E}\{(U - \mathcal{E}(U))(V - \mathcal{E}(V))\} = 0$, and we find for orthogonal variables that $\mathcal{E}(UV) = 0$. Uncorrelated variables have a zero second joint moment. In comparison, statistically independent variables have all joint moments equal to zero, so statistical independence is a much stronger statement about how uncoupled two variables are as compared with correlation.

The example provided in Figure G.2 shows 1,000 realizations of two statistically independent, zero-mean normal random variables whose joint distribution is

$$\mathbf{X} \sim \mathcal{N}(\mathbf{0}, \mathbf{K}) = \mathcal{N}\left(\begin{pmatrix} 0 \\ 0 \end{pmatrix}, \begin{pmatrix} 4 & 0 \\ 0 & 1 \end{pmatrix} \right).$$

That is, $\sigma_{X_1} = 2\sigma_{X_2}$ (see the following code). We know these variables are statistically independent because $\mathrm{cov}(X_1, X_2) = 0$ and they are normally distributed variables, but are they linearly independent? The question makes no sense, since statistical and

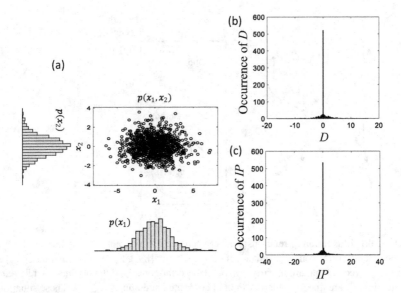

Figure G.2 (a) The joint pdf, $p(x_1, x_2)$, for statistically independent normal random variables X_1 and X_2 is shown for 1,000 realizations, along with its marginal pdfs $p(x_1)$ and $p(x_2)$. (b) A histogram of determinant values (D) found from pairs of (x_{1i}, x_{2j}), where $i \neq j$. Linearly dependent pairs have $D = 0$. (c) A histogram of inner products (IP) between vector pairs described in (b). Orthogonal pairs have $IP = 0$.

linear independence are incommensurate properties, but let's check using a numerical experiment. In the MATLAB code, we formed (x_1, x_2) vector pairs and computed the determinant of the 2×2 matrix whose columns are the vectors. The results are shown in Figure G.2b. While none of the D values is exactly zero, about half are close to zero. This shows that linear independence says nothing about statistical independence. Similarly, we provide a histogram of inner products of vector pairs (Figure G.2c) to test for orthogonality. Again, about half are orthogonal or nearly so. Statistical properties describe the random variables, while the algebraic properties discussed in the last section describe properties of small samples of deterministic or random variables. Statistical and linear independence are separate concepts that should not be confused. The term "uncoupled" generally refers to a lack of relationship with respect to variables describing state or statistical properties.

```
%%%%%%%%%%%%%%%%%%%%%%%%%%%%%  Code generating data in Fig. G.2  %%%%%%%%%%%%%%%%%%%%%%%%%%%%%%%
%
Np=1000;X=randn(Np,2);X(:,1)=2*X(:,1);          % create statistically independent X1 and X2
scatterhist(X(:,1),X(:,2));                      % histograms of joint pdf and marginals
D=zeros(1,Np-2);cntD=0;ip=zeros(1,Np-2);cntip=0; % initialize
for j=1:2:Np-2
    Xp=X(j:j+1,:);
    D(j)=det(Xp);                                % check for linear independence
    if abs(D(j))<0.1; cntD=cntD+1; end
    ip(j)=Xp(1,:)*Xp(2,:)';                      % check for orthogonality
    if abs(ip(j))<0.01; cntip=cntip+1; end
end
Nbins=200;
figure;h=histogram(D,Nbins);                     % histogram determinants
figure;histogram(ip,Nbins);                      % histogram inner produces
```

Bibliography

[1] CK Abbey, RJ Zemp, J Liu, KK Lindfors, MF Insana, "Observer efficiency in discrimination tasks simulating malignant and benign breast lesions with ultrasound," *IEEE Trans. Med. Imaging*, 25(2):198–209, 2006.

[2] CC Aggarwal, CK Reddy, eds., *Data Clustering: Algorithms and Applications*, CRC Press, Boca Raton, FL, 2014.

[3] RE Alvarez, A Macovski, "Energy-selective reconstructions in X-ray computerised tomography," *Phys. Med. Biol.*, 21(5):733–744, 1976.

[4] TW Anderson, *An Introduction to Multivariate Statistical Analysis*, John Wiley, New York, 1958.

[5] HC Andrews, BR Hunt, *Digital Image Processing*, Prentice-Hall, Engelwood Cliffs, NJ, 1977.

[6] D Arthur, S Vassilvitskii, "K-means++: the advantages of careful seeding," *SODA 07: Proc. 18th Annual ACM-SIAM Symp. Discrete Algor.*, 1027–1035, 2007.

[7] J Assländer, MA Cloos, F Knoll, DK Sodickson, J Hennig, R Lattanzi, "Low rank alternating direction method of multipliers reconstruction for MR fingerprinting," *Magn. Reson. Med.*, 79(1):83–96, 2018.

[8] RG Baraniuk, "Compressive sensing," *IEEE Sig. Proc. Mag.*, 24(4):118–124, 2007.

[9] HH Barrett, "Objective assessment of image quality: effects of quantum noise and object variability," *J. Opt. Soc. Am. A*, 7(7):1266–1278, 1990.

[10] HH Barrett, KJ Myers, *Foundations of Image Science*, Wiley-Interscience, Hoboken, NJ, 2004.

[11] RG Bartle, *The Elements of Integration and Lebesgue Measure*, Wiley Classics Library, vol. 56, Wiley, New York, 1995.

[12] A Beck, M Teboulle, "A fast iterative shrinkage-thresholding algorithm for linear inverse problems," *SIAM J. Imaging Sci.*, 2(1):183–202, 2009.

[13] JS Bendat, AG Piersol, *Random Data: Analysis and Measurement Procedures*, 2nd ed., Wiley-Interscience, New York, 1986.

[14] A Ben-Naim, *A Farewell to Entropy: Statistical Thermodynamics Based on Information*, World Scientific, Singapore, 2008.

[15] PR Bevington, DK Robinson, *Data Reduction and Error Analysis for the Physical Sciences*, 2nd ed., WCB/McGraw-Hill, Boston, 1992.

[16] EG Birgin, JM Martinez, *Practical Augmented Lagrangian Methods for Constrained Optimization: Fundamentals of Algorithms*, SIAM, Philadelphia, 2014.

[17] CM Bishop, *Neural Networks for Pattern Recognition*, Clarendon Press, Oxford, UK, 1995.

[18] M Born, E Wolf, *Principles of Optics: Electromagnetic Theory of Propagation, Interference and Diffraction of Light*, 7th ed., Cambridge University Press, Cambridge, UK, 1999.

[19] GEP Box, "Science and statistics," *J. Am. Stat. Assoc.*, 71(356):791–799, 1976.

[20] S Boyd, N Parikh, E Chu, B Peleato, J Eckstein, "Distributed optimization and statistical learning via the alternating direction method of multipliers," *Found. Trends Mach. Learn.*, 3(1):1–122, 2010.

[21] RN Bracewell, *The Fourier Transform and Its Applications*, 2nd ed., McGraw-Hill, New York, 1978.

[22] EN Bruce, *Biomedical Signal Processing and Signal Modeling*, Wiley & Sons, New York, 2001.

[23] JT Bushberg, JA Seibert, EM Leibholdt Jr, JM Boone, *The Essential Physics of Medical Imaging*, 3rd ed., Lippincott Williams & Wilkins, Philadelphia, 2012.

[24] E Candès, Y Plan, "Near-ideal model selection by ℓ_1 minimization," *Ann. Stat.*, 37(5A):2145–2177, 2009.

[25] EJ Candès, MB Wakin, "An introduction to compressive sampling," *IEEE Signal Proc. Mag.*, 25(2):21–30, 2008.

[26] GB Coleman, HC Andrews, "Image segmentation by clustering," *Proc. IEEE*, 67(5):773–785, 1979.

[27] TM Cover, "Learning in pattern recognition." In: *Methodologies of Pattern Recognition*, S Watanabe, ed., Academic Press, New York, 1969.

[28] TM Cover, JA Thomas, *Elements of Information Theory*, John Wiley & Sons, New York, 1991.

[29] D Donoho, "Compressed sensing," *IEEE Trans. Inf. Theory*, 52(4):1289–1306, 2006.

[30] DD Dorfman, E Alf Jr, "Maximum likelihood estimation of parameters of signal detection theory: a direct solution," *Psychometrika*, 33(1):117–124, 1968.

[31] RO Duda, PE Hart, DG Stark, *Pattern Classification*, 2nd ed., Wiley Interscience, John Wiley and Sons, New York, 2000.

[32] YC Eldar, G Kutyniok, *Compressed Sensing: Theory and Applications*, Cambridge University Press, Cambridge, UK, 2012.

[33] E Elshik, CR Bester, A Nel, "Appropriate solar spectrum usage: the novel design of a photovoltaic thermal system," 2016. www.researchgate.net/publication/299559828_Appropriate_Solar_Spectrum_Usage_The_Novel_Design_of_a_Photovoltaic_Thermal_System

[34] RD Evans, *The Atomic Nucleus*, McGraw-Hill, New York, 1955.

[35] SA Frank, "The common patterns of nature," *J. Evol. Biol.*, 22(8):1563–1585, 2009.

[36] E Fredenberg, "Spectral and dual-energy x-ray imaging for medical applications," *Nucl. Inst. Methods Phys. Res. A*, 878:7487, 2018.

[37] K Fukunaga, *Introduction to Statistical Pattern Recognition*, Academic Press, San Diego, 1990.

[38] D Gabor, "Theory of communication: part 1. The analysis of information," *J. Inst. Elect. Eng. Part III, Radio Commun.*, 93:429–441, 1946.

[39] BS Garra BS, MF Insana, TH Shawker, RF Wagner, M Bradford, MA Russell, "Quantitative ultrasonic detection and classification for diffuse liver disease: comparison with human observer performance," *Invest. Radiol.*, 24:196–203, 1989.

[40] J Ghaboussi, MF Insana, *Understanding Systems: A Grand Challenge for 21st Century Engineering,* World Scientific, Singapore, 2018.

[41] GH Golub, CF VanLoan, *Matrix Computations*, 4th ed., Johns Hopkins University Press, Baltimore, 2013.

[42] RC Gonzalez, RE Woods, *Digital Image Processing*, Addison-Wesley, Reading, MA, 1992.

[43] JW Goodman, *Introduction to Fourier Optics*, McGraw-Hill, San Francisco, 1968.

[44] JW Goodman, *Statistical Optics*, Wiley Interscience, New York, 1985.

[45] IS Gradshteyn, IM Ryzhik, *Table of Integrals, Series, and Products*, 5th ed., A Jeffrey, ed., Academic Press, San Diego, 1994.

[46] Data originally provided by W. B. Gratzer, Med. Res. Council Labs, Holly Hill, London, and N. Kollias, Wellman Laboratories, Harvard Medical School, Boston. See https://en.wikipedia.org/wiki/Pulse_oximetry.

[47] DM Green, JA Swets, *Signal Detection Theory and Psychophysics*, Wiley & Sons, New York, 1966.

[48] FG Guerrero, "A new look at the classical entropy of written English," *CoRR*, 2009. http://arxiv.org/abs/0911.2284.

[49] JP Haldar, D Hernando, ZP Liang, "Compressed-sensing MRI with random encoding," *IEEE Trans. Med. Imaging*, 30(4):893–903, 2011.

[50] T Hastie, R Tibshirani, J Friedman, *The Elements of Statistical Learning: Data Mining, Inference, and Prediction*, Springer, New York, 2009.

[51] JM Heffernan, RJ Smith, LM Wahl, "Perspectives on the basic reproductive ratio," *J. R. Soc. Interface*, 2:281–293, 2005.

[52] HW Hethcote, "The basic epidemiological models: models, expressions for R_0 parameter estimation, and applications," In: *Mathematical Understanding of Infectious Disease Dynamics*, S Ma, Y Xia, eds., World Scientific, Singapore, 2008, pp. 1–61.

[53] FC Hoppensteadt, CS Peskin, *Modeling and Simulation in Medicine and the Life Sciences,* Springer, New York, 2002.

[54] ICRU Report 54, *Medical Imaging: The Assessment of Image Quality*, 7910 Woodmont Avenue, Bethesda, MD, 20814, April 1996.

[55] MF Insana, TJ Hall, "Visual detection efficiency in ultrasonic imaging: a framework for objective assessment of image quality," *J. Acoust. Soc. Am.*, 95(4):2081–2090, 1994.

[56] MF Insana, RF Wagner, BS Garra, R Momenan, TH Shawker, "Pattern recognition methods for optimizing multivariate tissue signatures in diagnostic ultrasound," *Ultrasonic Imag.*, 8:165–180, 1986.

[57] MJ Jaroszeski, G Radcliff, "Fundamentals of flow cytometry," *Molec. Biotech.*, 11:37–53, 1999.

[58] AC Kak, M Slaney, *Principles of Computerized Tomographic Imaging*, SIAM, Philadelphia, 2001.

[59] MJ Keeling, P Rohani, *Modeling Infectious Diseases in Humans and Animals*, Princeton University Press, Princeton, NJ, 2007.

[60] EH Kerner, "Dynamical aspects of kinetics," *Bull. Math. Biophys.* 26, 333–349, 1964; also EH Kerner, "Note on Hamiltonian format of Lotka–Volterra dynamics," *Phys. Lett. A*, 151:401–402, 1990.

[61] EH Kerner, "Comment on Hamiltonian structures for the n-dimensional Lotka-Volterra equations," *J. Math. Phys.* 38(2):1218–1223, 1997.

[62] M-W Kim, Y Zhu, J Hedhli, LW Dobrucki, MF Insana, "Multi-dimensional clutter filter optimization for ultrasonic perfusion imaging," *IEEE Trans. Ultrason. Ferroelec. Freq. Control*, 65(11):2020–2029, 2018.

[63] S Kullback, *Information Theory and Statistics*, Dover, New York, 1997.

[64] JL Lancaster, B Hasegawa, *Fundamental Mathematics and Physics of Medical Imaging*, CRC Press, Boca Raton, FL, 2016.

[65] LA Lehmann, RE Alvarez, A Macovski, WR Brody, NJ Pelc, SJ Riederer, AL Hall, "Generalized imaging combinations in dual kVp digital radiography," *Med. Phys.*, 8(5):659–667, 1981.

[66] SP Lloyd, "Least squares quantization in PCM," *IEEE Trans. Inf. Theory*, 28(2):129–137, 1982.

[67] A Macovski, "Ultrasonic imaging using arrays," *Proc. IEEE*, 67:484–495, 1979.

[68] A Macovski, *Medical Imaging Systems*, Prentice-Hall, Upper Saddle River, NJ, 1983.

[69] SL Marple, Jr, *Digital Spectral Analysis with Applications*, Prentice-Hall, Englewood Cliffs, NJ, 1987.

[70] RN McDonough, AD Whalen, *Detection of Signals in Noise*, 2nd ed., Academic Press, San Diego, 1995.

[71] JD Murray, *Mathematical Biology*, 3rd ed., vol. 1, Springer, New York, 2007.

[72] K Nakamura, ed., *Ultrasonic Transducers: Materials and Design for Sensors, Actuators and Medical Applications*, Woodhead Publishing, Elsevier, 2012.

[73] NQ Nguyen, CK Abbey, MF Insana, "Objective assessment of sonographic quality I: task information," *IEEE Trans. Med. Imaging*, 32(4):683–690, 2013.

[74] DO North, "An analysis of the factors which determine signal/noise discrimination in pulse-carrier systems," Tech. Rep. PTR-6C, RCA Laboratories, June 1943. Reprinted in *Proc. IEEE*, 51:1016–1027, 1963.

[75] NA Obuchowski, D Katzman McClish, "Sample size determination for diagnostic accuracy studies involving binormal ROC curve indices," *Stat. Med.*, 16:1529–1542, 1997.

[76] EP Odum, *Fundamentals of Ecology*, W. B. Saunders, Philadelphia, 1953.

[77] AV Oppenheim, RW Schafer, JR Buck, *Discrete-Time Signal Processing*, 2nd ed., Prentice-Hall, Upper Saddle River, NJ, 1999.

[78] BØ Palsson, *Systems Biology: Properties of Reconstructed Networks*, Cambridge University Press, Cambridge UK, 2006.

[79] X Pan, CE Metz, "The 'proper' binormal model: parametric receiver operating characteristic curve estimation with degenerate data," *Acad. Radiol.*, 4:380–389, 1997.

[80] A Papoulis, *Probability, Random Variables, and Stochastic Processes*, 3rd ed., WCB McGraw-Hill, Boston, 1991.

[81] R Penrose, "A generalized inverse for matrices," *Math. Proc. Cambridge Phil. Soc.*, 51(3):406–413, 1955.

[82] M Plank, "Hamiltonian structures for the n-dimensional Lotka–Volterra equations," *J. Math. Phys.* 36(7): 3520–3534, 1995.

[83] SC Prasad, WR Hendee, PL Carson, "Intensity distribution, modulation transfer function, and the effective dimension of a line-focus x-ray focal spot," *Med. Phys.*, 3(4):217–223, 1976.

[84] A Rajkomar, J Dean, I Kohane, "Machine learning in medicine," *N. Engl. J. Med.*, 380:1347–1358, 2019.

[85] DP Aditya Rao, CH Renumadhavi, MG Chandra, R Srinivasan, "Compressed sensing methods for DNA microarrays, RNA interference, and metagenomics," *J. Comput. Biol.*, 22(2):145–158, 2015.

[86] S Ravishankar, Y Bresler, "Learning sparsifying transforms," *IEEE Trans. Sig. Proc.* 61(5):1072–1086, 2013.

[87] S Ravishankar, JC Ye, JA Fessler, "Image reconstruction: from sparsity to data-adaptive methods and machine learning," *Proc. IEEE*, 108(1):3–10, 2020.

[88] 3rd Baron Rayleigh, JW Strutt, "The problem of the random walk," *Nature*, 72(1866):318, 1905.

[89] JL Rodgers, WA Nicewander, L Toothaker, "Linearly independent, orthogonal, and uncorrelated variables," *Am. Statistician*, 38(2):133–134, 1984.

[90] S Ross, *A First Course in Probability*, Macmillan, New York, 1976.

[91] SI Rubinow, *Introduction to Mathematical Biology*, John Wiley & Sons, New York, 1975.

[92] MA Savageau, *Biochemical Systems Analysis: A Study of Function and Design in Molecular Biology*, CreateSpace Independent Publishing Platform, 2010.

[93] O Schade, *Image Quality: A Comparison of Photographic and Television Systems*, RCA Laboratories, Princeton, NJ, 1975.

[94] JL Semmlow, *Circuits Systems and Signals for Bioengineers: A Matlab-Based Introduction*, Elsevier, New York, 2005.

[95] CE Shannon, "A mathematical theory of communication," *Bell Syst. Tech. J.*, 27:379–423, 623–656, 1948.

[96] CE Shannon, "Prediction and entropy of printed English," *Bell Syst. Tech. J.*, 30:47–51, 1951.

[97] HM Shapiro, *Practical Flow Cytometry*, 4th ed., Wiley-Liss, Hoboken, NJ, 2003.

[98] R Shaw, "The equivalent quantum efficiency of the photographic process," *J. Photographic Sci.*, 11:199–204, 1963.

[99] W Simon, *Mathematical Techniques for Biology and Medicine*, Dover, New York, 1986.

[100] D Slepian, *Key Papers in the Development of Information Theory*, IEEE Press, Piscataway, NJ, 1974.

[101] S So, KK Paliwal, "Reconstruction of a signal from the real part of its discrete Fourier transform," *IEEE Sig. Proc. Mag.*, 35(2):162–164, 174, 2018.

[102] FG Sommer, RT Hoppe, L Fellingham, BA Carroll, H Solomon, S Yousem, "Spleen structure in Hodgkin disease: ultrasonic characterization: work in progress," *Radiology*, 153(1):219–222, 1984.

[103] A Statnikov, CF Aliferis, DP Hardin, I Guyon, *A Gentle Introduction to Support Vector Machines in Biomedicine*, vol. 1, World Scientific, Singapore, 2011.

[104] WP Tanner, Jr, TG Birdsall, "Definitions of d' and η as psychophysical measures," *J. Acoust. Soc. Am.*, 30(10):922–928, 1958.

[105] A Tharwat, "Linear vs. quadratic discriminant analysis classifier: a tutorial," *Int. J. Applied Pattern Recognition*, 3(2):145–180. 2016.

[106] KE Thomenius, "Evolution of ultrasound beamformers," *Proc. IEEE Ultrason. Symp.*, 1615–1622, 1996.

[107] R Tibshirani, "Regression shrinkage and selection via the lasso: a retrospective," *J. Royal Statist. Soc. B*, 73(pt 3):273–282, 2011.

[108] DJ Tward, JH Siewerdsen, "Noise aliasing and the 3D NEQ of flat-panel cone-beam CT: effect of 2D/3D apertures and sampling," *Med. Phys.*, 36(8):3830–3843, 2009.

[109] H Urkowitz, *Signal Theory and Random Processes*, Artech House, Norwood, MA, 1983.

[110] HL Van Trees, *Detection, Estimation, and Modulation Theory, Part I*, Wiley, New York, 1968. Note that a second edition to Part I, published in 2013 by Wiley with coauthors Kristine L. Bell and Zhi Tian, expands the development of optimal detectors.

[111] HL Van Trees, *Detection, Estimation, and Modulation Theory, Part IV: Optimum Array Processing*, Wiley, New York, 2002.

[112] F Verhulst, *Nonlinear Differential Equations and Dynamical Systems*, Springer, Berlin, 1996.

[113] RF Wagner, DG Brown, "Overview of a unified SNR analysis of medical imaging systems," *IEEE Trans. Med. Imaging*, 1(4):210–213, 1982.

[114] RF Wagner, DG Brown, "Unified SNR analysis of medical imaging systems," *Phys. Med. Biol.*, 30:489–518, 1985.

[115] RF Wagner, DG Brown, MS Pastel, "Application of information theory to the assessment of computed tomography," *Med. Phys.*, 6(2):83–94, 1979.

[116] RF Wagner, MF Insana, RJ Jennings, DG Brown, "Multivariate signal and texture discrimination in medical imaging," In: *Statistical Efficiency of Natural and Artificial Vision*, HG Barlow, DG Peli, eds., Rank Prize Fund Publishing, London, 1986.

[117] RF Wagner, MF Insana, SW Smith, "Fundamental correlation lengths of coherent speckle in medical ultrasonic images," *IEEE Trans. Ultrason. Ferro. Freq. Control*, 35:34–44, 1988.

[118] RF Wagner, KE Weaver, EW Denny, RG Bostrom, "Toward a unified view of radiological imaging systems. Part I: noiseless images," *Med. Phys.*, 1(1):11–24, 1974.

[119] M Wernick, JN Aarsvold, *Emission Tomography: The Fundamentals of PET and SPECT*, Elsevier/Academic Press, San Diego, 2004.

[120] LA Zadeh, "The determination of the impulsive response of variable networks," *J. Appl. Phys.*, 21:642–645, 1950.

[121] LA Zadeh, "Initial conditions in linear-varying parameter systems," *J. Appl. Phys.*, 22(6):782–786, 1951.

[122] RJ Zemp, CK Abbey, MF Insana, "Linear system models for ultrasonic imaging: application to signal statistics," *IEEE Trans. Ultrason. Ferro. Freq. Control*, 50:642–654, 2003.

[123] DP Zhou, W Peng, L Chen, X Bao, "Brillouin optical time-domain analysis via compressed sensing," *Opt. Lett.*, 43(22):5496–5499, 2018.

Index

Printed in the United States
by Baker & Taylor Publisher Services